Guidebook: Toxic Substances Control Act

Volume I

Editor

George Dominguez
Director
Government Relations
Safety, Health, and
Ecology Department
CIBA-GEIGY Corporation
Ardsley, New York

CRC Press is an imprint of the
Taylor & Francis Group, an **informa** business

CRC Press
Taylor & Francis Group
6000 Broken Sound Parkway NW, Suite 300
Boca Raton, FL 33487-2742

Reissued 2019 by CRC Press

A Library of Congress record exists under LC control number:

ISBN 13: 978-0-367-26301-0 (hbk)
ISBN 13: 978-0-367-26302-7 (pbk)
ISBN 13: 978-0-429-29244-6 (ebk)

Visit the Taylor & Francis Web site at http://www.taylorandfrancis.com and the CRC Press Web site at http://www.crcpress.com

EDITOR

George S. Dominguez, M.B.A., is Director, Government Relations, Safety, Health, and Ecology Department, CIBA-GEIGY Corporation, Ardsley, New York.

Mr. Dominguez received his B.S. and M.B.A. from Kentucky Christian University, Ashland. He is also a graduate of the Manhattan Medical and Dental Assistant School and the U.S. Army Intelligence School.

Mr. Dominguez is a member of the Board of Governors of the Synthetic Organic Chemical Manufacturers Association. He is a member of the Government Relations Committee and Chairman of the Chemical Regulations Advisory Committee of the Manufacturing Chemists Association. He is also a member of the Ecology Steering Committee and Chairman of the Toxic Substances Task Force of the National Association of Manufacturers.

Mr. Dominguez has had numerous publications and is the author of *Product Management; Marketing in a Shortage Economy; How to be a Successful Product Manager;* and *Business, Government, and the Public Interest.*

He has also taught and lectured widely, was a member of the faculty of New York University, Alamance Technical Institute, and is a frequent guest lecturer at the New York University Graduate School of Education.

ADVISORY BOARD

CONTRIBUTORS

Dr. John D. Behun
Corporate Manager of Toxic Substances Control
Environmental Affairs and Toxicology Department
Mobil Oil Corporation
New York, New York

Etcyl H. Blair, Ph.D.
Director, Health and Environmental Research
Dow Chemical Company
Midland, Michigan

Roger E. Copland, J.D.
Associate Director
Regulatory and Legislative Affairs
Chemical Specialties Manufacturers Association
Washington, D.C.

Theodore Ellison, Ph.D.
Senior Toxicologist
Mobil Chemical Company
Edison, New Jersey

Charles Ganz, Ph.D.
President and Technical Director
EN-CAS Analytical Laboratories
Winston-Salem, North Carolina

Fred D. Hoerger, Ph.D.
Manager, Regulatory and Legislative Issues
Health and Environmental Sciences
Dow Chemical Company
Midland, Michigan

Glenn E. Schweitzer, M.S.
Director, Office of Toxic Substances
Environmental Protection Agency
Washington, D.C.

Marvin E. Winquist, B.A.
Manager, Special Chemicals
Designed Products Department
Dow Chemical Company
Midland, Michigan

TABLE OF CONTENTS

Section I

Introduction

INTRODUCTION

THE TOXIC SUBSTANCES CONTROL ACT: A GUIDE TO THE GUIDEBOOK

A great deal has been said and written about the Toxic Substances Control Act (TSCA) both before and after it was signed into law on October 11, 1976 by President Ford. In light of the vast new regulatory authority given to EPA and the essential transformation of nonregulated sectors of the chemical industry to regulated ones, undoubtedly there remains much more to be said and written in the future. The full effect of this new legislation will only become apparent as the statute is interpreted and implemented by the U.S. Environmental Protection Agency (EPA) and as such implementation is translated into specific regulatory and compliance requirements. As this occurs, there will be, without question, many additional observations, interpretations, reviews, and commentaries by all concerned.

However, our purpose at this time is not to merely add to this body of literature. Rather, it is to provide what we perceive to be a very essential and important contribution to the chemical industry in general and to the smaller to medium-sized manufacturers, processors, formulators, or users in particular. The guidebook is not just another extensive review and analysis of the law itself (which, in fact, we also provide) but an exhaustive evaluation of:

1. Probable effects on chemical industry operations
2. Actions, costs, alternatives, and compliance programs
3. Recommendations for organizational preparation
4. Review of major rules and regulations (as they are proposed and promulgated)
5. Evaluation of the effects of this law on all affected sectors of the chemical industry (including users)
6. An introduction to toxicological and environmental effects testing
7. A resource list of available commercial testing laboratories
8. An extensive bibliography for further study
9. A permanent basis for a concise "on-source" reference on TSCA

In effect, it provides a complete "guidebook" not only to the law, but offers routes for preparation and response to its many requirements. The emphasis is on the practical, with clear and precise advice on anticipated compliance approaches and suggestions for early organizational preparation. This, augmented with an introduction to toxicological and environmental effects testing and a listing of available commercial testing sources (combined with a provision for periodic updating), should provide not only a true "guide to TSCA" but to company activity. As there are unquestioned effects on formulators and uses of chemical substances and mixtures, extensive interpretation of the consequences to these segments of the chemical industrial community have been carefully and fully analyzed and presented.

The following introductory remarks are intended to provide an additional frame of reference to better understand, not only the objectives of this publication and thus increase its utility, but to provide the proper perspective on those aspects, issues, and requirements which are of the greatest interest and concern to you and your company.

OBJECTIVES

In preparing this guidebook, the Editor and Advisory Board had several objectives constantly in mind. These were to provide:

1. A clear, complete, yet concise summary of the law itself, useful to both lawyers and nonlawyers
2. An exhaustive yet understandable analysis of the law from both the legal and business viewpoints
3. Practical recommendations for company compliance, planning activity, and testing
4. An updating program of important rules and regulations, if justified by the demand for this work, would be made available to all purchasers of this Guidebook.

FORMAT AND APPROACH

To satisfy the many and demanding objectives, we initially assembled an Advisory Board of expert practitioners, all of whom have had long and direct involvement with this legislation. As you will note, not only is the business community represented, but we are most pleased to include an EPA representative who, while not participating directly on behalf of EPA but rather through personal efforts, nevertheless does so with EPA permission. This provides not only a continuity of interaction with the agency, but clearly reflects the agency's interest in public communication, coordination, and assistance.

INTRODUCTION TO THE TOXIC SUBSTANCES CONTROL ACT (TSCA)

It is not my place as Editor to preempt the detailed and expert comments on this law and its probable effects which follow. However, a few brief introductory observations may help to place the importance of this new law and your reasons for being vitally concerned with it into perspective.

While this law became effective on January 1, 1977, there are many provisions which will become effective over the course of the next 2 years as provided by specific statutory requirements and deadlines. In addition, there are many requirements that will ensue as a result of agency rule making, some of which are mandatory (nondiscretionary) and some of which are discretionary. Therefore, we will, in effect, see two timetables: events which can be predicted by certain dates, as they are compulsory and relate to particular deadlines set forth in the law, e.g., promulgation of the initial inventory list (November 11, 1977) and the establishment of a premanufacturing notification system (December 12, 1977); while other important rules and regulations will occur based on an internal EPA-developed timetable, e.g., testing rules (Section 4 of the Act) or application of the law to "significant new uses" (Section 5 of the Act). In our summary of the law, these and various other statutory and discretionary deadlines (insofar as they have or can be estimated by the agency or the industry) are clearly indicated. The important point to bear in mind, from a business point of view, is that we are dealing with both. While it may be possible to dictate actions with some certainty relative to those deadlines and events, as stipulated in the law, it is a matter of considerable judgment as to the others, which proceed at the discretion of the Administrator. It is in regard to the latter that some of the specific advice contained in this guidebook will prove most useful, since it should assist you in determining your company's compliance programs. This is especially the case in business planning, organizational, and budgetary considerations.

In examining TSCA, it is very important to understand not only its specific requirements (the full text provided in Section II) but also to have some appreciation of the interpretation that experienced lawyers place on the many and varied sections and language of the law itself. This guidance is contained in the very extensive analysis that has been prepared by the expert and experienced Washington law firm of Cleary, Gottlieb, Steen, and Hamilton, reprinted here with the kind permission of the Synthetic Organic Chemical Manufacturers Association (SOCMA) for whom it was originally developed. The practical business insights provided by these interpretive statements are most helpful in understanding not only the language of the law, but provides a greater appreciation of its meaning as both a matter of law and in terms of potential compliance requirements and business effects.

In this context, comprehension of the intent of the law is vital: what Congress meant to accomplish with this legislation and what construction was placed not only on the entire law but on individual sections as it underwent legislative development. To gain this perspective, it must be remembered that this law has, in fact, been long in coming and has undergone considerable review and debate not only in the 94th Congress but in two preceding sessions as well. The law is the outgrowth of a recommendation made to former President Nixon by the Council on Environmental Quality who, in 1971, identified the manufacture and marketing of industrial chemicals without any mandated premarket health or environmental effects testing as a matter of critical national concern. The Congress was quick to respond to this perceived need and various bills were introduced and debated in the 92nd and 93rd Congresses. However, for many and varied reasons, none of these passed as final legislation, and therefore, it was not until the 94th session that a bill was voted out of both houses and eventually signed into law by President Ford on October 11, 1976. The basic purpose of the law as stated by the Congress and later somewhat restated by EPA in its draft

"Assessment and Control of Chemical Problems: An Approach to Implementing the Toxic Substances Control Act" (more briefly referred to as EPA's overall TSCA Strategy Document) is "to protect human health and the environment from unreasonable risk now and in future generations."

To enable one to better perceive the congressional intent, not only is the summary and analysis important, but so is the record of the joint Senate and House Conference Committee who met and finally developed the law as it was passed. For purposes of studying this critical document, it has been provided in its entirety (Section III) and you are urged to read it just as carefully as you do the law itself for from the "conference report" comes clarification of not only various Sections of the law, but a fuller appreciation of its conceptual foundations and the authority that Congress intended to give to EPA. This is important not only from the perspective of what authority it intended to extend to the agency, but also for clarification as to what authority it did not intend to give the agency.

Given these intentions, the statute grants extensive mandatory authority to EPA to require premarket notification on new chemicals and later to require similar premarket notification for what uses are yet to be classified as significant new uses. Broad authority to require testing of new and existing chemicals for health and environmental effects is granted the EPA. While the Act does not yet specify what tests will be required or how they should be conducted, control over chemicals that might involve chronic health effects such as carcinogenicity, mutagenicity, and teratogenicity is its primary objective. However, acute effects as well as behavioral and synergetic effects are also mentioned. In the area of environmental effects, the emphasis once again seems to be on long-term consideration and factors such as bioaccumulation, bioconcentration, persistence in the environmental and related phenomena are mentioned in the EPA Strategy Document; however, acute effects are no doubt also covered. It is important from the standpoint of analyzing the requirements of this legislation on the manufacture, formulation, or uses that it is realized just how costly and time consuming these tests can be. At this point, we cannot provide with certainty any delineation of the types of tests or the specific costs of testing that will be required. However, since this is such a vital area in planning new product and existing product activity as well as budgeting, we have provided an extensive discussion of the many types of tests that may be required in both fields (health and environmental effects) as well as listed resources (commercial testing laboratories who are able to undertake such tests). However, considering that the premarket notification of at least 90 days before marketing will commence as of December 12, 1977, and that certain of these tests require long lead times, the time for new-product introduction is now.

Insofar as the regulatory authority of the agency is concerned, it is important to recognize that both corporate and personal civil and criminal liability are associated with violations of several prohibited actions. These are clearly described in the Act and in the accompanying text. However, the key concept is that civil and criminal liability are provided. Insofar as regulatory authority is concerned, again, the agency has vast discretion and can take actions ranging from outright and total bans to imposition of quality-control procedures. These various options cannot only be taken with respect to individual substances, but on "categories" of substances and on a geographic or regional basis. All of which makes for a very complex matrix of possible regulatory action. Here, again, familiarization with the various regulatory positions is important since it is also the clear and declared intention of the Congress to implement this act "in a reasonable and prudent manner, and that the Administrator shall consider the environmental, economic impact of any action the Administrator takes or proposed to take under this Act." While this intent underlies the entire agency administration of the act, there is further explanation in Section 6 of the Act dealing with regulations which impose an additional restriction on the Administrator. When contemplating and taking regulatory action, he shall control unreasonable risk by "using the least burdensome requirements." The latter is most important to the industry and underscores the need for thorough knowledge of regulatory action in order to determine which is the least burdensome regulatory option.

Another area of concern is that of the relationship of this law to others, not only those administered by EPA but by other agencies such as the Occupational Safety and Health Administration (OSHA). Here again, the agency, while having some overlapping and seemingly duplicate authority, should use this act only if the issues cannot be addressed adequately by the other law and/or agency. Therefore, the Administrator must make such a determination and refer the issue to the other department or agency first. It is important once again from the business viewpoint to understand these limitations and to be in a position to assert the legitimate rights of the company in terms of such options.

There are many other important features of this legislation; however, these should suffice as an introduction to provide you with a "feel" for the law and the numerous reasons why you must understand not only its specific provisions, but the underlying intent and purpose. In this way, you can best exercise your rights if contentions later arise. By so doing, you will have insight into precautionary preparations that are wisely investigated even without the issuance of specific regulations and rules based on the aforementioned distinctions between statutory deadlines – actions that you can readily anticipate and plan on, and those that are discretionary – they will eventually occur but you cannot definitely predetermine when.

BUSINESS AND ECONOMIC EFFECTS

A law of this complexity and magnitude regulating a large and important segment of American industry necessarily has not only economic, but also other effects on the regulated industry. Several attempts have been made to assess these and range from an initial low EPA assessment of $40 million (later revised upward by the agency to its present estimate of $80 to 140 million) to an evaluation by GAO which calculated a cost to industry of $100 to 200 million. The most comprehensive effort was made by the industry itself through a study which was conducted by Foster D. Snell (a division of Booz, Allen, and Hamilton) for the Manufacturing Chemists Association. This study was completed in June 1975 and indicated the following:

1. Chemical industry cost increases ranging from $360 million to $1.3 billion/year in compliance and testing expenditures
2. A decline in the number of new products of anywhere from 10 to 30%
3. Potential job dislocation from 20 to 80 thousand by 1985
4. Potential additional research and development investments of $600 million/year to offset negative effects on innovation

From the range of estimates which have been indicated, it is apparent that at this time there is no unanimity of opinion as to complete economic impact on the one hand or on the resultant benefits of this legislation.

The law specifically stipulates that EPA must take economic factors into consideration and as stated in Section 2 of the Toxic Substances Control Act "it is the intent of Congress that the Administrator shall carry out this Act in a reasonable and prudent manner and that the Administrator shall consider the environmental, economic, and social impact of any action the Administrator takes or proposes to take under this Act." The need to consider economic effects is further reinforced in Section 6 of the Act (Regulation of Hazardous Chemical Substances and Mixtures) wherein the statute states that in exercising his regulatory authorities and options, the Administrator must take economic factors into consideration.

Regardless of which of these estimates is correct, and time will certainly tell, there is clear recognition on the part of all concerned – industry, Congress, and EPA, that despite the benefits that will unquestionably occur, there will be costs. Which will outweigh which and what will be the cost remain to be more precisely calculated. In this context, it should be urged that each company keep complete and accurate records of its costs associated with this legislation. Only in this way can a final objective assessment be undertaken at some later date.

It is clear from the Congressional Record that the Congress, while recognizing the potential value of this act, did not intend undue negative economic or other effects, such as reduction in innovation or chemical industry stultification, either in the passage of the law or its subsequent implementation. However, it is the obligation of the industry to assist both the Congress and the agency in making the precise determinations that will be necessary to objectively determine if this is what occurs. Conversely, we must also strive to quantify and understand the acknowledged benefits. Only by measuring both sides of the equation can optimum social benefit occur.

RECENT DEVELOPMENTS RELATIVE TO IMPLEMENTATION OF TSCA

Several months have elapsed since this statute took effect. During that time, one of the most significant developments was EPA's alteration in its approach in constructing the initial inventory required by the statute. The Agency had proposed a procedure in the March 9, 1977 *Federal Register* which has been completely superseded by reproposal rules in the *Federal Register* of August 2, 1977.

The new proposal is important not only in the fact that it expands the initial reporting requirements of chemical manufacturers, but that it alters the timetable for the establishment of the inventory itself and the initiation of premanufacturing notification requirements. The original requirement of the statute was to complete the inventory list by November 11, 1977 and to establish and require premanufacturing notification on December 12, 1977. By virtue of the new proposed regulations, a copy of which is provided here, the initial inventory list could not be completed until early 1978 and premanufacturing notification requirements therefore could not be required until mid to late 1978.

In reading the various timetables and comments that are contained in some of the following articles, please bear in mind that they were written prior to this development and still reflect the initial timetables stipulated in the statute. The reader should realize that these are now altered based on the proposed rules and should also be aware that they might change in the final rules which will probably be published in October 1977. Therefore, the *Federal Register* should be carefully monitored for the final rule making which will provide both the final reporting requirements and schedule.

USING THE GUIDEBOOK

While this, like any book, can be read cover-to-cover and ultimately should be, it is suggested that by reading it in the following sequence you may well derive far greater benefit:

1. Start with Section V – "Summary and Analysis of the Toxic Substances Control Act." This will serve to introduce you to the statute and all of its provisions without having to wade through the precise legal language of the statute which, to many lawyers, can be quite confusing and time consuming.

2. If you want to examine some of the specific provisions in the original, use the reference contained in the Cleary, Gottlieb summary and refer back to Section II which contains the statute in its entirety.

3. If you want to check on the intent of the Congress with regard to any point, refer to Section 3 which contains the full Conference Report, keyed section by section. In this way, any aspect of a Section that you might wish to refer to based on either your reading in the Cleary, Gottlieb summary or by virtue of reading the act itself can be easily located. The report will concentrate on the positions and interpretations of both the House of Representatives and the Senate, as well as on the final position taken in the ultimate bill.

4. After these, you should refer to "EPA's Implementation Plans" (Section IV) – "The Toxic Substances Control Act: Where are We Going?" and the Appendix for the latest February 17th Draft EPA document – "Assessment and Control of Chemical Problems – an Approach to Implementing the Toxic Substances Control Act." These two documents will give you an overview of EPA philosophy, intent and purpose, and planned implementation of this law.

5. With all of this background, you should now have a basic appreciation for the law and its effects on your company. However, for detailed examination of this, you should proceed to "Corporate Preparedness" (Section VII) – "Corporate Preparedness for the Toxic Substances Control Act" if you are a basic manufacturer. If you are a formulator, you should read "Industries' Assessment of the Toxic Substances Control Act" (Section VI) – "Toxic Substances Control Act – the Formulator's Viewpoint" first and then later, the previously mentioned two sections. These should still interest you since what affects your supplier will ultimately have some effect on you, not only in monetary terms but quite possibly in terms of availability of materials.

6. After reviewing these sections you should be in a position to start formulating your own compliance plans and programs. To assist in this regard, especially for those with little or no knowledge of health and environmental effects testing, we have prepared "Toxicological and Environmental Effects Testing" (Section VIII) – "Toxicological Effects Testing" under the skilled authority of Dr. Theodore Ellison.

This excellent summary provides a synopsis of major testing practices and should assist those unfamiliar with these terms and their meanings in better understanding this highly technical area.

7. Another section of the guidebook – "References and Resources" is self-explanatory and provided for your more specific use. The most significant section is Section IX – "Regulatory Review and Summary." This section now contains only one item – EPA reproposed rules on inventory reporting.

COMPLYING WITH TSCA

As a practical matter, there are obviously a number of questions that arise when trying to determine how to best comply with TSCA. These range all the way from broad policy determinations that have to be made with the company (such as reorganization) to the details of initial inventory preparation. To provide you with some expert guidance in this area, we have two sections: one dealing with corporate preparedness in general, "Corporate Preparedness for the Toxic Substances Control Act" and the other on how larger companies are organizing to comply "Chemical Industry's Assessment of the Toxic Substances Control Act." In addition, since the perspectives and concerns of formulators and users differ from those of primary manufacturers, we have also provided an article most generously prepared by the Chemical Specialties Manufacturers Association (CSMA) which presents detailed coverage of the formulator downstream viewpoints.

When reading these, it may be beneficial to keep in mind some of the distinctions that have already been made:

1. What are the requirements by statutes with precise deadlines for delineation?
2. What are the requirements that will come from discretionary authority within the agency as to deadlines?
3. What are the immediately effective provisions of the Act, e.g., Section 8(e) reporting of substantial risk, and what should you be doing about them now?
4. What is the overall strategy and implementation intention of the agency and how will this affect my compliance planning?
5. What are the informational and other requirements necessary to effectively assure compliance?
6. How do I look at the distinctions between short- and long-term company compliance activities?

Remembering these distinctions while reading the various sections of this book should assist in developing specific compliance programs involving organizational considerations and budgeting in your particular company.

ACKNOWLEDGMENTS

As Editor, my task was more than eased by the expertise and resourcefulness of my Advisory Board. My personal thanks goes to each and every one of them for their invaluable assistance in reviewing and commenting on this publication from its conceptual inception to final publication. Obviously, this board, regardless of skill or knowledge, could accomplish little or nothing were it not for the actual articles that represent the substance of the guidebook. I know that I speak for the entire board when I express my thanks, which is hardly sufficient for the actual work involved, to each of our authors who not only took the time from their busy schedules to prepare their manuscripts but, more importantly, were willing to share their knowledge with us in the preparation of this publication. And last, but not least, we must express our gratitude to the publisher and its internal Editorial Staff, especially Vicki Masseria, who certainly eased the Editor's burden by doing all of the hard work and leaving him the glory. May she and the others at CRC Press know that their hard work does not go unnoticed or unappreciated.

George S. Dominguez
Editor
July 26, 1977
Wilton, Connecticut

Section II

The Toxic Substances Control Act

S. 3149

Ninety-fourth Congress of the United States of America

AT THE SECOND SESSION

Begun and held at the City of Washington on Monday, the nineteenth day of January, one thousand nine hundred and seventy-six

An Act

To regulate commerce and protect human health and the environment by requiring testing and necessary use restrictions on certain chemical substances, and for other purposes.

Be it enacted by the Senate and House of Representatives of the United States of America in Congress assembled,

SECTION 1. SHORT TITLE AND TABLE OF CONTENTS.

This Act may be cited as the "Toxic Substances Control Act".

TABLE OF CONTENTS

SEC. 2. FINDINGS, POLICY, AND INTENT.

(a) FINDINGS.—The Congress finds that—

(1) human beings and the environment are being exposed each year to a large number of chemical substances and mixtures;

(2) among the many chemical substances and mixtures which are constantly being developed and produced, there are some whose manufacture, processing, distribution in commerce, use, or disposal may present an unreasonable risk of injury to health or the environment; and

(3) the effective regulation of interstate commerce in such chemical substances and mixtures also necessitates the regulation of intrastate commerce in such chemical substances and mixtures.

(b) POLICY.—It is the policy of the United States that—

(1) adequate data should be developed with respect to the effect of chemical substances and mixtures on health and the environ-

ment and that the development of such data should be the responsibility of those who manufacture and those who process such chemical substances and mixtures;

(2) adequate authority should exist to regulate chemical substances and mixtures which present an unreasonable risk of injury to health or the environment, and to take action with respect to chemical substances and mixtures which are imminent hazards; and

(3) authority over chemical substances and mixtures should be exercised in such a manner as not to impede unduly or create unnecessary economic barriers to technological innovation while fulfilling the primary purpose of this Act to assure that such innovation and commerce in such chemical substances and mixtures do not present an unreasonable risk of injury to health or the environment.

(c) Intent of Congress.—It is the intent of Congress that the Administrator shall carry out this Act in a reasonable and prudent manner, and that the Administrator shall consider the environmental, economic, and social impact of any action the Administrator takes or proposes to take under this Act.

SEC. 3. DEFINITIONS.

As used in this Act:

(1) the term "Administrator" means the Administrator of the Environmental Protection Agency.

(2)(A) Except as provided in subparagraph (B), the term "chemical substance" means any organic or inorganic substance of a particular molecular identity, including—

(i) any combination of such substances occurring in whole or in part as a result of a chemical reaction or occurring in nature, and

(ii) any element or uncombined radical.

(B) Such term does not include—

(i) any mixture,

(ii) any pesticide (as defined in the Federal Insecticide, Fungicide, and Rodenticide Act) when manufactured, processed, or distributed in commerce for use as a pesticide,

(iii) tobacco or any tobacco product,

(iv) any source material, special nuclear material, or byproduct material (as such terms are defined in the Atomic Energy Act of 1954 and regulations issued under such Act),

(v) any article the sale of which is subject to the tax imposed by section 4181 of the Internal Revenue Code of 1954 (determined without regard to any exemptions from such tax provided by section 4182 or 4221 or any other provision of such Code), and

(vi) any food, food additive, drug, cosmetic, or device (as such terms are defined in section 201 of the Federal Food, Drug, and Cosmetic Act) when manufactured, processed, or distributed in commerce for use as a food, food additive, drug, cosmetic, or device.

The term "food" as used in clause (vi) of this subparagraph includes poultry and poultry products (as defined in sections 4(e) and 4(f) of the Poultry Products Inspection Act), meat and meat food products (as defined in section 1(j) of the Federal Meat Inspection Act), and eggs and egg products (as defined in section 4 of the Egg Products Inspection Act).

(3) The term "commerce" means trade, traffic, transportation, or other commerce (A) between a place in a State and any place outside

of such State, or (B) which affects trade, traffic, transportation, or commerce described in clause (A).

(4) The terms "distribute in commerce" and "distribution in commerce" when used to describe an action taken with respect to a chemical substance or mixture or article containing a substance or mixture mean to sell, or the sale of, the substance, mixture, or article in commerce; to introduce or deliver for introduction into commerce, or the introduction or delivery for introduction into commerce of, the substance, mixture, or article; or to hold, or the holding of, the substance, mixture, or article after its introduction into commerce.

(5) The term "environment" includes water, air, and land and the interrelationship which exists among and between water, air, and land and all living things.

(6) The term "health and safety study" means any study of any effect of a chemical substance or mixture on health or the environment or on both, including underlying data and epidemiological studies, studies of occupational exposure to a chemical substance or mixture, toxicological, clinical, and ecological studies of a chemical substance or mixture, and any test performed pursuant to this Act.

(7) The term "manufacture" means to import into the customs territory of the United States (as defined in general headnote 2 of the Tariff Schedules of the United States), produce, or manufacture.

(8) The term "mixture" means any combination of two or more chemical substances if the combination does not occur in nature and is not, in whole or in part, the result of a chemical reaction; except that such term does include any combination which occurs, in whole or in part, as a result of a chemical reaction if none of the chemical substances comprising the combination is a new chemical substance and if the combination could have been manufactured for commercial purposes without a chemical reaction at the time the chemical substances comprising the combination were combined.

(9) The term "new chemical substance" means any chemical substance which is not included in the chemical substance list compiled and published under section 8(b).

(10) The term "process" means the preparation of a chemical substance or mixture, after its manufacture, for distribution in commerce—

(A) in the same form or physical state as, or in a different form or physical state from, that in which it was received by the person so preparing such substance or mixture, or

(B) as part of an article containing the chemical substance or mixture.

(11) The term "processor" means any person who processes a chemical substance or mixture.

(12) The term "standards for the development of test data" means a prescription of—

(A) the—

(i) health and environmental effects, and

(ii) information relating to toxicity, persistence, and other characteristics which affect health and the environment,

for which test data for a chemical substance or mixture are to be developed and any analysis that is to be performed on such data, and

(B) to the extent necessary to assure that data respecting such effects and characteristics are reliable and adequate—

(i) the manner in which such data are to be developed,

(ii) the specification of any test protocol or methodology to be employed in the development of such data, and

(iii) such other requirements as are necessary to provide such assurance.

(13) The term "State" means any State of the United States, the District of Columbia, the Commonwealth of Puerto Rico, the Virgin Islands, Guam, the Canal Zone, American Samoa, the Northern Mariana Islands, or any other territory or possession of the United States.

(14) The term "United States", when used in the geographic sense, means all of the States.

SEC. 4. TESTING OF CHEMICAL SUBSTANCES AND MIXTURES.

(a) TESTING REQUIREMENTS.—If the Administrator finds that—

(1)(A)(i) the manufacture, distribution in commerce, processing, use, or disposal of a chemical substance or mixture, or that any combination of such activities, may present an unreasonable risk of injury to health or the environment,

(ii) there are insufficient data and experience upon which the effects of such manufacture, distribution in commerce, processing, use, or disposal of such substance or mixture or of any combination of such activities on health or the environment can reasonably be determined or predicted, and

(iii) testing of such substance or mixture with respect to such effects is necessary to develop such data; or

(B)(i) a chemical substance or mixture is or will be produced in substantial quantities, and (I) it enters or may reasonably be anticipated to enter the environment in substantial quantities or (II) there is or may be significant or substantial human exposure to such substance or mixture,

(ii) there are insufficient data and experience upon which the effects of the manufacture, distribution in commerce, processing, use, or disposal of such substance or mixture or of any combination of such activities on health or the environment can reasonably be determined or predicted, and

(iii) testing of such substance or mixture with respect to such effects is necessary to develop such data; and

(2) in the case of a mixture, the effects which the mixture's manufacture, distribution in commerce, processing, use, or disposal or any combination of such activities may have on health or the environment may not be reasonably and more efficiently determined or predicted by testing the chemical substances which comprise the mixture;

the Administrator shall by rule require that testing be conducted on such substance or mixture to develop data with respect to the health and environmental effects for which there is an insufficiency of data and experience and which are relevant to a determination that the manufacture, distribution in commerce, processing, use, or disposal of such substance or mixture, or that any combination of such activities, does or does not present an unreasonable risk of injury to health or the environment.

(b)(1) TESTING REQUIREMENT RULE.—A rule under subsection (a) shall include—

(A) identification of the chemical substance or mixture for which testing is required under the rule,

(B) standards for the development of test data for such substance or mixture, and

(C) with respect to chemical substances which are not new chemical substances and to mixtures, a specification of the period (which period may not be of unreasonable duration) within which the persons required to conduct the testing shall submit to the Administrator data developed in accordance with the standards referred to in subparagraph (B).

In determining the standards and period to be included, pursuant to subparagraphs (B) and (C), in a rule under subsection (a), the Administrator's considerations shall include the relative costs of the various test protocols and methodologies which may be required under the rule and the reasonably foreseeable availability of the facilities and personnel needed to perform the testing required under the rule. Any such rule may require the submission to the Administrator of preliminary data during the period prescribed under subparagraph (C).

(2)(A) The health and environmental effects for which standards for the development of test data may be prescribed include carcinogenesis, mutagenesis, teratogenesis, behavioral disorders, cumulative or synergistic effects, and any other effect which may present an unreasonable risk of injury to health or the environment. The characteristics of chemical substances and mixtures for which such standards may be prescribed include persistence, acute toxicity, subacute toxicity, chronic toxicity, and any other characteristic which may present such a risk. The methodologies that may be prescribed in such standards include epidemiologic studies, serial or hierarchical tests, in vitro tests, and whole animal tests, except that before prescribing epidemiologic studies of employees, the Administrator shall consult with the Director of the National Institute for Occupational Safety and Health.

(B) From time to time, but not less than once each 12 months, the Administrator shall review the adequacy of the standards for development of data prescribed in rules under subsection (a) and shall, if necessary, institute proceedings to make appropriate revisions of such standards.

(3)(A) A rule under subsection (a) respecting a chemical substance or mixture shall require the persons described in subparagraph (B) to conduct tests and submit data to the Administrator on such substance or mixture, except that the Administrator may permit two or more of such persons to designate one such person or a qualified third party to conduct such tests and submit such data on behalf of the persons making the designation.

(B) The following persons shall be required to conduct tests and submit data on a chemical substance or mixture subject to a rule under subsection (a):

(i) Each person who manufactures or intends to manufacture such substance or mixture if the Administrator makes a finding described in subsection (a)(1)(A)(ii) or (a)(1)(B)(ii) with respect to the manufacture of such substance or mixture.

(ii) Each person who processes or intends to process such substance or mixture if the Administrator makes a finding described in subsection (a)(1)(A)(ii) or (a)(1)(B)(ii) with respect to the processing of such substance or mixture.

(iii) Each person who manufactures or processes or intends to manufacture or process such substance or mixture if the Administrator makes a finding described in subsection (a)(1)(A)(ii) or (a)(1)(B)(ii) with respect to the distribution in commerce, use, or disposal of such substance or mixture.

(4) Any rule under subsection (a) requiring the testing of and submission of data for a particular chemical substance or mixture shall expire at the end of the reimbursement period (as defined in subsection (c)(3)(B)) which is applicable to test data for such substance or mixture unless the Administrator repeals the rule before such date;

and a rule under subsection (a) requiring the testing of and submission of data for a category of chemical substances or mixtures shall expire with respect to a chemical substance or mixture included in the category at the end of the reimbursement period (as so defined) which is applicable to test data for such substance or mixture unless the Administrator before such date repeals the application of the rule to such substance or mixture or repeals the rule.

(5) Rules issued under subsection (a) (and any substantive amendment thereto or repeal thereof) shall be promulgated pursuant to section 553 of title 5, United States Code, except that (A) the Administrator shall give interested persons an opportunity for the oral presentation of data, views, or arguments, in addition to an opportunity to make written submissions; (B) a transcript shall be made of any oral presentation; and (C) the Administrator shall make and publish with the rule the findings described in paragraph (1)(A) or (1)(B) of subsection (a) and, in the case of a rule respecting a mixture, the finding described in paragraph (2) of such subsection.

(c) EXEMPTION.—(1) Any person required by a rule under subsection (a) to conduct tests and submit data on a chemical substance or mixture may apply to the Administrator (in such form and manner as the Administrator shall prescribe) for an exemption from such requirement.

(2) If, upon receipt of an application under paragraph (1), the Administrator determines that—

(A) the chemical substance or mixture with respect to which such application was submitted is equivalent to a chemical substance or mixture for which data has been submitted to the Administrator in accordance with a rule under subsection (a) or for which data is being developed pursuant to such a rule, and

(B) submission of data by the applicant on such substance or mixture would be duplicative of data which has been submitted to the Administrator in accordance with such rule or which is being developed pursuant to such rule,

the Administrator shall exempt, in accordance with paragraph (3) or (4), the applicant from conducting tests and submitting data on such substance or mixture under the rule with respect to which such application was submitted.

(3)(A) If the exemption under paragraph (2) of any person from the requirement to conduct tests and submit test data on a chemical substance or mixture is granted on the basis of the existence of previously submitted test data and if such exemption is granted during the reimbursement period for such test data (as prescribed by subparagraph (B)), then (unless such person and the persons referred to in clauses (i) and (ii) agree on the amount and method of reimbursement) the Administrator shall order the person granted the exemption to provide fair and equitable reimbursement (in an amount determined under rules of the Administrator)—

(i) to the person who previously submitted such test data, for a portion of the costs incurred by such person in complying with the requirement to submit such data, and

(ii) to any other person who has been required under this subparagraph to contribute with respect to such costs, for a portion of the amount such person was required to contribute.

In promulgating rules for the determination of fair and equitable reimbursement to the persons described in clauses (i) and (ii) for costs incurred with respect to a chemical substance or mixture, the Administrator shall, after consultation with the Attorney General and the Federal Trade Commission, consider all relevant factors, including the effect on the competitive position of the person required to provide reimbursement in relation to the person to be reimbursed and the share of the market for such substance or mixture of the person required to provide reimbursement in relation to the share of such market of the persons to be reimbursed. An order under this subparagraph shall, for purposes of judicial review, be considered final agency action.

(B) For purposes of subparagraph (A), the reimbursement period for any test data for a chemical substance or mixture is a period—

(i) beginning on the date such data is submitted in accordance with a rule promulgated under subsection (a), and

(ii) ending—

(I) five years after the date referred to in clause (i), or

(II) at the expiration of a period which begins on the date referred to in clause (i) and which is equal to the period which the Administrator determines was necessary to develop such data,

whichever is later.

(4)(A) If the exemption under paragraph (2) of any person from the requirement to conduct tests and submit test data on a chemical substance or mixture is granted on the basis of the fact that test data is being developed by one or more persons pursuant to a rule promulgated under subsection (a), then (unless such person and the persons referred to in clauses (i) and (ii) agree on the amount and method of reimbursement) the Administrator shall order the person granted the exemption to provide fair and equitable reimbursement (in an amount determined under rules of the Administrator)—

(i) to each such person who is developing such test data, for a portion of the costs incurred by each such person in complying with such rule, and

(ii) to any other person who has been required under this subparagraph to contribute with respect to the costs of complying with such rule, for a portion of the amount such person was required to contribute.

In promulgating rules for the determination of fair and equitable reimbursement to the persons described in clauses (i) and (ii) for costs incurred with respect to a chemical substance or mixture, the Administrator shall, after consultation with the Attorney General and the Federal Trade Commission, consider the factors described in the second sentence of paragraph (3)(A). An order under this subparagraph shall, for purposes of judicial review, be considered final agency action.

(B) If any exemption is granted under paragraph (2) on the basis of the fact that one or more persons are developing test data pursuant to a rule promulgated under subsection (a) and if after such exemption is granted the Administrator determines that no such person has complied with such rule, the Administrator shall (i) after providing written notice to the person who holds such exemption and an opportunity for a hearing, by order terminate such exemption, and (ii) notify in writing such person of the requirements of the rule with respect to which such exemption was granted.

(d) Notice.—Upon the receipt of any test data pursuant to a rule under subsection (a), the Administrator shall publish a notice of the receipt of such data in the Federal Register within 15 days of its receipt. Subject to section 14, each such notice shall (1) identify the chemical substance or mixture for which data have been received; (2) list the uses or intended uses of such substance or mixture and the information required by the applicable standards for the development of test data; and (3) describe the nature of the test data developed. Except as otherwise provided in section 14, such data shall be made available by the Administrator for examination by any person.

(e) PRIORITY LIST.—(1) (A) There is established a committee to make recommendations to the Administrator respecting the chemical substances and mixtures to which the Administrator should give priority consideration for the promulgation of a rule under subsection (a). In making such a recommendation with respect to any chemical substance or mixture, the committee shall consider all relevant factors, including—

(i) the quantities in which the substance or mixture is or will be manufactured,

(ii) the quantities in which the substance or mixture enters or will enter the environment,

(iii) the number of individuals who are or will be exposed to the substance or mixture in their places of employment and the duration of such exposure,

(iv) the extent to which human beings are or will be exposed to the substance or mixture,

(v) the extent to which the substance or mixture is closely related to a chemical substance or mixture which is known to present an unreasonable risk of injury to health or the environment,

(vi) the existence of data concerning the effects of the substance or mixture on health or the environment,

(vii) the extent to which testing of the substance or mixture may result in the development of data upon which the effects of the substance or mixture on health or the environment can reasonably be determined or predicted, and

(viii) the reasonably foreseeable availability of facilities and personnel for performing testing on the substance or mixture.

The recommendations of the committee shall be in the form of a list of chemical substances and mixtures which shall be set forth, either by individual substance or mixture or by groups of substances or mixtures, in the order in which the committee determines the Administrator should take action under subsection (a) with respect to the substances and mixtures. In establishing such list, the committee shall give priority attention to those chemical substances and mixtures which are known to cause or contribute to or which are suspected of causing or contributing to cancer, gene mutations, or birth defects. The committee shall designate chemical substances and mixtures on the list with respect to which the committee determines the Administrator should, within 12 months of the date on which such substances and mixtures are first designated, initiate a proceeding under subsection (a). The total number of chemical substances and mixtures on the list which are designated under the preceding sentence may not, at any time, exceed 50.

(B) As soon as practicable but not later than nine months after the effective date of this Act, the committee shall publish in the Federal Register and transmit to the Administrator the list and designations required by subparagraph (A) together with the reasons for the committee's inclusion of each chemical substance or mixture on the list. At least every six months after the date of the transmission to the Administrator of the list pursuant to the preceding sentence, the committee shall make such revisions in the list as it determines to be necessary and shall transmit them to the Administrator together with the committee's reasons for the revisions. Upon receipt of any such revision, the Administrator shall publish in the Federal Register the list with such revision, the reasons for such revision, and the designations made under subparagraph (A). The Administrator shall provide reasonable opportunity to any interested person to file with the Administrator written comments on the committee's list, any revision of such list by the committee, and designations made by the committee, and shall make such comments available to the public. Within the 12-month period beginning on the date of the first inclusion on the list of a chemical substance or mixture designated by the committee under subparagraph (A) the Administrator shall with respect to such chemical substance or mixture either initiate a rulemaking proceeding under subsection (a) or if such a proceeding is not initiated within such period, publish in the Federal Register the Administrator's reason for not initiating such a proceeding.

(2) (A) The committee established by paragraph (1) (A) shall consist of eight members as follows:

(i) One member appointed by the Administrator from the Environmental Protection Agency.

(ii) One member appointed by the Secretary of Labor from officers or employees of the Department of Labor engaged in the Secretary's activities under the Occupational Safety and Health Act of 1970.

(iii) One member appointed by the Chairman of the Council on Environmental Quality from the Council or its officers or employees.

(iv) One member appointed by the Director of the National Institute for Occupational Safety and Health from officers or employees of the Institute.

(v) One member appointed by the Director of the National Institute of Environmental Health Sciences from officers or employees of the Institute.

(vi) One member appointed by the Director of the National Cancer Institute from officers or employees of the Institute.

(vii) One member appointed by the Director of the National Science Foundation from officers or employees of the Foundation.

(viii) One member appointed by the Secretary of Commerce from officers or employees of the Department of Commerce.

(B) (i) An appointed member may designate an individual to serve on the committee on the member's behalf. Such a designation may be made only with the approval of the applicable appointing authority and only if the individual is from the entity from which the member was appointed.

(ii) No individual may serve as a member of the committee for more than four years in the aggregate. If any member of the committee leaves the entity from which the member was appointed, such member may not continue as a member of the committee, and the member's position shall be considered to be vacant. A vacancy in the committee shall be filled in the same manner in which the original appointment was made.

(iii) Initial appointments to the committee shall be made not later than the 60th day after the effective date of this Act. Not later than the 90th day after such date the members of the committee shall hold a meeting for the selection of a chairperson from among their number.

(C) (i) No member of the committee, or designee of such member, shall accept employment or compensation from any person subject to any requirement of this Act or of any rule promulgated or order issued thereunder, for a period of at least 12 months after termination of service on the committee.

(ii) No person, while serving as a member of the committee, or designee of such member, may own any stocks or bonds, or have any pecuniary interest, of substantial value in any person engaged in the manufacture, processing, or distribution in commerce of any chemical substance or mixture subject to any requirement of this Act or of any rule promulgated or order issued thereunder.

(iii) The Administrator, acting through attorneys of the Environmental Protection Agency, or the Attorney General may bring an action in the appropriate district court of the United States to restrain any violation of this subparagraph.

(D) The Administrator shall provide the committee such administrative support services as may be necessary to enable the committee to carry out its function under this subsection.

(f) REQUIRED ACTIONS.—Upon the receipt of—

(1) any test data required to be submitted under this Act, or

(2) any other information available to the Administrator,

which indicates to the Administrator that there may be a reasonable basis to conclude that a chemical substance or mixture presents or will present a significant risk of serious or widespread harm to human beings from cancer, gene mutations, or birth defects, the Administrator shall, within the 180-day period beginning on the date of the receipt of such data or information, initiate appropriate action under section 5, 6, or 7 to prevent or reduce to a sufficient extent such risk or publish in the Federal Register a finding that such risk is not unreasonable. For good cause shown the Administrator may extend such period for an additional period of not more than 90 days. The Administrator shall publish in the Federal Register notice of any such extension and the reasons therefor. A finding by the Administrator that a risk is not unreasonable shall be considered agency action for purposes of judicial review under chapter 7 of title 5, United States Code. This subsection shall not take effect until two years after the effective date of this Act.

(g) PETITION FOR STANDARDS FOR THE DEVELOPMENT OF TEST DATA.—A person intending to manufacture or process a chemical substance for which notice is required under section 5(a) and who is not required under a rule under subsection (a) to conduct tests and submit data on such substance may petition the Administrator to prescribe standards for the development of test data for such substance. The Administrator shall by order either grant or deny any such petition within 60 days of its receipt. If the petition is granted, the Administrator shall prescribe such standards for such substance within 75 days of the date the petition is granted. If the petition is denied, the Administrator shall publish, subject to section 14, in the Federal Register the reasons for such denial.

SEC. 5. MANUFACTURING AND PROCESSING NOTICES.

(a) IN GENERAL.—(1) Except as provided in subsection (h), no person may—

(A) manufacture a new chemical substance on or after the 30th day after the date on which the Administrator first publishes the list required by section 8(b), or

(B) manufacture or process any chemical substance for a use which the Administrator has determined, in accordance with paragraph (2), is a significant new use,

unless such person submits to the Administrator, at least 90 days before such manufacture or processing, a notice, in accordance with subsection (d), of such person's intention to manufacture or process such substance and such person complies with any applicable requirement of subsection (b).

(2) A determination by the Administrator that a use of a chemical substance is a significant new use with respect to which notification is required under paragraph (1) shall be made by a rule promulgated after a consideration of all relevant factors, including—

(A) the projected volume of manufacturing and processing of a chemical substance,

(B) the extent to which a use changes the type or form of exposure of human beings or the environment to a chemical substance,

(C) the extent to which a use increases the magnitude and duration of exposure of human beings or the environment to a chemical substance, and

(D) the reasonably anticipated manner and methods of manufacturing, processing, distribution in commerce, and disposal of a chemical substance.

(b) Submission of Test Data.—(1) (A) If (i) a person is required by subsection (a) (1) to submit a notice to the Administrator before beginning the manufacture or processing of a chemical substance, and (ii) such person is required to submit test data for such substance pursuant to a rule promulgated under section 4 before the submission of such notice, such person shall submit to the Administrator such data in accordance with such rule at the time notice is submitted in accordance with subsection (a) (1).

(B) If—

(i) a person is required by subsection (a) (1) to submit a notice to the Administrator, and

(ii) such person has been granted an exemption under section 4(c) from the requirements of a rule promulgated under section 4 before the submission of such notice,

such person may not, before the expiration of the 90 day period which begins on the date of the submission in accordance with such rule of the test data the submission or development of which was the basis for the exemption, manufacture such substance if such person is subject to subsection (a) (1) (A) or manufacture or process such substance for a significant new use if the person is subject to subsection (a) (1) (B).

(2) (A) If a person—

(i) is required by subsection (a) (1) to submit a notice to the Administrator before beginning the manufacture or processing of a chemical substance listed under paragraph (4), and

(ii) is not required by a rule promulgated under section 4 before the submission of such notice to submit test data for such substance,

such person shall submit to the Administrator data prescribed by subparagraph (B) at the time notice is submitted in accordance with subsection (a) (1).

(B) Data submitted pursuant to subparagraph (A) shall be data which the person submitting the data believes show that—

(i) in the case of a substance with respect to which notice is required under subsection (a) (1) (A), the manufacture, processing, distribution in commerce, use, and disposal of the chemical substance or any combination of such activities will not present an unreasonable risk of injury to health or the environment, or

(ii) in the case of a chemical substance with respect to which notice is required under subsection (a) (1) (B), the intended significant new use of the chemical substance will not present an unreasonable risk of injury to health or the environment.

(3) Data submitted under paragraph (1) or (2) shall be made available, subject to section 14, for examination by interested persons.

(4)(A)(i) The Administrator may, by rule, compile and keep current a list of chemical substances with respect to which the Administrator finds that the manufacture, processing, distribution in commerce, use, or disposal, or any combination of such activities, presents or may present an unreasonable risk of injury to health or the environment.

(ii) In making a finding under clause (i) that the manufacture, processing, distribution in commerce, use, or disposal of a chemical substance or any combination of such activities presents or may present an unreasonable risk of injury to health or the environment, the Administrator shall consider all relevant factors, including—

(I) the effects of the chemical substance on health and the magnitude of human exposure to such substance; and

(II) the effects of the chemical substance on the environment and the magnitude of environmental exposure to such substance.

(B) The Administrator shall, in prescribing a rule under subparagraph (A) which lists any chemical substance, identify those uses, if any, which the Administrator determines, by rule under subsection (a)(2), would constitute a significant new use of such substance.

(C) Any rule under subparagraph (A), and any substantive amendment or repeal of such a rule, shall be promulgated pursuant to the procedures specified in section 553 of title 5, United States Code, except that (i) the Administrator shall give interested persons an opportunity for the oral presentation of data, views, or arguments, in addition to an opportunity to make written submissions, (ii) a transcript shall be kept of any oral presentation, and (iii) the Administrator shall make and publish with the rule the finding described in subparagraph (A).

(c) Extension of Notice Period.—The Administrator may for good cause extend for additional periods (not to exceed in the aggregate 90 days) the period, prescribed by subsection (a) or (b) before which the manufacturing or processing of a chemical substance subject to such subsection may begin. Subject to section 14, such an extension and the reasons therefor shall be published in the Federal Register and shall constitute a final agency action subject to judicial review.

(d) Content of Notice; Publications in the Federal Register.—(1) The notice required by subsection (a) shall include—

(A) insofar as known to the person submitting the notice or insofar as reasonably ascertainable, the information described in subparagraphs (A), (B), (C), (D), (F), and (G) of section 8(a)(2), and

(B) in such form and manner as the Administrator may prescribe, any test data in the possession or control of the person giving such notice which are related to the effect of any manufacture, processing, distribution in commerce, use, or disposal of such substance or any article containing such substance, or of any combination of such activities, on health or the environment, and

(C) a description of any other data concerning the environmental and health effects of such substance, insofar as known to the person making the notice or insofar as reasonably ascertainable.

Such a notice shall be made available, subject to section 14, for examination by interested persons.

(2) Subject to section 14, not later than five days (excluding Saturdays, Sundays and legal holidays) after the date of the receipt of a

notice under subsection (a) or of data under subsection (b), the Administrator shall publish in the Federal Register a notice which—

(A) identifies the chemical substance for which notice or data has been received;

(B) lists the uses or intended uses of such substance; and

(C) in the case of the receipt of data under subsection (b), describes the nature of the tests performed on such substance and any data which was developed pursuant to subsection (b) or a rule under section 4.

A notice under this paragraph respecting a chemical substance shall identify the chemical substance by generic class unless the Administrator determines that more specific identification is required in the public interest.

(3) At the beginning of each month the Administrator shall publish a list in the Federal Register of (A) each chemical substance for which notice has been received under subsection (a) and for which the notification period prescribed by subsection (a), (b), or (c) has not expired, and (B) each chemical substance for which such notification period has expired since the last publication in the Federal Register of such list.

(e) Regulation Pending Development of Information.—(1) (A) If the Administrator determines that—

(i) the information available to the Administrator is insufficient to permit a reasoned evaluation of the health and environmental effects of a chemical substance with respect to which notice is required by subsection (a); and

(ii) (I) in the absence of sufficient information to permit the Administrator to make such an evaluation, the manufacture, processing, distribution in commerce, use, or disposal of such substance, or any combination of such activities, may present an unreasonable risk of injury to health or the environment, or

(II) such substance is or will be produced in substantial quantities, and such substance either enters or may reasonably be anticipated to enter the environment in substantial quantities or there is or may be significant or substantial human exposure to the substance,

the Administrator may issue a proposed order, to take effect on the expiration of the notification period applicable to the manufacturing or processing of such substance under subsection (a), (b), or (c), to prohibit or limit the manufacture, processing, distribution in commerce, use, or disposal of such substance or to prohibit or limit any combination of such activities.

(B) A proposed order may not be issued under subparagraph (A) respecting a chemical substance (i) later than 45 days before the expiration of the notification period applicable to the manufacture or processing of such substance under subsection (a), (b), or (c), and (ii) unless the Administrator has, on or before the issuance of the proposed order, notified, in writing, each manufacturer or processor, as the case may be, of such substance of the determination which underlies such order.

(C) If a manufacturer or processor of a chemical substance to be subject to a proposed order issued under subparagraph (A) files with the Administrator (within the 30-day period beginning on the date such manufacturer or processor received the notice required by subparagraph (B)(ii)) objections specifying with particularity the provisions of the order deemed objectionable and stating the grounds therefor, the proposed order shall not take effect.

(2)(A)(i) Except as provided in clause (ii), if with respect to a chemical substance with respect to which notice is required by subsection (a), the Administrator makes the determination described in paragraph (1)(A) and if—

(I) the Administrator does not issue a proposed order under paragraph (1) respecting such substance, or

(II) the Administrator issues such an order respecting such substance but such order does not take effect because objections were filed under paragraph (1)(C) with respect to it,

the Administrator, through attorneys of the Environmental Protection Agency, shall apply to the United States District Court for the District of Columbia or the United States district court for the judicial district in which the manufacturer or processor, as the case may be, of such substance is found, resides, or transacts business for an injunction to prohibit or limit the manufacture, processing, distribution in commerce, use, or disposal of such substance (or to prohibit or limit any combination of such activities).

(ii) If the Administrator issues a proposed order under paragraph (1)(A) respecting a chemical substance but such order does not take effect because objections have been filed under paragraph (1)(C) with respect to it, the Administrator is not required to apply for an injunction under clause (i) respecting such substance if the Administrator determines, on the basis of such objections, that the determinations under paragraph (1)(A) may not be made.

(B) A district court of the United States which receives an application under subparagraph (A)(i) for an injunction respecting a chemical substance shall issue such injunction if the court finds that—

(i) the information available to the Administrator is insufficient to permit a reasoned evaluation of the health and environmental effects of a chemical substance with respect to which notice is required by subsection (a); and

(ii)(I) in the absence of sufficient information to permit the Administrator to make such an evaluation, the manufacture, processing, distribution in commerce, use, or disposal of such substance, or any combination of such activities, may present an unreasonable risk of injury to health or the environment, or

(II) such substance is or will be produced in substantial quantities, and such substance either enters or may reasonably be anticipated to enter the environment in substantial quantities or there is or may be significant or substantial human exposure to the substance.

(C) Pending the completion of a proceeding for the issuance of an injunction under subparagraph (B) respecting a chemical substance, the court may, upon application of the Administrator made through attorneys of the Environmental Protection Agency, issue a temporary restraining order or a preliminary injunction to prohibit the manufacture, processing, distribution in commerce, use, or disposal of such a substance (or any combination of such activities) if the court finds that the notification period applicable under subsection (a), (b), or (c) to the manufacturing or processing of such substance may expire before such proceeding can be completed.

(D) After the submission to the Administrator of test data sufficient to evaluate the health and environmental effects of a chemical substance subject to an injunction issued under subparagraph (B) and the evaluation of such data by the Administrator, the district court of the United States which issued such injunction shall, upon petition, dissolve the injunction unless the Administrator has initiated a proceeding for the issuance of a rule under section 6(a) respecting the substance. If such a proceeding has been initiated, such court shall continue the injunction in effect until the effective date of the rule promulgated in such proceeding or, if such proceeding is terminated without the promulgation of a rule, upon the termination of the proceeding, whichever occurs first.

(f) PROTECTION AGAINST UNREASONABLE RISKS.—(1) If the Administrator finds that there is a reasonable basis to conclude that the manufacture, processing, distribution in commerce, use, or disposal of a chemical substance with respect to which notice is required by subsection (a), or that any combination of such activities, presents or will present an unreasonable risk of injury to health or environment before a rule promulgated under section 6 can protect against such risk, the Administrator shall, before the expiration of the notification period applicable under subsection (a), (b), or (c) to the manufacturing or processing of such substance, take the action authorized by paragraph (2) or (3) to the extent necessary to protect against such risk.

(2) The Administrator may issue a proposed rule under section 6(a) to apply to a chemical substance with respect to which a finding was made under paragraph (1)—

(A) a requirement limiting the amount of such substance which may be manufactured, processed, or distributed in commerce,

(B) a requirement described in paragraph (2), (3), (4), (5), (6), or (7) of section 6(a), or

(C) any combination of the requirements referred to in subparagraph (B).

Such a proposed rule shall be effective upon its publication in the Federal Register. Section 6(d)(2)(B) shall apply with respect to such rule.

(3)(A) The Administrator may—

(i) issue a proposed order to prohibit the manufacture, processing, or distribution in commerce of a substance with respect to which a finding was made under paragraph (1), or

(ii) apply, through attorneys of the Environmental Protection Agency, to the United States District Court for the District of Columbia or the United States district court for the judicial district in which the manufacturer, or processor, as the case may be, of such substance, is found, resides, or transacts business for an injunction to prohibit the manufacture, processing, or distribution in commerce of such substance.

A proposed order issued under clause (i) respecting a chemical substance shall take effect on the expiration of the notification period applicable under subsection (a), (b), or (c) to the manufacture or processing of such substance.

(B) If the district court of the United States to which an application has been made under subparagraph (A)(ii) finds that there is a reasonable basis to conclude that the manufacture, processing, distribution in commerce, use, or disposal of the chemical substance with respect to which such application was made, or that any combination of such activities, presents or will present an unreasonable risk of injury to health or the environment before a rule promulgated under section 6 can protect against such risk, the court shall issue an injunction to prohibit the manufacture, processing, or distribution in commerce of such substance or to prohibit any combination of such activities.

(C) The provisions of subparagraphs (B) and (C) of subsection (e)(1) shall apply with respect to an order issued under clause (i) of subparagraph (A); and the provisions of subparagraph (C) of subsection (e)(2) shall apply with respect to an injunction issued under subparagraph (B).

(D) If the Administrator issues an order pursuant to subparagraph (A)(i) respecting a chemical substance and objections are filed in accordance with subsection (e)(1)(C), the Administrator shall seek an injunction under subparagraph (A)(ii) respecting such substance unless the Administrator determines, on the basis of such objections, that such substance does not or will not present an unreasonable risk of injury to health or the environment.

(g) STATEMENT OF REASONS FOR NOT TAKING ACTION.—If the Administrator has not initiated any action under this section or section 6 or 7 to prohibit or limit the manufacture, processing, distribution in commerce, use, or disposal of a chemical substance, with respect to which notification or data is required by subsection (a)(1)(B) or (b), before the expiration of the notification period applicable to the manufacturing or processing of such substance, the Administrator shall publish a statement of the Administrator's reasons for not initiating such action. Such a statement shall be published in the Federal Register before the expiration of such period. Publication of such statement in accordance with the preceding sentence is not a prerequisite to the manufacturing or processing of the substance with respect to which the statement is to be published.

(h) EXEMPTIONS.—(1) The Administrator may, upon application, exempt any person from any requirement of subsection (a) or (b) to permit such person to manufacture or process a chemical substance for test marketing purposes—

(A) upon a showing by such person satisfactory to the Administrator that the manufacture, processing, distribution in commerce, use, and disposal of such substance, and that any combination of such activities, for such purposes will not present any unreasonable risk of injury to health or the environment, and

(B) under such restrictions as the Administrator considers appropriate.

(2)(A) The Administrator may, upon application, exempt any person from the requirement of subsection (b)(2) to submit data for a chemical substance. If, upon receipt of an application under the preceding sentence, the Administrator determines that—

(i) the chemical substance with respect to which such application was submitted is equivalent to a chemical substance for which data has been submitted to the Administrator as required by subsection (b)(2), and

(ii) submission of data by the applicant on such substance would be duplicative of data which has been submitted to the Administrator in accordance with such subsection,

the Administrator shall exempt the applicant from the requirement to submit such data on such substance. No exemption which is granted under this subparagraph with respect to the submission of data for a chemical substance may take effect before the beginning of the reimbursement period applicable to such data.

(B) If the Administrator exempts any person, under subparagraph (A), from submitting data required under subsection (b)(2) for a chemical substance because of the existence of previously submitted data and if such exemption is granted during the reimbursement period for such data, then (unless such person and the persons referred to in clauses (i) and (ii) agree on the amount and method of reimbursement) the Administrator shall order the person granted the exemption to provide fair and equitable reimbursement (in an amount determined under rules of the Administrator)—

(i) to the person who previously submitted the data on which the exemption was based, for a portion of the costs incurred by such person in complying with the requirement under subsection (b)(2) to submit such data, and

(ii) to any other person who has been required under this subparagraph to contribute with respect to such costs, for a portion of the amount such person was required to contribute.

In promulgating rules for the determination of fair and equitable reimbursement to the persons described in clauses (i) and (ii) for costs incurred with respect to a chemical substance, the Administrator shall, after consultation with the Attorney General and the Federal Trade Commission, consider all relevant factors, including the effect on the competitive position of the person required to provide reimbursement in relation to the persons to be reimbursed and the share of the market for such substance of the person required to provide reimbursement in relation to the share of such market of the persons to be reimbursed. For purposes of judicial review, an order under this subparagraph shall be considered final agency action.

(C) For purposes of this paragraph, the reimbursement period for any previously submitted data for a chemical substance is a period—

(i) beginning on the date of the termination of the prohibition, imposed under this section, on the manufacture or processing of such substance by the person who submitted such data to the Administrator, and

(ii) ending—

(I) five years after the date referred to in clause (i), or

(II) at the expiration of a period which begins on the date referred to in clause (i) and is equal to the period which the Administrator determines was necessary to develop such data,

whichever is later.

(3) The requirements of subsections (a) and (b) do not apply with respect to the manufacturing or processing of any chemical substance which is manufactured or processed, or proposed to be manufactured or processed, only in small quantities (as defined by the Administrator by rule) solely for purposes of—

(A) scientific experimentation or analysis, or

(B) chemical research on, or analysis of such substance or another substance, including such research or analysis for the development of a product,

if all persons engaged in such experimentation, research, or analysis for a manufacturer or processor are notified (in such form and manner as the Administrator may prescribe) of any risk to health which the manufacturer, processor, or the Administrator has reason to believe may be associated with such chemical substance.

(4) The Administrator may, upon application and by rule, exempt the manufacturer of any new chemical substance from all or part of the requirements of this section if the Administrator determines that the manufacture, processing, distribution in commerce, use, or disposal of such chemical substance, or that any combination of such activities, will not present an unreasonable risk of injury to health or the environment. A rule promulgated under this paragraph (and any substantive amendment to, or repeal of, such a rule) shall be promulgated in accordance with paragraphs (2) and (3) of section 6(c).

(5) The Administrator may, upon application, make the requirements of subsections (a) and (b) inapplicable with respect to the manufacturing or processing of any chemical substance (A) which exists temporarily as a result of a chemical reaction in the manufacturing or processing of a mixture or another chemical substance, and (B) to which there is no, and will not be, human or environmental exposure.

(6) Immediately upon receipt of an application under paragraph (1) or (5) the Administrator shall publish in the Federal Register notice of the receipt of such application. The Administrator shall give interested persons an opportunity to comment upon any such application and shall, within 45 days of its receipt, either approve or deny the application. The Administrator shall publish in the Federal Register notice of the approval or denial of such an application.

(i) Definition.—For purposes of this section, the terms "manufacture" and "process" mean manufacturing or processing for commercial purposes.

SEC. 6. REGULATION OF HAZARDOUS CHEMICAL SUBSTANCES AND MIXTURES.

(a) Scope of Regulation.—If the Administrator finds that there is a reasonable basis to conclude that the manufacture, processing, distribution in commerce, use, or disposal of a chemical substance or mixture, or that any combination of such activities, presents or will present an unreasonable risk of injury to health or the environment, the Administrator shall by rule apply one or more of the following requirements to such substance or mixture to the extent necessary to protect adequately against such risk using the least burdensome requirements:

(1) A requirement (A) prohibiting the manufacturing, processing, or distribution in commerce of such substance or mixture, or (B) limiting the amount of such substance or mixture which may be manufactured, processed, or distributed in commerce.

(2) A requirement—

(A) prohibiting the manufacture, processing, or distribution in commerce of such substance or mixture for (i) a particular use or (ii) a particular use in a concentration in excess of a level specified by the Administrator in the rule imposing the requirement, or

(B) limiting the amount of such substance or mixture which may be manufactured, processed, or distributed in commerce for (i) a particular use or (ii) a particular use in a concentration in excess of a level specified by the Administrator in the rule imposing the requirement.

(3) A requirement that such substance or mixture or any article containing such substance or mixture be marked with or accompanied by clear and adequate warnings and instructions with respect to its use, distribution in commerce, or disposal or with respect to any combination of such activities. The form and content of such warnings and instructions shall be prescribed by the Administrator.

(4) A requirement that manufacturers and processors of such substance or mixture make and retain records of the processes used to manufacture or process such substance or mixture and monitor or conduct tests which are reasonable and necessary to assure compliance with the requirements of any rule applicable under this subsection.

(5) A requirement prohibiting or otherwise regulating any manner or method of commercial use of such substance or mixture.

(6)(A) A requirement prohibiting or otherwise regulating any manner or method of disposal of such substance or mixture, or of any article containing such substance or mixture, by its manufacturer or processor or by any other person who uses, or disposes of, it for commercial purposes.

(B) A requirement under subparagraph (A) may not require any person to take any action which would be in violation of any law or requirement of, or in effect for, a State or political subdivision, and shall require each person subject to it to notify each State and political subdivision in which a required disposal may occur of such disposal.

(7) A requirement directing manufacturers or processors of such substance or mixture (A) to give notice of such unreasonable risk of injury to distributors in commerce of such substance or mixture and, to the extent reasonably ascertainable, to other persons in possession of such substance or mixture or exposed to such substance or mixture, (B) to give public notice of such risk of injury, and (C) to replace or repurchase such substance or mixture as elected by the person to which the requirement is directed.

Any requirement (or combination of requirements) imposed under this subsection may be limited in application to specified geographic areas.

(b) Quality Control.—If the Administrator has a reasonable basis to conclude that a particular manufacturer or processor is manufacturing or processing a chemical substance or mixture in a manner which unintentionally causes the chemical substance or mixture to present or which will cause it to present an unreasonable risk of injury to health or the environment—

(1) the Administrator may by order require such manufacturer or processor to submit a description of the relevant quality control procedures followed in the manufacturing or processing of such chemical substance or mixture; and

(2) if the Administrator determines—

(A) that such quality control procedures are inadequate to prevent the chemical substance or mixture from presenting such risk of injury, the Administrator may order the manufacturer or processor to revise such quality control procedures to the extent necessary to remedy such inadequacy; or

(B) that the use of such quality control procedures has resulted in the distribution in commerce of chemical substances or mixtures which present an unreasonable risk of injury to health or the environment, the Administrator may order the manufacturer or processor to (i) give notice of such risk to processors or distributors in commerce of any such substance or mixture, or to both, and, to the extent reasonably ascertainable, to any other person in possession of or exposed to any such substance, (ii) to give public notice of such risk, and (iii) to provide such replacement or repurchase of any such substance or mixture as is necessary to adequately protect health or the environment.

A determination under subparagraph (A) or (B) of paragraph (2) shall be made on the record after opportunity for hearing in accordance with section 554 of title 5, United States Code. Any manufacturer or processor subject to a requirement to replace or repurchase a chemical substance or mixture may elect either to replace or repurchase the substance or mixture and shall take either such action in the manner prescribed by the Administrator.

(c) PROMULGATION OF SUBSECTION (a) RULES.—(1) In promulgating any rule under subsection (a) with respect to a chemical substance or mixture, the Administrator shall consider and publish a statement with respect to—

(A) the effects of such substance or mixture on health and the magnitude of the exposure of human beings to such substance or mixture,

(B) the effects of such substance or mixture on the environment and the magnitude of the exposure of the environment to such substance or mixture,

(C) the benefits of such substance or mixture for various uses and the availability of substitutes for such uses, and

(D) the reasonably ascertainable economic consequences of the rule, after consideration of the effect on the national economy, small business, technological innovation, the environment, and public health.

If the Administrator determines that a risk of injury to health or the environment could be eliminated or reduced to a sufficient extent by actions taken under another Federal law (or laws) administered in whole or in part by the Administrator, the Administrator may not promulgate a rule under subsection (a) to protect against such risk of injury unless the Administrator finds, in the Administrator's discretion, that it is in the public interest to protect against such risk under this Act. In making such a finding the Administrator shall consider (i) all relevant aspects of the risk, as determined by the Administrator in the Administrator's discretion, (ii) a comparison of the estimated costs of complying with actions taken under this Act and under such law (or laws), and (iii) the relative efficiency of actions under this Act and under such law (or laws) to protect against such risk of injury.

(2) When prescribing a rule under subsection (a) the Administrator shall proceed in accordance with section 553 of title 5, United States Code (without regard to any reference in such section to sections 556 and 557 of such title), and shall also (A) publish a notice of proposed rulemaking stating with particularity the reason for the proposed rule; (B) allow interested persons to submit written data, views, and arguments, and make all such submissions publicly available; (C) provide an opportunity for an informal hearing in accordance with paragraph (3); (D) promulgate, if appropriate, a final rule based on the matter in the rulemaking record (as defined in section 19(a)), and (E) make and publish with the rule the finding described in subsection (a).

(3) Informal hearings required by paragraph (2)(C) shall be conducted by the Administrator in accordance with the following requirements:

(A) Subject to subparagraph (B), an interested person is entitled—

(i) to present such person's position orally or by documentary submissions (or both), and

(ii) if the Administrator determines that there are disputed issues of material fact it is necessary to resolve, to present such rebuttal submissions and to conduct (or have conducted under subparagraph (B)(ii)) such cross-examination of persons as the Administrator determines (I) to be appropriate, and (II) to be required for a full and true disclosure with respect to such issues.

(B) The Administrator may prescribe such rules and make such rulings concerning procedures in such hearings to avoid unnecessary costs or delay. Such rules or rulings may include (i) the imposition of reasonable time limits on each interested person's oral presentations, and (ii) requirements that any cross-examination to which a person may be entitled under subparagraph (A) be conducted by the Administrator on behalf of that person in such manner as the Administrator determines (I) to be appropriate, and (II) to be required for a full and true disclosure with respect to disputed issues of material fact.

(C) (i) Except as provided in clause (ii), if a group of persons each of whom under subparagraphs (A) and (B) would be entitled to conduct (or have conducted) cross-examination and who are determined by the Administrator to have the same or similar interests in the proceeding cannot agree upon a single representative of such interests for purposes of cross-examination, the Administrator may make rules and rulings (I) limiting the representation of such interest for such purposes, and (II) governing the manner in which such cross-examination shall be limited.

(ii) When any person who is a member of a group with respect to which the Administrator has made a determination under clause (i) is unable to agree upon group representation with the other members of the group, then such person shall not be denied under the authority of clause (i) the opportunity to conduct (or have conducted) cross-examination as to issues affecting the person's particular interests if (I) the person satisfies the Administrator that the person has made a reasonable and good faith effort to reach agreement upon group representation with the other members of the group and (II) the Administrator determines that there are substantial and relevant issues which are not adequately presented by the group representative.

(D) A verbatim transcript shall be taken of any oral presentation made, and cross-examination conducted in any informal hearing under this subsection. Such transcript shall be available to the public.

(4) (A) The Administrator may, pursuant to rules prescribed by the Administrator, provide compensation for reasonable attorneys' fees, expert witness fees, and other costs of participating in a rulemaking proceeding for the promulgation of a rule under subsection (a) to any person—

(i) who represents an interest which would substantially contribute to a fair determination of the issues to be resolved in the proceeding, and

(ii) if—

(I) the economic interest of such person is small in comparison to the costs of effective participation in the proceeding by such person, or

(II) such person demonstrates to the satisfaction of the Administrator that such person does not have sufficient resources adequately to participate in the proceeding without compensation under this subparagraph.

In determining for purposes of clause (i) if an interest will substantially contribute to a fair determination of the issues to be resolved in a proceeding, the Administrator shall take into account the number and complexity of such issues and the extent to which representation of such interest will contribute to widespread public participation in the proceeding and representation of a fair balance of interests for the resolution of such issues.

(B) In determining whether compensation should be provided to a person under subparagraph (A) and the amount of such compensation, the Administrator shall take into account the financial burden which will be incurred by such person in participating in the rulemaking proceeding. The Administrator shall take such action as may be necessary to ensure that the aggregate amount of compensation paid under this paragraph in any fiscal year to all persons who, in rulemaking proceedings in which they receive compensation, are persons who either—

(i) would be regulated by the proposed rule, or

(ii) represent persons who would be so regulated,

may not exceed 25 per centum of the aggregate amount paid as compensation under this paragraph to all persons in such fiscal year.

(5) Paragraph (1), (2), (3), and (4) of this subsection apply to the promulgation of a rule repealing, or making a substantive amendment to, a rule promulgated under subsection (a).

(d) Effective Date.—(1) The Administrator shall specify in any rule under subsection (a) the date on which it shall take effect, which date shall be as soon as feasible.

(2) (A) The Administrator may declare a proposed rule under subsection (a) to be effective upon its publication in the Federal Register and until the effective date of final action taken, in accordance with subparagraph (B), respecting such rule if—

(i) the Administrator determines that—

(I) the manufacture, processing, distribution in commerce, use, or disposal of the chemical substance or mixture subject to such proposed rule or any combination of such activities is likely to result in an unreasonable risk of serious or widespread injury to health or the environment before such effective date; and

(II) making such proposed rule so effective is necessary to protect the public interest; and

(ii) in the case of a proposed rule to prohibit the manufacture, processing, or distribution of a chemical substance or mixture because of the risk determined under clause (i)(I), a court has in an action under section 7 granted relief with respect to such risk associated with such substance or mixture.

Such a proposed rule which is made so effective shall not, for purposes of judicial review, be considered final agency action.

(B) If the Administrator makes a proposed rule effective upon its publication in the Federal Register, the Administrator shall, as expeditiously as possible, give interested persons prompt notice of such action, provide reasonable opportunity, in accordance with paragraphs (2) and (3) of subsection (c), for a hearing on such rule, and either promulgate such rule (as proposed or with modifications) or revoke it; and if such a hearing is requested, the Administrator shall commence the hearing within five days from the date such request is made unless the Administrator and the person making the request agree upon a later date for the hearing to begin, and after the hearing is concluded the Administrator shall, within ten days of the conclusion of the hearing, either promulgate such rule (as proposed or with modifications) or revoke it.

(e) Polychlorinated Biphenyls.—(1) Within six months after the effective date of this Act the Administrator shall promulgate rules to—

(A) prescribe methods for the disposal of polychlorinated biphenyls, and

(B) require polychlorinated biphenyls to be marked with clear and adequate warnings, and instructions with respect to their processing, distribution in commerce, use, or disposal or with respect to any combination of such activities.

Requirements prescribed by rules under this paragraph shall be consistent with the requirements of paragraphs (2) and (3).

(2)(A) Except as provided under subparagraph (B), effective one year after the effective date of this Act no person may manufacture, process, or distribute in commerce or use any polychlorinated biphenyl in any manner other than in a totally enclosed manner.

(B) The Administrator may by rule authorize the manufacture, processing, distribution in commerce or use (or any combination of such activities) of any polychlorinated biphenyl in a manner other than in a totally enclosed manner if the Administrator finds that such manufacture, processing, distribution in commerce, or use (or combination of such activities) will not present an unreasonable risk of injury to health or the environment.

(C) For the purposes of this paragraph, the term "totally enclosed manner" means any manner which will ensure that any exposure of human beings or the environment to a polychlorinated biphenyl will be insignificant as determined by the Administrator by rule.

(3)(A) Except as provided in subparagraphs (B) and (C)—

(i) no person may manufacture any polychlorinated biphenyl after two years after the effective date of this Act, and

(ii) no person may process or distribute in commerce any polychlorinated biphenyl after two and one-half years after such date.

(B) Any person may petition the Administrator for an exemption from the requirements of subparagraph (A), and the Administrator may grant by rule such an exemption if the Administrator finds that—

(i) an unreasonable risk of injury to health or environment would not result, and

(ii) good faith efforts have been made to develop a chemical substance which does not present an unreasonable risk of injury to health or the environment and which may be substituted for such polychlorinated biphenyl.

An exemption granted under this subparagraph shall be subject to such terms and conditions as the Administrator may prescribe and shall be in effect for such period (but not more than one year from the date it is granted) as the Administrator may prescribe.

(C) Subparagraph (A) shall not apply to the distribution in commerce of any polychlorinated biphenyl if such polychlorinated biphenyl was sold for purposes other than resale before two and one half years after the date of enactment of this Act.

(4) Any rule under paragraph (1), (2)(B), or (3)(B) shall be promulgated in accordance with paragraphs (2), (3), and (4) of subsection (c).

(5) This subsection does not limit the authority of the Administrator, under any other provision of this Act or any other Federal law, to take action respecting any polychlorinated biphenyl.

SEC. 7. IMMINENT HAZARDS.

(a) Actions Authorized and Required.—(1) The Administrator may commence a civil action in an appropriate district court of the United States—

(A) for seizure of an imminently hazardous chemical substance or mixture or any article containing such a substance or mixture,

(B) for relief (as authorized by subsection (b)) against any person who manufactures, processes, distributes in commerce, or uses, or disposes of, an imminently hazardous chemical substance or mixture or any article containing such a substance or mixture, or

(C) for both such seizure and relief.

A civil action may be commenced under this paragraph notwithstanding the existence of a rule under section 4, 5, or 6 or an order under section 5, and notwithstanding the pendency of any administrative or judicial proceeding under any provision of this Act.

(2) If the Administrator has not made a rule under section 6(a) immediately effective (as authorized by subsection 6(d)(2)(A)(i)) with respect to an imminently hazardous chemical substance or mixture, the Administrator shall commence in a district court of the United States with respect to such substance or mixture or article containing such substance or mixture a civil action described in subparagraph (A), (B), or (C) of paragraph (1).

(b) Relief Authorized.—(1) The district court of the United States in which an action under subsection (a) is brought shall have jurisdiction to grant such temporary or permanent relief as may be necessary to protect health or the environment from the unreasonable risk associated with the chemical substance, mixture, or article involved in such action.

(2) In the case of an action under subsection (a) brought against a person who manufactures, processes, or distributes in commerce a chemical substance or mixture or an article containing a chemical substance or mixture, the relief authorized by paragraph (1) may include the issuance of a mandatory order requiring (A) in the case of purchasers of such substance, mixture, or article known to the defendant, notification to such purchasers of the risk associated with it; (B) public notice of such risk; (C) recall; (D) the replacement or repurchase of such substance, mixture, or article; or (E) any combination of the actions described in the preceding clauses.

(3) In the case of an action under subsection (a) against a chemical substance, mixture, or article, such substance, mixture, or article may be proceeded against by process of libel for its seizure and condemnation. Proceedings in such an action shall conform as nearly as possible to proceedings in rem in admiralty.

(c) Venue and Consolidation.—(1)(A) An action under subsection (a) against a person who manufactures, processes, or distributes a chemical substance or mixture or an article containing a chemical substance or mixture may be brought in the United States District Court for the District of Columbia or for any judicial district in which any of the defendants is found, resides, or transacts business; and process in such an action may be served on a defendant in any other district in which such defendant resides or may be found. An action under subsection (a) against a chemical substance, mixture, or article may be brought in any United States district court within the jurisdiction of which the substance, mixture, or article is found.

(B) In determining the judicial district in which an action may be brought under subsection (a) in instances in which such action may be brought in more than one judicial district, the Administrator shall take into account the convenience of the parties.

(C) Subpeonas requiring attendance of witnesses in an action brought under subsection (a) may be served in any judicial district.

(2) Whenever proceedings under subsection (a) involving identical chemical substances, mixtures, or articles are pending in courts in two or more judicial districts, they shall be consolidated for trial by order of any such court upon application reasonably made by any party in interest, upon notice to all parties in interest.

(d) ACTION UNDER SECTION 6.—Where appropriate, concurrently with the filing of an action under subsection (a) or as soon thereafter as may be practicable, the Administrator shall initiate a proceeding for the promulgation of a rule under section 6(a).

(e) REPRESENTATION.—Notwithstanding any other provision of law, in any action under subsection (a), the Administrator may direct attorneys of the Environmental Protection Agency to appear and represent the Administrator in such an action.

(f) DEFINITION.—For the purposes of subsection (a), the term "imminently hazardous chemical substance or mixture" means a chemical substance or mixture which presents an imminent and unreasonable risk of serious or widespread injury to health or the environment. Such a risk to health or the environment shall be considered imminent if it is shown that the manufacture, processing, distribution in commerce, use, or disposal of the chemical substance or mixture, or that any combination of such activities, is likely to result in such injury to health or the environment before a final rule under section 6 can protect against such risk.

SEC. 8. REPORTING AND RETENTION OF INFORMATION.

(a) REPORTS.—(1) The Administrator shall promulgate rules under which—

(A) each person (other than a small manufacturer or processor) who manufactures or processes or proposes to manufacture or process a chemical substance (other than a chemical substance described in subparagraph (B)(ii)) shall maintain such records, and shall submit to the Administrator such reports, as the Administrator may reasonably require, and

(B) each person (other than a small manufacturer or processor) who manufactures or processes or proposes to manufacture or process—

(i) a mixture, or

(ii) a chemical substance in small quantities (as defined by the Administrator by rule) solely for purposes of scientific experimentation or analysis or chemical research on, or analysis of, such substance or another substance, including any such research or analysis for the development of a product,

shall maintain records and submit to the Administrator reports but only to the extent the Administrator determines the maintenance of records or submission of reports, or both, is necessary for the effective enforcement of this Act.

The Administrator may not require in a rule promulgated under this paragraph the maintenance of records or the submission of reports with respect to changes in the proportions of the components of a mixture unless the Administrator finds that the maintenance of such records or the submission of such reports, or both, is necessary for the effective enforcement of this Act. For purposes of the compilation of the list of chemical substances required under subsection (b), the Administrator shall promulgate rules pursuant to this subsection not later than 180 days after the effective date of this Act.

(2) The Administrator may require under paragraph (1) maintenance of records and reporting with respect to the following insofar as known to the person making the report or insofar as reasonably ascertainable:

(A) The common or trade name, the chemical identity, and the molecular structure of each chemical substance or mixture for which such a report is required.

(B) The categories or proposed categories of use of each such substance or mixture.

(C) The total amount of each such substance and mixture manufactured or processed, reasonable estimates of the total amount to be manufactured or processed, the amount manufactured or processed for each of its categories of use, and reasonable estimates of the amount to be manufactured or processed for each of its categories of use or proposed categories of use.

(D) A description of the byproducts resulting from the manufacture, processing, use, or disposal of each such substance or mixture.

(E) All existing data concerning the environmental and health effects of such substance or mixture.

(F) The number of individuals exposed, and reasonable estimates of the number who will be exposed, to such substance or mixture in their places of employment and the duration of such exposure.

(G) In the initial report under paragraph (1) on such substance or mixture, the manner or method of its disposal, and in any subsequent report on such substance or mixture, any change in such manner or method.

To the extent feasible, the Administrator shall not require under paragraph (1), any reporting which is unnecessary or duplicative.

(3)(A)(i) The Administrator may by rule require a small manufacturer or processor of a chemical substance to submit to the Administrator such information respecting the chemical substance as the Administrator may require for publication of the first list of chemical substances required by subsection (b).

(ii) The Administrator may by rule require a small manufacturer or processor of a chemical substance or mixture—

(I) subject to a rule proposed or promulgated under section 4, 5(b)(4), or 6, or an order in effect under section 5(e), or

(II) with respect to which relief has been granted pursuant to a civil action brought under section 5 or 7,

to maintain such records on such substance or mixture, and to submit to the Administrator such reports on such substance or mixture, as the Administrator may reasonably require. A rule under this clause requiring reporting may require reporting with respect to the matters referred to in paragraph (2).

(B) The Administrator, after consultation with the Administrator of the Small Business Administration, shall by rule prescribe standards for determining the manufacturers and processors which qualify as small manufacturers and processors for purposes of this paragraph and paragraph (1).

(b) Inventory.—(1) The Administrator shall compile, keep current, and publish a list of each chemical substance which is manufactured or processed in the United States. Such list shall at least include each chemical substance which any person reports, under section 5 or subsection (a) of this section, is manufactured or processed in the United States. Such list may not include any chemical substance which was not manufactured or processed in the United States within three years before the effective date of the rules promulgated pursuant to the last sentence of subsection (a)(1). In the case of a chemical substance for which a notice is submitted in accordance with section 5, such chemical substance shall be included in such list as of the earliest date (as determined by the Administrator) on which such substance was manufactured or processed in the United States. The Administrator shall first publish such a list not later than 315 days after the effective date of this Act. The Administrator shall not include in such list any chemical substance which is manufactured or processed only in small quantities (as defined by the Administrator by rule) solely for purposes of scientific experimentation or analysis or chemical research on, or analysis of, such substance or another substance, including such research or analysis for the development of a product.

(2) To the extent consistent with the purposes of this Act, the Administrator may, in lieu of listing, pursuant to paragraph (1), a chemical substance individually, list a category of chemical substances in which such substance is included.

(c) RECORDS.—Any person who manufactures, processes, or distributes in commerce any chemical substance or mixture shall maintain records of significant adverse reactions to health or the environment, as determined by the Administrator by rule, alleged to have been caused by the substance or mixture. Records of such adverse reactions to the health of employees shall be retained for a period of 30 years from the date such reactions were first reported to or known by the person maintaining such records. Any other record of such adverse reactions shall be retained for a period of five years from the date the information contained in the record was first reported to or known by the person maintaining the record. Records required to be maintained under this subsection shall include records of consumer allegations of personal injury or harm to health, reports of occupational disease or injury, and reports or complaints of injury to the environment submitted to the manufacturer, processor, or distributor in commerce from any source. Upon request of any duly designated representative of the Administrator, each person who is required to maintain records under this subsection shall permit the inspection of such records and shall submit copies of such records.

(d) HEALTH AND SAFETY STUDIES.—The Administrator shall promulgate rules under which the Administrator shall require any person who manufactures, processes, or distributes in commerce or who proposes to manufacture, process, or distribute in commerce any chemical substance or mixture (or with respect to paragraph (2), any person who has possession of a study) to submit to the Administrator—

(1) lists of health and safety studies (A) conducted or initiated by or for such person with respect to such substance or mixture at any time, (B) known to such person, or (C) reasonably ascertainable by such person, except that the Administrator may exclude certain types or categories of studies from the requirements of this subsection if the Administrator finds that submission of lists of such studies are unnecessary to carry out the purposes of this Act; and

(2) copies of any study contained on a list submitted pursuant to paragraph (1) or otherwise known by such person.

(e) NOTICE TO ADMINISTRATOR OF SUBSTANTIAL RISKS.—Any person who manufactures, processes, or distributes in commerce a chemical substance or mixture and who obtains information which reasonably supports the conclusion that such substance or mixture presents a substantial risk of injury to health or the environment shall immediately inform the Administrator of such information unless such person has actual knowledge that the Administrator has been adequately informed of such information.

(f) DEFINITIONS.—For purposes of this section, the terms "manufacture" and "process" mean manufacture or process for commercial purposes.

SEC. 9. RELATIONSHIP TO OTHER FEDERAL LAWS.

(a) LAWS NOT ADMINISTERED BY THE ADMINISTRATOR.—(1) If the Administrator has reasonable basis to conclude that the manufacture, processing, distribution in commerce, use, or disposal of a chemical substance or mixture, or that any combination of such activities, presents or will present an unreasonable risk of injury to health or the environment and determines, in the Administrator's discretion, that such risk may be prevented or reduced to a sufficient extent by action taken under a Federal law not administered by the Administrator, the Administrator shall submit to the agency which administers such law a report which describes such risk and includes in such description a specification of the activity or combination of activities which the Administrator has reason to believe so presents such risk. Such report shall also request such agency—

(A)(i) to determine if the risk described in such report may be prevented or reduced to a sufficient extent by action taken under such law, and

(ii) if the agency determines that such risk may be so prevented or reduced, to issue an order declaring whether or not the activity or combination of activities specified in the description of such risk presents such risk; and

(B) to respond to the Administrator with respect to the matters described in subparagraph (A).

Any report of the Administrator shall include a detailed statement of the information on which it is based and shall be published in the Federal Register. The agency receiving a request under such a report shall make the requested determination, issue the requested order, and make the requested response within such time as the Administrator specifies in the request, but such time specified may not be less than 90 days from the date the request was made. The response of an agency shall be accompanied by a detailed statement of the findings and conclusions of the agency and shall be published in the Federal Register.

(2) If the Administrator makes a report under paragraph (1) with respect to a chemical substance or mixture and the agency to which such report was made either—

(A) issues an order declaring that the activity or combination of activities specified in the description of the risk described in the report does not present the risk described in the report, or

(B) initiates, within 90 days of the publication in the Federal Register of the response of the agency under paragraph (1), action under the law (or laws) administered by such agency to protect against such risk associated with such activity or combination of activities,

the Administrator may not take any action under section 6 or 7 with respect to such risk.

(3) If the Administrator has initiated action under section 6 or 7 with respect to a risk associated with a chemical substance or mixture which was the subject of a report made to an agency under paragraph (1), such agency shall before taking action under the law (or laws)

administered by it to protect against such risk consult with the Administrator for the purpose of avoiding duplication of Federal action against such risk.

(b) Laws Administered by the Administrator.—The Administrator shall coordinate actions taken under this Act with actions taken under other Federal laws administered in whole or in part by the Administrator. If the Administrator determines that a risk to health or the environment associated with a chemical substance or mixture could be eliminated or reduced to a sufficient extent by actions taken under the authorities contained in such other Federal laws, the Administrator shall use such authorities to protect against such risk unless the Administrator determines, in the Administrator's discretion, that it is in the public interest to protect against such risk by actions taken under this Act. This subsection shall not be construed to relieve the Administrator of any requirement imposed on the Administrator by such other Federal laws.

(c) Occupational Safety and Health.—In exercising any authority under this Act, the Administrator shall not, for purposes of section 4(b)(1) of the Occupational Safety and Health Act of 1970, be deemed to be exercising statutory authority to prescribe or enforce standards or regulations affecting occupational safety and health.

(d) Coordination.—In administering this Act, the Administrator shall consult and coordinate with the Secretary of Health, Education, and Welfare and the heads of any other appropriate Federal executive department or agency, any relevant independent regulatory agency, and any other appropriate instrumentality of the Federal Government for the purpose of achieving the maximum enforcement of this Act while imposing the least burdens of duplicative requirements on those subject to the Act and for other purposes. The Administrator shall, in the report required by section 30, report annually to the Congress on actions taken to coordinate with such other Federal departments, agencies, or instrumentalities, and on actions taken to coordinate the authority under this Act with the authority granted under other Acts referred to in subsection (b).

SEC. 10. RESEARCH, DEVELOPMENT, COLLECTION, DISSEMINATION, AND UTILIZATION OF DATA.

(a) Authority.—The Administrator shall, in consultation and cooperation with the Secretary of Health, Education, and Welfare and with other heads of appropriate departments and agencies, conduct such research, development, and monitoring as is necessary to carry out the purposes of this Act. The Administrator may enter into contracts and may make grants for research, development, and monitoring under this subsection. Contracts may be entered into under this subsection without regard to sections 3648 and 3709 of the Revised Statutes (31 U.S.C. 529, 14 U.S.C. 5).

(b) Data Systems.—(1) The Administrator shall establish, administer, and be responsible for the continuing activities of an interagency committee which shall design, establish, and coordinate an efficient and effective system, within the Environmental Protection Agency, for the collection, dissemination to other Federal departments and agencies, and use of data submitted to the Administrator under this Act.

(2)(A) The Administrator shall, in consultation and cooperation with the Secretary of Health, Education, and Welfare and other heads of appropriate departments and agencies design, establish, and coordinate an efficient and effective system for the retrieval of toxicological and other scientific data which could be useful to the Administrator in carrying out the purposes of this Act. Systematized retrieval shall be developed for use by all Federal and other departments and agencies with responsibilities in the area of regulation or study of chemical substances and mixtures and their effect on health or the environment.

(B) The Administrator, in consultation and cooperation with the Secretary of Health, Education, and Welfare, may make grants and enter into contracts for the development of a data retrieval system described in subparagraph (A). Contracts may be entered into under this subparagraph without regard to sections 3648 and 3709 of the Revised Statutes (31 U.S.C. 529, 41 U.S.C. 5).

(c) SCREENING TECHNIQUES.—The Administrator shall coordinate, with the Assistant Secretary for Health of the Department of Health, Education, and Welfare, research undertaken by the Administrator and directed toward the development of rapid, reliable, and economical screening techniques for carcinogenic, mutagenic, teratogenic, and ecological effects of chemical substances and mixtures.

(d) MONITORING.—The Administrator shall, in consultation and cooperation with the Secretary of Health, Education, and Welfare, establish and be responsible for research aimed at the development, in cooperation with local, State, and Federal agencies, of monitoring techniques and instruments which may be used in the detection of toxic chemical substances and mixtures and which are reliable, economical, and capable of being implemented under a wide variety of conditions.

(e) BASIC RESEARCH.—The Administrator shall, in consultation and cooperation with the Secretary of Health, Education, and Welfare, establish research programs to develop the fundamental scientific basis of the screening and monitoring techniques described in subsections (c) and (d), the bounds of the reliability of such techniques, and the opportunities for their improvement.

(f) TRAINING.—The Administrator shall establish and promote programs and workshops to train or facilitate the training of Federal laboratory and technical personnel in existing or newly developed screening and monitoring techniques.

(g) EXCHANGE OF RESEARCH AND DEVELOPMENT RESULTS.—The Administrator shall, in consultation with the Secretary of Health, Education, and Welfare and other heads of appropriate departments and agencies, establish and coordinate a system for exchange among Federal, State, and local authorities of research and development results respecting toxic chemical substances and mixtures, including a system to facilitate and promote the development of standard data format and analysis and consistent testing procedures.

SEC. 11. INSPECTIONS AND SUBPOENAS.

(a) IN GENERAL.—For purposes of administering this Act, the Administrator, and any duly designated representative of the Administrator, may inspect any establishment, facility, or other premises in which chemical substances or mixtures are manufactured, processed, stored, or held before or after their distribution in commerce and any conveyance being used to transport chemical substances, mixtures, or such articles in connection with distribution in commerce. Such an inspection may only be made upon the presentation of appropriate credentials and of a written notice to the owner, operator, or agent in charge of the premises or conveyance to be inspected. A separate notice shall be given for each such inspection, but a notice shall not be required for each entry made during the period covered by the inspection. Each such inspection shall be commenced and completed with reasonable promptness and shall be conducted at reasonable times, within reasonable limits, and in a reasonable manner.

(b) SCOPE.—(1) Except as provided in paragraph (2), an inspection conducted under subsection (a) shall extend to all things within the premises or conveyance inspected (including records, files, papers, processes, controls, and facilities) bearing on whether the requirements of this Act applicable to the chemical substances or mixtures within such premises or conveyance have been complied with.

(2) No inspection under subsection (a) shall extend to—

(A) financial data,

(B) sales data (other than shipment data),

(C) pricing data,

(D) personnel data, or

(E) research data (other than data required by this Act or under a rule promulgated thereunder),

unless the nature and extent of such data are described with reasonable specificity in the written notice required by subsection (a) for such inspection.

(c) SUBPOENAS.—In carrying out this Act, the Administrator may by subpoena require the attendance and testimony of witnesses and the production of reports, papers, documents, answers to questions, and other information that the Administrator deems necessary. Witnesses shall be paid the same fees and mileage that are paid witnesses in the courts of the United States. In the event of contumacy, failure, or refusal of any person to obey any such subpoena, any district court of the United States in which venue is proper shall have jurisdiction to order any such person to comply with such subpoena. Any failure to obey such an order of the court is punishable by the court as a contempt thereof.

SEC. 12. EXPORTS.

(a) IN GENERAL.—(1) Except as provided in paragraph (2) and subsection (b), this Act (other than section 8) shall not apply to any chemical substance, mixture, or to an article containing a chemical substance or mixture, if—

(A) it can be shown that such substance, mixture, or article is being manufactured, processed, or distributed in commerce for export from the United States, unless such substance, mixture, or article was, in fact, manufactured, processed, or distributed in commerce, for use in the United States, and

(B) such substance, mixture, or article (when distributed in commerce), or any container in which it is enclosed (when so distributed), bears a stamp or label stating that such substance, mixture, or article is intended for export.

(2) Paragraph (1) shall not apply to any chemical substance, mixture, or article if the Administrator finds that the substance, mixture, or article will present an unreasonable risk of injury to health within the United States or to the environment of the United States. The Administrator may require, under section 4, testing of any chemical substance or mixture exempted from this Act by paragraph (1) for the purpose of determining whether or not such substance or mixture presents an unreasonable risk of injury to health within the United States or to the environment of the United States.

(b) NOTICE.—(1) If any person exports or intends to export to a foreign country a chemical substance or mixture for which the submission of data is required under section 4 or 5(b), such person shall notify the Administrator of such exportation or intent to export and the Administrator shall furnish to the government of such country notice of the availability of the data submitted to the Administrator under such section for such substance or mixture.

(2) If any person exports or intends to export to a foreign country a chemical substance or mixture for which an order has been issued

under section 5 or a rule has been proposed or promulgated under section 5 or 6, or with respect to which an action is pending, or relief has been granted under section 5 or 7, such person shall notify the Administrator of such exportation or intent to export and the Administrator shall furnish to the government of such country notice of such rule, order, action, or relief.

SEC. 13. ENTRY INTO CUSTOMS TERRITORY OF THE UNITED STATES.

(a) In General.—(1) The Secretary of the Treasury shall refuse entry into the customs territory of the United States (as defined in general headnote 2 to the Tariff Schedules of the United States) of any chemical substance, mixture, or article containing a chemical substance or mixture offered for such entry if—

(A) it fails to comply with any rule in effect under this Act, or

(B) it is offered for entry in violation of section 5 or 6, a rule or order under section 5 or 6, or an order issued in a civil action brought under section 5 or 7.

(2) If a chemical substance, mixture, or article is refused entry under paragraph (1), the Secretary of the Treasury shall notify the consignee of such entry refusal, shall not release it to the consignee, and shall cause its disposal or storage (under such rules as the Secretary of the Treasury may prescribe) if it has not been exported by the consignee within 90 days from the date of receipt of notice of such refusal, except that the Secretary of the Treasury may, pending a review by the Administrator of the entry refusal, release to the consignee such substance, mixture, or article on execution of bond for the amount of the full invoice of such substance, mixture, or article (as such value is set forth in the customs entry), together with the duty thereon. On failure to return such substance, mixture, or article for any cause to the custody of the Secretary of the Treasury when demanded, such consignee shall be liable to the United States for liquidated damages equal to the full amount of such bond. All charges for storage, cartage, and labor on and for disposal of substances, mixtures, or articles which are refused entry or release under this section shall be paid by the owner or consignee, and in default of such payment shall constitute a lien against any future entry made by such owner or consignee.

(b) Rules.—The Secretary of the Treasury, after consultation with the Administrator, shall issue rules for the administration of subsection (a) of this section.

SEC. 14. DISCLOSURE OF DATA.

(a) In General.—Except as provided by subsection (b), any information reported to, or otherwise obtained by, the Administrator (or any representative of the Administrator) under this Act, which is exempt from disclosure pursuant to subsection (a) of section 552 of title 5, United States Code, by reason of subsection (b)(4) of such section, shall, notwithstanding the provisions of any other section of this Act, not be disclosed by the Administrator or by any officer or employee of the United States, except that such information—

(1) shall be disclosed to any officer or employee of the United States—

(A) in connection with the official duties of such officer or employee under any law for the protection of health or the environment, or

(B) for specific law enforcement purposes;

(2) shall be disclosed to contractors with the United States and employees of such contractors if in the opinion of the Administra-

tor such disclosure is necessary for the satisfactory performance by the contractor of a contract with the United States entered into on or after the date of enactment of this Act for the performance of work in connection with this Act and under such conditions as the Administrator may specify;

(3) shall be disclosed if the Administrator determines it necessary to protect health or the environment against an unreasonable risk of injury to health or the environment; or

(4) may be disclosed when relevant in any proceeding under this Act, except that disclosure in such a proceeding shall be made in such manner as to preserve confidentiality to the extent practicable without impairing the proceeding.

In any proceeding under section 552(a) of title 5, United States Code, to obtain information the disclosure of which has been denied because of the provisions of this subsection, the Administrator may not rely on section 552(b)(3) of such title to sustain the Administrator's action.

(b) DATA FROM HEALTH AND SAFETY STUDIES.—(1) Subsection (a) does not prohibit the disclosure of—

(A) any health and safety study which is submitted under this Act with respect to—

(i) any chemical substance or mixture which, on the date on which such study is to be disclosed has been offered for commercial distribution, or

(ii) any chemical substance or mixture for which testing is required under section 4 or for which notification is required under section 5, and

(B) any data reported to, or otherwise obtained by, the Administrator from a health and safety study which relates to a chemical substance or mixture described in clause (i) or (ii) of subparagraph (A).

This paragraph does not authorize the release of any data which discloses processes used in the manufacturing or processing of a chemical substance or mixture or, in the case of a mixture, the release of data disclosing the portion of the mixture comprised by any of the chemical substances in the mixture.

(2) If a request is made to the Administrator under subsection (a) of section 552 of title 5, United States Code, for information which is described in the first sentence of paragraph (1) and which is not information described in the second sentence of such paragraph, the Administrator may not deny such request on the basis of subsection (b)(4) of such section.

(c) DESIGNATION AND RELEASE OF CONFIDENTIAL DATA.—(1) In submitting data under this Act, a manufacturer, processor, or distributor in commerce may (A) designate the data which such person believes is entitled to confidential treatment under subsection (a), and (B) submit such designated data separately from other data submitted under this Act. A designation under this paragraph shall be made in writing and in such manner as the Administrator may prescribe.

(2)(A) Except as provided by subparagraph (B), if the Administrator proposes to release for inspection data which has been designated under paragraph (1)(A), the Administrator shall notify, in writing and by certified mail, the manufacturer, processor, or distributor in commerce who submitted such data of the intent to release such data. If the release of such data is to be made pursuant to a request made under section 552(a) of title 5, United States Code, such notice shall be given immediately upon approval of such request by the Administrator. The Administrator may not release such data until the expiration of 30 days after the manufacturer, processor, or distributor in commerce submitting such data has received the notice required by this subparagraph.

(B)(i) Subparagraph (A) shall not apply to the release of information under paragraph (1), (2), (3), or (4) of subsection (a), except that the Administrator may not release data under paragraph (3) of subsection (a) unless the Administrator has notified each manufacturer, processor, and distributor in commerce who submitted such data of such release. Such notice shall be made in writing by certified mail at least 15 days before the release of such data, except that if the Administrator determines that the release of such data is necessary to protect against an imminent, unreasonable risk of injury to health or the environment, such notice may be made by such means as the Administrator determines will provide notice at least 24 hours before such release is made.

(ii) Subparagraph (A) shall not apply to the release of information described in subsection (b)(1) other than information described in the second sentence of such subsection.

(d) CRIMINAL PENALTY FOR WRONGFUL DISCLOSURE.—(1) Any officer or employee of the United States or former officer or employee of the United States, who by virtue of such employment or official position has obtained possession of, or has access to, material the disclosure of which is prohibited by subsection (a), and who knowing that disclosure of such material is prohibited by such subsection, willfully discloses the material in any manner to any person not entitled to receive it, shall be guilty of a misdemeanor and fined not more than $5,000 or imprisoned for not more than one year, or both. Section 1905 of title 18, United States Code, does not apply with respect to the publishing, divulging, disclosure, or making known of, or making available, information reported or otherwise obtained under this Act.

(2) For the purposes of paragraph (1), any contractor with the United States who is furnished information as authorized by subsection (a)(2), and any employee of any such contractor, shall be considered to be an employee of the United States.

(e) ACCESS BY CONGRESS.—Notwithstanding any limitation contained in this section or any other provision of law, all information reported to or otherwise obtained by the Administrator (or any representative of the Administrator) under this Act shall be made available, upon written request of any duly authorized committee of the Congress, to such committee.

SEC. 15. PROHIBITED ACTS.

It shall be unlawful for any person to—

(1) fail or refuse to comply with (A) any rule promulgated or order issued under section 4, (B) any requirement prescribed by section 5 or 6, or (C) any rule promulgated or order issued under section 5 or 6;

(2) use for commercial purposes a chemical substance or mixture which such person knew or had reason to know was manufactured, processed, or distributed in commerce in violation of section 5 or 6, a rule or order under section 5 or 6, or an order issued in action brought under section 5 or 7;

(3) fail or refuse to (A) establish or maintain records, (B) submit reports, notices, or other information, or (C) permit access to or copying of records, as required by this Act or a rule thereunder; or

(4) fail or refuse to permit entry or inspection as required by section 11.

SEC. 16. PENALTIES.

(a) CIVIL.—(1) Any person who violates a provision of section 15 shall be liable to the United States for a civil penalty in an amount not to exceed $25,000 for each such violation. Each day such a violation continues shall, for purposes of this subsection, constitute a separate violation of section 15.

(2) (A) A civil penalty for a violation of section 15 shall be assessed by the Administrator by an order made on the record after opportunity (provided in accordance with this subparagraph) for a hearing in accordance with section 554 of title 5, United States Code. Before issuing such an order, the Administrator shall give written notice to the person to be assessed a civil penalty under such order of the Administrator's proposal to issue such order and provide such person an opportunity to request, within 15 days of the date the notice is received by such person, such a hearing on the order.

(B) In determining the amount of a civil penalty, the Administrator shall take into account the nature, circumstances, extent, and gravity of the violation or violations and, with respect to the violator, ability to pay, effect on ability to continue to do business, any history of prior such violations, the degree of culpability, and such other matters as justice may require.

(C) The Administrator may compromise, modify, or remit, with or without conditions, any civil penalty which may be imposed under this subsection. The amount of such penalty, when finally determined, or the amount agreed upon in compromise, may be deducted from any sums owing by the United States to the person charged.

(3) Any person who requested in accordance with paragraph (2) (A) a hearing respecting the assessment of a civil penalty and who is aggrieved by an order assessing a civil penalty may file a petition for judicial review of such order with the United States Court of Appeals for the District of Columbia Circuit or for any other circuit in which such person resides or transacts business. Such a petition may only be filed within the 30-day period beginning on the date the order making such assessment was issued.

(4) If any person fails to pay an assessment of a civil penalty—

(A) after the order making the assessment has become a final order and if such person does not file a petition for judicial review of the order in accordance with paragraph (3), or

(B) after a court in an action brought under paragraph (3) has entered a final judgment in favor of the Administrator,

the Attorney General shall recover the amount assessed (plus interest at currently prevailing rates from the date of the expiration of the 30-day period referred to in paragraph (3) or the date of such final judgment, as the case may be) in an action brought in any appropriate district court of the United States. In such an action, the validity, amount, and appropriateness of such penalty shall not be subject to review.

(b) CRIMINAL.—Any person who knowingly or willfully violates any provision of section 15 shall, in addition to or in lieu of any civil penalty which may be imposed under subsection (a) of this section for such violation, be subject, upon conviction, to a fine of not more than $25,000 for each day of violation, or to imprisonment for not more than one year, or both.

SEC. 17. SPECIFIC ENFORCEMENT AND SEIZURE.

(a) SPECIFIC ENFORCEMENT.—(1) The district courts of the United States shall have jurisdiction over civil actions to—

(A) restrain any violation of section 15,

(B) restrain any person from taking any action prohibited by section 5 or 6 or by a rule or order under section 5 or 6,

(C) compel the taking of any action required by or under this Act, or

(D) direct any manufacturer or processor of a chemical substance or mixture manufactured or processed in violation of section 5 or 6 or a rule or order under section 5 or 6 and distributed in commerce, (i) to give notice of such fact to distributors in commerce of such substance or mixture and, to the extent reasonably ascertainable, to other persons in possession of such substance or mixture or exposed to such substance or mixture, (ii) to give public notice of such risk of injury, and (iii) to either replace or repurchase such substance or mixture, whichever the person to which the requirement is directed elects.

(2) A civil action described in paragraph (1) may be brought—

(A) in the case of a civil action described in subparagraph (A) of such paragraph, in the United States district court for the judicial district wherein any act, omission, or transaction constituting a violation of section 15 occurred or wherein the defendant is found or transacts business, or

(B) in the case of any other civil action described in such paragraph, in the United States district court for the judicial district wherein the defendant is found or transacts business.

In any such civil action process may be served on a defendant in any judicial district in which a defendant resides or may be found. Subpoenas requiring attendance of witnesses in any such action may be served in any judicial district.

(b) Seizure.—Any chemical substance or mixture which was manufactured, processed, or distributed in commerce in violation of this Act or any rule promulgated or order issued under this Act or any article containing such a substance or mixture shall be liable to be proceeded against, by process of libel for the seizure and condemnation of such substance, mixture, or article, in any district court of the United States within the jurisdiction of which such substance, mixture, or article is found. Such proceedings shall conform as nearly as possible to proceedings in rem in admiralty.

SEC. 18. PREEMPTION.

(a) Effect on State Law.—(1) Except as provided in paragraph (2), nothing in this Act shall affect the authority of any State or political subdivision of a State to establish or continue in effect regulation of any chemical substance, mixture, or article containing a chemical substance or mixture.

(2) Except as provided in subsection (b)—

(A) if the Administrator requires by a rule promulgated under section 4 the testing of a chemical substance or mixture, no State or political subdivision may, after the effective date of such rule, establish or continue in effect a requirement for the testing of such substance or mixture for purposes similar to those for which testing is required under such rule; and

(B) if the Administrator prescribes a rule or order under section 5 or 6 (other than a rule imposing a requirement described in subsection (a)(6) of section 6) which is applicable to a chemical substance or mixture, and which is designed to protect against a risk of injury to health or the environment associated with such substance or mixture, no State or political subdivision of a State may, after the effective date of such requirement, establish or continue in effect, any requirement which is applicable to such substance or mixture, or an article containing such substance or mixture, and which is designed to protect against such risk unless such requirement (i) is identical to the requirement prescribed by the Administrator, (ii) is adopted under the authority of the Clean Air Act or any other Federal law, or (iii) prohibits the use of such substance or mixture in such State or political subdivision (other than its use in the manufacture or processing of other substances or mixtures).

(b) Exemption.—Upon application of a State or political subdivision of a State the Administrator may by rule exempt from subsection (a)(2), under such conditions as may be prescribed in such rule, a requirement of such State or political subdivision designed to protect against a risk of injury to health or the environment associated with a chemical substance, mixture, or article containing a chemical substance or mixture if—

(1) compliance with the requirement would not cause the manufacturing, processing, distribution in commerce, or use of the substance, mixture, or article to be in violation of the applicable requirement under this Act described in subsection (a)(2), and

(2) the State or political subdivision requirement (A) provides a significantly higher degree of protection from such risk than the requirement under this Act described in subsection (a)(2) and (B) does not, through difficulties in marketing, distribution, or other factors, unduly burden interstate commerce.

SEC. 19. JUDICIAL REVIEW.

(a) In General.—(1)(A) Not later than 60 days after the date of the promulgation of a rule under section 4(a), 5(a)(2), 5(b)(4), 6(a), 6(e), or 8, any person may file a petition for judicial review of such rule with the United States Court of Appeals for the District of Columbia Circuit or for the circuit in which such person resides or in which such person's principal place of business is located. Courts of appeals of the United States shall have exclusive jurisdiction of any action to obtain judicial review (other than in an enforcement proceeding) of such a rule if any district court of the United States would have had jurisdiction of such action but for this subparagraph.

(B) Courts of appeals of the United States shall have exclusive jurisdiction of any action to obtain judicial review (other than in an enforcement proceeding) of an order issued under subparagraph (A) or (B) of section 6(b)(1) if any district court of the United States would have had jurisdiction of such action but for this subparagraph.

(2) Copies of any petition filed under paragraph (1)(A) shall be transmitted forthwith to the Administrator and to the Attorney General by the clerk of the court with which such petition was filed. The provisions of section 2112 of title 28, United States Code, shall apply to the filing of the rulemaking record of proceedings on which the Administrator based the rule being reviewed under this section and to the transfer of proceedings between United States courts of appeals.

(3) For purposes of this section, the term "rulemaking record" means—

(A) the rule being reviewed under this section;

(B) in the case of a rule under section 4(a), the finding required by such section, in the case of a rule under section 5(b)(4), the finding required by such section, in the case of a rule under section 6(a) the finding required by section 5(f) or 6(a), as the case may be, in the case of a rule under section 6(a), the statement required by section 6(c)(1), and in the case of a rule under section 6(e), the findings required by paragraph (2)(B) or (3)(B) of such section, as the case may be;

(C) any transcript required to be made of oral presentations made in proceedings for the promulgation of such rule;

(D) any written submission of interested parties respecting the promulgation of such rule; and

(E) any other information which the Administrator considers to be relevant to such rule and which the Administrator identified, on or before the date of the promulgation of such rule, in a notice published in the Federal Register.

(b) ADDITIONAL SUBMISSIONS AND PRESENTATIONS; MODIFICATIONS.—If in an action under this section to review a rule the petitioner or the Administrator applies to the court for leave to make additional oral submissions or written presentations respecting such rule and shows to the satisfaction of the court that such submissions and presentations would be material and that there were reasonable grounds for the submissions and failure to make such submissions and presentations in the proceeding before the Administrator, the court may order the Administrator to provide additional opportunity to make such submissions and presentations. The Administrator may modify or set aside the rule being reviewed or make a new rule by reason of the additional submissions and presentations and shall file such modified or new rule with the return of such submissions and presentations. The court shall thereafter review such new or modified rule.

(c) STANDARD OF REVIEW.—(1)(A) Upon the filing of a petition under subsection (a)(1) for judicial review of a rule, the court shall have jurisdiction (i) to grant appropriate relief, including interim relief, as provided in chapter 7 of title 5, United States Code, and (ii) except as otherwise provided in subparagraph (B), to review such rule in accordance with chapter 7 of title 5, United States Code.

(B) Section 706 of title 5, United States Code, shall apply to review of a rule under this section, except that—

(i) in the case of review of a rule under section 4(a), 5(b)(4), 6(a), or 6(e), the standard for review prescribed by paragraph (2)(E) of such section 706 shall not apply and the court shall hold unlawful and set aside such rule if the court finds that the rule is not supported by substantial evidence in the rulemaking record (as defined in subsection (a)(3)) taken as a whole;

(ii) in the case of review of a rule under section 6(a), the court shall hold unlawful and set aside such rule if it finds that—

(I) a determination by the Administrator under section 6(c)(3) that the petitioner seeking review of such rule is not entitled to conduct (or have conducted) cross-examination or to present rebuttal submissions, or

(II) a rule of, or ruling by, the Administrator under section 6(c)(3) limiting such petitioner's cross-examination or oral presentations,

has precluded disclosure of disputed material facts which was necessary to a fair determination by the Administrator of the rulemaking proceeding taken as a whole; and section 706(2)(D) shall not apply with respect to a determination, rule, or ruling referred to in subclause (I) or (II); and

(iii) the court may not review the contents and adequacy of—

(I) any statement required to be made pursuant to section 6(c)(1), or

(II) any statement of basis and purpose required by section 553(c) of title 5, United States Code, to be incorporated in the rule

except as part of a review of the rulemaking record taken as a whole.

The term "evidence" as used in clause (i) means any matter in the rulemaking record.

(C) A determination, rule, or ruling of the Administrator described in subparagraph (B)(ii) may be reviewed only in an action under this section and only in accordance with such subparagraph.

(2) The judgment of the court affirming or setting aside, in whole or in part, any rule reviewed in accordance with this section shall be final, subject to review by the Supreme Court of the United States upon certiorari or certification, as provided in section 1254 of title 28, United States Code.

(d) FEES AND COSTS.—The decision of the court in an action commenced under subsection (a), or of the Supreme Court of the United States on review of such a decision, may include an award of costs of suit and reasonable fees for attorneys and expert witnesses if the court determines that such an award is appropriate.

(e) OTHER REMEDIES.—The remedies as provided in this section shall be in addition to and not in lieu of any other remedies provided by law.

SEC. 20. CITIZENS' CIVIL ACTIONS.

(a) IN GENERAL.—Except as provided in subsection (b), any person may commence a civil action—

(1) against any person (including (A) the United States, and (B) any other governmental instrumentality or agency to the extent permitted by the eleventh amendment to the Constitution) who is alleged to be in violation of this Act or any rule promulgated under section 4, 5, or 6 or order issued under section 5 to restrain such violation, or

(2) against the Administrator to compel the Administrator to perform any act or duty under this Act which is not discretionary.

Any civil action under paragraph (1) shall be brought in the United States district court for the district in which the alleged violation occurred or in which the defendant resides or in which the defendant's principal place of business is located. Any action brought under paragraph (2) shall be brought in the United States District Court for the District of Columbia, or the United States district court for the judicial district in which the plaintiff is domiciled. The district courts of the United States shall have jurisdiction over suits brought under this section, without regard to the amount in controversy or the citizenship of the parties. In any civil action under this subsection process may be served on a defendant in any judicial district in which the defendant resides or may be found and subpoenas for witnesses may be served in any judicial district.

(b) LIMITATION.—No civil action may be commenced—

(1) under subsection (a)(1) to restrain a violation of this Act or rule or order under this Act—

(A) before the expiration of 60 days after the plaintiff has given notice of such violation (i) to the Administrator, and (ii) to the person who is alleged to have committed such violation, or

(B) if the Administrator has commenced and is diligently prosecuting a proceeding for the issuance of an order under section 16(a)(2) to require compliance with this Act or with such rule or order or if the Attorney General has commenced and is diligently prosecuting a civil action in a court of the United States to require compliance with this Act or with such rule or order, but if such proceeding or civil action is commenced after the giving of notice, any person giving such notice may intervene as a matter of right in such proceeding or action; or

(2) under subsection (a)(2) before the expiration of 60 days after the plaintiff has given notice to the Administrator of the alleged failure of the Administrator to perform an act or duty which is the basis for such action or, in the case of an action under such subsection for the failure of the Administrator to file an action under section 7, before the expiration of ten days after such notification.

Notice under this subsection shall be given in such manner as the Administrator shall prescribe by rule.

(c) GENERAL.—(1) In any action under this section, the Administrator, if not a party, may intervene as a matter of right.

(2) The court, in issuing any final order in any action brought pursuant to subsection (a), may award costs of suit and reasonable fees for attorneys and expert witnesses if the court determines that such an award is appropriate. Any court, in issuing its decision in an action brought to review such an order, may award costs of suit and reasonable fees for attorneys if the court determines that such an award is appropriate.

(3) Nothing in this section shall restrict any right which any person (or class of persons) may have under any statute or common law to seek enforcement of this Act or any rule or order under this Act or to seek any other relief.

(d) CONSOLIDATION.—When two or more civil actions brought under subsection (a) involving the same defendant and the same issues or violations are pending in two or more judicial districts, such pending actions, upon application of such defendants to such actions which is made to a court in which any such action is brought, may, if such court in its discretion so decides, be consolidated for trial by order (issued after giving all parties reasonable notice and opportunity to be heard) of such court and tried in—

(1) any district which is selected by such defendant and in which one of such actions is pending,

(2) a district which is agreed upon by stipulation between all the parties to such actions and in which one of such actions is pending, or

(3) a district which is selected by the court and in which one of such actions is pending.

The court issuing such an order shall give prompt notification of the order to the other courts in which the civil actions consolidated under the order are pending.

SEC. 21. CITIZENS' PETITIONS.

(a) IN GENERAL.—Any person may petition the Administrator to initiate a proceeding for the issuance, amendment, or repeal of a rule under section 4, 6, or 8 or an order under section 5(e) or (6)(b)(2).

(b) PROCEDURES.—(1) Such petition shall be filed in the principal office of the Administrator and shall set forth the facts which it is claimed establish that it is necessary to issue, amend, or repeal a rule under section 4, 6, or 8 or an order under section 5(e), 6(b)(1)(A), or 6(b)(1)(B).

(2) The Administrator may hold a public hearing or may conduct such investigation or proceeding as the Administrator deems appropriate in order to determine whether or not such petition should be granted.

(3) Within 90 days after filing of a petition described in paragraph (1), the Administrator shall either grant or deny the petition. If the Administrator grants such petition, the Administrator shall promptly

commence an appropriate proceeding in accordance with section 4, 5, 6, or 8. If the Administrator denies such petition, the Administrator shall publish in the Federal Register the Administrator's reasons for such denial.

(4)(A) If the Administrator denies a petition filed under this section (or if the Administrator fails to grant or deny such petition within the 90-day period) the petitioner may commence a civil action in a district court of the United States to compel the Administrator to initiate a rulemaking proceeding as requested in the petition. Any such action shall be filed within 60 days after the Administrator's denial of the petition or, if the Administrator fails to grant or deny the petition within 90 days after filing the petition, within 60 days after the expiration of the 90-day period.

(B) In an action under subparagraph (A) respecting a petition to initiate a proceeding to issue a rule under section 4, 6, or 8 or an order under section 5(e) or 6(b)(2), the petitioner shall be provided an opportunity to have such petition considered by the court in a de novo proceeding. If the petitioner demonstrates to the satisfaction of the court by a preponderance of the evidence that—

(i) in the case of a petition to initiate a proceeding for the issuance of a rule under section 4 or an order under section 5(e)—

(I) information available to the Administrator is insufficient to permit a reasoned evaluation of the health and environmental effects of the chemical substance to be subject to such rule or order; and

(II) in the absence of such information, the substance may present an unreasonable risk to health or the environment, or the substance is or will be produced in substantial quantities and it enters or may reasonably be anticipated to enter the environment in substantial quantities or there is or may be significant or substantial human exposure to it; or

(ii) in the case of a petition to initiate a proceeding for the issuance of a rule under section 6 or 8 or an order under section 6(b)(2), there is a reasonable basis to conclude that the issuance of such a rule or order is necessary to protect health or the environment against an unreasonable risk of injury to health or the environment.

the court shall order the Administrator to initiate the action requested by the petitioner. If the court finds that the extent of the risk to health or the environment alleged by the petitioner is less than the extent of risks to health or the environment with respect to which the Administrator is taking action under this Act and there are insufficient resources available to the Administrator to take the action requested by the petitioner, the court may permit the Administrator to defer initiating the action requested by the petitioner until such time as the court prescribes.

(C) The court in issuing any final order in any action brought pursuant to subparagraph (A) may award costs of suit and reasonable fees for attorneys and expert witnesses if the court determines that such an award is appropriate. Any court, in issuing its decision in an action brought to review such an order, may award costs of suit and reasonable fees for attorneys if the court determines that such an award is appropriate.

(5) The remedies under this section shall be in addition to, and not in lieu of, other remedies provided by law.

SEC. 22. NATIONAL DEFENSE WAIVER.

The Administrator shall waive compliance with any provision of this Act upon a request and determination by the President that the requested waiver is necessary in the interest of national defense. The Administrator shall maintain a written record of the basis upon which such waiver was granted and make such record available for in camera examination when relevant in a judicial proceeding under this Act. Upon the issuance of such a waiver, the Administrator shall publish in the Federal Register a notice that the waiver was granted for national defense purposes, unless, upon the request of the President, the Administrator determines to omit such publication because the publication itself would be contrary to the interests of national defense, in which event the Administrator shall submit notice thereof to the Armed Services Committees of the Senate and the House of Representatives.

SEC. 23. EMPLOYEE PROTECTION.

(a) In General.—No employer may discharge any employee or otherwise discriminate against any employee with respect to the employee's compensation, terms, conditions, or privileges of employment because the employee (or any person acting pursuant to a request of the employee) has—

(1) commenced, caused to be commenced, or is about to commence or cause to be commenced a proceeding under this Act;

(2) testified or is about to testify in any such proceeding; or

(3) assisted or participated or is about to assist or participate in any manner in such a proceeding or in any other action to carry out the purposes of this Act.

(b) Remedy.—(1) Any employee who believes that the employee has been discharged or otherwise discriminated against by any person in violation of subsection (a) of this section may, within 30 days after such alleged violation occurs, file (or have any person file on the employee's behalf) a complaint with the Secretary of Labor (hereinafter in this section referred to as the "Secretary") alleging such discharge or discrimination. Upon receipt of such a complaint, the Secretary shall notify the person named in the complaint of the filing of the complaint.

(2)(A) Upon receipt of a complaint filed under paragraph (1), the Secretary shall conduct an investigation of the violation alleged in the complaint. Within 30 days of the receipt of such complaint, the Secretary shall complete such investigation and shall notify in writing the complainant (and any person acting on behalf of the complainant) and the person alleged to have committed such violation of the results of the investigation conducted pursuant to this paragraph. Within ninety days of the receipt of such complaint the Secretary shall, unless the proceeding on the complaint is terminated by the Secretary on the basis of a settlement entered into by the Secretary and the person alleged to have committed such violation, issue an order either providing the relief prescribed by subparagraph (B) or denying the complaint. An order of the Secretary shall be made on the record after notice and opportunity for agency hearing. The Secretary may not enter into a settlement terminating a proceeding on a complaint without the participation and consent of the complainant.

(B) If in response to a complaint filed under paragraph (1) the Secretary determines that a violation of subsection (a) of this section has occurred, the Secretary shall order (i) the person who committed such violation to take affirmative action to abate the violation, (ii) such person to reinstate the complainant to the complainant's former position together with the compensation (including back pay), terms, conditions, and privileges of the complainant's employment, (iii) compensatory damages, and (iv) where appropriate, exemplary damages. If such an order issued, the Secretary, at the request of the complainant, shall assess against the person against whom the order is issued a sum equal to the aggregate amount of all costs and expenses (including attorney's fees) reasonably incurred, as determined by the Secretary, by the complainant for, or in connection with, the bringing of the complaint upon which the order was issued.

(c) REVIEW.—(1) Any employee or employer adversely affected or aggrieved by an order issued under subsection (b) may obtain review of the order in the United States Court of Appeals for the circuit in which the violation, with respect to which the order was issued, allegedly occurred. The petition for review must be filed within sixty days from the issuance of the Secretary's order. Review shall conform to chapter 7 of title 5 of the United States Code.

(2) An order of the Secretary, with respect to which review could have been obtained under paragraph (1), shall not be subject to judicial review in any criminal or other civil proceeding.

(d) ENFORCEMENT.—Whenever a person has failed to comply with an order issued under subsection (b)(2), the Secretary shall file a civil action in the United States district court for the district in which the violation was found to occur to enforce such order. In actions brought under this subsection, the district courts shall have jurisdiction to grant all appropriate relief, including injunctive relief and compensatory and exemplary damages. Civil actions brought under this subsection shall be heard and decided expeditiously.

(e) EXCLUSION.—Subsection (a) of this section shall not apply with respect to any employee who, acting without direction from the employee's employer (or any agent of the employer), deliberately causes a violation of any requirement of this Act.

SEC. 24. EMPLOYMENT EFFECTS.

(a) IN GENERAL.—The Administrator shall evaluate on a continuing basis the potential effects on employment (including reductions in employment or loss of employment from threatened plant closures) of—

(1) the issuance of a rule or order under section 4, 5, or 6, or

(2) a requirement of section 5 or 6.

(b)(1) INVESTIGATIONS.—Any employee (or any representative of an employee) may request the Administrator to make an investigation of—

(A) a discharge or layoff or threatened discharge or layoff of the employee, or

(B) adverse or threatened adverse effects on the employee's employment,

allegedly resulting from a rule or order under section 4, 5, or 6 or a requirement of section 5 or 6. Any such request shall be made in writing, shall set forth with reasonable particularity the grounds for the request, and shall be signed by the employee, or representative of such employee, making the request.

(2)(A) Upon receipt of a request made in accordance with paragraph (1) the Administrator shall (i) conduct the investigation requested, and (ii) if requested by any interested person, hold public hearings on any matter involved in the investigation unless the Administrator, by order issued within 45 days of the date such hearings are requested, denies the request for the hearings because the Administrator determines there are no reasonable grounds for holding such hearings. If the Administrator makes such a determination, the Administrator shall notify in writing the person requesting the hearing of the determination and the reasons therefor and shall publish the determination and the reasons therefor in the Federal Register.

(B) If public hearings are to be held on any matter involved in an investigation conducted under this subsection—

(i) at least five days' notice shall be provided the person making the request for the investigation and any person identified in such request,

(ii) such hearings shall be held in accordance with section 6(c)(3), and

(iii) each employee who made or for whom was made a request for such hearings and the employer of such employee shall be required to present information respecting the applicable matter referred to in paragraph (1)(A) or (1)(B) together with the basis for such information.

(3) Upon completion of an investigation under paragraph (2), the Administrator shall make findings of fact, shall make such recommendations as the Administrator deems appropriate, and shall make available to the public such findings and recommendations.

(4) This section shall not be construed to require the Administrator to amend or repeal any rule or order in effect under this Act.

SEC. 25. STUDIES.

(a) Indemnification Study.—The Administrator shall conduct a study of all Federal laws administered by the Administrator for the purpose of determining whether and under what conditions, if any, indemnification should be accorded any person as a result of any action taken by the Administrator under any such law. The study shall—

(1) include an estimate of the probable cost of any indemnification programs which may be recommended;

(2) include an examination of all viable means of financing the cost of any recommended indemnification; and

(3) be completed and submitted to Congress within two years from the effective date of enactment of this Act.

The General Accounting Office shall review the adequacy of the study submitted to Congress pursuant to paragraph (3) and shall report the results of its review to the Congress within six months of the date such study is submitted to Congress.

(b) Classification, Storage, and Retrieval Study.—The Council on Environmental Quality, in consultation with the Administrator, the Secretary of Health, Education, and Welfare, the Secretary of Commerce, and the heads of other appropriate Federal departments or agencies, shall coordinate a study of the feasibility of establishing (1) a standard classification system for chemical substances and related substances, and (2) a standard means for storing and for obtaining rapid access to information respecting such substances. A report on such study shall be completed and submitted to Congress not later than 18 months after the effective date of enactment of this Act.

SEC. 26. ADMINISTRATION OF THE ACT.

(a) Cooperation of Federal Agencies.—Upon request by the Administrator, each Federal department and agency is authorized—

(1) to make its services, personnel, and facilities available (with or without reimbursement) to the Administrator to assist the Administrator in the administration of this Act; and

(2) to furnish to the Administrator such information, data, estimates, and statistics, and to allow the Administrator access to all information in its possession as the Administrator may reasonably determine to be necessary for the administration of this Act.

(b) FEES.—(1) The Administrator may, by rule, require the payment of a reasonable fee from any person required to submit data under section 4 or 5 to defray the cost of administering this Act. Such rules shall not provide for any fee in excess of $2,500 or, in the case of a small business concern, any fee in excess of $100. In setting a fee under this paragraph, the Administrator shall take into account the ability to pay of the person required to submit the data and the cost to the Administrator of reviewing such data. Such rules may provide for sharing such a fee in any case in which the expenses of testing are shared under section 4 or 5.

(2) The Administrator, after consultation with the Administrator of the Small Business Administration, shall by rule prescribe standards for determining the persons which qualify as small business concerns for purposes of paragraph (1).

(c) ACTION WITH RESPECT TO CATEGORIES.—(1) Any action authorized or required to be taken by the Administrator under any provision of this Act with respect to a chemical substance or mixture may be taken by the Administrator in accordance with that provision with respect to a category of chemical substances or mixtures. Whenever the Administrator takes action under a provision of this Act with respect to a category of chemical substances or mixtures, any reference in this Act to a chemical substance or mixture (insofar as it relates to such action) shall be deemed to be a reference to each chemical substance or mixture in such category.

(2) For purposes of paragraph (1):

(A) The term "category of chemical substances" means a group of chemical substances the members of which are similar in molecular structure, in physical, chemical, or biological properties, in use, or in mode of entrance into the human body or into the environment, or the members of which are in some other way suitable for classification as such for purposes of this Act, except that such term does not mean a group of chemical substances which are grouped together solely on the basis of their being new chemical substances.

(B) The term "category of mixtures" means a group of mixtures the members of which are similar in molecular structure, in physical, chemical, or biological properties, in use, or in the mode of entrance into the human body or into the environment, or the members of which are in some other way suitable for classification as such for purposes of this Act.

(d) ASSISTANCE OFFICE.—The Administrator shall establish in the Environmental Protection Agency an identifiable office to provide technical and other nonfinancial assistance to manufacturers and processors of chemical substances and mixtures respecting the requirements of this Act applicable to such manufacturers and processors, the policy of the Agency respecting the application of such requirements to such manufacturers and processors, and the means and methods by which such manufacturers and processors may comply with such requirements.

(e) FINANCIAL DISCLOSURES.—(1) Except as provided under paragraph (3), each officer or employee of the Environmental Protection Agency and the Department of Health, Education, and Welfare who—

(A) performs any function or duty under this Act, and

(B) has any known financial interest (i) in any person subject to this Act or any rule or order in effect under this Act, or (ii) in any person who applies for or receives any grant or contract under this Act,

shall, on February 1, 1978, and on February 1 of each year thereafter, file with the Administrator or the Secretary of Health, Education, and Welfare (hereinafter in this subsection referred to as the "Secretary"), as appropriate, a written statement concerning all such interests held by such officer or employee during the preceding calendar year. Such statement shall be made available to the public.

(2) The Administrator and the Secretary shall—

(A) act within 90 days of the effective date of this Act—

(i) to define the term "known financial interests" for purposes of paragraph (1), and

(ii) to establish the methods by which the requirement to file written statements specified in paragraph (1) will be monitored and enforced, including appropriate provisions for review by the Administrator and the Secretary of such statements; and

(B) report to the Congress on June 1, 1978, and on June 1 of each year thereafter with respect to such statements and the actions taken in regard thereto during the preceding calendar year.

(3) The Administrator may by rule identify specific positions with the Environmental Protection Agency, and the Secretary may by rule identify specific positions with the Department of Health, Education, and Welfare, which are of a nonregulatory or nonpolicymaking nature, and the Administrator and the Secretary may by rule provide that officers or employees occupying such positions shall be exempt from the requirements of paragraph (1).

(4) This subsection does not supersede any requirement of chapter 11 of title 18, United States Code.

(5) Any officer or employee who is subject to, and knowingly violates, this subsection or any rule issued thereunder, shall be fined not more than $2,500 or imprisoned not more than one year, or both.

(f) Statement of Basis and Purpose.—Any final order issued under this Act shall be accompanied by a statement of its basis and purpose. The contents and adequacy of any such statement shall not be subject to judicial review in any respect.

(g) Assistant Administrator.—(1) The President, by and with the advice and consent of the Senate, shall appoint an Assistant Administrator for Toxic Substances of the Environmental Protection Agency. Such Assistant Administrator shall be a qualified individual who is, by reason of background and experience, especially qualified to direct a program concerning the effects of chemicals on human health and the environment. Such Assistant Administrator shall be responsible for (A) the collection of data, (B) the preparation of studies, (C) the making of recommendations to the Administrator for regulatory and other actions to carry out the purposes and to facilitate the administration of this Act, and (D) such other functions as the Administrator may assign or delegate.

(2) The Assistant Administrator to be appointed under paragraph (1) shall (A) be in addition to the Assistant Administrators of the Environmental Protection Agency authorized by section 1(d) of Reorganization Plan No. 3 of 1970, and (B) be compensated at the rate of pay authorized for such Assistant Administrators.

SEC. 27. DEVELOPMENT AND EVALUATION OF TEST METHODS.

(a) IN GENERAL.—The Secretary of Health, Education, and Welfare, in consultation with the Administrator and acting through the Assistant Secretary for Health, may conduct, and make grants to public and nonprofit private entities and enter into contracts with public and private entities for, projects for the development and evaluation of inexpensive and efficient methods (1) for determining and evaluating the health and environmental effects of chemical substances and mixtures, and their toxicity, persistence, and other characteristics which affect health and the environment, and (2) which may be used for the development of test data to meet the requirements of rules promulgated under section 4. The Administrator shall consider such methods in prescribing under section 4 standards for the development of test data.

(b) APPROVAL BY SECRETARY.—No grant may be made or contract entered into under subsection (a) unless an application therefor has been submitted to and approved by the Secretary. Such an application shall be submitted in such form and manner and contain such information as the Secretary may require. The Secretary may apply such conditions to grants and contracts under subsection (a) as the Secretary determines are necessary to carry out the purposes of such subsection. Contracts may be entered into under such subsection without regard to sections 3648 and 3709 of the Revised Statutes (31 U.S.C. 529; 41 U.S.C. 5).

(c) ANNUAL REPORTS.—(1) The Secretary shall prepare and submit to the President and the Congress on or before January 1 of each year a report of the number of grants made and contracts entered into under this section and the results of such grants and contracts.

(2) The Secretary shall periodically publish in the Federal Register reports describing the progress and results of any contract entered into or grant made under this section.

SEC. 28. STATE PROGRAMS.

(a) IN GENERAL.—For the purpose of complementing (but not reducing) the authority of, or actions taken by, the Administrator under this Act, the Administrator may make grants to States for the establishment and operation of programs to prevent or eliminate unreasonable risks within the States to health or the environment which are associated with a chemical substance or mixture and with respect to which the Administrator is unable or is not likely to take action under this Act for their prevention or elimination. The amount of a grant under this subsection shall be determined by the Administrator, except that no grant for any State program may exceed 75 per centum of the establishment and operation costs (as determined by the Administrator) of such program during the period for which the grant is made.

(b) APPROVAL BY ADMINISTRATOR.—(1) No grant may be made under subsection (a) unless an application therefor is submitted to and approved by the Administrator. Such an application shall be submitted in such form and manner as the Administrator may require and shall—

(A) set forth the need of the applicant for a grant under subsection (a),

(B) identify the agency or agencies of the State which shall establish or operate, or both, the program for which the application is submitted,

(C) describe the actions proposed to be taken under such program,

(D) contain or be supported by assurances satisfactory to the Administrator that such program shall, to the extent feasible, be integrated with other programs of the applicant for environmental and public health protection,

(E) provide for the making of such reports and evaluations as the Administrator may require, and

(F) contain such other information as the Administrator may prescribe.

(2) The Administrator may approve an application submitted in accordance with paragraph (1) only if the applicant has established to the satisfaction of the Administrator a priority need, as determined under rules of the Administrator, for the grant for which the application has been submitted. Such rules shall take into consideration the seriousness of the health effects in a State which are associated with chemical substances or mixtures, including cancer, birth defects, and gene mutations, the extent of the exposure in a State of human beings and the environment to chemical substances and mixtures, and the extent to which chemical substances and mixtures are manufactured, processed, used, and disposed of in a State.

(c) ANNUAL REPORTS.—Not later than six months after the end of each of the fiscal years 1979, 1980, and 1981, the Administrator shall submit to the Congress a report respecting the programs assisted by grants under subsection (a) in the preceding fiscal year and the extent to which the Administrator has disseminated information respecting such programs.

(d) AUTHORIZATION.—For the purpose of making grants under subsection (a) there are authorized to be appropriated $1,500,000 for the fiscal year ending September 30, 1977, $1,500,000 for the fiscal year ending September 30, 1978, and $1,500,000 for the fiscal year ending September 30, 1979. Sums appropriated under this subsection shall remain available until expended.

SEC. 29. AUTHORIZATION FOR APPROPRIATIONS.

There are authorized to be appropriated to the Administrator for purposes of carrying out this Act (other than sections 27 and 28 and subsections (a) and (c) through (g) of section 10 thereof) $10,100,000 for the fiscal year ending September 30, 1977, $12,625,000 for the fiscal year ending September 30, 1978, $16,200,000 for the fiscal year ending September 30, 1979. No part of the funds appropriated under this section may be used to construct any research laboratories.

SEC. 30. ANNUAL REPORT.

The Administrator shall prepare and submit to the President and the Congress on or before January 1, 1978, and on or before January 1 of each succeeding year a comprehensive report on the administration of this Act during the preceding fiscal year. Such report shall include—

(1) a list of the testing required under section 4 during the year for which the report is made and an estimate of the costs incurred during such year by the persons required to perform such tests;

(2) the number of notices received during such year under section 5, the number of such notices received during such year under such section for chemical substances subject to a section 4 rule, and a summary of any action taken during such year under section 5(g);

(3) a list of rules issued during such year under section 6;

(4) a list, with a brief statement of the issues, of completed or pending judicial actions under this Act and administrative actions under section 16 during such year;

(5) a summary of major problems encountered in the administration of this Act; and

(6) such recommendations for additional legislation as the Administrator deems necessary to carry out the purposes of this Act.

SEC. 31. EFFECTIVE DATE.

Except as provided in section 4(f), this Act shall take effect on January 1, 1977.

Speaker of the House of Representatives.

Vice President of the United States and
President of the Senate.

Section III

The Conference Report

94TH CONGRESS } HOUSE OF REPRESENTATIVES { REPORT
2d Session } { No. 94–1679

TOXIC SUBSTANCES CONTROL ACT

SEPTEMBER 23, 1976.—Ordered to be printed

Mr. STAGGERS, from the committee of conference, submitted the following

CONFERENCE REPORT

[To accompany S. 3149]

The committee of conference on the disagreeing votes of the two Houses on the amendment of the House to the bill (S. 3149) to regulate commerce and protect human health and the environment by requiring testing and necessary use restrictions on certain chemical substances, and for other purposes, having met, after full and free conference, have agreed to recommend and do recommend to their respective Houses as follows:

That the Senate recede from its disagreement to the amendment of the House and agree to the same with an amendment as follows:

In lieu of the matter proposed to be inserted by the House amendment insert the following:

SECTION 1. SHORT TITLE AND TABLE OF CONTENTS.

This Act may be cited as the "Toxic Substances Control Act".

SECTION 1. This Act may be cited as the "Toxic Substances Control Act".

TABLE OF CONTENTS

76–804 O

SEC. 2. FINDINGS, POLICY, AND INTENT.

(a) FINDINGS.—The Congress finds that—

(1) human beings and the environment are being exposed each year to a large number of chemical substances and mixtures;

(2) among the many chemical substances and mixtures which are constantly being developed and produced, there are some whose manufacture, processing, distribution in commerce, use, or disposal may present an unreasonable risk of injury to health or the environment; and

(3) the effective regulation of interstate commerce in such chemical substances and mixtures also necessitates the regulation of intrastate commerce in such chemical substances and mixtures.

(b) POLICY.—It is the policy of the United States that—

(1) adequate data should be developed with respect to the effect of chemical substances and mixtures on health and the environment and that the development of such data should be the responsibility of those who manufacture and those who process such chemical substances and mixtures;

(2) adequate authority should exist to regulate chemical substances and mixtures which present an unreasonable risk of injury to health or the environment, and to take action with respect to chemical substances and mixtures which are imminent hazards; and

(3) authority over chemical substances and mixtures should be exercised in such a manner as not to impede unduly or create unnecessary economic barriers to technological innovation while fulfilling the primary purpose of this Act to assure that such innovation and commerce in such chemical substances and mixtures do not present an unreasonable risk of injury to health or the environment.

(c) INTENT OF CONGRESS.—It is the intent of Congress that the Administrator shall carry out this Act in a reasonable and prudent manner, and that the Administrator shall consider the environmental, economic, and social impact of any action the Administrator takes or proposes to take under this Act.

SEC. 3. DEFINITIONS.

As used in this Act:

(1) The term "Administrator" means the Administrator of the Environmental Protection Agency.

(2)(A) Except as provided in subparagraph (B), the term "chemical substance" means any organic or inorganic substance of a particular molecular identity, including—

(i) any combination of such substances occurring in whole or in part as a result of a chemical reaction or occurring in nature, and

(ii) any element or uncombined radical.

(B) Such term does not include—

(i) any mixture,

(ii) any pesticide (as defined in the Federal Insecticide, Fungicide, and Rodenticide Act) when manufactured, processed, or distributed in commerce for use as a pesticide,

(iii) tobacco or any tobacco product,

(iv) any source material, special nuclear material, or byproduct material (as such terms are defined in the Atomic Energy Act of 1954 and regulations issued under such Act),

(v) any article the sale of which is subject to the tax imposed by section 4181 of the Internal Revenue Code of 1954 (determined without regard to any exemptions from such tax provided by section 4182 or 4221 or any other provision of such Code), and

(vi) any food, food additive, drug, cosmetic, or device (as such terms are defined in section 201 of the Federal Food, Drug, and Cosmetic Act) when manufactured, processed, or distributed in commerce for use as a food, food additive, drug, cosmetic, or device.

The term "food" as used in clause (vi) of this subparagraph includes poultry and poultry products (as defined in sections 4(e) and 4(f) of the Poultry Products Inspection Act), meat and meat food products (as defined in section 1(j) of the Federal Meat Inspection Act), and eggs and egg products (as defined in section 4 of the Egg Products Inspection Act).

(3) The term "commerce" means trade, traffic, transportation, or other commerce (A) between a place in a State and any place outside of such State, or (B) which affects trade, traffic, transportation, or commerce described in clause (A).

(4) The terms "distribute in commerce" and "distribution in commerce" when used to describe an action taken with respect to a chemical substance or mixture or article containing a substance or mixture mean to sell, or the sale of, the substace, mixture, or article in commerce: to introduce or deliver for introduction into commerce, or the introduction or delivery for introduction into commerce of, the substance, mixture, or article; or to hold, or the holding of, the substance, mixture, or article after its introduction into commerce.

(5) The term "environment" includes water, air, and land and the interrelationship which exists among and between water, air, and land and all living things.

(6) The term "health and safety study" means any study of any effect of a chemical substance or mixture on health or the environment or on both, including underlying data and epidemiological studies, studies of occupational exposure to a chemical substance or mixture, toxicological, clinical, and ecological studies of a chemical substance or mixture, and any test performed pursuant to this Act.

(7) The term "manufacture" means to import into the customs territory of the United States (as defined in general headnote 2 of the Tariff Schedules of the United States), produce, or manufacture.

(8) The term "mixture" means any combination of two or more chemical substances if the combination does not occur in nature and is not, in whole or in part, the result of a chemical reaction; except that such terms does include any combination which occurs, in whole or in part, as a result of a chemical reaction if none of the chemical substances comprising the combination is a new chemical substance and if the combination could have been manufactured for commercial purposes without a chemical reaction at the time the chemical substances comprising the combination were combined.

(9) The term "new chemical substance" means any chemical substance which is not included in the chemical substance list compiled and published under section 8(b).

(10) The term "process" means the preparation of a chemical substance or mixture, after its manufacture, for distribution in commerce—

(A) in the same form or physical state as, or in a different form or physical state from, that in which it was received by the person so preparing such substance or mixture, or

(B) as part of an article containing the chemical substance or mixture.

(11) The term "processor" means any person who processes a chemical substance or mixture.

(12) The term "standards for the development of test data" means a prescription of—

(A) the—

(i) health and environmental effects, and

(ii) information relating to toxicity, persistence, and other characteristics which affect health and the environment,

for which test data for a chemical substance or mixture are to be developed and any analysis that is to be performed on such data, and

(B) to the extent necessary to assure that data respecting such effects and characteristics are reliable and adequate—

(i) the manner in which such data are to be developed,

(ii) the specification of any test protocol or methodology to be employed in the development of such data, and

(iii) such other requirements as are necessary to provide such assurance.

(13) The term "State" means any State of the United States, the District of Columbia, the Commonwealth of Puerto Rico, the Virgin Islands, Guam, the Canal Zone, American Samoa, the Northern Mariana Islands, or any other territory or possession of the United States.

(14) The term "United States", when used in the geographic sense, means all of the States.

SEC. 4. TESTING OF CHEMICAL SUBSTANCES AND MIXTURES.

(a) TESTING REQUIREMENTS.—If the Administrator finds that—

(1)(A)(i) the manufacture, distribution in commerce, processing, use, or disposal of a chemical substance or mixture, or that any combination of such activities, may present an unreasonable risk of injury to health or the environment,

(ii) there are insufficient data and experience upon which the effects of such manufacture, distribution in commerce, processing, use, or disposal of such substance or mixture or of any combination of such activities on health or the environment can reasonably be determined or predicted, and

(iii) testing of such substance or mixture with respect to such effects is necessary to develop such data; or

(B)(i) a chemical substance or mixture is or will be produced in substantial quantities, and (I) it enters or may reasonably be anticipated to enter the environment in substantial quantities or (II) there is or may be significant or substantial human exposure to such substance or mixture,

(ii) there are insufficient data and experience upon which the effects of the manufacture, distribution in commerce, processing, use, or disposal of such substance or mixture or of any combination of such activities on health or the environment can reasonably be determined or predicted, and

(iii) testing of such substance or mixture with respect to such effects is necessary to develop such data; and

(2) in the case of a mixture, the effects which the mixture's manufacture, distribution in commerce, processing, use, or disposal or any combination of such activities may have on health or the environment may not be reasonably and more efficiently determined or predicted by testing the chemical substances which comprise the mixture;

the Administrator shall by rule require that testing be conducted on such substance or mixture to develop data with respect to the health and environmental effects for which there is an insufficiency of data and experience and which are relevant to a determination that the manufacture, distribution in commerce, processing, use, or disposal of such substance or mixture, or that any combination of such activities, does or does not present an unreasonable risk of injury to health or the environment.

(b)(1) TESTING REQUIREMENT RULE.—A rule under subsection (a) shall include—

(A) identification of the chemical substance or mixture for which testing is required under the rule.

(B) standards for the development of test data for such substance or mixture, and

(C) with respect to chemical substances which are not new chemical substances and to mixtures, a specification of the period (which period may not be of unreasonable duration) within which the persons required to conduct the testing shall submit to the Administrator data developed in accordance with the standards referred to in subparagraph (B).

In determining the standards and period to be included, pursuant to subparagraphs (B) and (C), in a rule under subsection (a), the Administrator's considerations shall include the relative costs of the various test protocols and methodologies which may be required under the rule and the reasonably foreseeable availability of the facilities and personnel needed to perform the testing required under the rule. Any such rule may require the submission to the Administrator of preliminary data during the period prescribed under subparagraph (C).

(2)(A) The health and environmental effects for which standards for the development of test data may be prescribed include carcinogenesis, mutagenesis, teratogenesis, behavioral disorders, cumulative or synergistic effects, and any other effect which may present an unreasonable risk of injury to health or the environment. The characteristics of chemical substances and mixtures for which such standards may be prescribed include persistence, acute toxicity, subacute toxicity, chronic toxicity, and any other characteristic which may present such a risk. The methodologies that may be prescribed in such standards include epidemiologic studies, serial or hierarchical tests, in vitro tests, and whole animal tests, except that before prescribing epidemiologic studies of employees, the Administrator shall consult with the Director of the National Institute for Occupational Safety and Health.

(B) From time to time, but not less than once each 12 months, the Administrator shall review the adequacy of the standards for development of data prescribed in rules under subsection (a) and shall, if necessary, institute proceedings to make appropriate revisions of such standards.

(3)(A) A rule under subsection (a) respecting a chemical substance or mixture shall require the persons described in subparagraph (B) to conduct tests and submit data to the Administrator on such substance or mixture, except that the Administrator may permit two or more of such persons to designate one such person or a qualified third party to conduct such tests and submit such data on behalf of the persons making the deignation.

(B) The following persons shall be required to conduct tests and submit data on a chemical substance or mixture subject to a rule under subsection (a):

(i) Each person who manufactures or intends to manufacture such substance or mixture if the Administrator makes a finding described in subsection (a)(1)(A)(ii) or (a)(1)(B)(ii) with respect to the manufacture of such substance or mixture.

(ii) Each person who processes or intends to process such substance or mixture if the Administrator makes a finding described in subsection (a)(1)(A)(ii) or (a)(1)(B)(ii) with respect to the processing of such substance or mixture.

(iii) Each person who manufactures or processes or intends to manufacture or process such substance or mixture if the Administrator makes a finding described in subsection (a)(1)(A)(ii) or (a)(1)(B)(ii) with respect to the distribution in commerce, use, or disposal of such substance or mixture.

(4) Any rule under subsection (a) requiring the testing of and submission of data for a particular chemical substance or mixture shall expire at the end of the reimbursement period (as defined in subsection

(c)(3)(B)) which is applicable to test data for such substance or mixture unless the Administrator repeals the rule before such date; and a rule under subsection (a) requiring the testing of and submission of data for a category of chemical substances or mixtures shall expire with respect to a chemical substance or mixture included in the category at the end of the reimbursement period (as so defined) which is applicable to test data for such substance or mixture unless the Administrator before such date repeals the application of the rule to such substance or mixture or repeals the rule.

(5) Rules issued under subsection (a) (and any substantive amendment thereto or repeal thereof) shall be promulgated pursuant to section 553 of title 5, United States Code, except that (A) the Administrator shall give interested persons an opportunity for the oral presentation of data, views, or arguments, in addition to an opportunity to make written submissions; (B) a transcript shall be made of any oral presentation; and (C) the Administrator shall make and publish with the rule the findings described in paragraph (1)(A) or (1)(B) of subsection (a) and, in the case of a rule respecting a mixture, the finding described in paragraph (2) of such subsection.

(c) EXEMPTION.—(1) Any person required by a rule under subsection (a) to conduct tests and submit data on a chemical substance or mixture may apply to the Administrator (in such form and manner as the Administrator shall prescribe) for an exemption from such requirement.

(2) If, upon receipt of an application under paragraph (1), the Administrator determines that—

(A) the chemical substance or mixture with respect to which such application was submitted is equivalent to a chemical substance or mixture for which data has been submitted to the Administrator in accordance with a rule under subsection (a) or for which data is being developed pursuant to such a rule, and

(B) submission of data by the applicant on such substance or mixture would be duplicative of data which has been submitted to the Administrator in accordance with such rule or which is being developed pursuant to such rule,

the Administrator shall exempt, in accordance with paragraph (3) or (4), the applicant from conducting tests and submitting data on such substance or mixture under the rule with respect to which such application was submitted.

(3)(A) If the exemption under paragraph (2) of any person from the requirement to conduct tests and submit test data on a chemical substance or mixture is granted on the basis of the existence of previously submitted test data and if such exemption is granted during the reimbursement period for such test data (as prescribed by subparagraph (B)), then (unless such person and the persons referred to in clauses (i) and (ii) agree on the amount and method of reimbursement) the Administrator shall order the person granted the exemption to provide fair and equitable reimbursement (in an amount determined under rules of the Administrator)—

(i) to the person who previously submitted such test data, for a portion of the costs incurred by such person in complying with the requirement to submit such data, and

(ii) to any other person who has been required under this subparagraph to contribute with respect to such costs, for a portion of the amount such person was required to contribute.

In promulgating rules for the determination of fair and equitable reimbursement to the persons described in clauses (i) and (ii) for costs incurred with respect to a chemical substance or mixture, the Administrator shall, after consultation with the Attorney General and the Federal Trade Commission, consider all relevant factors, including the effect on the competitive position of the person required to provide reimbursement in relation to the person to be reimbursed and the share of the market for such substance or mixture of the person required to provide reimbursement in relation to the share of such market of the persons to be reimbursed. An order under this subparagraph shall, for purposes of judicial review, be considered final agency action.

(B) For purposes of subparagraph (A), the reimbursement period for any test data for a chemical substance or mixture is a period—

(i) beginning on the date such data is submitted in accordance with a rule promulgated under subsection (a), and

(ii) ending—

(I) five years after the date referred to in clause (i), or

(II) at the expiration of a period which begins on the date referred to in clause (i) and which is equal to the period which the Administrator determines was necessary to develop such data, whichever is later.

(4)(A) If the exemption under paragraph (2) of any person from the requirement to conduct tests and submit test data on a chemical substance or mixture is granted on the basis of the fact that test data is being developed by one or more persons pursuant to a rule promulgated under subsection (a), then (unless such person and the persons referred to in clauses (i) and (ii) agree on the amount and method of reimbursement) the Administrator shall order the person granted the exemption to provide fair and equitable reimbursement (in an amount determined under rules of the Administrator)—

(i) to each such person who is developing such test data, for a portion of the costs incurred by each such person in complying with such rule, and

(ii) to any other person who has been required under this subparagraph to contribute with respect to the costs of complying with such rule, for a portion of the amount such person was required to contribute.

In promulgating rules for the determination of fair and equitable reimbursement to the persons described in clauses (i) and (ii) for costs incurred with respect to a chemical substance or mixture, the Administrator shall, after consultation with the Attorney General and the Federal Trade Commission, consider the factors described in the second sentence of paragraph (3)(A). An order under this subparagraph shall, for purposes of judicial review, be considered final agency action.

(B) If any exemption is granted under paragraph (2) on the basis of the fact that one or more persons are developing test data pursuant to a rule promulgated under subsection (a) and if after such exemp-

tion is granted the Administrator determines that no such person has complied with such rule, the Administrator shall (i) after providing written notice to the person who holds such exemption and an opportunity for a hearing, by order terminate such exemption, and (ii) notify in writing such person of the requirements of the rule with respect to which such exemption was granted.

(d) NOTICE.—Upon the receipt of any test data pursuant to a rule under subsection (a), the Administrator shall publish a notice of the receipt of such data in the Federal Register within 15 days of its receipt. Subject to section 14, each such notice shall (1) identify the chemical substance or mixture for which data have been recived; (2) list the uses or intended uses of such substance or mixture and the information required by the applicable standards for the development of test data; and (3) describe the nature of the test data developed. Except as otherwise provided in section 14, such data shall be made available by the Administrator for examination by any person.

(e) PRIORITY LIST.—(1)(A) There is established a committee to make recommendations to the Administrator respecting the chemical substances and mixtures to which the Administrator should give priority consideration for the promulgation of a rule under subsection (a). In making such a recommendation with respect to any chemical substance or mixture, the committee shall consider all relevant factors, including—

(i) the quantities in which the substance or mixture is or will be manufactured,

(ii) the quantities in which the substance or mixture enters or will enter the environment,

(iii) the number of individuals who are or will be exposed to the substance or mixture in their places of employment and the duration of such exposure,

(iv) the extent to which human beings are or will be exposed to the substance or mixture,

(v) the extent to which the substance or mixture is closely related to a chemical substance or mixture which is known to present an unreasonable risk of injury to health or the environment,

(vi) the existence of data concerning the effects of the substance or mixture on health or the environment,

(vii) the extent to which testing of the substance or mixture may result in the development of data upon which the effects of the substance or mixture on health or the environment can reasonably be determined or predicted, and

(viii) the reasonably foreseeable availability of facilities and personnel for performing testing on the substance or mixture.

The recommendations of the committee shall be in the form of a list of chemical substances and mixtures which shall be set forth, either by individual substance or mixture or by groups of substances or mixtures, in the order in which the committee determines the Administrator should take action under subsection (a) with respect to the substances and mixtures. In establishing such list, the committee shall give priority attention to those chemical substances and mixtures which are known to cause or contribute to or which are suspected of causing or contributing to cancer, gene mutations, or birth defects. The committee shall designate chemical substances and mixtures on the list with respect to which the committee determines the Administrator should, within 12 months of the date on which such substances and mixtures are first designated, initiate a proceeding under subsection (a). The total number of chemical substances and mixtures on the list which are designated under the preceding sentence may not, at any time, exceed 50.

(B) As soon as practicable but not later than nine months after the effective date of this Act, the committee shall publish in the Federal Register and transmit to the Administer the list and designations required by subparagraph (A) together with the reasons for the committee's inclusion of each chemical substance or mixture on the list. At least every six months after the date of the transmission to the Administrator of the list pursuant to the preceding sentence, the committee shall make such revisions in the list as it determines to be necessary and shall transmit them to the Administrator together with the committee's reasons for the revisions. Upon receipt of any such revision, the Administrator shall publish in the Federal Register the list with such revision, the reasons for such revision, and the designations made under subparagraph (A). The Administrator shall provide reasonable opportunity to any interested person to file with the Administrator written comments on the committee's list, any revision of such list by the committee, and designations made by the committee, and shall make such comments available to the public. Within the 12-month period beginning on the date of the first inclusion on the list of a chemical substance or mixture designated by the committee under subparagraph (A) the Administrator shall with respect to such chemical substance or mixture either initiate a rulemaking proceeding under subsection (a) or if such a proceeding is not initiated within such period, publish in the Federal Register the Administrator's reason for not initiating such a proceeding.

(2) (A) The committee established by paragraph (1) (A) shall consist of eight members as follows:

(i) One member appointed by the Administrator from the Environmental Protection Agency.

(ii) One member appointed by the Secretary of Labor from officers or employees of the Department of Labor engaged in the Secretary's activities under the Occupational Safety and Health Act of 1970.

(iii) One member appointed by the Chairman of the Council on Environmental Quality from the Council or its officers or employees.

(iv) One member appointed by the Director of the National Institute for Occupational Safety and Health from officers or employees of the Institute.

(v) One member appointed by the Director of the National Institute of Environmental Health Sciences from officers or employees of the Institute.

(vi) One member appointed by the Director of the National Cancer Institute from officers or employees of the Institute.

(vii) One member appointed by the Director of the National Science Foundation from officers or employees of the Foundation.

(viii) One member appointed by the Secretary of Commerce from officers or employees of the Department of Commerce.

(B) (i) An appointed member may designate an individual to serve on the committee on the member's behalf. Such a designation may be made only with the approval of the applicable appointing authority and only if the individual is from the entity from which the member was appointed.

(ii) No individual may serve as a member of the committee for more than four years in the aggregate. If any member of the committee leaves the entity from which the member was appointed, such member may not continue as a member of the committee, and the member's position shall be considered to be vacant. A vacancy in the committee shall be filled in the same manner in which the original appointment was made.

(iii) Initial appointments to the committee shall be made not later than the 60th day after the effective date of this Act. Not later than the 90th day after such date the members of the committee shall hold a meeting for the selection of a chairperson from among their number.

(C) (i) No member of the committee, or designee of such member, shall accept employment or compensation from any person subject to any requirement of this Act or of any rule promulgated or order issued thereunder, for a period of at least 12 months after termination of service on the committe.

(ii) No person, while serving as a member of the committee, or designee of such member, may own any stocks or bonds, or have any pecuniary interest, of substantial value in any person engaged in the manufacture, processing, or distribution in commerce of any chemical substance or mixture subject to any requirement of this Act or of any rule promulgated or order issued thereunder.

(iii) The Administrator, acting through attorneys of the Environmental Protection Agency, or the Attorney General may bring an action in the appropriate district court of the United States to restrain any violation of this subparagraph.

(D) The Administrator shall provide the committee such administrative support services as may be necessary to enable the committee to carry out its function under this subsection.

(f) Required Actions.—Upon the receipt of—

(1) any test data required to be submitted under this Act, or

(2) any other information available to the Administrator,

which indicates to the Administrator that there may be a reasonable basis to conclude that a chemical substance or mixture presents or will present a significant risk of serious or widespread harm to human beings from cancer, gene mutations, or birth defects, the Administrator shall, within the 180-day period beginning on the date of the receipt of such data or information, initiate appropriate action under section 5, 6, or 7 prevent or reduce to a sufficient extent such risk or publish in the Federal Register a finding that such risk is not unreasonable. For good cause shown the Administrator may extend such period for an additional period of not more than 90 days. The Administrator shall publish in the Federal Register notice of any such extension and the reasons therefor. A finding by the Administrator that a risk is not unreasonable shall be considered agency action for purposes of judicial review under chapter 7 of title 5, United States Code. This subsection shall not take effect until two years after the effective date of this Act.

(g) Petition for Standards for the Development of Test Data.—*A person intending to manufacture or process a chemical substance for which notice is required under section 5(a) and who is not required under a rule under subsection (a) to conduct tests and submit data on such substance may petition the Administrator to prescribe standards for the development of test data for such substance. The Administrator shall by order either grant or deny any such petition within 60 days of its receipt. If the petition is granted, the Administrator shall prescribe such standards for such substance with 75 days of the date the petition is granted. If the petition is denied, the Administrator shall publish, subject to section 14, in the Federal Register the reasons for such denial.*

SEC. 5. MANUFACTURING AND PROCESSING NOTICES.

Sec. 5 (a) In General.—*(1) Except as provided in subsection (h), no person may*—

(A) manufacture a new chemical substance on or after the 30th day after the date on which the Administrator first publishes the list required by section 8(b), or

(B) manufacture or process any chemical substance for a use which the Administrator has determined, in accordance with paragraph (2), is a significant new use,

unless such person submits to the Administrator, at least 90 days before such manufacture or processing, a notice, in accordance with subsection (d), of such person's intention to manufacture or process such substance and such person complies with any applicable requirement of subsection (b).

(2) A determination by the Administrator that a use of a chemical substance is a significant new use with respect to which notification is required under paragraph (1) shall be made by a rule promulgated after a consideration of all relevant factors, including—

(A) the projected volume of manufacturing and processing of a chemical substance,

(B) the extent to which a use changes the type or form of exposure of human beings or the environment to a chemical substance,

(C) the extent to which a use increases the magnitude and duration of exposure of human beings or the environment to a chemical substance, and

(D) the reasonably anticipated manner and methods of manufacturing, processing, distribution in commerce, and disposal of a chemical substance.

(b) Submission of Test Data.—*(1)(A) If (i) a person is required by subsection (a)(1) to submit a notice to the Administrator before beginning the manufacture or processing of a chemical substance, and (ii) such person is required to submit test data for such substance pursuant to a rule promulgated under section 4 before the submission of such notice, such person shall submit to the Administrator such data in accordance with such rule at the time notice is submitted in accordance with subsection (a)(1).*

(B) If—

(i) a person is required by subsection (a)(1) to submit a notice to the Administrator, and

(ii) such person has been granted an exemption under section

4(c) from the requirements of a rule promulgated under section 4 before the submission of such notice,

such person may not, before the expiration of the 90 day period which begins on the date of the submission in accordance with such rule of the test data the submission or development of which was the basis for the exemption, manufacture such substance if such person is subject to subsection (a)(1)(A) or manufacture or process such substance for a significant new use if the person is subject to subsection (a)(1)(B).

(2)(A) If a person—

(i) is required by subsection (a)(1) to submit a notice to the Administrator before beginning the manufacture or processing of a chemical substance listed under paragraph (4), and

(ii) is not required by a rule promulgated under section 4 before the submission of such notice to submit test data for such substance,

such person shall submit to the Administrator data prescribed by subparagraph (B) at the time notice is submitted in accordance with subsection (a)(1).

(B) Data submitted pursuant to subparagraph (A) shall be data which the person submitting the data believes show that—

(i) in the case of a substance with respect to which notice is required under subsection (a)(1)(A), the manufacture, processing, distribution in commerce, use, and disposal of the chemical substance or any combination of such activities will not present an unreasonable risk of injury to health or the environment, or

(ii) in the case of a chemical substance with respect to which notice is required under subsection (a)(1)(B), the intended significant new use of the chemical substance will not present an unreasonable risk of injury to health or the environment.

(3) Data submitted under paragraph (1) or (2) shall be made available, subject to section 14, for examination by interested persons.

(4)(A)(i) The Administrator may, by rule, compile and keep current a list of chemical substances with respect to which the Administrator finds that the manufacture, processing, distribution in commerce, use, or disposal, or any combination of such activities, presents or may present an unreasonable risk of injury to health or the environment.

(ii) In making a finding under clause (i) that the manufacture, processing, distribution in commerce, use, or disposal of a chemical substance or any combination of such activities presents or may present an unreasonable risk of injury to health or the environment, the Administrator shall consider all relevant factors, including—

(I) the effects of the chemical substance on health and the magnitude of human exposure to such substance; and

(II) the effects of the chemical substance on the environment and the magnitude of environmental exposure to such substance.

(B) The Administrator shall, in prescribing a rule under subparagraph (A) which lists any chemical substance, identify those uses, if any, which the Administrator determines, by rule under subsection (a)(2), would constitute a significant new use of such substance.

(C) Any rule under subparagraph (A), and any substantive amendment or repeal of such a rule, shall be promulgated pursuant to the

procedures specified in section 355 of title 5, United States Code, except that (i) the Administrator shall give interested persons an opportunity for the oral presentation of data, views, or arguments, in addition to an opportunity to make written submissions, (ii) a transcript shall be kept of any oral presentation, and (iii) the Administrator shall make and publish with the rule the finding described in subparagraph (A).

(c) EXTENSION OF NOTICE PERIOD.—The Administrator may for good cause extend for additional periods (not to exceed in the aggregate 90 days) the period, prescribed by subsection (a) or (b) before which the manufacturing or processing of a chemical substance subject to such subsection may begin. Subject to section 14, such an extension and the reasons therefor shall be published in the Federal Register and shall constitute a final agency action subject to judicial review.

(d) CONTENT OF NOTICE; PUBLICATIONS IN THE FEDERAL REGISTER.—(1) The notice required by subsection (a) shall include—

(A) insofar as known to the person submitting the notice or insofar as reasonably ascertainable, the information described in subparagraphs (A), (B), (C), (D), (F), and (G) of section 8(a)(2), and

(B) in such form and manner as the Administrator may prescribe, any test data in the possession or control of the person giving such notice which are related to the effect of any manufacture, processing, distribution in commerce, use, or disposal of such substance or any article containing such substance, or of any combination of such activities, on health or the environment, and

(C) a description of any other data concerning the environmental and health effects of such substance, insofar as known to the person making the notice or insofar as reasonably ascertainable.

Such a notice shall be made available, subject to section 14, for examination by interested persons.

(2) Subject to section 14, not later than five days (excluding Saturdays, Sundays and legal holidays) after the date of the receipt of a notice under subsection (a) or of data under subsection (b), the Administrator shall publish in the Federal Register a notice which—

(A) identifies the chemical substance for which notice or data has been received;

(B) lists the uses or intended uses of such substance; and

(C) in the case of the receipt of data under subsection (b), describes the nature of the tests performed on such substance and any data which was developed pursuant to subsection (b) or a rule under section 4.

A notice under this paragraph respecting a chemical substance shall identify the chemical substance by generic class unless the Administrator determines that more specific identification is required in the public interest.

(3) At the beginning of each month the Administrator shall publish a list in the Federal Register of (A) each chemical substance for which notice has been received under subsection (a) and for which the notification period prescribed by subsection (a), (b), or (c) has not expired, and (B) each chemical substance for which such notifica-

tion period has expired since the last publication in the Federal Register of such list.

(e) Regulation Pending Development of Information.—(1)(A) If the Administrator determines that—

(i) the information available to the Administrator is insufficient to permit a reasoned evaluation of the health and environmental effects of a chemical substance with respect to which notice is required by subsection (a); and

(ii)(I) in the absence of sufficient information to permit the Administrator to make such an evaluation, the manufacture, processing, distribution in commerce, use, or disposal of such substance, or any combination of such activities, may present an unreasonable risk of injury to health or the environment, or

(II) such substance is or will be produced in substantial quantities, and such substance either enters or may reasonably be anticipated to enter the environment in substantial quantities or there is or may be significant or substantial human exposure to the substance,

the Administrator may issue a proposed order, to take effect on the expiration of the notification period applicable to the manufacturing or processing of such substance under subsection (a), (b), or (c), to prohibit or limit the manufacture, processing, distribution in commerce, use, or disposal of such substance or to prohibit or limit any combination of such activities.

(B) A proposal order may not be issued under subparagraph (A) respecting a chemical substance (i) later than 45 days before the expiration of the notification peroid applicable to the manufacture or processing of such substance under subsection (a), (b), or (c), and (ii) unless the Administrator has, on or before the issuance of the proposed order, notified, in writing, each manufacturer or processor, as the case may be, of such substance of the determination which underlie such order.

(C) If a manufacturer or processor of a chemcial substance to be subject to a proposed order issued under subparagraph (A) files with the Administrator (within the 30-day period beginning on the date such maunfacturer or processor received the notice required by subparagraph (B)(ii)) objections specifying with particularity the provisions of the order deemed objectionable and stating the grounds therefor, the proposed order shall not take effect.

(2)(A)(i) Except as provided in clause (ii), if with respect to a chemical subtance with respect to which notice is required by subsection (a), the Administrator makes the determination described in paragraph (1)(A) and if—

(I) the Administrator does not issue a proposed order under paragraph (1) respecting such substance, or

(II) the Administrator issues such an order respecting such substance but such order does not take effect because objections were filed under paragraph (1)(C) with respect to it,

the Administrator, through attorneys of the Environmental Protection Agency, shall apply to the United States District Court for the District of Columbia or the United States district court for the judicial district in which the manufacturer or processor, as the case may be, of such substance is found, resides, or transacts business for an injunc-

tion to prohibit or limit the manufacture, processing, distribution in commerce, use, or disposal of such substance (or to prohibit or limit any combination of such activities).

(ii) If the Administrator issues a proposed order under paragraph (1)(A) respecting a chemical substance but such order does not take effect because objections have been filed under paragraph (1)(C) with respect to it, the Administrator is not required to apply for an injunction under clause (i) respecting such substance if the Administrator determines, on the basis of such objections, that the determinations under paragraph (1)(A) may not be made.

(B) A district court of the United States which receives an application under subparagraph (A)(i) for an injunction respecting a chemical substance shall issue such injunction if the court finds that—

(i) the information available to the Administrator is insufficient to permit a reasoned evaluation of the health and environmental effects of a chemical substance with respect to which notice is required by subsection (a); and

(ii)(I) in the absence of sufficient information to permit the Administrator to make such an evaluation, the manufacture, processing, distribution in commerce, use, or disposal of such substance, or any combination of such activities, may present an unreasonable risk of injury to health or the environment, or

(II) such substance is or will be produced in substantial quantities, and such substance either enters or may reasonably be anticipated to enter the environment in substantial quantities or there is or may be significant or substantial human exposure to the substance.

(C) Pending the completion of a proceeding for the issuance of an injunction under subparagraph (B) respecting a chemical substance, the court may, upon application of the Administrator made through attorneys of the Environmental Protection Agency, issue a temporary restraining order or a preliminary injunction to prohibit the manufacture, processing, distribution in commerce, use, or disposal of such a substance (or any combination of such activities) if the court finds that the notification period applicable under subsection (a) (b), or (c) to the manufacturing or processing of such substance may expire before such proceeding can be completed.

(D) After the submission to the Administrator of test data sufficient to evaluate the health and environmental effects of a chemical substance subject to an injunction issued under subparagraph (B) and the evaluation of such data by the Administrator, the district court of the United States which issued such injunction shall, upon petition, disolve the injunction unless the Administrator has initiated a proceeding for the issuance of a rule under section 6(a) respecting the substance. If such a proceeding has been initiated, such court shall continue the injunction in effect until the effective date of the rule promulgated in such proceeding or, if such proceeding is terminated without the promulgation of a rule, upon the termination of the proceeding, whichever occurs first.

(f) PROTECTION AGAINST UNREASONABLE RISKS.—(1) If the Administrator finds that there is a reasonable basis to conclude that the manufacture, processing, distribution in commerce, use, or disposal of a

chemical substance with respect to which notice is required by subsection (a), or that any combination of such activities, presents or will present an unreasonable risk of injury to health or environment before a rule promulgated under section 6 can protect against such risk, the Administrator shall, before the expiration of the notification period applicable under subsection (a), (b), or (c) to the manufacturing or processing of such substance, take the action authorized by paragraph (2) or (3) to the extent necessary to protect against such risk.

(2) The Administrator may issue a proposed rule under section 6(a) to apply to a chemical substance with respect to which a finding was made under paragraph (1)—

(A) a requirement limiting the amount of such substance which may be manufactured, processed, or distributed in commerce,

(B) a requirement described in paragraph (2), (3), (4), (5), (6), or (7) of section 6(a), or

(C) any combination of the requirements referred to in subparagraph (B).

Such a proposed rule shall be effective upon its publication in the Federal Register. Section 6(d)(2)(B) shall apply with respect to such rule.

(3)(A) The Administrator may—

(i) issue a proposed order to prohibit the manufacture, processing, or distribution in commerce of a substance with respect to which a finding was made under paragraph (1), or

(ii) apply, through attorneys of the Environmental Protection Agency, to the United States District Court for the District of Columbia or the United States district court for the judicial district in which the manufacturer, or processor, as the case may be, of such substance, is found, resides, or transacts business for an injunction to prohibit the manufacture, processing, or distribution in commerce of such substance.

A proposed order issued under clause (i) respecting a chemical substance shall take effect on the expiration of the notification period applicable under subsection (a), (b), or (c) to the manufacture or processing of such substance.

(B) If the district court of the United States to which an application has been made under subparagraph (A)(ii) finds that there is a reasonable basis to conclude that the manufacture, processing, distribution in commerce, use, or disposal of the chemical substance with respect to which such application was made, or that any combination of such activities, presents or will present an unreasonable risk of injury to health or the environment before a rule promulgated under section 6 can protect against such risk, the court shall issue an injunction to prohibit the manufacture, processing, or distribution in commerce of such substance or to prohibit any combination of such activities.

(C) The provisions of subparagraphs (B) and (C) of subsection (e)(1) shall apply with respect to an order issued under clause (i) of subparagraph (A); and the provisions of subparagraph (C) of subsection (e)(2) shall apply with respect to an injunction issued under subparagraph (B).

(D) If the Administrator issues an order pursuant to subparagraph (A)(i) respecting a chemical substance and objections are filed in

accordance with subsection (e)(1)(C), the Administrator shall seek an injunction under subparagraph (A)(ii) respecting such substance unless the Administrator determines, on the basis of such objections, that such substance does not or will not present an unreasonable risk of injury to health or the environment.

(g) STATEMENT OF REASONS FOR NOT TAKING ACTION.—If the Administrator has not initiated any action under this section or section 6 or 7 to prohibit or limit the manufacture, processing, distribution in commerce, use, or disposal of a chemical substance, with respect to which notification or data is required by subsection (a)(1)(B) or (b), before the expiration of the notification period applicable to the manufacturing or processing of such substance, the Administrator shall publish a statement of the Administrator's reasons for not initiating such action. Such a statement shall be published in the Federal Register before the expiration of such period. Publication of such statement in accordance with the preceding sentence is not a prerequisite to the manufacturing or processing of the substance with respect to which the statement is to be published.

(h) EXEMPTIONS.—(1) The Administrator may, upon application, exempt any person from any requirement of subsection (a) or (b) to permit such person to manufacture or process a chemical substance for test marketing purposes—

(A) upon a showing by such person satisfactory to the Administrator that the manufacture, processing, distribution in commerce, use, and disposal of such substance, and that any combination of such activities, for such purposes will not present any unreasonable risk of injury to health or the environment, and

(B) under such restrictions as the Administrator considers appropriate.

(2)(A) The Administrator may, upon application, exempt any person from the requirement of subsection (b)(2) to submit data for a chemical substance. If upon receipt of an application under the preceding sentence, the Administrator determines that—

(i) the chemical substance with respect to which such application was submitted is equivalent to a chemical substance for which data has been submitted to the Administrator as required by subsection (b)(2), and

(ii) submission of data by the applicant on such substance would be duplicative of data which has been submitted to the Administrator in accordance with such subsection,

the Administrator shall exempt the applicant from the requirement to submit such data on such substance. No exemption which is granted under this subparagraph with respect to the submission of data for a chemical substance may take effect before the beginning of the reimbursement period applicable to such data.

(B) If the Administrator exempts any person, under subparagraph (A), from submitting data required under subsection (b)(2) for a chemical substance because of the existence of previously submitted data and if such exemption is granted during the reimbursement period for such data, then (unless such person and the persons referred to in clauses (i) and (ii) agree on the amount and method of reimbursement) the Administrator shall order the person granted the ex-

emption to provide fair and equitable reimbursement (in an amount determined under rules of the Administrator)—

(i) to the person who previously submitted the data on which the exemption was based, for a portion of the costs incurred by such person in complying with the requirement under subsection (b)(2) to submit such data, and

(ii) to any other person who has been required under this subparagraph to contribute with respect to such costs, for a portion of the amount such person was required to contribute.

In promulgating rules for the determination of fair and equitable reimbursement to the persons described in clauses (i) and (ii) for costs incurred with respect to a chemical substance, the Administrator shall, after consultation with the Attorney General and the Federal Trade Commission, consider all relevant factors, including the effect on the competitive position of the person required to provide reimbursement in relation to the persons to be reimbursed and the share of the market for such substance of the person required to provide reimbursement in relation to the share of such market of the persons to be reimbursed. For purposes of judicial review, an order under this subparagraph shall be considered final agency action.

(C) For purposes of this paragraph, the reimbursement period for any previously submitted data for a chemical substance is a period—

(i) beginning on the date of the termination of the prohibition, imposed under this section, on the manufacture or processing of such substance by the person who submitted such data to the Administrator, and

(ii) ending—

(I) five years after the date referred to in clause (i), or

(II) at the expiration of a period which begins on the date referred to in clause (i) and is equal to the period which the Administrator determines was necessary to develop such data,

whichever is later.

(3) The requirements of subsections (a) and (b) do not apply with respect to the manufacturing or processing of any chemical substance which is manufactured or processed, or proposed to be manufactured or processed, only in small quantities (as defined by the Administrator by rule) solely for purposes of—

(A) scientific experimentation or analysis, or

(B) chemical research on, or analysis of such substance or another substance, including such research or analysis for the development of a product,

if all persons engaged in such experimentation, research, or analysis for a manufacturer or processor are notified (in such form and manner as the Administrator may prescribe) of any risk to health which the manufacturer, processor, or the Administrator has reason to believe may be associated with such chemical substance.

(4) The Administrator may, upon application and by rule, exempt the manufacturer of any new chemical substance from all or part of the requirements of this section if the Administrator determines that the manufacture, processing, distribution in commerce, use, or disposal of such chemical substance, or that any combination of such

activities, will not present an unreasonable risk of injury to health or the environment. A rule promulgated under this paragraph (and any substantive amendment to, or repeal of, such a rule) shall be promulgated in accordance with paragraphs (2) and (3) of section 6(c).

(5) The Administrator may, upon application, make the requirements of subsections (a) and (b) inapplicable with respect to the manufacturing or processing of any chemical substance (A) which exists temporarily as a result of a chemical reaction in the manufacturing or processing of a mixture or another chemical substance, and (B) to which there is no, and will not be, human or environmental exposure.

(6) Immediately upon receipt of an application under paragraph (1) or (5) the Administrator shall publish in the Federal Register notice of the receipt of such application. The Administrator shall give interested persons an opportunity to comment upon any such application and shall, within 45 days of its receipt, either approve or deny the application. The Administrator shall publish in the Federal Register notice of the approval or denial of such an application.

(i) DEFINITION.—For purposes of this section, the terms "manufacture" and "process" mean manufacturing or processing for commercial purposes.

SEC. 6. REGULATION OF HAZARDOUS CHEMICAL SUBSTANCES AND MIXTURES.

(a) SCOPE OF REGULATION.—If the Administrator finds that there is a reasonable basis to conclude that the manufacture, processing, distribution in commerce, use, or disposal of a chemical substance or mixture, or that any combination of such activities, presents or will present an unreasonable risk of injury to health or the environment, the Administrator shall by rule apply one or more of the following requirements to such substance or mixture to the extent necessary to protect adequately against such risk using the least burdensome requirements:

(1) A requirement (A) prohibiting the manufacturing, processing, or distribution in commerce of such substance or mixture, or (B) limiting the amount of such substance or mixture which may be manufactured, processed, or distributed in commerce.

(2) A requirement—

(A) prohibiting the manufacture, processing, or distribution in commerce of such substance or mixture for (i) a particular use or (ii) a particular use in a concentration in excess of a level specified by the Administrator in the rule imposing the requirement, or

(B) limiting the amount of such substance or mixture which may be manufactured, processed, or distributed in commerce for (i) a particular use or (ii) a particular use in a concentration in excess of a level specified by the Administrator in the rule imposing the requirement.

(3) A requirement that such substance or mixture or any article containing such substance or mixture be marked with or accompanied by clear and adequate warnings and instructions with respect to its use, distribution in commerce, or disposal or with respect to any combination of such activities. The form and

content of such warnings and instructions shall be prescribed by the Administrator.

(4) A requirement that manufacturers and processors of such substance or mixture make and retain records of the processes used to manufacture or process such substance or mixture and monitor or conduct tests which are reasonable and necessary to assure compliance with the requirements of any rule applicable under this subsection.

(5) A requirement prohibiting or otherwise regulating any manner or method of commercial use of such substance or mixture.

(6)(A) A requirement prohibiting or otherwise regulating any manner or method of disposal of such substance or mixture, or of any article containing such substance or mixture, by its manufacturer or processor or by any other person who uses, or disposes of, it for commercial purposes.

(B) A requirement under subparagraph (A) may not require any person to take any action which would be in violation of any law or requirement of, or in effect for, a State or political subdivision, and shall require each person subject to it to notify each State and political subdivision in which a required disposal may occur of such disposal.

(7) A requirement directing manufacturers or processors of such substance or mixture (A) to give notice of such unreasonable risk of injury to distributors in commerce of such substance or mixture and, to the extent reasonably ascertainable, to other persons in possession of such substance or mixture or exposed to such substance or mixture, (B) to give public notice of such risk of injury, and (C) to replace or repurchase such substance or mixture as elected by the person to which the requirement is directed.

Any requirement (or combination of requirements) imposed under this subsection may be limited in application to specified geographic areas.

(b) Quality Control.—If the Administrator has a reasonable basis to conclude that a particular manufacturer or processor is manufacturing or processing a chemical substance or mixture in a manner which unintentionally causes the chemical substance or mixture to present or which will cause it to present an unreasonable risk of injury to health or the environment—

(1) the Administrator may by order require such manufacturer or processor to submit a description of the relevant quality control procedures followed in the manufacturing or processing of such chemical substance or mixture; and

(2) if the Administrator determines—

(A) that such quality control procedures are inadequate to prevent the chemical substance or mixture from presenting such risk of injury, the Administrator may order the manufacturer or processor to revise such quality control procedures to the extent necessary to remedy such inadequacy; or

(B) that the use of such quality control procedures has resulted in the distribution in commerce of chemical substances or mixtures which present an unreasonable risk of injury to health or the environment, the Administrator may order the

manufacturer or processor to (i) give notice of such risk to processors or distributors in commerce of any such substance or mixture, or to both, and, to the extent reasonably ascertainable, to any other person in possession of or exposed to any such substance, (ii) to give public notice of such risk, and (iii) to provide such replacement or repurchase of any such substance or mixture as is necessary to adequately protect health or the environment.

A determination under subparagraph (A) or (B) of paragraph (2) shall be made on the record after opportunity for hearing in accordance with section 554 of title 5, United States Code. Any manufacturer or processor subject to a requirement to replace or repurchase a chemical substance or mixture may elect either to replace or repurchase the substance or mixture and shall take either such action in the manner prescribed by the Administrator.

(c) PROMULGATION OF SUBSECTION (a) RULES.—(1) In promulgating any rule under subsection (a) with respect to a chemical substance or mixture, the Administrator shall consider and publish a statement with respect to—

(A) the effects of such substance or mixture on health and the magnitude of the exposure of human beings to such substance or mixture,

(B) the effects of such substance or mixture on the environment and the magnitude of the exposure of the environment to such substance or mixture,

(C) the benefits of such substance or mixture for various uses and the availability of substitutes for such uses, and

(D) the reasonably ascertainable economic consequences of the rule, after consideration of the effect on the national economy, small business, technological innovation, the environment, and public health.

If the Administrator determines that a risk of injury to health or the environment could be eliminated or reduced to a sufficient extent by actions taken under another Federal law (or laws) administered in whole or in part by the Administrator, the Administrator may not promulgate a rule under subsection (a) to protect against such risk of injury unless the Administrator finds, in the Administrator's discretion, that it is in the public interest to protect against such risk under this Act. In making such a finding the Administrator shall consider (i) all relevant aspects of the risk, as determined by the Administrator in the Administrator's discretion, (ii) a comparison of the estimated costs of complying with actions under this Act and under such law (or laws), and (iii) the relative efficiency of actions under this Act and under such law (or laws) to protect against such risk of injury.

(2) When prescribing a rule under subsection (a) the Administrator shall proceed in accordance with section 553 of title 5, United States Code (without regard to any reference in such section to sections 556 and 557 of such title), and shall also (A) publish a notice of proposed rulemaking stating with particularity the reason for the proposed rule; (B) allow interested persons to submit written data, views, and arguments, and make all such submissions publicly available; (C)

provide an opportunity for an informal hearing in accordance with paragraph (3); (D) promulgate, if appropriate, a final rule based on the matter in the rulemaking record (as defined in section 19(a)), and (E) make and publish with the rule the finding described in subsection (a).

(3) Informal hearings required by paragraph (2)(C) shall be conducted by the Administrator in accordance with the following requirements:

(A) Subject to subparagraph (B), an interested person is entitled—

(i) to present such person's position orally or by documentary submissions (or both), and

(ii) if the Administrator determines that there are disputed issues of material fact it is necessary to resolve, to present such rebuttal submissions and to conduct (or have conducted under subparagraph (B)(ii)) such cross-examination of persons as the Administrator determines (I) to be appropriate, and (II) to be required for a full and true disclosure with respect to such issues.

(B) The Administrator may prescribe such rules and make such rulings concerning procedures in such hearings to avoid unnecessary costs or delay. Such rules or rulings may include (i) the imposition of reasonable time limits on each interested person's oral presentations, and (ii) requirements that any cross-examination to which a person may be entitled under subparagraph (A) be conducted by the Administrator on behalf of that person in such manner as the Administrator determines (I) to be appropriate, and (II) to be required for a full and true disclosure with respect to disputed issues of material fact.

(C)(i) Except as provided in clause (ii), if a group of persons each of whom under subparagraphs (A) and (B) would be entitled to conduct (or have conducted) cross-examination and who are determined by the Administrator to have the same or similar interests in the proceeding cannot agree upon a single representative of such interests for purposes of cross-examination, the Administrator may make rules and rulings (I) limiting the representation of such interest for such purposes, and (II) governing the manner in which such cross-examination shall be limited.

(ii) When any person who is a member of a group with respect to which the Administrator has made a determination under clause (i) is unable to agree upon group representation with the other members of the group, then such person shall not be denied under the authority of clause (i) the opportunity to conduct (or have conducted) cross-examination as to issues affecting the person's particular interests if (I) the person satisfies the Administrator that the person has made a reasonable and good faith effort to reach agreement upon group representation with the other members of the group and (II) the Administrator determines that there are substantial and relevant issues which are not adequately presented by the group representative.

(D) A verbatim transcript shall be taken of any oral presentation made, and cross-examination conducted in any informal hear-

ing under this subsection. Such transcript shall be available to the public.

(4)(A) The Administrator may, pursuant to rules prescribed by the Administrator, provide compensation for reasonable attorneys' fees, expert witness fees, and other costs of participating in a rule-making proceeding for the promulgation of a rule under subsection (a) to any person—

(i) who represents an interest which would substantially contribute to a fair determination of the issues to be resolved in the proceeding, and

(ii) if—

(I) the economic interest of such person is small in comparison to the costs of effective participation in the proceeding by such person, or

(II) such person demonstrates to the satisfaction of the Administrator that such person does not have sufficient resources adequately to participate in the proceeding without compensation under this subparagraph.

In determining for purposes of clause (i) if an interest will substantially contribute to a fair determination of the issues to be resolved in a proceeding, the Administrator shall take into account the number and complexity of such issues and the extent to which representation of such interest will contribute to widespread public participation in the proceeding and representation of a fair balance of interests for the resolution of such issues.

(B) In determining whether compensation should be provided to a person under subparagraph (A) and the amount of such compensation, the Administrator shall take into account the financial burden which will be incurred by such person in participating in the rule-making proceeding. The Administrator shall take such action as may be necessary to ensure that the aggregate amount of compensation paid under this paragraph in any fiscal year to all persons who, in rulemaking proceedings in which they receive compensation, are persons who either—

(i) would be regulated by the proposed rule, or

(ii) represent persons who would be so regulated,

may not exceed 25 per centum of the aggregate amount paid as compensation under this paragraph to all persons in such fiscal year.

(5) Paragraphs (1), (2), (3), and (4) of this subsection apply to the promulgation of a rule repealing, or making a substantive amendment to, a rule promulgated under subsection (a).

(d) EFFECTIVE DATE.—(1) The Administrator shall specify in any rule under subsection (a) the date on which it shall take effect, which date shall be as soon as feasible.

(2)(A) The Administrator may declare a proposed rule under subsection (a) to be effective upon its publication in the Federal Register and until the effective date of final action taken, in accordance with subparagraph (B), respecting such rule if—

(i) the Administrator determines that—

(I) the manufacture, processing, distribution in commerce, use, or disposal of the chemical substance or mixture subject to such proposed rule or any combination of such activities is likely to result in an unreasonable risk of serious or wide-

spread injury to health or the environment before such effective date; and

(II) making such proposed rule so effective is necessary to protect the public interest; and

(ii) in the case of a proposed rule to prohibit the manufacture, processing, or distribution of a chemical substance or mixture because of the risk determined under clause (i)(I), a court has in an action under section 7 granted relief with respect to such risk associated with such substance or mixture.

Such a proposed rule which is made so effective shall not, for purposes of judicial review, be considered final agency action.

(B) If the Administrator makes a proposed rule effective upon its publication in the Federal Register, the Administrator shall, as expeditiously as possible, give interested persons prompt notice of such action, provide reasonable opportunity, in accordance with paragraphs (2) and (3) of subsection (c), for a hearing on such rule, and either promulgate such rule (as proposed or with modifications) or revoke it; and if such a hearing is requested, the Administrator shall commence the hearing within five days from the date such request is made unless the Administrator and the person making the request agree upon a later date for the hearing to begin, and after the hearing is concluded the Administrator shall, within ten days of the conclusion of the hearing, either promulgate such rule (as proposed or with modifications) or revoke it.

(e) POLYCHLORINATED BIPHENYLS.—(1) Within six months after the effective date of this Act the Administrator shall promulgate rules to—

(A) prescribe methods for the disposal of polychlorinated biphenyls, and

(B) require polychlorinated biphenyls to be marked with clear and adequate warnings and instructions with respect to their processing, distribution in commerce, use, or disposal or with respect to any combination of such activities.

Requirements prescribed by rules under this paragraph shall be consistent with the requirements of paragraphs (2) and (3).

(2)(A) Except as provided under subparagraph (B), effective one year after the effective date of this Act no person may manufacture, process, or distribute in commerce or use any polychlorinated biphenyl in any manner other than in a totally enclosed manner.

(B) The Administrator may by rule authorize the manufacture, processing, distribution in commerce, or use (or any combination of such activities) of any polychlorinated biphenyl in a manner other than in a totally enclosed manner if the Administrator finds that such manufacture, processing, distribution in commerce, or use (or combination of such activities) will not present an unreasonable risk of injury to health or the environment.

(C) For the purposes of this paragraph, the term "totally enclosed manner" means any manner which will ensure that any exposure of human beings or the environment to a polychlorinated biphenyl will be insignificant as determined by the Administrator by rule.

(3)(A) Except as provided in subparagraphs (B) and (C)—

(i) no person may manufacture any polychlorinated biphenyl after two years after the effective date of this Act, and

(ii) no person may process or distribute in commerce any polychlorinated biphenyl after two and one-half years after such date.

(B) Any person may petition the Administrator for an exemption from the requirements of subparagraph (A), and the Administrator may grant by rule such an exemption if the Administrator finds that—

(i) an unreasonable risk of injury to health or environment would not result, and

(ii) good faith efforts have been made to develop a chemical substance which does not present an unreasonable risk of injury to health or the environment and which may be substituted for such polychlorinated biphenyl.

An exemption granted under this subparagraph shall be subject to such terms and conditions as the Administrator may prescribe and shall be in effect for such period (but not more than one year from the date is it granted) as the Administrator may prescribe.

(C) Subparagraph (A) shall not apply to the distribution in commerce of any polychlorinated biphenyl if such polychlorinated biphenyl was sold for purposes other than resale before two and one half years after the date of enactment of this Act.

(4) Any rule under paragraph (1), (2)(B), or (3)(B) shall be promulgated in accordance with paragraphs (2), (3), and (4) of subsection (c).

(5) This subsection does not limit the authority of the Administrator, under any other provision of this Act or any other Federal law, to take action respecting any polychlorinated biphenyl.

SEC. 7. IMMINENT HAZARDS.

(a) Actions Authorized and Required.—(1) The Administrator may commence a civil action in an appropriate district court of the United States—

(A) for seizure of an imminently hazardous chemical substance or mixture or any article containing such a substance or mixture,

(B) for relief (as authorized by subsection (b)) against any person who manufactures, processes, distributes in commerce, or uses, or disposes of, an imminently hazardous chemical substance or mixture or any article containing such a substance or mixture, or

(C) for both such seizure and relief.

A civil action may be commenced under this paragraph notwithstanding the existence of a rule under section 4, 5, or 6 or an order under section 5, and notwithstanding the pendency of any administrative or judicial proceeding under any provision of this Act.

(2) If the Administrator has not made a rule under section 6(a) immediately effective (as authorized by subsection 6(d)(2)(A)(i)) with respect to an imminently hazardous chemical substance or mixture, the Administrator shall commence in a district court of the United States with respect to such substance or mixture or article containing such substance or mixture a civil action described in subparagraph (A), (B), or (C) of paragraph (1).

(b) Relief Authorized.—(1) The district court of the United States in which an action under subsection (a) is brought shall have

jurisdiction to grant such temporary or permanent relief as may be necessary to protect health or the environment from the unreasonable risk associated with the chemical substance, mixture, or article involved in such action.

(2) In the case of an action under subsection (a) brought against a person who manufactures, processes, or distributes in commerce a chemical substance or mixture or an article containing a chemical substance or mixture, the relief authorized by paragraph (1) may include the issuance of a mandatory order requiring (A) in the case of purchasers of such substance, mixture, or article known to the defendant, notification to such purchasers of the risk associated with it; (B) public notice of such risk; (C) recall; (D) the replacement or repurchase of such substance, mixture, or article; or (E) any combination of the actions described in the preceding clauses.

(3) In the case of an action under subsection (a) against a chemical substance, mixture, or article, such substance, mixture, or article may be proceeded against by process of libel for its seizure and condemnation. Proceedings in such an action shall conform as nearly as possible to proceedings in rem in admiralty.

(c) VENUE AND CONSOLIDATION.—(1)(A) An action under subsection (a) against a person who manufactures, processes, or distributes a chemical substance or mixture or an article containing a chemical substance or mixture may be brought in the United States District Court for the District of Columbia or for any judicial district in which any of the defendants is found, resides, or transacts business; and process in such an action may be served on a defendant in any other district in which such defendant resides or may be found. An action under subsection (a) against a chemical substance, mixture, or article may be brought in any United States district court within the jurisdiction of which the substance, mixture, or article is found.

(B) In determining the judicial district in which an action may be brought under subsection (a) in instances in which such action may be brought in more than one judicial district, the Administrator shall take into account the convenience of the parties.

(C) Subpeonas requiring attendance of witnesses in an action brought under subsection (a) may be served in any judicial district.

(2) Whenever proceedings under subsection (a) involving identical chemical substances, mixtures, or articles are pending in courts in two or more judicial districts, they shall be consolidated for trial by order of any such court upon application reasonably made by any party in interest, upon notice to all parties in interest.

(d) ACTION UNDER SECTION 6.—Where appropriate, concurrently with the filing of an action under subsection (a) or as soon thereafter as may be practicable, the Administrator shall initiate a proceeding for the promulgation of a rule under section 6(a).

(e) REPRESENTATION.—Notwithstanding any other provision of law, in any action under subsection (a), the Administrator may direct attorneys of the Environmental Protection Agency to appear and represent the Administrator in such an action.

(f) DEFINITION.—For the purposes of subsection (a), the term "imminently hazardous chemical substance or mixture" means a chemical substance or mixture which presents an imminent and unreason-

able risk of serious or widespread injury to health or the environment. Such a risk to health or the environment shall be considered imminent if it is shown that the manufacture, processing, distribution in commerce, use, or disposal of the chemical substance or mixture, or that any combination of such activities, is likely to result in such injury to health or the environment before a final rule under section 6 can protect against such risk.

SEC. 8. REPORTING AND RETENTION OF INFORMATION.

(a) Reports.—(1) The Administrator shall promulgate rules under which—

(A) each person (other than a small manufacturer or processor) who manufactures or processes or proposes to manufacture or process a chemical substance (other than a chemical substance described in subparagraph (B)(ii)) shall maintain such records, and shall submit to the Administrator such reports, as the Administrator may reasonably require, and

(B) each person (other than a small manufacturer or processor) who manufactures or processes or proposes to manufacture or process—

(i) a mixture, or

(ii) a chemical substance in small quantities (as defined by the Administrator by rule) solely for purposes of scientific experimentation or analysis or chemical research on, or analysis of, such substance or another substance, including such research or analysis for the development of a product,

shall maintain records and submit to the Administrator reports but only to the extent the Administrator determines the maintenance of records or submission of reports, or both, is necessary for the effective enforcement of this Act.

The Administrator may not require in a rule promulgated under this paragraph the maintenance of records or the submission of reports with respect to changes in the proportions of the components of a mixture unless the Administrator finds that the maintenance of such records or the submission of such reports, or both, is necessary for the effective enforcement of this Act. For purposes of the compilation of the list of chemical substances required under subsection (b), the Administrator shall promulgate rules pursuant to this subsection not later than 180 days after the effective date of this Act.

(2) The Administrator may require under paragraph (1) maintenance of records and reporting with respect to the following insofar as known to the person making the report or insofar as reasonably ascertainable:

(A) The common or trade name, the chemical identity, and the molecular structure of each chemical substance or mixture for which such a report is required.

(B) The categories or proposed categories of use of each such substance or mixture.

(C) The total amount of each such substance and mixture manufactured or processed, reasonable estimates of the total amount to be manufactured or processed, the amount manufactured or processed for each of its categories of use, and reasonable estimates of the amount to be manufactured or processed for each of its categories of use or proposed categories of use.

(D) A description of the byproducts resulting from the manufacture, processing, use, or disposal of each such substance or mixture.

(E) All existing data concerning the environmental and health effects of such substance or mixture.

(F) The number of individuals exposed, and reasonable estimates of the number who will be exposed, to such substance or mixture in their places of employment and the duration of such exposure.

(G) In the initial report under paragraph (1) on such substance or mixture, the manner or method of its disposal, and in any subsequent report on such substance or mixture, any change in such manner or method.

To the extent feasible, the Administration shall not require under paragraph (1) any reporting which is unnecessary or duplicative.

(3)(A)(i) The Administrator may by rule require a small manufacturer or processor of a chemical substance to submit to the Administrator such information respecting the chemical substance as the Administrator may require for publication of the first list of chemical substances required by subsection (b).

(ii) The Administrator may by rule require a small manufacturer or processor of a chemical substance or mixture—

(I) subject to a rule proposed or promulgated under section 4, 5(b)(4), or 6, or an order in effect under section 5(e), or

(II) with respect to which relief has been granted pursuant to a civil action brought under section 5 or 7,

to maintain such records on such substance or mixture, and to submit to the Administrator such reports on such substance or mixture, as the Administrator may reasonably require. A rule under this clause requiring reporting may require reporting with respect to the matters referred to in paragraph (2).

(B) The Administrator, after consultation with the Administrator of the Small Business Administration, shall by rule prescribe standards for determining the manufacturers and processors which qualify as small manufacturers and processors for purposes of this paragraph and paragraph (1).

(b) INVENTORY.—(1) The Administrator shall compile, keep current, and publish a list of each chemical substance which is manufactured or processed in the United States. Such list shall at least include each chemical substance which any person reports, under section 5 or subsection (a) of this section, is manufactured or processed in the United States. Such list may not include any chemical substance which was not manufactured or processed in the United States within three years before the effective date of the rules promulgated pursuant to the last sentence of subsection (a)(1). In the case of a chemical substance for which a notice is submitted in accordance with section 5, such chemical substance shall be included in such list as of the earliest date (as determined by the Administrator) on which such substance was manufactured or processed in the United States. The Administrator shall first publish such a list not later than 315 days after the effective date of this Act. The Administrator shall not include in such list any chemical substance which is manufactured

or processed only in small quantities (as defined by the Administrator by rule) solely for purposes of scientific experimentation or analysis or chemical research on, or analysis of, such substance or another substance, including such research or analysis for the development of a product,

(2) To the extent consistent with the purposes of this Act, the Administrator may, in lieu of listing, pursuant to paragraph (1), a chemical substance individually, list a category of chemical substances in which such substance is included.

(c) RECORDS.—Any person who manufactures, processes, or distributes in commerce any chemical substance or mixture shall maintain records of significant adverse reactions to health or the environvent, as determined by the Administrator by rule, alleged to have been caused by the substance or mixture. Records of such adverse reactions to the health of employees shall be retained for a period of 30 years from the date such reactions were first reported to or known by the person maintining such records. Any other record of such adverse reactions shall be retained for a period of five years from the date the information contained in the record was first reported to or known by the person maintaining the record. Records required to be maintained under this subsection shall include records of consumer allegations of personal injury or harm to health, reports of occupational disease or injury, and reports or complaints of injury to the environment submitted to the manufacturer, processor, or distributor in commerce from any source. Upon request of any duly designated representative of the Administrator, each person who is required to maintain records under this subsection shall permit the inspection of such records and shall submit copies of such records.

(d) HEALTH AND SAFETY STUDIES.—The Administrator shall promulgate rules under which the Administrator shall require any person who manufactures, processes, or distributes in commerce or who proposes to manufacture, process, or distribute in commerce any chemical substance or mixture (or with respect to paragraph (2), any person who has possession of a study) to submit to the Administrator—

(1) lists of health and safety studies (A) conducted or initiated by or for such person with respect to such substance or mixture at any time, (B) known to such person, or (C) reasonably ascertainable by such person, except that the Administrator may exclude certain types or categories of studies from the requirements of this subsection if the Administrator finds that submission of lists of such studies are unnecessary to carry out the purposes of this Act; and

(2) copies of any study contained on a list submitted pursuant to paragraph (1) or otherwise known by such person.

(e) NOTICE TO ADMINISTRATOR OF SUBSTANTIAL RISKS.—Any person who manufactures, processes, or distributes in commerce a chemical substance or mixture and who obtains information which reasonably supports the conclusion that such substance or mixture presents a substantial risk of injury to health or the environment shall immediately inform the Administrator of such information unless such person has actual knowledge that the Administrator has been adequately informed of such information.

(f) *Definitions.—For purposes of this section, the terms "manufacture" and "process" mean manufacture or process for commercial purposes.*

SEC. 9. RELATIONSHIP TO OTHER FEDERAL LAWS.

(a) Laws Not Administered by the Administrator.—(1) If the Administrator has reasonable basis to conclude that the manufacture, processing, distribution in commerce, use, or disposal of a chemical substance or mixture, or that any combination of such activities, presents or will present an unreasonable risk of injury to health or the environment and determines, in the Administrator's discretion, that such risk may be prevented or reduced to a sufficient extent by action taken under a Federal law not administered by the Administrator, the Administrator shall submit to the agency which administers such law a report which describes such risk and includes in such description a specification of the activity or combination of activities which the Administrator has reason to believe so presents such risk. Such report shall also request such agency—

(A)(i) to determine if the risk described in such report may be prevented or reduced to a sufficient extent by action taken under such law, and

(ii) if the agency determines that such risk may be so prevented or reduced, to issue an order declaring whether or not the activity or combination of activities specified in the description of such risk presents such risk; and

(B) to respond to the Administrator with respect to the matters described in subparagraph (A).

Any report of the Administrator shall include a detailed statement of the information on which it is based and shall be published in the Federal Register. The agency receiving a request under such a report shall make the requested determination, issue the requested order, and make the requested response within such time as the Administrator specifies in the request, but such time specified may not be less than 90 days from the date the request was made. The response of an agency shall be accompanied by a detailed statement of the findings and conclusions of the agency and shall be published in the Federal Register.

(2) If the Administrator makes a report under paragraph (1) with respect to a chemical substance or mixture and the agency to which such report was made either—

(A) issues an order declaring that the activity or combination of activities specified in the description of the risk described in the report does not present the risk described in the report, or

(B) initiates, within 90 days of the publication in the Federal Register of the response of the agency under paragraph (1), action under the law (or laws) administered by such agency to protect against such risk associated with such activity or combination of activities,

the Administrator may not take any action under section 6 or 7 with respect to such risk.

(3) If the Administrator has initiated action under section 6 or 7 with respect to a risk associated with a chemical substance or mixture which was the subject of a report made to an agency under paragraph

(1), such agency shall before taking action under the law (or laws) administered by it to protect against such risk consult with the Administrator for the purpose of avoiding duplication of Federal action against such risk.

(b) Laws Administered by the Administrator.—The Administrator shall coordinate actions taken under this Act with actions taken under other Federal laws administered in whole or in part by the Administrator. If the Administrator determines that a risk to health or the environment associated with a chemical substance or mixture could be eliminated or reduced to a sufficient extent by actions taken under the authorities contained in such other Federal laws, the Administrator shall use such authorities to protect against such risk unless the Administrator determines, in the Administrator's discretion, that it is in the public interest to protect against such risk by actions taken under this Act. This subsection shall not be construed to relieve the Administrator of any requirement imposed on the Administrator by such other Federal laws.

(c) Occupational Safety and Health.—In exercising any authority under this Act, the Administrator shall not, for purposes of section 4(b)(1) of the Occupational Safety and Health Act of 1970, be deemed to be exercising statutory authority to prescribe or enforce standards or regulations affecting occupational safety and health.

(d) Coordination.—In administering this Act, the Administrator shall consult and coordinate with the Secretary of Health, Education, and Welfare and the heads of any other appropriate Federal executive department or agency, any relevant independent regulatory agency, and any other appropriate instrumentality of the Federal Government for the purpose of achieving the maximum enforcement of this Act while imposing the least burdens of duplicative requirements on those subject to the Act and for other purposes. The Administrator shall, in the report required by section 30, report annually to the Congress on actions taken to coordinate with such other Federal departments, agencies, or instrumentalities, and on actions taken to coordinate the authority under this Act with the authority granted under other Acts referred to in subsection (b).

SEC. 10. RESEARCH, DEVELOPMENT, COLLECTION, DISSEMINATION, AND UTILIZATION OF DATA.

(a) Authority.—The Administrator shall, in consultation and cooperation with the Secretary of Health, Education, and Welfare and with other heads of appropriate departments and agencies, conduct such research, development, and monitoring as is necessary to carry out the purposes of this Act. The Administrator may enter into contracts and may make grants for research, development, and monitoring under this subsection. Contracts may be entered into under this subsection without regard to sections 3648 and 3709 of the Revised Statutes (31 U.S.C. 529, 14 U.S.C. 5).

(b) Data Systems.—(1) The Administrator shall establish, administer, and be responsible for the continuing activities of an interagency committee which will design, establish, and coordinate an efficient and effective system, within the Environmental Protection Agency, for the collection, dissemination to other Federal departments and agencies, and use of data submitted to the Administrator under this Act.

(2)(A) The Administrator shall, in consultation and cooperation with the Secretary of Health, Education, and Welfare and other heads of appropriate departments and agencies design, establish, and coordinate an efficient and effective system for the retrieval of toxicological and other scientific data which could be useful to the Administrator in carrying out the purposes of this Act. Systematized retrieval shall be developed for use by all Federal and other departments and agencies with responsibilities in the area of regulation or study of chemical substances and mixtures and their effect on health or the environment.

(B) The Administrator, in consultation and cooperation with the Secretary of Health, Education, and Welfare may make grants and enter into contracts for the development of a data retrieval system described in subparagraph (A). Contracts may be entered into under this subparagraph wtihout regard to sections 3648 and 3709 of the Revised Statutes (31 U.S.C. 529, 41 U.S.C. 5).

(c) SCREENING TECHNIQUES.—The Administrator shall coordinate, with the Assistant Secretary for Health of the Department of Health, Education, and Welfare, research undertaken by the Administrator and directed toward the development of rapid, reliable, and economical screening techniques for carcinogenic, mutagenic, teratogenic, and ecological effects of chemical substances and mixtures.

(d) MONITORING.—The Administrator shall, in consultation and cooperation with the Secretary of Health, Education, and Welfare, establish and be responsible for research aimed at the development, in cooperation with local, State, and Federal agencies, of monitoring techniques and instruments which may be used in the detection of toxic chemical substances and mixtures and which are reliable, economical, and capable of being implemented under a wide variety of conditions.

(e) BASIC RESEARCH.—The Administrator shall, in consultation and cooperation with the Secretary of Health, Education, and Welfare, establish research programs to develop the fundamental scientific basis of the screening and monitoring techniques described in subsections (c) and (d), the bounds of the reliability of such techniques, and the oppotunities for their improvement.

(f) TRAINING.—The Administrator shall establish and promote programs and workshops to train or facilitate the training of Federal laboratory and technical personnel in existing or newly developed screening and monitoring techniques.

(g) EXCHANGE OF RESEARCH AND DEVELOPMENT RESULTS.—The Administrator shall, in consultation with the Secretary of Health, Education, and Welfare and other heads of appropriate departments and agencies, establish and coordinate a system for exchange among Federal, State, and local authorities of research and development results respecting toxic chemical substances and mixtures, including a system to facilitate and promote the development of standard data format and analysis and consistent testing procedures.

SEC. 11. INSPECTIONS AND SUBPOENAS.

SEC. 11. (a) IN GENERAL.—For purposes of administering this

(a) IN GENERAL.—For purposes of administering this Act, the Administrator, and any duly designated representative of the Administra-

tor may inspect any establishment, facility, or other premises in which chemical substances or mixtures are manufactured, processed, stored, or held before or after their distribution in commerce and any conveyance being used to transport chemical substances, mixtures, or such articles in connection with distribution in commerce. Such an inspection may only be made upon the presentation of appropriate credentials and of a written notice to the owner, operator, or agent in charge of the premises or conveyance to be inspected. A separate notice shall be given for each such inspection, but a notice shall not be required for each entry made during the period covered by the inspection. Each such inspection shall be commenced and completed with reasonable promptness and shall be conducted at reasonable times, within reasonable limits, and in a reasonable manner.

(b) Scope.—(1) Except as provided in paragraph (2), an inspection conducted under subsection (a) shall extend to all things within the premises or conveyance inspected (including records, files, papers, processes, controls, and facilities) bearing on whether the requirements of this Act applicable to the chemical substances or mixtures within such premises or conveyance have been complied with.

(2) No inspection under subsection (a) shall extend to—

(A) financial data,

(B) sales data (other than shipment data),

(C) pricing data,

(D) personnel data, or

(E) research data (other than data required by this Act or under a rule promulgated thereunder),

unless the nature and extent of such data are described with reasonable specificity in the written notice required by subsection (a) for such inspection.

(c) Subpoenas.—In carrying out this Act, the Administrator may by subpoena require the attendance and testimony of witnesses and the production of reports, papers, documents, answers to questions, and other information that the Administrator deems necessary. Witnesses shall be paid the same fees and mileage that are paid witnesses in the courts of the United States. In the event of contumacy, failure, or refusal of any person to obey any such subpoena, any district court of the United States in which venue is proper shall have jurisdiction to order any such person to comply with such subpoena. Any failure to obey such an order of the court is punishable by the court as a contempt thereof.

SEC. 12. EXPORTS.

(a) In General.—(1) Except as provided in paragraph (2) and subsection (b), this Act (other than section 8) shall not apply to any chemical substance, mixture, or to an article containing a chemical substance or mixture, if—

(A) it can be shown that such substance, mixture, or article is being manufactured, processed, or distributed in commerce for export from the United States, unless such substance, mixture, or article was, in fact, manufactured, processed, or distributed in commerce, for use in the United States, and

(B) such substance, mixture, or article (when distributed in commerce), or any container in which it is enclosed (when so

distributed), *bears a stamp or label stating that such substance, mixture, or article is intended for export.*

(*2*) *Paragraph* (*1*) *shall not apply to any chemical substance, mixture, or article if the Administrator finds that the substance, mixture, or article will present an unreasonable risk of injury to health within the United States or to the environment of the United States. The Administrator may require, under section 4, testing of any chemical substance or mixture exempted from this Act by paragraph* (*1*) *for the purpose of determining whether or not such substance or mixture presents an unreasonable risk of injury to health within the United States or to the environment of the United States.*

(*b*) *Notice.*—(*1*) *If any person exports or intends to export to a foreign country a chemical substance or mixture for which the submission of data is required under section 4 or 5*(*b*), *such person shall notify the Administrator of such exportation or intent to export and the Administrator shall furnish to the government of such country notice of the availability of the data submitted to the Administrator under such section for such substance or mixture.*

(*2*) *If any person exports or intends to export to a foreign country a chemical substance or mixture for which an order has been issued under section 5 or a rule has been proposed or promulgated under section 5 or 6, or with respect to which an action is pending, or relief has been granted under section 5 or 7, such person shall notify the Administrator of such exportation or intent to export and the Administrator shall furnish to the government of such country notice of such rule, order, action, or relief.*

SEC. 13. ENTRY INTO CUSTOMS TERRITORY OF THE UNITED STATES.

(*a*) *In General.*—(*1*) *The Secretary of the Treasury shall refuse entry into the customs territory of the United States* (*as defined in general headnote 2 to the Tariff Schedules of the United States*) *of any chemical substance, mixture, or article containing a chemical substance or mixture offered for such entry if*—

(*A*) *it fails to comply with any rule in effect under this Act, or*

(*B*) *it is offered for entry in violation of section 5 or 6, a rule or order under section 5 or 6, or an order issued in a civil action brought under section 5 or 7.*

(*2*) *If a chemical substance, mixture, or article is refused entry under paragraph* (*1*), *the Secretary of the Treasury shall notify the consignee of such entry refusal, shall not release it to the consignee, and shall cause its disposal or storage* (*under such rules as the Secretary of the Treasury may prescibe*) *if it has not been exported by the consignee within 90 days from the date of receipt of notice of such refusal, except that the Secretary of the Treasury may, pending a review by the Administrator of the entry refusal, release to the consignee such substance, mixture, or article on execution of bond for the amount of the full invoice of such substance, mixture, or article* (*as such value is set forth in the customs entry*), *together with the duty thereon. On failure to return such substance, mixture, or article for any cause to the custody of the Secretary of the Treasury when demanded, such consignee shall be liable to the United States for liquidated damages equal to the full amount of such bond. All charges for storage, cartage, and labor on and for disposal of substances, mixtures, or articles which are refused entry or release under this section shall be paid by the owner or consignee, and in default of such payment shall constitute a lien against any future entry made by such owner or consignee.*

(b) Rules.—The Secretary of the Treasury, after consultation with the Administrator, shall issue rules for the administration of subsection (a) of this section.

SEC. 14. DISCLOSURE OF DATA.

(a) In General.—Except as provided by subsection (b) any information reported to, or otherwise obtained by, the Administrator (or any representative of the Administrator) under this Act, which is exempt from disclosure pursuant to subsection (a) of section 552 of title 5, United States Code, by reason of subsection (b)(4) of such section, shall, notwithstanding the provisions of any other section of this Act, not be disclosed by the Administrator or by any officer or employee of the United States, except that such information—

(1) shall be disclosed to any officer or employee of the United States—

(A) in connection with the official duties of such officer or employee under any law for the protection of health or the environment, or

(B) for specific law enforcement purposes;

(2) shall be disclosed to contractors with the United States and employees of such contractors if in the opinion of the Administrator such disclosure is necessary for the satisfactory performance by the contractor of a contract with the United States entered into on or after the date of enactment of this Act for the performance of work in connection with this Act and under such conditions as the Administrator may specify;

(3) shall be disclosed if the Administrator determines it necessary to protect health or the environment against an unreasonable risk of injury to health or the environment; or

(4) may be disclosed when relevant in any proceeding under this Act, except that disclosure in such a proceeding shall be made in such manner as to preserve confidentiality to the extent practicable without impairing the proceeding.

In any proceeding under section 552(a) of title 5, United States Code, to obtain information the disclosure of which has been denied because of the provisions of this subsection, the Administrator may not rely on section 552(b)(3) of such title to sustain the Administrator's action.

(b) Data From Health and Safety Studies.—(1) Subsection (a) does not prohibit the disclosure of—

(A) any health and safety study which is submitted under this Act with respect to—

(i) any chemical substance or mixture which, on the date on which such study is to be disclosed, has been offered for commercial distribution, or

(ii) any chemical substance or mixture for which testing is required under section 4 or for which notification is required under section 5, and

(B) any data reported to, or otherwise obtained by, the Administrator from a health and safety study which relates to a chemical

substance or mixture described in clause (i) or (ii) of subparagraph (A)

This paragraph does not authorize the release of any data which discloses processes used in the manufacturing or processing of a chemical substance or mixture or, in the case of a mixture, the release of data disclosing the portion of the mixture comprised by any of the chemical substances in the mixture.

(2) If a request is made to the Administrator under subsection (a) of section 552 of title 5, United States Code, for information which is described in the first sentence of paragraph (1) and which is not information described in the second sentence of such paragraph, the Administrator may not deny such request on the basis of subsection (b)(4) of such section.

(c) DESIGNATION AND RELEASE OF CONFIDENTIAL DATA.—(1) In submitting data under this Act, a manufacturer, processor, or distributor in commerce may (A) designate the data which such person believes is entitled to confidential treatment under subsection (a), and (B) submit such designated data separately from other data submitted under this Act. A designation under this paragraph shall be made in writing and in such manner as the Administrator may prescribe.

(2)(A) Except as provided by subparagraph (B), if the Administrator proposes to release for inspection data which has been designated under paragraph (1)(A), the Administrator shall notify, in writing and by certified mail, the manufacturer, processor, or distributor in commerce who submitted such data of the intent to release such data. If the release of such data is to be made pursuant to a request made under section 552(a) of title 5, United States Code, such notice shall be given immediately upon approval of such request by the Administrator. The Administrator may not release such data until the expiration of 30 days after the manufacturer, processor, or distributor in commerce submitting such data has received the notice required by this subparagraph.

(B)(i) Subparagraph (A) shall not apply to the release of information under paragraph (1), (2), (3), or (4) of subsection (a), except that the Administrator may not release data under paragraph (3) of subsection (a) unless the Administrator has notified each manufacturer, processor, and distributor in commerce who submitted such data of such release. Such notice shall be made in writing by certified mail at least 15 days before the release of such data, except that if the Administrator determines that the release of such data is necessary to protect against an imminent, unreasonable risk of injury to health or the environment, such notice may be made by such means as the Administrator determines will provide notice at least 24 hours before such release is made.

(ii) Subparagraph (A) shall not apply to the release of information described in the second sentence of such subsection.

(d) CRIMINAL PENALTY FOR WRONGFUL DISCLOSURE.—(1) Any officer or employee of the United States or former officer or employee of the United States, who by virtue of such employment or official position has obtained possession of, or has access to, material the disclosure of which is prohibited by subsection (a), and who knowing that disclosure of such material is prohibited by such subsection, willfully discloses the material in any manner to any person not entitled to receive it, shall be guilty of a mis-

demeanor and fined not more than $5,000 or imprisoned for not more than one year, or both. Section 1905 of title 18, United States Code, does not apply with respect to the publishing, divulging, disclosure, or making known of, or making available, information reported or otherwise obtained under this Act.

(2) For the purposes of paragraph (1), any contractor with the United States who is furnished information as authorized by subsection (a)(2), and any employee of any such contractor, shall be considered to be an employee of the United States.

(e) ACCESS BY CONGRESS.—Notwithstanding any limitation contained in this section or any other provision of law, all information reported to or otherwise obtained by the Administrator (or any representative of the Administrator) under this Act shall be made available, upon written request of any duly authorized committee of the Congress, to such committee.

SEC. 15. PROHIBITED ACTS.

It shall be unlawful for any person to—

(1) fail or refuse to comply with (A) any rule promulgated or order issued under section 4, (B) any requirement prescribed by section 5 or 6, or (C) any rule promulgated or order issued under section 5 or 6;

(2) use for commercial purposes a chemical substance or mixture which such person knew or had reason to know was manufactured, processed, or distributed in commerce in violation of section 5 or 6, a rule or order under section 5 or 6, or an order issued in an action brought under section 5 or 7;

(3) fail or refuse to (A) establish or maintain records, (B) submit reports, notices, or other information, or (C) permit access to or copying of records, as required by this Act or a rule thereunder; or

(4) fail or refuse to permit entry or inspection as required by section 11.

SEC. 16. PENALTIES.

(a) CIVIL.—(1) Any person who violates a provision of section 15 shall be liable to the United States for a civil penalty in an amount not to exceed $25,000 for each such violation. Each day such a violation continues shall, for purposes of this subsection, constitute a separate violation of section 15.

(2)(A) A civil penalty for a violation of section 15 shall be assessed by the Administrator by an order made on the record after opportunity (provided in accordance with this subparagraph) for a hearing in accordance with section 554 of title 5, United States Code. Before issuing such an order, the Administrator shall give written notice to the person to be assessed a civil penalty under such order of the Administrator's proposal to issue such order and provide such person an opportunity to request, within 15 rays of the date the notice is received by such person, such a hearing on the order.

(B) In determining the amount of a civil penalty, the Administrator shall take into account the nature, circumstances, extent, and gravity of the violation or violations and, with respect to the violator, ability to pay, effect on ability to continue to do business, any history of prior such violations, the degree of culpability, and such other matters as justice may require.

(C) The Administrator may compromise, modify, or remit, with or without conditions, any civil penalty which may be imposed under this subsection. The amount of such penalty, when finally determined, or the amount agreed upon in compromise, may be deducted from any sums owing by the United States to the person charged.

(3) Any person who requested in accordance with paragraph (2)(A) a hearing respecting the assessment of a civil penalty and who is aggrieved by an order assessing a civil penalty may file a petition for judicial review of such order with the United States Court of Appeals for the District of Columbia Circuit or for any other circuit in which such person resides or transacts business. Such a petition may only be filed within the 30-day period beginning on the date the order making such assessment was issued.

(4) If any person fails to pay an assessment of a civil penalty—

(A) after the order making the assessment has become a final order and if such person does not file a petition for judicial review of the order in accordance with paragraph (3), or

(B) after a court in an action brought under paragraph (3) has entered a final judgment in favor of the Administrator,

the Attorney General shall recover the amount assessed (plus interest at currently prevailing rates from the date of the expiration of the 30-day period referred to in paragraph (3) or the date of such final judgment, as the case may be) in an action brought in any appropriate district court of the United States. In such an action, the validity, amount, and appropriateness of such penalty shall not be subject to review.

(b) CRIMINAL.—Any person who knowingly or willfully violates any provision of section 15 shall, in addition to or in lieu of any civil penalty which may be imposed under subsection (a) of this section for such violation, be subject, upon conviction, to a fine of not more than $25,000 for each day of violation, or to imprisonment for not more than one year, or both.

SEC. 17. SPECIFIC ENFORCEMENT AND SEIZURE.

(a) SPECIFIC ENFORCEMENT.—(1) The district courts of the United States shall have jurisdiction over civil actions to—

(A) restrain any violation of section 15,

(B) restrain any person from taking any action prohibited by section 5 or 6 or by a rule or order under section 5 or 6,

(C) compel the taking of any action required by or under this Act, or

(D) direct any manufacturer or processor of a chemical substance or mixture manufactured or processed in violation of section 5 or 6 or a rule or order under section 5 or 6 and distributed in commerce, (i) to give notice of such fact to distributors in commerce of such substance or mixture and, to the extent reasonably ascertainable, to other persons in possession of such substance or mixture or exposed to such substance or mixture, (ii) to give public notice of such risk of injury, and (iii) to either replace or repurchase such substance or mixture, whichever the person to which the requirement is directed elects.

(2) A civil action described in paragraph (1) may be brought—

(A) in the case of a civil action described in subparagraph (A) of such paragraph, in the United States district court for the judicial

district wherein any act, omission, or transaction constituting a violation of section 15 occurred or wherein the defendant is found or transacts business, or

(B) in the case of any other civil action described in such paragraph, in the United States district court for the judicial district wherein the defendant is found or transacts business.

In any such civil action process may be served on a defendant in any judicial district in which a defendant resides or may be found. Subpoenas requiring attendance of witnesses in any such action may be served in any judicial district.

(b) Seizure.—Any chemical substance or mixture which was manufactured, processed, or distributed in commerce in violation of this Act or any rule promulgated or order issued under this Act or any article containing such a substance or mixture shall be liable to be proceeded against, by process of libel for the seizure and condemnation of such substance, mixture, or article, in any district court of the United States within the jurisdiction of which such substance, mixture, or article is found. Such proceedings shall conform as nearly as possible to proceedings in rem in admiralty.

SEC. 18. PREEMPTION.

(a) Effect on State Law.—(1) Except as provided in paragraph (2), nothing in this Act shall affect the authority of any State or political subdivision of a State to establish or continue in effect regulation of any chemical substance, mixture, or article containing a chemical substance or mixture.

(2) Except as provided in subsection (b)—

(A) if the Administrator requires by a rule promulgated under section 4 the testing of a chemical substance or mixture, no State or political subdivision may, after the effective date of such rule, establish or continue in effect a requirement for the testing of such substance or mixture for purposes similar to those for which testing is required under such rule; and

(B) if the Administrator prescribes a rule or order under section 5 or 6 (other than a rule imposing a requirement described in subsection (a)(6) of section 6) which is applicable to a chemical substance or mixture, and which is designed to protect against a risk of injury to health or the environment associated with such substance or mixture, no State or political subdivision of a State may, after the effective date of such requirement, establish or continue in effect, any requirement which is applicable to such substance or mixture, or an article containing such substance or mixture, and which is designed to protect against such risk unless such requirement (i) is identical to the requirement prescribed by the Administrator, (ii) is adopted under the authority of the Clean Air Act or any other Federal law, or (iii) prohibits the use of such substance or mixture in such State or political subdivision (other than its use in the manufacture of processing of other substances of mixtures).

(b) Exemption.—Upon application of a State or political subdivision of a State the Administrator may by rule exempt from subsection (a)(2), under such conditions as may be prescribed in such rule, a requirement of such State or political subdivision designed to protect against a risk

of injury to health or the environment associated with a chemical substance, mixture, or article containing a chemical substance or mixture if—

(1) compliance with the requirement would not cause the manufacturing, processing, distribution in commerce, or use of the substance, mixture, or article to be in violation of the applicable requirement under this Act described in subsection (a)(2), and

(2) the State or political subdivision requirement (A) provides a significantly higher degree of protection from such risk than the requirement under this Act described in subsection (a)(2) and (B) does not, through difficulties in marketing, distribution, or other factors, unduly burden interstate commerce.

SEC. 19. JUDICIAL REVIEW.

(a) IN GENERAL.—(1)(A) Not later than 60 days after the date of the promulgation of a rule under section 4(a), 5(a)(2), 5(b)(4), 6(a) 6(e), or 8, any person may file a petition for judicial review of such rule with the United States Court of Appeals for the District of Columbia Circuit or for the circuit in which such person resides or in which such person's principal place of business is located. Courts of appeals of the United States shall have exclusive jurisdiction of any action to obtain judicial review (other than in an enforcement proceeding) of such a rule if any district court of the United States would have had jurisdiction of such action but for this subparagraph.

(B) Courts of appeals of the United States shall have exclusive jurisdiction of any action to obtain judicial review (other than in an enforcement proceeding) of an order issued under subparagraph (A) or (B) of section 6(b)(1) if any district court of the United States would have had jurisdiction of such action but for this subparagraph.

(2) Copies of any petition filed under paragraph (1)(A) shall be transmitted forthwith to the Administrator and to the Attorney General by the clerk of the court with which such petition was filed. The provisions of section 2112 of title 28, United States Code, shall apply to the filing of the rulemaking record of proceedings on which the Administrator based the rule being reviewed under this section and to the transfer of proceedings between United States courts of appeals.

(3) For purposes of this section, the term "rulemaking record" means—

(A) the rule being reviewed under this section;

(B) in the case of a rule under section 4(a), the finding required by such section, in the case of a rule under section 5(b)(4), the finding required by such section, in the case of a rule under section 6(a) the finding required by section 5(f) or 6(a), as the case may be, in the case of a rule under section 6(a), the statement required by section 6(c)(1), and in the case of a rule under section 6(e), the findings required by paragraph (2)(B) or (3)(B) of such section, as the case may be;

(C) any transcript required to be made of oral presentations made in proceedings for the promulgation of such rule;

(D) any written submission of interested parties respecting the promulgation of such rule; and

(E) any other information which the Administrator considers to be relevant to such rule and which the Administrator identi-

fied, on or before the date of the promulgation of such rule, in a notice published in the Federal Register.

(*b*) *Additional Submissions and Presentations; Modifications.—If in an action under this section to review a rule the petitioner or the Administrator applies to the court for leave to make additional oral submissions or written presentations respecting such rule and shows to the satisfaction of the court that such submissions and presentations would be material and that there were reasonable grounds for the submissions and failure to make such submissions and presentations in the proceeding before the Administrator, the court may order the Administrator to provide additional opportunity to make such submissions and presentations. The Administrator may modify or set aside the rule being reviewed or make a new rule by reason of the additional submissions and presentations and shall file such modified or new rule with the return of such submissions and presentations. The court shall thereafter review such new or modified rule.*

(*c*) *Standard of Review.—(1)(A) Upon the filing of a petition under subsection (a)(1) for judicial review of a rule, the court shall have jurisdiction (i) to grant appropriate relief, including interim relief, as provided in chapter 7 of title 5, United States Code, and (ii) except as otherwise provided in subparagraph (B), to review such rule in accordance with chapter 7 of title 5, United States Code.*

(B) Section 706 of title 5, United States Code, shall apply to review of a rule under this section, except that—

(i) in the case of review of a rule under section 4(a), 5(b)(4), 6(a), or 6(e), the standard for review prescribed by paragraph (2)(E) of such section 706 shall not apply and the court shall hold unlawful and set aside such rule if the court finds that the rule is not supported by substantial evidence in the rulemaking record (as defined in subsection (a)(3)) taken as a whole;

(ii) in the case of review of a rule under section 6(a), the court shall hold unlawful and set aside such rule if it finds that—

(I) a determination by the Administrator under section 6(c)(3) that the petitioner seeking review of such rule is not entitled to conduct (or have conducted) cross-examination or to present rebuttal submissions, or

(II) a rule of, or ruling by, the Administrator under section 6(c)(3) limiting such petitioner's cross-examination or oral presentations,

has precluded disclosure of disputed material facts which was necessary to a fair determination by the Administrator of the rulemaking proceeding taken as a whole; and section 706(2)(D) shall not apply with respect to a determination, rule, or ruling referred to in subclause (I) or (II); and

(iii) the court may not review the contents and adequacy of—

(I) any statement required to be made pursuant to section 6(c)(1), or

(II) any statement of basis and purpose required by section 553(c) of title 5, United States Code, to be incorporated in the rule

except as part of a review of the rulmaking record taken as a whole.

The term "evidence" as used in clause (i) means any matter in the rulmaking record.

(C) A determination, rule, or ruling of the Administrator described in subparagraph (B)(ii) may be reviewed only in an action under this section and only in accordance with such subparagraph.

(2) The judgment of the court affirming or setting aside, in whole or in part, any rule reviewed in accordance with this section shall be final, subject to review by the Supreme Court of the United States upon certiorari or certification, as provided in section 1254 of title 28, United States Code.

(d) FEES AND COSTS.—The decision of the court in an action commenced under subsection (a), or of the Supreme Court of the United States on review of such a decision, may include an award of costs of suit and reasonable fees for attorneys and expert witnesses if the court determines that such an award is appropriate.

(e) OTHER REMEDIES.—The remedies provided in this section shall be in addition to and not in lieu of any other remedies provided by law.

SEC. 20. CITIZENS' CIVIL ACTIONS.

(a) IN GENERAL.—Except as provided in subsection (b), any person may commence a civil action—

(1) against any person (including (A) the United States, and (B) any other governmental instrumentality or agency to the extent permitted by the eleventh amendment to the Constitution) who is alleged to be in violation of this Act or any rule promulgated under section 4, 5, or 6 or order issued under section 5 to restrain such violation, or

(2) against the Administrator to compel the Administrator to perform any act or duty under this Act which is not discretionary.

Any civil action under paragraph (1) shall be brought in the United States district court for the district in which the alleged violation occurred or in which the defendant resides or in which the defendant's principal place of business is located. Any action brought under paragraph (2) shall be brought in the United States District Court for the District of Columbia, or the United States district court for the judicial district in which the plaintiff is domiciled. The district courts of the United States shall have jurisdiction over suits brought under this section, without regard to the amount in controversy or the citizenship of the parties. In any civil action under this subsection process may be served on a defendant in any judicial district in which the defendant resides or may be found and subpoenas for witnesses may be served in any judicial district.

(b) LIMITATION.—No civil action may be commenced—

(1) under subsection (a)(1) to restrain a violation of this Act or rule or order under this Act—

(A) before the expiration of 60 days after the plaintiff has given notice of such violation (i) to the Administrator, and (ii) to the person who is alleged to have committed such violation, or

(B) If the Administrator has commenced and is diligently prosecuting a proceeding for the issuance of an order under section 16(a)(2) to require compliance with this Act or with such rule or order or if the Attorney General has commenced and is diligently prosecuting a civil action in a court

of the United States to require compliance with this Act or with such rule or order, but if such proceeding or civil action is commenced after the giving of notice, any person giving such notice may intervene as a matter of right in such proceeding or action; or

(2) under subsection (a)(2) before the expiration of 60 days after the plaintiff has given notice to the Administrator of the alleged failure of the Administrator to perform an act or duty which is the basis for such action or, in the case of an action under such subsection for the failure of the Administrator to file an action under section 7, before the expiration of ten days after such notification.

Notice under this subsection shall be given in such manner as the Administrator shall prescribe by rule.

(c) General.—(1) In any action under this section, the Administrator, if not a party, may intervene as a matter of right.

(2) The court, in issuing any final order in any action brought pursuant to subsection (a), may award costs of suit and reasonable fees for attorneys and expert witnesses if the court determines that such an award is appropriate. Any court, in issuing its decision in an action brought to review such an order, may award costs of suit and reasonable fees for attorneys if the court determines that such an award is appropriate.

(3) Nothing in this section shall restrict any right which any person (or class of persons) may have under any statute or common law to seek enforcement of this Act or any rule or order under this Act or to seek any other relief.

(d) Consolidation.—When two or more civil actions brought under subsection (a) involving the same defendant and the same issues or violations are pending in two or more judicial districts, such pending actions, upon application of such defendants to such actions which is made to a court in which any such action is brought, may, if such court in its discretion so decides, be consolidated for trial by order (issued after giving all parties reasonable notice and opportunity to be heard) of such court and tried in—

(1) any district which is selected by such defendant and in which one of such actions is pending,

(2) a district which is agreed upon by stipulation between all the parties to such actions and in which one of such actions is pending, or

(3) a district which is selected by the court and in which one of such actions is pending.

The court issuing such an order shall give prompt notification of the order to the other courts in which the civil actions consolidated under the order are pending.

SEC. 21. CITIZENS' PETITIONS.

(a) In General.—Any person may petition the Administrator to initiate a proceeding for the issuance, amendment, or repeal of a rule under section 4, 6, or 8 or an order under section 5(e) or (6)(b)(2).

(b) Procdeures.—(1) Such petition shall be filed in the principal office of the Administrator and shall set forth the facts which it is claimed establish that it is necessary to issue, amend, or repeal a rule under section 4, 6, or 8 or an order under section 5(e), 6(b)(1)(A), or 6(b)(1)(B).

(2) The Administrator may hold a public hearing or may conduct such investigation or proceeding as the Administrator deems appropriate in order to determine whether or not such petition should be granted.

(3) Within 90 days after filing of a petition described in paragraph (1), the Administrator shall either grant or deny the petition. If the Administrator grants such petition, the Administrator shall promptly commence an appropriate proceeding in accordance with section 4, 5, 6, or 8. If the Administrator denies such petition, the Administrator shall publish in the Federal Register the Administrator's reasons for such denial.

(4)(A) If the Administrator denies a petition filed under this section (or if the Administrator fails to grant or deny such petition within the 90-day period) the petitioner may commence a civil action in a district court of the United States to compel the Administrator to initiate a rulemaking proceeding as requested in the petition. Any such action shall be filed within 60 days after the Administrator's denial of the petition or, if the Administrator fails to grant or deny the petition within 90 days after filing the petition, within 60 days after the expiration of the 90-day period.

(B) In an action under subparagraph (A) respecting a petition to initiate a proceeding to issue a rule under section 4, 6, or 8 or an order under section 5(e) or 6(b)(2), the petitioner shall be provided an opportunity to have such petition considered by the court in a de novo proceeding. If the petitioner demonstrates to the satisfaction of the court by a preponderance of the evidence that—

(i) in the case of a petition to initiate a proceeding for the issuance of a rule under section 4 or an order under section 5(e)—

(I) information available to the Administrator is insufficient to permit a reasoned evaluation of the health and environmental effects of the chemical substance to be subject to such rule or order; and

(II) in the absence of such information, the substance may present an unreasonable risk to health or the environment, or the substance is or will be produced in substantial quantities and it enters or may reasonably be anticipated to enter the environment in substantial quantities or there is or may be significant or substantial human exposure to it; or

(ii) in the case of a petition to initiate a proceeding for the issuance of a rule under section 6 or 8 or an order under section 6(b)(2), there is a reasonable basis to conclude that the issuance of such a rule or order is necessary to protect health or the environment against an unreasonable risk of injury to health or the environment.

the court shall order the Administrator to initiate the action requested by the petitioner. If the court finds that the extent of the risk to health or the environment alleged by the petitioner is less than the extent of risks to health or the environment with respect to which the Administrator is taking action under this Act and there are insufficient resources available to the Administrator to take the action requested by the petitioner, the court may permit the Administrator to defer initiating the action requested by the petitioner until such time as the court prescribes.

(C) The court in issuing any final order in any action brought pursuant to subparagraph (A) may award costs of suit and reasonable fees for at-

toneys and expert witnesses if the court determines that such an award is appropriate. Any court, in issuing its decision in an action brought to review such an order, may award costs of suit and reasonable fees for attorneys if the court determines that such an award is appropriate.

(5) The remedies under this section shall be in addition to, and not in lieu of, other remedies provided by law.

NATIONAL DEFENSE WAIVER

SEC. 22. The Administrator shall waive compliance with any provision of this Act upon a request and determination by the President that the requested waiver is necessary in the interest of national defense. The Administrator shall maintain a written record of the basis upon which such waiver was granted and make such record available for in camera examination when relevant in a judicial proceeding under this Act. Upon the issuance of such a waiver, the Administrator shall publish in the Federal Register a notice that the waiver was granted for national defense purposes, unless, upon the request of the President, the Administrator determines to omit such publication because the publication itself would be contrary to the interests of national defense, in which event the Administrator shall submit notice thereof to the Armed Services Committees of the Senate and the House of Representatives.

EMPLOYEE PROTECTION

SEC. 23. (a) IN GENERAL.—No employer may discharge any employee or otherwise discriminate against any employee with respect to the employee's compensation, terms, conditions, or privileges of employment because the employee (or any person acting pursuant to a request of the employee) has—

(1) commenced, caused to be commenced, or is about to commence or cause to be commenced a proceeding under this Act;

(2) testified or is about to testify in any such proceeding; or

(3) assisted or participated or is about to assist or participate in any manner in such a proceeding or in any other action to carry out the purposes of this Act.

(b) REMEDY.—(1) Any employee who believes that the employee has been discharged or otherwise discriminated against by any person in violation of subsection (a) of this section may, within 30 days after such alleged violation occurs, file (or have any person file on the employee's behalf) a complaint with the Secretary of Labor (hereinafter in this section referred to as the "Secretary") alleging such discharge or discrimination. Upon receipt of such a complaint, the Secretary shall notify the person named in the complaint of the filing of the complaint.

(2)(A) Upon receipt of a complaint filed under paragraph (1), the Secretary shall conduct an investigation of the violation alleged in the complaint. Within 30 days of the receipt of such complaint, the Secretary shall complete such investigation and shall notify in writing the complainant (and any person acting on behalf of the complainant) and the person alleged to have committed such violation of the results of the investigation conducted pursuant to this paragraph. Within ninety days of the receipt of such complaint the Secretary shall, unless the proceeding

on the complaint is terminated by the Secretary on the basis of a settlement entered into by the Secretary and the person alleged to have committed such violation, issue an order either providing the relief prescribed by subparagraph (B) or denying the complaint. An order of the Secretary shall be made on the record after notice and opportunity for agency hearing. The Secretary may not enter into a settlement terminating a proceeding on a complaint without the participation and consent of the complainant.

(B) If in response to a complaint filed under paragraph (1) the Secretary determines that a violation of subsection (a) of this section has occurred, the Secretary shall order (i) the person who committed such violation to take affirmative action to abate the violation, (ii) such person to reinstate the complainant to the complainant's former position together with the compensation (including back pay), terms, conditions, and privileges of the complainant's employment, (iii) compensatory damages, and (iv) where appropriate, exemplary damages. If such an order issued, the Secretary, at the request of the complainant, shall assess against the person against whom the order is issued a sum equal to the aggregate amount of all costs and expenses (including attorney's fees) reasonably incurred, as determined by the Secretary, by the complainant for, or in connection with, the bringing of the complaint upon which the order was issued.

(c) Review.—Any employee or employer adversely affected or aggrieved by an order issued under subsection (b) may obtain review of the order in the United States Court of Appeals for the circuit in which the violation, with respect to which the order was issued, allegedly occurred. The petition for review must be filed within sixty days from the issuance of the Secretary's order. Review shall conform to chapter 7 of title 5 of the United States Code.

(2) An order of the Secretary, with respect to which review could have been obtained under paragraph (1), shall not be subject to judicial review in any criminal or other civil proceeding.

(d) Enforcement.—Whenever a person has failed to comply with an order issued under subsection (b)(2), the Secretary shall file a civil action in the United States district court for the district in which the violation was found to occur to enforce such order. In actions brought under this subsection, the district courts shall have jurisdiction to grant all appropriate relief, including injunctive relief and compensatory and exemplary damages. Civil actions brought under this subsection shall be heard and decided expeditiously.

(e) Exclusion.—Subsection (a) of this section shall not apply with respect to any employee who, acting without direction from the employee's employer (or any agent of the employer), deliberately causes a violation of any requirement of this Act.

EMPLOYMENT EFFECTS

Sec. 24. (a) In General.—The Administrator shall evaluate on a continuing basis the potential effects on employment (including reductions in employment or loss of employment from threatened plant closures) of—

(1) the issuance of a rule or order under section 4, 5, or 6, or

(2) a requirement of section 5 or 6.

(b)(1) Investigations.—Any employee (or any representative of an employee) may request the Administrator to make an investigation of—

(A) a discharge or layoff or threatened discharge or layoff of the employee, or

(B) adverse or threatened adverse effects on the employee's employment,

allegedly resulting from a rule or order under section 4, 5, or 6 or a requirement of section 5 or 6. Any such request shall be made in writing, shall set forth with reasonable particularity the grounds for the request, and shall be signed by the employee, or representative of such employee, making the request.

(2)(A) Upon receipt of a request made in accordance with paragraph (1) the Administrator shall (i) conduct the investigation requested, and (ii) if requested by any interested person, hold public hearings on any matter involved in the investigation unless the Administrator, by order issued within 45 days of the date such hearings are requested, denies the request for the hearings because the Administrator determines there are no reasonable grounds for holding such hearings. If the Administrator makes such a determination, the Administrator shall notify in writing the person requesting the hearing of the determination and the reasons therefor and shall publish the determination and the reasons therefor in the Federal Register.

(B) If public hearings are to be held on any matter involved in an investigation conducted under this subsection—

(i) at least five days' notice shall be provided the person making the request for the investigation and any person identified in such request,

(ii) such hearings shall be held in accordance with section 6(c)(3), and

(iii) each employee who made or for whom was made a request for such hearings and the employer of such employee shall be required to present information respecting the applicable matter referred to in paragraph (1)(A) or (1)(B) together with the basis for such information.

(3) Upon completion of an investigation under paragraph (2), the Administrator shall make findings of fact, shall make such recommendations as the Administrator deems appropriate, and shall make available to the public such findings and recommendations.

(4) This section shall not be construed to require the Administrator to amend or repeal any rule or order in effect under this Act.

SEC. 25. STUDIES.

(a) Indemnification Study.—The Administrator shall conduct a study of all Federal laws administered by the Administrator for the purpose of determining whether and under what conditions, if any, indemnification should be accorded any person as a result of any action taken by the Administrator under any such law. The study shall—

(1) include an estimate of the probable cost of any indemnification programs which may be recommended;

(2) include an examination of all viable means of financing the cost of any recommended indemnification; and

(3) be completed and submitted to Congress within two years from the effective date of enactment of this Act.

The General Accounting Office shall review the adequacy of the study submitted to Congress pursuant to paragraph (3) and shall report the results of its review to the Congress within six months of the date such study is submitted to Congress.

(b) CLASSIFICATION, STORAGE, AND RETRIEVAL STUDY.—The Council on Environmental Quality, in consultation with the Administrator, the Secretary of Health, Education, and Welfare, the Secretary of Commerce, and the heads of other appropriate Federal departments or agencies, shall coordinate a study of the feasibility of establishing (1) a standard classification system for chemical substances and related substances, and (2) a standard means for storing and for obtaining rapid access to information respecting such substances. A report on such study shall be completed and submitted to Congress not later than 18 months after the effective date of enactment of this Act.

SEC. 26. ADMINISTRATION OF THE ACT.

(a) COOPERATION OF FEDERAL AGENCIES.—Upon request by the Administrator, each Federal department and agency is authorized—

(1) to make its services, personnel, and facilities available (with or without reimbursement) to the Administrator to assist the Administrator in the administration of this Act; and

(2) to furnish to the Administrator such information, data, estimates, and statistics, and to allow the Administrator access to all information in its possession as the Administrator may reasonably determine to be necessary for the administration of this Act.

(b) FEES.—(1) The Administrator may, by rule, require the payment of a reasonable fee from any person required to submit data under section 4 or 5 to defray the cost of administering this Act. Such rules shall not provide for any fee in excess of $2,500 or, in the case of a small business concern, any fee in excess of $100. In setting a fee und erthis paragraph, the Administrator shall take into account the ability to pay of the person required to submit the data and the cost to the Administrator of reviewing such data. Such rules may provide for sharing such a fee in any case in which the expenses of testing are shared under section 4 or 5.

(2) The Administrator, after consultation with the Administrator of the Small Business Administration, shall by rule prescribe standards for determining the persons which qualify as small business concerns for purposes of paragraph (1).

(c) ACTION WITH RESPECT TO CATEGORIES.—(1) Any action authorized or required to be taken by the Administrator under any provision of this Act with respect to a chemical substance or mixture may be taken by the Administrator in accordance with that provision with respect to a category of chemical substances or mixtures. Whenever the Administrator takes action under a provision of this Act with respect to a category of chemical substances or mixtures, any reference in this Act to a chemical substance or mixture (insofar as it relates to such action) shall be deemed to be a reference to each chemical substance or mixture in such category.

(2) For purposes of paragraph (1):

(A) The term "category of chemical substances" means a group of chemical substances the members of which are similar in molecular structure, in physical, chemical, or biological properties, in use, or in mode of entrance into the human body or into the environment, or the members of which are in some other way suitable for classification as

such for purposes of this Act, except that such term does not mean a group of chemical substances which are grouped together solely on the basis of their being new chemical substances.

(B) The term "category of mixtures" means a group of mixtures the members of which are similar in molecular structure, in physical, chemical, or biological properties, in use, or in the mode of entrance into the human body or into the environment, or the members of which are in some other way suitable for classification as such for purposes of this Act.

(d) Assistance Office.—The Administrator shall establish in the Environmental Protection Agency an identifiable office to provide technical and other nonfinancial assistance to manufacturers and processors of chemical substances and mixtures respecting the requirements of this Act applicable to such manufacturers and processors, the policy of the Agency respecting the application of such requirements to such manufacturers and processors, and the means and methods by which such manufacturers and processors may comply with such requirements.

(e) Financial Disclosures.—(1) Except as provided under paragraph (3), each officer or employee of the Environmental Protection Agency and the Department of Health, Education, and Welfare who—

(A) performs any function or duty under this Act, and

(B) has any known financial interest (i) in any person subject to this Act or any rule or order in effect under this Act, or (ii) in any person who applies for or receives any grant or contract under this Act,

shall, on February 1, 1978, and on February 1 of each year thereafter, file with the Administrator or the Secretary of Health, Education, and Welfare (hereinafter in this subsection referred to as the "Secretary"), as appropriate, a written statement concerning all such interests held by such officer or employee during the preceding calendar year. Such statement shall be made available to the public.

(2) The Administrator and the Secretary shall—

(A) act within 90 days of the effective date of this Act—

(i) to define the term "known financial interests" for purposes of paragraph (1), and

(ii) to establish the methods by which the requirement to file written statements specified in paragraph (1) will be monitored and enforced, including appropriate provisions for review by the Administrator and the Secretary of such statements; and

(B) report to the Congress on June 1, 1978, and on June 1 of each year thereafter with respect to such statements and the actions taken in regard thereto during the preceding calendar year.

(3) The Administrator may by rule identify specific positions with the Environmental Protection Agency, and the Secretary may by rule identify specific positions with the Department of Health, Education, and Welfare, which are of a nonregulatory or nonpolicymaking nature, and the Administrator and the Secretary may by rule provide that officers or employees occupying such positions shall be exempt from the requirements of paragraph (1).

(4) This subsection does not supersede any requirement of chapter 11 of title 18, United States Code.

(5) Any officer or employee who is subject to, and knowingly violates, this subsection or any rule issued thereunder, shall be fined not more than $2,500 or imprisoned not more than one year, or both.

(*f*) *Statement of Basis and Purpose.—Any final order issued under this Act shall be accompanied by a statement of its basis and purpose. The contents and adequacy of any such statement shall not be subject to judicial review in any respect.*

(*g*) *Assistant Administrator.—(1) The President, by and with the advice and consent of the Senate, shall appoint an Assistant Administrator for Toxic Substances of the Environmental Protection Agency. Such Assistant Administrator shall be a qualified individual who is, by reason of background and experience, especially qualified to direct a program concerning the effects of chemicals on human health and the environment. Such Assistant Administrator shall be responsible for (A) the collection of data, (B) the preparation of studies, (C) the making of recommendations to the Administrator for regulatory and other actions to carry out the purposes and to facilitate the administration of this Act, and (D) such other functions as the Administrator may assign or delegate.*

(2) The Assistant Administrator to be appointed under paragraph (1) shall (A) be in addition to the Assistant Administrators of the Environmental Protection Agency authorized by section 1(d) of Reorganization Plan No. 3 of 1970, and (B) be compensated at the rate of pay authorized for such Assistant Administrators.

SEC. 27. DEVELOPMENT AND EVALUATION OF TEST METHODS.

(*a*) *In General.—The Secretary of Health, Education, and Welfare, in consultation with the Administrator and acting through the Assistant Secretary for Health, may conduct, and make grants to public and nonprofit private entities and enter into contracts with public and private entities for, projects for the development and evaluation of inexpensive and efficient methods (1) for determining and evaluating the health and environmental effects of chemical substances and mixtures, and their toxicity, persistence, and other characteristics which affect health and the environment, and (2) which may be used for the development of test data to meet the requirements of rules promulgated under section 4. The Administrator shall consider such methods in prescribing under section 4 standards for the development of test data.*

(*b*) *Approval by Secretary.—No grant may be made or contract entered into under subsection (a) unless an application therefor has been submitted to and approved by the Secretary. Such an application shall be submitted in such form and manner and contain such information as the Secretary may require. The Secretary may apply such conditions to grants and contracts under subsection (a) as the Secretary determines are necessary to carry out the purposes of such subsection. Contracts may be entered into under such subsection without regard to sections 3648 and 3709 of the Revised Statutes (31 U.S.C. 529; 41 U.S.C. 5).*

(c) *Annual Reports.—(1) The Secretary shall prepare and submit to the President and the Congress on or before January 1 of each year a report of the number of grants made and contracts entered into under this section and the results of such grants and contracts.*

(2) The Secretary shall periodically publish in the Federal Register reports describing the progress and results of any contract entered into or grant made under this section.

SEC. 28. STATE PROGRAMS.

(a) *In General.—For the purpose of complementing (but not reducing) the authority of, or actions taken by, the Administrator under this Act, the Administrator may make grants to States for the establishment and operation of programs to prevent or eliminate unreasonable risks within the States to health or the environment which are associated with a chemical substance or mixture and with respect to which the Administrator is unable or is not likely to take action under this Act for their prevention or elimination. The amount of a grant under this subsection shall be determined by the Administrator, except that no grant for any State program may exceed 75 per centum of the establishment and operation costs (as determined by the Administrator) of such program during the period for which the grant is made.*

(b) *Approval by Administrator.—(1) No grant may be made under subsection (a) unless an application therefor is submitted to and approved by the Administrator. Such an application shall be submitted in such form and manner as the Administrator may require and shall—*

(A) set forth the need of the applicant for a grant under subsection (a),

(B) identify the agency or agencies of the State which shall establish or operate, or both, the program for which the application is submitted,

(C) describe the actions proposed to be taken under such program,

(D) contain or be supported by assurances satisfactory to the Administrator that such program shall, to the extent feasible, be integrated with other programs of the applicant for environmental and public health protection,

(E) provide for the making of such reports and evaluations as the Administrator may require, and

(F) contain such other information as the Administrator may prescribe.

(2) The Administrator may approve an application submitted in accordance with paragraph (1) only if the applicant has established to the satisfaction of the Administrator a priority need, as determined under rules of the Administrator, for the grant for which the application has been submitted. Such rules shall take into consideration the seriousness of the health effects in a State which are associated with chemical substances or mixtures, including cancer, birth defects, and gene mutations, the extent of the exposure in a State of human beings and the environment to chemical substances and mixtures, and the extent to which chemical substances and mixtures are manufactured, processed, used, and disposed of in a State.

(c) *Annual Reports.—Not later than six months after the end of each of the fiscal years 1979, 1980, and 1981, the Administrator shall submit to the Congress a report respecting the programs assisted by grants under subsection (a) in the preceding fiscal year and the extent to which the Administrator has disseminated information respecting such programs.*

(d) *Authorization.—For the purpose of making grants under subsection (a) there are authorized to be appropriated $1,500,000 for the fiscal year ending September 30, 1977, $1,500,000 for the fiscal year ending September 30, 1978, and $1,500,000 for the fiscal year ending*

September 30, 1979. Sums appropriated under this subsection shall remain available until expended.

SEC. 29. AUTHORIZATION FOR APPROPRIATIONS.

There are authorized to be appropriated to the Administrator for purposes of carrying out this Act (other than sections 27 and 28 and subsections (a) and (c) through (g) of section 10 thereof) $10,100,000 for the fiscal year ending September 30, 1977, $12,625,000 for the fiscal year ending September 30, 1978, $16,200,000 for the fiscal year ending September 30, 1979. No part of the funds appropriated under this section may be used to construct any research laboratories.

SEC. 30. ANNUAL REPORT.

The Administrator shall prepare and submit to the President and the Congress on or before January 1, 1978, and on or before January 1 of each succeeding year a comprehensive report on the administration of this Act during the preceding fiscal year. Such report shall include—

(1) a list of the testing required under section 4 during the year for which the report is made and an estimate of the costs incurred during such year by the persons required to perform such tests;

(2) the number of notices received during such year under section 5, the number of such notices received during such year under such section for chemical substances subject to a section 4 rule, and a summary of any action taken during such year under section 5(g);

(3) a list of rules issued during such year under section 6;

(4) a list, with a brief statement of the issues, of completed or pending judicial actions under this Act and administrative actions under section 16 during such year;

(5) a summary of major problems encountered in the administration of this Act; and

(6) such recommendations for additional legislation as the Administrator deems necessary to carry out the purposes of this Act.

SEC. 31. EFFECTIVE DATE.

Except as provided in section 4(f), this Act shall take effect on January 1, 1977.

And the House agree to the same.

HARLEY O. STAGGERS,
JOHN M. MURPHY,
W. S. STUCKEY,
BOB ECKHARDT,
RALPH H. METCALFE,
WILLIAM BRODHEAD,
JAMES H. SCHEUER,
SAMUEL L. DEVINE,
JAMES T. BROYHILL,
MATTHEW J. RINALD,
Managers on the Part of the House.

WARREN G. MAGNUSON,
VANCE HARTKE,
PHILIP A. HART,
JOHN A. DURKIN,
JOHN V. TUNNEY,
HOWARD BAKER,
TED STEVENS,
Managers on the Part of the Senate.

JOINT EXPLANATORY STATEMENT OF THE COMMITTEE OF CONFERENCE

The managers on the part of the House and the Senate at the conference on the disagreeing votes of the two Houses on the amendment of the House to the bill (S. 3149) to regulate commerce and protect human health and the environment by requiring testing and necessary use restrictions on certain chemical substances, and for other purposes, submit the following joint statement to the House and the Senate in explanation of the effect of the action agreed upon by the managers and recommended in the accompanying conference report:

The House amendment struck out all of the Senate bill after the enacting clause and inserted a substitute text.

The Senate recedes from its disagreement to the amendment of the House with an amendment which is a substitute for the Senate bill and the House amendment. The differences between the Senate bill, the House amendment, and the substitute agreed to in conference are noted below, except for clerical corrections, conforming changes made necessary by agreements reached by the conferees, and minor drafting and clarifying changes.

FINDINGS, POLICY, AND INTENT

Senate bill (section 2)

Section 2(a) outlines Congressional policy underlying the Toxic Substances Control Act. Congress finds that: human beings and the environment are exposed to numerous chemical substances and mixtures; some of these may cause or contribute to an unreasonable risk of injury to health or the environment; and the effective regulation of such substances and mixtures in interstate commerce necessitates regulation of intrastate commerce as well.

Subsection (b) sets forth that it is the policy of the United States that adequate data on the health and environmental effects of such chemical substances and mixtures should be developed. Such data development should be the responsibility of those who manufacture and process such chemical substances and mixtures. Adequate authority should exist to regulate chemical substances and mixtures, but the exercise of such authority should not unduly impede technological innovation.

Subsection (c) contains a declaration of Congressional intent as to how the Administrator shall fulfill the responsibilities under this Act. The Administrator shall carry out this Act in a reasonable and prudent manner and consider the environmental, economic, and social impact of any action taken or proposed under this Act.

House amendment (section 2)

The House amendment is nearly identical to the Senate bill. However, the House amendment confines its data development mandate to hazardous or potentially hazardous substances and mixtures, in contrast to the broader mandate contained in the Senate bill.

Conference substitute (section 2)

The conference substitute follows the Senate provision. Adequate data should be developed concerning the health and environmental effects of chemical substances and mixtures. Such data development should be the responsibility of those who manufacture or process such substances and mixtures. Adequate authority should exist to regulate chemical substances and mixtures which present an unreasonable risk of injury to health or the environment and to take action with respect to chemical substances and mixtures which are imminent hazards.

DEFINITIONS

Senate bill (section 3)

The Senate bill includes definitions for the Act, the principal ones of which are as follows:

1. Chemical substance is defined as (i) any organic or inorganic substance of a particular molecular identity including a combination of such substances occurring as a result of a chemical reaction, or (ii) any element or uncombined radical. The term specifically excludes any mixture; any pesticide; tobacco and tobacco products; special nuclear materials or by-product materials as defined in the Atomic Energy Act of 1954; articles which if sold would be subject to the tax imposed by section 4181 of the Internal Revenue Code of 1954 (i.e., pistols, firearms, revolvers, shells and cartridges); any substance found in or on any food, drug, cosmetic or device and any substance produced for research and development purposes intended only for use in or on any food, drug cosmetic or device.
2. The term "environment" includes human beings and their environment, water, atmosphere, and land and the interrelationships which exist among and between these.
3. "Manufacture" means to import, produce, or manufacture for commercial purposes.
4. The term "mixture" means any combination of two or more chemical substances if such substances do not react chemically with each other and if the combination is not a result of the chemical reaction. Mixture also includes combinations of substances which occur in nature.

The Senate bill authorizes the Administrator to exclude from coverage of the Act any chemical substance or mixture if the Administrator determines, by rule, that the substance or mixture does not present an unreasonable risk of injury to health or the environment. However, the exclusion shall not apply to section 7 or section 8(e) of the Senate bill.

House amendment (section 3)

1. The House definition of "chemical substance" is similar to that of the Senate bill, except that the term includes organic or inorganic substances or combinations of such substances occurring in nature. The exclusions from the definition of chemical substances are similar to the Senate bill, although the House amendment specifically excludes food additives along with foods, drugs, cosmetics, and devices.

2. The House amendment defines "environment" to include water, air and land, and the interrelationships which exist among and between water, air and land, and all living things.

3. The House amendment, like the Senate bill, defines manufacture to mean import, produce, or manufacture, but the definition is not limited to such activities done for commercial purposes.

4. The term "mixture' is defined to mean any combination of two or more chemical substances if the combination does not occur in nature and is not, in whole or in part, the result of a chemical reaction. However, certain reaction-produced combinations are included in the term "mixture" in order to prevent disparate treatment of identical combinations simply because of the number of steps used in the manufacture of the combination. If each of the chemical substances comprising the combination is not a new chemical substance and if the combination could have been manufactured for commercial purposes without a chemical reaction occurring at the time the substances comprising the combination were combined, then the combination is included within the term "mixture".

The House bill does not contain general authority for the Administrator to exclude any chemical substance or mixture from the provisions of the bill. However, section 5(i)(5) of the House amendment authorizes the Administrator to exclude any chemcial substance from the notification requirements of section 5 if the Administrator determines, by rule, that the substance will not cause or significantly contribute to an unreasonable risk to health or the environment.

Conference substitute (section 3)

The conference substitute adopts the definitions contained in the House amendment.

The conferees recognize that virtually no chemical substance exists in a completely pure state and intend that any reference to a chemical substance includes all impurities and concomitant products, including incidental reaction products, contaminants, co-products, and trace materials. Thus the definition of term "chemical substance" shall be applied to chemical substances as actually produced and marketed. For example, when the Administrator promulgates a rule under section 6(a) to regulate a particular substance, such rule will apply to the identified substance, including all the impurities and other concomitant products, without explicitly identifying such impurities and concomitant products within the rule.

It is expected that the Administrator will develop guidelines for the purpose of clarifying the extent to which impurities and concomitant products will be included within a reference to "chemical substance" as it relates to the various provisions of the Act. While impurities and concomitant products are included within references to a "chemical substance" under the Act, the Administrator is obviously authorized to move against them separately under the applicable provisions of the Act.

TESTING OF CHEMICAL SUBSTANCES AND MIXTURES

The term "health and safety study" is important as it describes information to which various provisions of the Act are applicable. For example, section 8(d) requires manufacturers, processors, and distributors in commerce to list such studies with the Administrator. Moreover, section 14(b) contains provisions concerning the availability of health and safety studies to the public.

The conference substitute defines health and safety study to mean any study of any effect of a chemical substance or mixture on health or the environment, including underlying data and epidemiological studies, studies of occupational exposure to a chemical substance or mixture, toxicological, chemical, and ecological studies of a chemical substance or mixture, and any test performed pursuant to this Act.

It is intended that the term be interpreted broadly. Not only is information which arises as a result of a formal, disciplined study included but other information relating to the effects of a chemcial substance or mixture on health and the environment is also included. Any data which bears on the effects of a chemical substance on health or the environment would be included.

HARLEY O. STAGGERS,
JOHN M. MURPHY,
W. S. STUCKEY,
BOB ECKHARDT,
RALPH H. METCALFE,
WILLIAM BRODHEAD,
JAMES H. SCHEUER,
SAMUEL L. DEVINE,
JAMES T. BROYHILL,
MATTHEW J. RINALDO,
Managers on the Part of the House.
WARREN G. MAGUNSON,
VANCE HARTKE,
PHILIP A. HART,
JOHN A. DURKIN,
JOHN V. TUNNEY,
HOWARD BAKER,
T. E. STEVENS,
Managers on the Part of the Senate.

Senate bill (section 4)

Section 4 authorizes the Administrator to require testing of chemical substances and mixtures to ascertain potential effects on human health and the environment. Under subsection (a), the Administrator must, by rule, require the testing of a chemical substance or mixture if the Administrator finds (1) that the chemical substance or mixture may present an unreasonable risk of injury to health or the environment, or that there may be significant human or environmental exposure because substantial quantities will be produced and such substance or mixture may perhaps present an adverse effect on health or the environment; (2) there are insufficient data or experience to reasonably determine or predict its health and environmental effects; and (3) testing is necessary to develop such data. If no reliable data is available to the Administrator, the finding that such substance or mixture may perhaps present an adverse effect on health or the environment shall be presumed. In the case of a mixture, the bill requires an additional finding that testing the chemical substances which comprise the mixture is not a more efficient and reasonable method to determine effects on health and the environment. When requiring tests under subsection (a), the Administrator shall consider reasonably ascertainable costs and other burdens associated with conducting tests and publish such considerations in the Federal Register.

Subsection (b) sets forth the required contents of the testing rule and provides an illustrative list of health and environmental effects for which test standards may be required. This subsection also describes some of the methodologies which the testing rule may prescribe. In addition, it describes which manufacturers and processors will be required to conduct the testing.

The Administrator shall review and, if appropriate, revise the standards for development of data at least once per year. Testing rules shall be issued in accordance with the rulemaking procedures of section 553, of title 5, United States Code, except that the Administrator shall allow interested persons the opportunity to make oral presentations of data, views, or arguments in addition to written submissions. A transcript of such oral presentations is required.

Subsection (c) provides a procedure whereby persons may apply to the Administrator for an exemption from a testing requirement rule in order to avoid submission of duplicative data. If an exemption is granted, a cost-sharing procedure is provided. A person providing reimbursement may have access to test data, subject to the confidentiality provisions of section 14.

Subsection (d) requires the Administrator to publish a notice of receipt of test data in the Federal Register and to make the data available to the public within 15 days of receipt.

Subsection (e) establishes an interagency advisory committee comprised of qualified and appropriate Federal officials to make recommendations to the Administrator regarding testing priorities.

The committee shall submit a list of chemical substances and mixtures in the order in which the committee determines the Administrator should promulgate testing rules under subsection (a). Within 12 months after the inclusion of a chemical on such list, the Administrator shall either initiate a rulemaking proceeding under subsection (a) or publish reasons for not initiating a proceeding in the Federal Register. Subsection (e) also contains specific conflict of interest provisions applicable to members of the interagency advisory committee.

Subsection (f) specifies required actions by the Administrator in response to test data or other information which indicate that a substance or mixture has a potential to induce cancer, gene mutations, or birth defects at levels for which human exposure exists or may exist with appropriate safety margins. The Administrator shall either take appropriate action under section 5(e), 6(a), or 7 within 180 days after the date of receipt of such data or information or publish in the Federal Register a finding that no unreasonable risk of injury is presented and reasons for making such a finding. Such requirement shall not take effect until two years after enactment.

House amendment (section 4)

Like the Senate bill, the House amendment requires that the Administrator find that there are insufficient data and experience upon which to determine or predict the effects of the manufacture, distribution in commerce, processing, use, or disposal of a chemical substance or mixture. It also requires a finding that testing of the substance or mixture is necessary to develop such data. However, the House bill differs from the Senate bill in that it requires a finding that a chemical substance or mixture may "cause or significantly contribute" to an unreasonable risk, whereas the Senate bill requires a finding that the substance or mixture may "present" an unreasonable risk.

Section 4(a)(1)(B) of the House amendment sets forth a second set of conditions under which the Administrator shall require testing. If the Administrator finds (1)(A) that a chemical substance or mixture is or will be produced in substantial quantities; and (B)(i) either that it enters or may reasonably be anticipated to enter the environment in substantial quantities or (ii) there is or may be significant or substantial human exposure to the substance or mixture, (2) there is an insufficiency of data, and (3) testing is necessary to develop the data, the Administrator shall, by rule, require testing.

The House amendment requires the Administrator to consult with the Director of the National Institute for Occupational Safety and Health before prescribing epidemiologic studies under the testing requirement rule. The House bill requires the Administrator to make and publish findings under subsection (a)(1)(A) or (a)(1)(B) before the issuance of a rule ordering persons to conduct tests. The House amendment also provides for the expiration of a testing requirement rule at the end of the reimbursement period.

The House amendment provides for exceptions from the testing rule in order to avoid submission of duplicative data. If an exemption is granted, reimbursement requirements similar to those in the Senate bill apply, except that the reimbursement period may last as long as five years, instead of the two-year period in the Senate bill. In promulgating rules to use in determining fair and equitable reimbursement, the House amendment does not require the Administrator to consult with the Attorney General and the Federal Trade Commission.

With respect to the interagency committee's priority list submitted to the Administrator, the House amendment does not require the Administrator either to initiate a rulemaking proceeding or to publish in the Federal Register the Administrator's reasons for not initiating such a proceeding. The House amendment does not include the conflict of interest provisions found in the Senate bill relating to members of the interagency advisory committee.

Conference substitute (section 4)

The conference substitute is similar to the House amendment with respect to the findings which the Administrator must make in order to require a manufacturer or processor to test a chemical substance or mixture, except that the term "presents" is used in lieu of "cause or significantly contribute to". The conference substitute includes this term throughout the bill when speaking of a risk.

In using the term, the conferees intend that the Administrator be able to address substances and mixtures which indirectly present unreasonable risks, as well as those which directly present such risks. Further, the conferees do not intend that a substance or mixture must be the single factor which results in the presentation of the risk.

Oftentimes an unreasonable risk will be presented because of the interrelationship or cumulative impact of a number of different substances or mixtures. The conferees intend that the Administrator have authority to protect health and the environment in such situations.

In following the House language, the conference substitute requires testing not only (1) in situations in which a substance or mixture may present an unreasonable risk, but also (2) in situations in which there may be substantial environmental or significant or substantial human exposure to a substance or mixture about which there is inadequate information to predict effects on health or the environment.

In the first situation, the conferees intend to focus the Administrator's attention on those chemical substances and mixtures about which there is a basis for concern, but about which there is inadequate information to reasonably predict or determine their effects on health or the environment. The Administrator need not show that the substance or mixture does or will present a risk.

The second situation reflects the conferees' recognition that there are certain situations in which testing should be conducted even though there is an absence of information indicating that the substance or mixture *per se* may be hazardous.

The conference substitute follows the House amendment with respect to the contents of the testing rule. The Senate provision concerning which manufacturers and processors are required to conduct the testing and submit test data is included. Like both the Senate bill and the House amendment, the conference substitute permits the Administrator to grant exemptions from a testing rule. To grant an exemption, the Administrator must determine whether the chemical substance or mixture is equivalent to a chemical substance or mixture for which test data is already being developed. In making this determination the conferees expect the Administrator to look at any contaminants in the chemical substance or mixture for which an exemption is being sought and ascertain whether any contaminants present might cause differences in test data which would be significant and which would, therefore, cause the Administrator to determine that the chemical substances or mixtures in that instance were not equivalent. It also follows the House amendment concerning reimbursement, except that the Administrator must consult with the Attorney General and the Federal Trade Commission in issuing rules which establish the general criteria for determining reimbursement.

The conference substitute retains a modified version of the Senate provision on the interagency committee which is established to make recommendations concerning testing priorities to the Administrator.

The Administrator shall provide administrative services to support such activities. These services shall encompass such things as clerical staff assistance and supplies. The conferees recognize the importance of the interagency committee recommendations and expect the interagency committee to deliberate with care; therefore, the conferees intend that members of the interagency committee shall be given adequate support services by EPA. They also shall be relieved of responsibilities within the entity they represent to the extent necessary to carry out their duties to the committee. Each entity represented shall provide its member with professional and research services. Here, as in all places in the bill where specific officers of the Federal Government are referred to by title, the conferees intend that such references be construed to mean successors to such offices as affected by any reorganization plan or the like.

The interagency committee may designate a maximum of 50 substances or mixtures with respect to which the Administrator should initiate a testing rulemaking proceeding within a year. No more than 50 substances or mixtures may be so designated at any one time. If the Administrator does not take such action within a year, the Administrator must publish in the Federal Register an explanation as to why such action has not been taken.

The conferees have given discretion to the interagency committee as to how many substances or mixtures should be designated. Although the committee may designate up to 50 substances or mixtures at any one time, the conferees wish to stress that the committee need not designate the maximum number. While it is intended that the recommendations of the interagency committee be given great weight by the Administrator, it should be emphasized that the decision to require testing rests with the Administrator.

The conferees do not intend that, in complying with the requirements of the statute, the Administrator divert from the regulatory activities of the Agency an inordinate amount of resources to justify the failure to require testing.

If the Administrator receives information which indicates to the Administrator that there may be a reasonable basis to conclude that a substance or mixture presents or will present a significant risk of serious or widespread harm to human beings from cancer, gene mutations, or birth defects, the Administrator shall initiate appropriate action under section 5, 6, or 7 to protect against the risk or publish in the Federal Register a finding that the risk is not unreasonable. Such action must be taken within 180 days of the receipt of the data, except that the Administrator may extend that period for an additional 90 days for good cause. This requirement does not take effect until two years after the date of enactment.

The conference substitute adopts a provision contained in the House amendment which enables any person who intends to manufacture or process a chemical substance which is not subject to a rule under section 4(a) to petition the Administrator to prescribe standards for the development of test data for such substance. The Administrator must grant or deny the petition within sixty days. If the petition is granted, the Administrator shall prescribe such standards within seventy-five days of the date on which the petition is granted. Any denial of such a petition must be published in the Federal Register.

MANUFACTURING AND PROCESSING NOTICES

Senate bill (section 5)

The Senate bill requires any manufacturer of a new chemical substance to notify the Administrator at least ninety days prior to the commencement of commercial manufacture of the new substance. The notice is to include the common and the trade name of the substance, its chemical identity and molecular structure, categories or proposed categories of use, reasonable estimates of the amount to be manufactured or processed, a description of the by-products, estimates of the number of people who will be exposed to the substance in their places of employment, and existing data concerning environmental and health effects of the substance.

In addition, if a testing rule applicable to the substance is in effect prior to submission of the notice, the manufacturer is required to submit along with the notification, the test data required to be developed by the testing rule.

The ninety day notification period may be extended by the Administrator for an additional ninety days for good cause shown. Notice of such extension must be published in the Federal Register.

Similar notification is required of any person intending to manufacture or process a chemical substance for a distribution in commerce, use, or disposal that has been identified by the Administrator, by rule, as a significant new use, distribution, or disposal. The Administrator is, by rule, to establish criteria defining a significant new distribution in commerce, use, or disposal. In establishing such criteria, the Administrator is to take into account the projected volume of production and category or categories of uses, increases in magnitude and duration of human or environmental exposure, routes of exposure, and the human health and environmental effects.

The Senate bill authorizes the Administrator to issue immediately effective administrative orders during the notification period to halt or limit the manufacture, processing, distribution in commerce, use or disposal of new chemical substances subject to the notification requirements in two situations.

First, the Administrator is required to issue such an order if the Administrator finds that a testing requirement under section 4(a) should be established or modified. The order is to remain in effect until an expedited rulemaking proceeding under section 4(a) to require testing can be completed, the testing performed and the test data submitted. Second, if the Administrator finds that a new substance presents or is likely to present an unreasonable risk of injury to health or the environment, the Administrator is also required to issue an immediately effective order. If such an order is issued, the Administrator must conduct an expedited rulemaking proceeding in accordance with the provisions of section 6(c) (2) and (3).

If the Administrator does not take action to prohibit or limit the manufacture, processing, distribution in commerce, use or disposal of a new substance during the notification period, the Administrator is required to publish a statement of reasons in the Federal Register for not taking such action. The statement must be published prior to the expiration of the notification period. Manufacture or processing of the new substance may commence following publication of the Administrator's statement. Failure to take action against a substance is an action subject to judicial review.

The Senate bill provides certain exemptions from the requirement to submit manufacturing and processing notice. The Administrator is authorized to grant such an exemption for test marketing or other specially limited purposes. In addition, the Administrator may exempt chemical substances which are intermediate reaction products formed during the manufacture of another chemical substance and to which there is no human or environmental exposure. In addition, the notification provisions of the Senate bill do not apply to any chemical substance manufactured in small quantities solely for scientific experimentation or analysis or for chemical research or analysis, including such research or analysis for the development of a product. However, the Administrator may, by rule, require notification prior to the manufacture or processing of such a substance upon a finding that the substance may cause or contribute to an unreasonable risk of injury to health or the environment. Although the section, by its terms, does not apply to mixtures, the Administrator is authorized to specify any mixture which shall be subject to the provisions of the section.

House amendment (section 5)

Like the Senate bill, the House amendment requires manufacturers of new chemical substance to notify the Administrator ninety days prior to the commencement of commercial manufacture of such new substance. The notice is to include information similar to that required by the Senate bill, including test data required to be developed by any applicable testing rule which has been promulgated under section 4 prior to the submission of the notice.

In addition, the House amendment requires the Administrator to compile and maintain a list of chemical substances which cause or significantly contribute to or may cause or significantly contribute to an unreasonable risk to health or the environment. If a person intends to manufacture a new chemical substance included on this list and if no testing rule applicable to the substance has been issued under section 4, the person must submit to the Administrator information which the person believes indicates that the chemical substance will not cause or significantly contribute to an unreasonable risk. Such information must be submitted along with the notice.

The House amendment also requires manufacturers or processors of an existing chemical substance for a new use which has been designated by the Administrator, by rule, as a significant new use, to provide similar notice ninety days prior to such manufacture or processing. The House amendment does not require notification prior to the manufacture or processing of a chemical substance for a significant new distribution in commerce or disposal.

In instances in which there is inadequate information to evaluate the effects of a new substance or of an existing substance for a significant new use, the Administrator is authorized to seek a court injunction to halt manufacture, processing or distribution in commerce. The Federal district courts are empowered to grant injunctions if the court finds that (1) there is inadequate information to reasonably evaluate the health and environmental effects of the new substance and (2) in the absence of such information, the substance may cause or significantly contribute to an unreasonable risk. If an injunction is granted, the Administrator shall conduct an expedited rulemaking proceeding to determine if a lesser restriction (rather than a total halt of manufacture, processing or distribution) would be adequate to protect health or the environment until adequate test data is developed and evaluated.

The House amendment does not require the Administrator to publish a statement of reasons for not taking action during the notification period to prohibit or limit the manufacture, processing, distribution, use or disposal of a new substance or of an existing substance manufactured or processed for a significant new use.

The House amendment also provides for exemptions from the notification requirements. The Administrator is authorized to provide an exemption for the manufacture and processing of a substance for test marketing purposes. The House bill specifically exempts from the notification requirements those chemical substances manufactured or processed in small quantities for scientific experimentation or analysis or for chemical research or analysis on such substance or another substance, including research and analysis for the development of a substance or another substance into a commercial product. However, all persons engaged in such experimentation, research or analysis for a manufacturer or processor must be notified of any risk to health which the manufacturer or processor has reason to believe may be associated with the substance.

The House amendment authorizes the Administrator, by rule, to exempt a manufacturer or processor of any new chemical substance from all or part of the requirements of this section if the Administrator determines that such chemical substance will not cause or significantly contribute to an unreasonable risk to health or the environment. The House amendment also contains an exemption clarifying that a chemical substance which, except for its inert ingredients, is identical to a chemical substance contained on the section 8(b) inventory will not be treated as a new chemical substance.

The House amendment does not authorize the Administrator to specify that a manufacturer of a mixture shall be subject to the notification requirements.

Conference substitute (section 5)

In general.—Section 5 sets out the notification requirements with which manufacturers of new chemical substances and manufacturers and processors of existing substances for significant new uses must comply. The requirements are intended to provide the Administrator with an opportunity to review and evaluate information with respect to the substance to determine if manufacture, processing, distribution in commerce, use or disposal should be limited, delayed or prohibited because data is insufficient to evaluate the health and environmental effects or because the substance or the new use presents or will present an unreasonable risk of injury to health or the environment.

The provisions of the section reflect the conferees recognition that the most desirable time to determine the health and environmental effects of a substance, and to take action to protect against any potential adverse effects, occurs before commercial production begins. Not only is human and environmental harm avoided or alleviated, but the cost of any regulatory action in terms of loss of jobs and capital investment is minimized. For these reasons the conferees have given the Administrator broad authority to act during the notification period.

Any person who intends to manufacture a new chemical substance or manufacture or process a chemical substance for a use which the Administrator, by rule, has determined is a significant new use, must give the Administrator at least 90 days notice before beginning such manufacture or processing, The 90-day period shall begin upon receipt of the notice by the Administrator or the Administrator's duly designated representative.

The conferees have not included the Senate provision which requires notification of significant new distributions or disposals. However, the conference substitute requires that the Administrator consider the reasonably anticipated manner and method of manufacturing, processing, distribution in commerce and disposal of a substance in determining when a use will be considered a significant new use. Thus, the conferees intend that any potential threats to health or the environment from the manufacture, processing, distribution in commerce, or disposal of a substance associated with a new use be considered by the Administrator when determining the significance of a new use. In addition, the Administrator shall consider the projected volume of manufacturing and processing of the substance for a use, the extent to which a use changes the type or form of exposure of human beings or the environment to a substance, and the extent to which such use increases the magnitude and duration of human or environmental exposure to a substance. Thus, a significant increase in the projected volume of manufacture or processing for a substance, a significant change in the type or form of human or environmental exposure, or a significant increase in the magnitude or duration of human or environmental exposure could be the basis for determining that a use is a significant new use.

Submission of test data.—Subsection (b) describes the instances in which a person subject to a notification requirement with respect to a chemical substance under subsection (a) must submit test data to the Administrator before manufacture of the substance or manufacture or processing of the substance for a significant new use can begin. If a rule under section 4 respecting a substance has been promulgated before submission of the notice required by subsection (a), then a person who is required by the section 4 rule to submit test data for the substance must submit such test data at the time the notice is submitted in accordance with subsection (a). This assures that the Administrator will have at least 90 days to evaluate the test data before the manufacture or processing begins. If a person has been granted an exemption from a testing rule under section 4 applicable to a new substance or to a significant new use of an existing substance, such person shall not begin manufacture or processing until 90 days after the date of submission of the test data on which the exemption is based.

It should be noted that if a testing rule under section 4 respecting a substance has not been promulgated prior to the submission of a notice required by section 5, the Administrator may promulgate a testing rule under section 4 for such substance without taking separate action under this section. However, such a rule would not delay the manufacture or processing of the substance.

The conferees adopted a provision from the House bill to insure that information respecting the health and environmental effects of any chemical substance which the Administrator has identified as a suspect chemical substance is submitted at the time of notification. Under the conference substitute the Administrator may, by rule, compile a list of chemical substances the manufacture, processing, distribution in commerce, use or disposal of which presents or may present an unreasonable risk of injury to health or environment. If a testing rule under section 4 has not been promulgated with respect to such substance before the submission of the notice, then the person submitting the notice must submit to the Administrator data which the person believes shows the manufacture, processing, distribution in commerce, use and disposal of the substance or any combination of such activities will not present an unreasonable risk to health or the environment.

Extension of notice period.—The Administrator, for good cause, may extend the 90 day notification period for additional periods not to exceed in the aggregate 90 days. Notice of any extension together with the reasons for it shall be published in the Federal Register and shall constitute final agency action subject to judical review.

The conferees intend that the Administrator have a large degree of flexibility in extending the notification period, so that manufacture or processing may begin as soon as the Administrator has sufficient information to evaluate the substance. For example, if the Administrator expects that sufficient data will be available 30 days after the original notification period will expire, then the conferees expect that the Administrator will settle on an extension period which will reasonably accommodate production of that data and time for administrative consideration. If production of the data is delayed, of course the Administrator may extend the original extension period. However, in no case may the extensions exceed a total of 90 days. Every time that the notification period is extended, the Administrator must publish notice of the extension in the Federal Register along with reasons therefor.

Content of notice; publications in the Federal Register.—The conference substitute requires the notice required under subsection (a) to include certain information described in section 8(a) (Reporting and Retention of Information) whether or not the Administrator has required its submission under that section; any test data in the possession or control of the person giving the notice which is related to the effects on health or the environment of any manufacture, processing, distribution in commerce, use or disposal of the substance; and a description of any other data concerning the health and environmental effects of the substance, insofar as known to the person making the report or insofar as reasonably ascertainable. The notice shall be made available, subject to section 14, for examination by interested persons.

In order that the public receive timely notification of any new chemical substance or any significant new use of an existing chemical substance, the conference substitute includes a provision which requires the Administrator to publish in the Federal Register a notice which identifies the chemical substance, lists the uses or intended uses of the substance, and describes the nature of tests performed on such substance and any data developed pursuant to subsection (b) or a rule under section 4. Such publication must occur within 5 days after the Administrator receives notice from the person who intends to manufacture or process.

The conference substitute also requires the Administrator to publish monthly a list of each chemical substance for which notice has been received under subsection (a) and for which the notification period has not expired. The Administrator must also include on the list those substances for which the notification period has expired since the last monthly publication.

Regulation pending development of information.—Subsection (e) sets out the procedures under which the Administrator can halt or limit the manufacture, processing, distribution in commerce, use, or disposal of a new substance or an existing substance for a significant new use when there is insufficient information to evaluate the health and environmental effects of the substance.

Action to prohibit or limit manufacture, processing, distribution in commerce, use, or disposal is required in instances in which the Administrator determines that:

(1) Information available to the Administrator is insufficient to permit a reasoned evaluation of the health and environmental effects of the substance, and

(2)(a). In the absence of information sufficient to permit the Administrator to make such an evaluation, the manufacture, processing, distribution in commerce, use, or disposal of the substance may present an unreasonable risk of injury to health or the environment, or

(b). The substance is or will be produced in substantial quantities, and (i) enters or may reasonably be anticipated to enter the environment in substantial quantities or (ii) there is or may be significant or substantial human exposure to the substance.

If the Administrator makes the above determination at least 45 days before the expiration of the notification period, then the Administrator may issue a proposed order to prohibit or limit the manufacture, processing, distribution in commerce, use or disposal of the substance. A limitation on manufacture or processing could, of course, include a labeling requirement. The proposed order will take effect upon the expiration of the notification period unless the manufacturer or processor subject to the order files objections with the Administrator, specifying with particularity the provision of the order deemed objectionable and stating the grounds for the objection. To prevent the order from becoming effective, the objections must be filed within 30 days after the manufacturer or processor has received in writing from the Administrator a notice of the proposed order. The conferees wish to stress that the Administrator must provide actual notice in writing to the manufacturer or processor who will be subject to the order. Notice is not to be published in the Federal Register, but is, of course, available to the public if it is not prohibited from disclosure under section 14.

This provision thus represents a melding of the Senate bill and the House amendment. In order to insure that timely action may be taken by the Administrator, the conference substitute authorizes the Administrator to issue an administrative order to take effect immediately upon the expiration of the notification period. However, to protect against unilateral action by the administrator without an adequate basis for action, the conference substitute borrows the procedure from section 701(e) of the Federal Food, Drug, and Cosmetic Act which permits the filing of objections by manufacturers and processors specifying with particularity the provisions of the order deemed objectionable and stating the grounds for the objections.

If such objections are filed, then the Administrator is instructed to seek an injunction in a Federal district court to prohibit or limit the manufacture, processing, distribution in commerce, use or disposal of the substance. Of course, if the objections filed with the Administrator indicate to the Administrator that the injunction is not necessary, then the Administrator is not required to seek the injunction.

If the court finds in such injunction action that (1) information available to the Administrator is insufficient to permit a reasoned evaluation of the health and environmental effects of the substance, and (2) (A) in the absence of such information, the manufacture, processing, distribution in commerce, use, or disposal of the substance may present an unreasonable risk of injury to health or the environment or (B) such substance is or will be produced in substantial quantities and it enters or may reasonably be anticipated to enter the environment in substantial quantities or there is or may be significant or substantial human exposure to the substance, then the court shall grant an injunction. The conferees intend that this two-part standard totally supplant the traditional elements which a party must ordinarily show before a court will exercise its equitable jurisdiction to grant an injunction. The conferees do not intend that the Administrator be required to make any showing other than that which is required for the court to make the two findings described above. Application of any other standard by the court would frustrate the purposes of this section that suspect chemicals be adequately tested to determine their health and environmental effects before commercial manufacture or processing begins.

The conference substitute authorizes such courts to grant a temporary restraining order or a preliminary injunction to prohibit manufacture, processing or distribution of a new substance or of an existing substance for a significant new use if the court finds that the notification period may expire before the action for an injunction can be completed. The conferees recognize that a manufacturer or processor, merely by beginning to manufacture, process, or distribute a new chemical or an existing chemical for a significant new use, could defeat the objective of section 5 to totally prevent environmental or human exposure to suspect new chemical substances or significant new uses of existing chemical substances until adequate testing can be performed and the data evaluated. Therefore, the conferees intend that the court freely exercise the authority to grant preliminary relief as is necessary to preserve the status quo in order to insure that the policy of this section can be fulfilled.

After submission of adequate test data to the Administrator and evaluation of such data, the district courts may, upon petition, dissolve the injunction unless the Administrator has initiated a proceeding under section 6(a) with respect to the substance. In such a situation, the injunction shall remain in effect until the effective date of a rule under section 6(a) or until the section 6 proceeding is terminated, whichever occurs first.

Protection against unreasonable risk.—Section 5(f) of the conference substitute requires the Administrator to take immediate action to prohibit or limit human or environmental exposure to a new chemical substance or to an existing chemical substance for a significant new use in certain situations. In section 5(f) the conference substitute authorizes the Administrator to issue a proposed rule under section 6(a), but such rule is to be effective upon its publication in the Federal Register. Such action is authorized in instances in which there is a reasonable basis to conclude that the manufacture, processing, distribution in commerce, use, or disposal of the substance presents or will present an unreasonable risk of injury to health or the environment before a rule promulgated under section 6(a) could protect against the risk. The conferees recognize, of course, that there is authority in section 6(d) under which the Administrator may make a proposed section 6(a) rule immediately effective. However, to invoke the section 6(d) authority the Administrator must find an imminent, unreasonable risk of serious or widespread injury. With respect to new chemical substances or substances for significant new uses, immediate action is authorized under section 5(f) when there is an imminent, unreasonable risk of injury, regardless of whether the injury will be serious or widespread.

The section 6(a) rule proposed and made immediately effective pursuant to the authority of this section may (A) limit the amount of a substance which may be manufactured, processed, or distributed in commerce, (B) contain any of the requirements described in paragraph (2), (3), (4), (5), (6), or (7) of subsection 6(a), or (C) contain any combination of the requirements described in clauses (A) and (B). Immediately following the publication in the Federal Register of a section 6(a) rule as authorized by this section, the Administrator must conduct an expedited rulemaking proceeding in accordance with the provisions of section 6(d)(2).

A rule under section 6(a) authorized by this section may not totally prohibit the manufacture, processing, or distribution in commerce of a new substance or an existing substance for a significant new use. In order to totally prohibit the manufacture, processing, or distribution in commerce of a new substance, or of an existing substance for a significant new use, the Administrator must either issue a proposed order which shall be subject to all the procedures applicable to the situation when there is an insufficiency of information, as described above, or obtain a court injunction. If the court finds that there is a reasonable basis to conclude that the manufacture, processing, distribution in commerce, use or disposal of a new substance or of an existing substance for a significant new use presents or will present an unreasonable risk of injury to health or the environment, the court shall issue an injunction. Again, the conferees intend that the court will not use the normal equity standard to determine if an injunction should be issued. Instead, the standard set out in section 5(f)(3)(B) of the conference substitute is intended to totally replace the normal injunction standard.

The conferees recognize that there will be instances in which there are a limited number of practical uses for a chemical substance and that by issuing an immediately effective proposed rule prohibiting those uses, the Administrator could effectively prohibit manufacture or processing altogether. The conferees view such a prohibition as a total prohibition of manufacture or processing and intend that the Administrator comply with the procedures of section 5 (f)(3) in order to obtain a total prohibition on manufacture or processing. The authority to issue an immediately effective rule to prohibit manufacture or processing for a use should be utilized only when there is more than one practical use of a substance and when the prohibition does not effectively ban all such uses. Likewise, the conferees do not intend that the Administrator utilize the authority to issue an immediately effective proposed rule so severely limiting the amount of a substance which may be manufactured, processed, or distributed in commerce as to effectively prohibit manufacture, processing, or distribution.

Statement of reasons for not taking action.—If, within the notification period, the Administrator has not initiated action under this section or section 6 or 7 to prohibit or limit the manufacture, processing, distribution in commerce, use or disposal of certain new chemical substances or of existing chemical substances for significant new uses, then subsection (g) requires the Administrator to issue a statement of reasons in the Federal Register for not initiating such action. The statement must be published prior to the expiration of the notification period. The chemical substances for which such a statement is required are those for which the Administrator, because of prior administrative action with respect to such chemical substances, has indicated there may be particular cause for concern. Specifically, a statement of reasons for not initiating action is required if a testing rule under section 4 applies to a new substance or an existing substance for a significant new use, or if a substance is listed under section 5(b)(4). In addition, if notification is required because a use constitutes a significant new use and, if no action is initiated during the notification period, the Administrator must issue a statement of reasons for not initiating such action.

Publication of the statement of reasons in accordance with this subsection is not a prerequisite to the manufacture or processing of the substance with respect to which the statement is to be published. Thus, the Administrator, merely by not issuing the statement of reasons, cannot delay the beginning of manufacture or processing of a new substance or a substance for a significant new use. Nonetheless, the Administrator must perform this non-discretionary duty and will subject himself not only to criticism by the Congress for not doing so, but may also subject himself to suit under section 20 of this Act or other provisions of law relating to the required performance of non-discretionary duties.

Exemptions.—Subsection (h) describes the situations in which a chemical substance may be manufactured or processed without regard to the notice and test data submission requirements of subsections (a) and (b). Paragraph (1) provides the authority for granting an exemption for test marketing purposes. Under paragraph (2) an exemption from the test data submission requirements of subsection (b) may be obtained if submission of data for the substance to be exempted would be duplicative of data already submitted to the Administrator. Paragraph (3) adopts the language in the House amendment specifically exempting from the notification requirements those chemical substances manufactured or processed or proposed to be manufactured or processed in small quantities (as defined by the

Administrator by rule) for scientific experimentation or analysis or for chemical research or analysis, including research and analysis for the development of the substance or another chemical substance into a commercial product. All persons engaged in such experimentation, research, or analysis for a manufacturer or processor must be notified or any risk to health which the manufacturer or processor or Administrator has reason to believe may be associated with the substance.

Under paragraph (4), the Administrator may, upon application and by rule, exempt the manufacturer of a new chemical substance from all or part of the requirements of this section if the Administrator determines that the manufacture, processing, distribution in commerce, use or disposal of the substance will not present an unreasonable risk of injury to health or the environment. A rule granting such an exemption must be promulgated in accordance with paragraphs (2), (3), and (4) of section 6(c).

Paragraph (5) authorizes the Administrator to make the requirements of subsection (a) and (b) inapplicable with respect to the manufacture or processing of a chemical substance which may temporarily exist as a result of a chemical reaction in the manufacture or processing of a mixture or another chemical substance and to which there is not or will not be any human or environmental exposure.

The conference substitute deletes the provision in the House amendment which clarified that a chemical substance is not to be treated as a new chemical substance solely because of the change in proportions of inert ingredients. This provision of the House amendment was deleted because under the definition of "mixture" in section 3 of the conference substitute the same result would occur, as any change in inert ingredients would constitute a new mixture not a new chemical substance. Mixtures are not covered by section 5.

Definition.—The terms "manufacture" and "process" as used in this section mean to manufacture or to process for commercial purposes. Since the term "manufacture" is defined to include import, persons who intend to import substances for commercial purposes will be treated in the same manner as domestic manufacturers under section 5.

REGULATION OF CHEMICAL SUBSTANCES AND MIXTURES

Senate bill (section 6)

The Senate bill requires the Administrator to impose restrictions on a chemical substance or mixture if the Administrator finds that the manufacture, processing, distribution in commerce, use, or disposal presents or is likely to present an unreasonable risk of injury to health or the environment. The Administrator shall impose one or more of several specified requirements as is necessary to adequately protect against the risk, using the least burdensome of effective controls.

A range of requirements is provided, from complete prohibitions on the manufacturing, processing, or distribution in commerce to labeling requirements. Among these is the authority to regulate the manner or method of use or disposal of such substance or mixture and the authority to require manufacturers or processors to replace or repurchase substances or mixtures found to present unreasonable risks.

The Senate bill also contains the authority to limit the amount of a substance or mixture which may be manufactured, processed, or distributed in commerce, or which may be manufactured, processed, or distributed in commerce for a particular use. A procedure for assigning permissible quotas if the applicable parties are unable to agree is provided. Supervision by the Attorney General and the Federal Trade Commission is provided for any voluntary efforts to establish quotas.

The Senate bill authorizes the Administrator to order manufacturers or processors to submit descriptions of relevant quality control procedures if the Administrator has good cause to believe that the manufacture or processing causes the adulteration of a chemical substance or mixture. If the Administrator determines that the quality control procedures of the manufacturer or processor are inadequate, the Administrator may order revisions in the quality control procedures to the extent necessary to remedy the inadequacy.

The Senate bill also requires the Administrator to consider relevant factors in imposing restrictions and to make findings with respect to certain factors.

The Senate bill contains a specific rulemaking procedure for rules imposing restrictions under this section. The procedure is an informal one, similar to the procedures in section 553 of title 5, United States Code, but there are exceptions, including an opportunity for an informal hearing. An opportunity for appropriate cross-examination in the hearing is provided under the supervision of the Administrator. Participants in a rulemaking proceeding may be compensated by the Administrator under specified criteria.

The Administrator may specify the date on which a rule under this section becomes effective, which shall be as soon as administratively feasible.

The Administrator may waive the required notice and comment period in those situations where compliance with the rulemaking provisions would present an unreasonable risk of death, serious or substantial personal injury, or serious or substantial environmental harm.

Finally, the Senate bill provides for the control of polychlorinated biphenyls (PCBs). Effective one year after the date of enactment of the Act, PCBs may not be used in any manner other than a totally enclosed manner, except that the Administrator may, by rule, authorize exceptions if the Administrator finds that no unreasonable risk of injury to health or the environment is presented. Effective two years after the date of enactment, the manufacture of any polychlorinated biphenyl would become unlawful. Effective two and one-half years after such date, the processing or distribution in commerce of PCB's would become unlawful, except that the Administrator may make exceptions if no unreasonable risk of injury to health or the environment is presented. Disposal regulations concerning PCB's shall be promulgated within six months after the date of enactment.

House amendment (section 6)

The standard for taking action against unreasonable risks under this section in the House amendment is slightly different from that contained in the Senate bill. If the Administrator finds that there is a reasonable basis on which to conclude that the manufacture, processing, distribution in commerce, use, or disposal of a chemical substance or mixture, or any combination of such actions causes or significantly contributes to, or will cause or significantly contribute to, an unreasonable risk to health or the environment, the Administrator shall impose requirements as necessary to protect against the risk. The requirements generally are similar to the requirements of the Senate bill, with several exceptions. The Administrator may not regulate the manner or method of use. Nor may the Administrator impose replacement or repurchase requirements for substances regulated. Requirements regulating the manner or method of disposal may not require any person to take an action in violation of a State or local law. Also, a person subject to a disposal requirement shall notify the State in which a required disposal may occur of such requirement. The House amendment does not contain authority to impose manufacturing or processing quotas.

The House amendment provides a procedure for protecting against unintentional contamination of a chemical substance due to the manner of manufacturing or processing. Quality control procedures may be required to be submitted. If found inadequate, the Administrator may order such procedures to be changed. In addition, if the quality control procedures have resulted in the distribution in commerce of substances or mixtures which cause or significantly contribute to an unreasonable risk to health or the environment, the Administrator may require manufacturers or processors to give notice and to replace or repurchase any such substance or mixture as is necessary to protect health or the environment.

In promulgating any rule under section 6(a), the Administrator shall consider all relevant factors and make findings with respect to certain factors.

The Administrator shall not promulgate a rule under this section if the risk could be eliminated or reduced to a sufficient extent under another federal law administered by the Administrator unless the Administrator makes a finding that it is in the public interest to do so, taking into consideration a number of enumerated factors.

The rulemaking procedures of the House amendment are generally similar to those contained in the Senate bill.

The Administrator may make a rule immediately effective if an unreasonable risk of serious or widespread harm to health or the environment will occur prior to the completion of the rulemaking proceedings, and making the rule so effective is necessary to protect the public interest. If a proposed rule totally prohibits the manufacture, processing or distribution of a chemical substance or mixture, a court must have previously taken action against a substance or mixture in an imminent hazard proceeding under section 7. An expedited rulemaking procedure is provided for immediately effective rules.

The provision of the House amendment relating to PCBs is similar to the Senate bill with a few exceptions. For example, the prohibitions effective in one year apply only to the manufacture, processing or distribution in commerce for a use other than a use in a totally enclosed manner. In addition, exemptions from the prohibitions relating to the manufacture, processing, and distribution of PCB's may be granted, if the Administrator determines that the exemption is necessary to protect health or the environment, and good faith efforts have been made to develop substitutes.

Conference substitute (section 6)

The conference substitute requires the Administrator to take action under this section against chemical substances or mixtures for which there is a reasonable basis to conclude that the manufacture, processing, distribution in commerce, use, or disposal of such chemical substances or mixtures, or any combination of such activities, presents or will present an unreasonable risk of injury to health or the environment. Requirements shall be imposed to the extent necessary to protect against the risk.

The requirements must be the least burdensome feasible for those subject to the requirement and for society while providing for an adequate margin of protection against the unreasonable risk.

The requirements which may be imposed are similar to those included in both the Senate bill and the House amendment. The Administrator may impose requirements regulating the manner or method of the commercial use of a substance a mixture and also requirements regulating the manner or method of disposal of a substance or mixture or an article containing a substance or mixture by the manufacturer or processor or by any other person who uses or disposes of it for commercial purposes. The provision of the House amendmnt that a disposal requirement may not require any person to take action in violation of any State law or political subdivision is included. The conference substitute also includes the Senate provision which authorizes the Administrator to require replacement or repurchase by manufacturers or processors of substances or mixtures with respect to which action has been taken under this section.

The provisions of the House bill relating to quality control are included.

The provisions of the Senate bill which authorize the Administrator to assign manufacturing or processing quotas are not included.

The conferees appreciate that if the Administrator chooses to impose a production limitation on any chemical substance, such limitation, if not carefully drawn, could produce monopoly profits. The conferees believe that the Administrator should consult with the Attorney General and the Federal Trade Commission in order to avoid any anticompetitive consequences.

The conference substitute requires the Administrator to consider certain enumerated factors and to publish a statement in the Federal Register with respect to them at the time of promulgation of a rule under section 6(a). Specifically, the Administrator must consider and publish a statement concerning the effects of the substance or mixture on health and the magnitude of human exposure to such substance or mixture; the effects of the substance or mixture on the environment and the magnitude of environmental exposure to such substance or mixture; the benefits of such substance or mixture for various uses and the availability of substitutes for such uses; and the reasonably ascertainable economic consequences of the rule, after consideration of the effects on the national economy, small business, innovation, the environment, and the public health. This requirement was contained in both the Senate bill and the House amendment. The purpose in requiring such a statement is to assure that the basis for the Administrator's rule are publicly enumerated. By requiring the statements, the conferees intend to emphasize key considerations which must be addressed. The conferees do not intend that the statement be detailed or voluminous. A succinct and precise statement of these key considerations will suffice. Of course, the statements will provide part of the rulemaking record for judicial review of a rule promulgated under section 6(a).

Moreover, if the Administrator determines that a risk may be eliminated or reduced to a sufficient extent by actions taken under another Federal law administered by the Administrator, action may not be taken under this section unless the Administrator finds, in the Administrator's discretion, that it is in the public interest to take action under this Act. By committing such determination to the Administrator's discretion, the conferees intend that such determination not be subject to judicial review.

The House provision relating to compensation for the costs of participating in a rulemaking proceeding and the provisions relating to the effective date of a rule are included. Generally rules are to be effective as soon as procedurally and administratively feasible. However, proposed rules may be declared to be immediately effective by the Administrator in certain instances. The rule may be declared immediately effective if the Administrator determines that the manufacture, processing, distribution in commerce, use, or disposal of a substance or mixture is likely to result in an unreasonable risk of serious or widespread injury to health or the environment before the normal effective date and making the rule immediately effective is necessary to protect the public interest. In the case of a rule to prohibit the manufacture, processing, or the distribution, a court must have granted relief in an imminent hazard action under section 7. An expedited rulemaking procedure is provided if a rule is made immediately effective.

The conference substitute includes procedures contained in both bills for prescribing rules. In general, rules under this section are to be prescribed in accordance with the informal rulemaking procedures of section 553 of title 5, United States Code, except that an opportunity for an oral hearing and for limited cross-examination is provided. The procedures are patterned after those contained in section 18 of the Magnuson-Moss Warranty-Federal Trade Commission Improvements Act.

The Senate bill, unlike the House amendment, specifically authorizes the Administrator to conduct cross-examination on behalf of the participants to the proceeding. Although the conference substitute retains the Senate provision, the conferees expect that in most instances the participants themselves would conduct the cross-examination, subject to the Administrator's time limitations and other rules.

When the Act states that the transcripts shall be available to the public, the conferees intend that such availability be construed in a reasonable manner. No person may be denied access to such information, but at the same time the Administrator shall not be required to assume the burden of copying what may be a formidable amount of material. Therefore, the conferees intend that the Administrator furnish copies of transcripts as long as a supply exists within EPA. However, if the amount of material is vast, or if EPA has run out of copies, then the person may inspect the transcript which shall be available at every regional office of EPA. EPA may photocopy a reasonable number of pages, such as 200, but shall in all cases afford any person the opportunity to photocopy as much of the transcript as the person desires. The cost of copying pages beyond a reasonable number shall be borne by the person; however, the Administrator shall not charge an unreasonable fee per page.

Generally, the provisions of the Senate bill relating to the control of polychlorinated biphenyls are included. The standard that must be satisfied before exemptions from the complete ban on polychlorinated biphenyls are granted contains elements of both the House and Senate provisions. Exemptions may be granted only if the Administrator finds that there is no unreasonable risk to health or the environment, and that good faith efforts have been made to develop a substitute. So that existing PCBs may be reused rather than disposed of, the prohibitions do not apply to distributions in commerce of PCBs sold for purposes other than resale before the effective date of the prohibition on distribution of PCBs.

IMMINENT HAZARDS

Senate bill (section 7)

The Senate bill authorizes the Administrator to initiate a judicial proceeding against an imminently hazardous chemical substance or mixture or against any person who manufactures, processes, distributes in commerce, uses or disposes of such substance or mixture or against both. The court is authorized to grant such temporary or permanent relief as is necessary to protect against the hazard. Such relief may include seizure and condemnation of the imminently hazardous substance or mixture. Further, the court is specifically authorized to require manufacturers, processors, or distributors to provide notice of the hazard to purchasers of the substance or mixture and to the public, and to recall and replace or repurchase the substance or mixture. Under the Senate bill an imminent hazard is considered to exist when the evidence is sufficient to show that a situation exists in which the continued use of a substance or mixture would be likely to result in unreasonable adverse effects on the environment or an unreasonable hazard to the survival of an endangered species. An unreasonable adverse effect is defined to mean an unreasonable risk to man or the environment taking into account the economic, social, and environmental costs and benefits of the use of the substance or mixture.

House amendment (section 7)

The House amendment differs from the Senate bill in three ways. First, in addition to authorizing action against imminently hazardous substances and mixtures, the House amendment explicitly authorizes actions against articles containing such substances or mixtures. Second, if the Administrator has not acted under section 6(d) of the House amendment which authorizes immediate administrative action against an imminent hazard, the Administrator is required to take action under section 7. Third, the House amendment differs from the Senate bill in its definition of an imminent hazard. Under the House amendment an imminently hazardous chemical substance or mixture is one which causes or significantly contributes to an imminent and unreasonable risk of serious or widespread harm to health or the environment. Such risk shall be considered imminent if it is shown that the manufacture, processing, distribution in commerce, use or disposal of a substance or mixture is likely to result in an unreasonable risk of serious or widespread harm to health or the environment before a final rule under section 6 can protect against the risk.

Conference substitute (section 7)

The conference substitute follows the House language with a clarification (contained in the Senate bill) that relief is authorized against persons who use or dispose of an imminently hazardous substance or mixture in addition to persons who manufacture, process, or distribute in commerce such substances or mixtures. If the Administrator has not used the authority provided in section 6(d)(2)(A)(i) to make a section 6(a) rule immediately effective in order to protect against an imminently hazardous substance or mixture, the Administrator must bring an action under section 7. The conferees have imposed such a nondiscretionary duty upon the Administrator to insure that protection is provided against imminently hazardous substances, mixtures, and articles containing such substances and mixtures.

The conferees wish to note that while the unreasonable risk of injury must be imminent, the physical manifestations of the injury itself need not be. Rather, an imminent hazard may be found at any point in the chain of events which may ultimately result in injury to health or the environment. The observance of actual injury is not essential to establish that an imminent hazard exists. The conferees intend that action under the imminent hazard section be able to occur early enough to prevent the final injury from materializing. In using the term "widespread injury" the conferees do not intend that the imminent hazard authority with respect to widespread harm be limited to instances in which the risk of injury is geographically widespread. Rather an unreasonable risk of harm affecting a substantial number of people, even though it is within a rather limited geographic area, should be deemed adequate to satisfy the requirement of an unreasonable risk of widespread injury to health. Of course if the risk of injury to health or environment is serious, it need not be widespread.

REPORTING AND RETENTION OF INFORMATION

Senate bill (section 8)

Section 8 sets forth requirements for reporting and retention of information. Under subsection (a) the Administrator shall issue rules which require each person who manufactures or processes a chemical substance or mixture to maintain records and to make such reports as the Administrator may reasonably require. Such rules shall require manufacturers or processors of chemical substances or mixtures who produce such substances or mixtures in small quantities solely for scientific experimentation or analysis for chemical research or analysis to maintain records and to submit reports only to the extent necessary for the effective enforcement of the Act. This subsection also contains an illustrative list of the kinds of information which the Administrator may require from manufacturers or processors of chemical substances.

To determine which substances are new chemical substances for the purpose of the pre-market notification provisions of section 5, subsection (b) requires the Administrator to publish an inventory of existing chemical substances or mixtures which any person report to be commercially manufactured or processed within the United States under subsection (a) or under section 5(a). The Administrator shall publish such list not later than 270 days after the date of enactment.

Subsection (c) requires any person who manufactures, processes, or distributes chemical substances or mixtures to maintain records of adverse reactions to health or the environment alleged to have been caused by any such substance or mixture. These records shall be maintained for 5 years from the date the information was reported to such person, except that reports dealing with adverse reactions of employees shall be retained for 30 years.

Subsection (d) requires the Administrator to promulgate rules with respect to the submission of lists of health and safety studies conducted or initiated by any manufacturer, processor, or distributor in commerce of any chemical substance or mixture. The Administrator may require the submission of any study appearing on the list.

Subsection (e) requires manufacturers, processors, or distributors in commerce of a chemical substance or mixture as well as their liability insurers to inform the Administrator when they receive information which supports the conclusion that such substance or mixture causes or contributes to an unreasonable risk of injury to health or the environment. Such persons are relieved of such requirement when they have reason to believe that the Administrator has been adequately informed of the risk.

House amendment (section 8)

Subsection (a) of the House amendment is substantially similar to the Senate bill except that it exempts small manufacturers or processors from the reporting requirements. The Administrator may, by rule, require such persons to maintain records and submit reports on a chemical substance or mixture subject to a rule or a proposed rule under section 4, 5(c), 5(g), or 6. In addition, if relief has been granted in an imminent hazard proceeding under section 7, the Administrator may, by rule, require a small manufacturer or processor to maintain records and submit reports. After consultation with the Administrator of the Small Business Administration, the Administrator, shall, by rule, prescribe standards for determining which manufacturers and processors will be considered "small" manufacturers and processors.

As a further limitation, section 8(a)(1)(B) specifies that the Administrator may not require the maintenance of records or the submission of reports with respect to changes in the proportions of the components of a mixture, unless the Administrator finds that such recordkeeping or reporting is necessary for the effective enforcement of the Act.

With respect to the inventory of existing chemical substances required by subsection (b), the House amendment provides that the inventory shall include at least each chemical substance which any person reports under section 5 or under section 8(a) was commercially manufactured or processed in the United States within 3 years before the effective date of the rules promulgated under section 8(a). The House amendment requires the publication of such inventory within 1 year after the effective date of the Act.

Subsection (c) differs from the Senate bill in that it allows the Administrator to determine, by rule, the requirements respecting the maintenance of records of adverse reactions to health or the environment alleged to have been caused by a substance or mixture. The Administrator may require that records relating to adverse reactions to employee health be retained for up to 50 years.

Subsection (d) concerning submission of lists of health and safety studies is similar to the Senate bill.

Subsection (e) of the House amendment does not require liability insurers to report to the Administrator information which supports the conclusion that a substance or mixture may cause or significantly contribute to an unreasonable risk of injury. Manufacturers, processors, and distributors must report information relating to a substantial risk to health or the environment unless they have actual knowledge that the Administrator has been adequately informed of such risk.

Subsection (f) of the House amendment provides definitions of "manufacture" and "process" for the purposes of section 8.

Conference substitute (section 8)

The conference substitute follows with some modification the House amendment of section 8 which outlines the policies and procedures for reporting and retention of information. Subsection (a) identifies which persons must, pursuant to rules promulgated by the Administrator, maintain records and make reports. The conference substitute provides an illustrative list of the kinds of activities for which recordkeeping and reporting may be required. The list includes such information as the identity of the chemical, categories of use, amounts manufactured or processed, by products, existing data, employees exposed, and the manner or method of disposal. The information specified may be required by the Administrator "insofar as known to the person making the report or insofar as reasonably ascertainable". The conferees intend that the "reasonably ascertainable" standard be an objective, rather than a subjective one. Thus, the manufacturer or processor must provide information of which a reasonable person similarly situated might be expected to have knowledge.

The conference substitute retains the exemptions in the House amendment relating to reporting by small businesses. The intent of the conferees is to protect small manufacturers and processors from unreasonably burdensome reporting requirements. However, the conferees do not intend to deny the Administrator access to information which may be necessary either to determine whether a rule or order should be promulgated or to enforce a final rule or order. Therefore, the conferees have specifically authorized the Administrator to obtain reports from small manufacturers and processors of a chemical substance or mixture with respect to which a rule has been proposed or promulgated under section 4, 5(b) (4), or 6, or with respect to which an order or rule is in effect under section 5(e) or 5(f). Thus, once a rule has been proposed, the Administrator may, by rule, issued in accordance with the informal rulemaking procedures of section 553 of title 5, United States Code, require reporting from small manufacturers and processors. Under such procedures, the Administrator will be able to obtain timely access to needed information. Similarly, reporting may be obtained from small manufacturers and processors of a substance or mixture with respect to which relief has been granted in a civil action under section 5 or 7.

The conference substitute adopts, with some clarification, the House amendment in subsection (b) which requires the Administrator to compile, keep current, and publish an inventory of chemical substances and mixtures manufactured or processed in the United States. The conference committee compromised on the date that the Administrator shall first publish the inventory, which publication shall take place 315 days after the effective date of the Act.

The conference substitute accepts the substance of the Senate bill in subsection (c), which states that records of significant adverse reactions (as defined by the Administrator by rule) shall be retained for five years after such reactions are reported. Under this provision an officer or employee designated by the Administrator may inspect the records maintained on adverse reactions. The conferees intend that persons under contract with the Administrator be considered employees of the Administrator. Such contractors and their employees may have access to records for purposes of this section and throughout the Act. The conferees recognize the special dangers presented to persons who are exposed to substances on a daily basis; therefore, records of adverse occupational effects must be retained for thirty years.

The seriousness, duration, and the frequency of reactions should be taken into account in establishing what constitutes a significant adverse reaction. For example, if an individual reports that a chemical substance causes his or her eyes to become inflamed and to tear, such reaction may be attributed to an isolated allergic reaction. However, if several persons report a similar reaction, then the reaction may indeed be significant. Because the ultimate significance of adverse reactions is difficult to predict, the conferees intend that the requirement to retain records err on the side of safety. Some very serious neurological disorders, for instance, at first present what appear to be trifling symptons.

The conference substitute includes the Senate version of subsection (d) concerning health and safety studies with slight modifications. As with the provision concerning adverse reactions, the conferees emphasize the importance of gaining information which errs on the side of too much rather than too little. Of course, the Administrator is to avoid imposing unnecessary or overly burdensome reporting requirements. In cases where test results are submitted, supporting data and the sources for such data must be included.

The conference substitute follows the House amendment for subsection (e) which provides that any manufacturer, processor, or distributor of a chemical substance or mixture who obtains information supporting the conclusion that such substance or mixture presents a substantial risk of injury to health or the environment shall notify the Administrator, unless such person has actual knowledge that the Administrator already possesses the information.

Senate bill (section 9)

Section 9(a) of the Senate bill provides that if the Administrator (A) has reason to believe that the manufacture, processing, distribution in commerce, use, or disposal of a chemical substance or mixture causes or contributes to, or is likely to cause or contribute to an unreasonable risk of injury to health or the environment, and (B) determines, in the Administrator's discretion, that such risk may be prevented or reduced to a sufficient extent by action taken under a Federal law not administered by EPA, then the Administrator must request the agency which administers such law to issue an order. Such agency shall consider all data submitted by the Administrator and issue an order declaring whether or not the manufacture, processing, distribution in commerce, use, or disposal of such substance or mixture causes or contributes to or is likely to cause or contribute to such a risk. If such agency makes such determination it shall also determine if such risk may be prevented or reduced to a sufficient extent by action taken under the law (or laws) administered by the agency.

The Administrator may specify the time within which the other agency must issue the order, but such time may not be less than 90 days from the date the request was made. The other agency must issue a report including a detailed statement of its findings and conclusions in response to the Administrator's request.

The Administrator shall not take any action under section 6 or 7 of this Act if such other agency (A) issues an order declaring that there is no unreasonable risk of injury, or (B) initiates action under the law (or laws) administered by such agency within 90 days of publication in the Federal Register of its report in response to the EPA request.

Section 9(a) of the Senate bill also states that nothing in this section shall prevent the Administrator from making any subsequent request or taking subsequent action under the Toxic Substances Control Act with respect to such risks if the requirements of section 9(a) are satisfied.

Section 9(a) of the Senate bill provides that if the Administrator has initiated action under section 6 or 7 of this bill with respect to a risk of injury which is the subject of a request to another agency, such other agency must consult with the Administrator to avoid duplication of Federal action against such risk before taking action under the law or laws it administers.

Section 9(b) of the Senate bill directs the Administrator to coordinate actions taken under this bill with actions taken under other Federal laws administered wholly or partially by the Administrator. The Administrator must use the authorities contained in such other Federal laws to protect against any risk to health or the environment associated with a chemical substance or mixture unless the Administrator, in the Administrator's discretion, determines that such risk might be more appropriately protected against under this Act. Section 9(b) does not relieve the Administrator of any duties or responsibilities imposed by other Federal law. Nor does section 9(b) affect any final action taken under such other Federal law or the extent to which human health or the environment is protected under such other law.

Section 9(c) of the Senate bill states that, in exercising any authority under this bill, the Administrator shall not, for purposes of section 4(b)(1) of the Occupational Safety and Health Act of 1970, be deemed to be exercising statutory authority to prescribe or enforce standards or regulations affecting occupational safety and health.

Section 9(d) of the Senate bill requires the Administrator to consult and coordinate with the Secretary of Health, Education, and Welfare and the heads of other appropriate Federal agencies, departments or instrumentalities for the purpose of achieving the maximum enforcement of this legislation while imposing the least burdens of duplicative requirements on those subject to the bill, and for other purposes. The Administrator shall report annually to the Congress on actions taken to so coordinate authority under this bill with the authority granted under other EPA-administered laws and laws administered by other Federal agencies.

Section 9(e) of the Senate bill provides that nothing in section 9 limits any requirement of section 4, 5 (other than section 5(e)(2)), or 8, or rules promulgated thereunder.

House amendment (section 9)

Section 9(a) of the House bill is similar to section 9(a) of the Senate bill; however, there are certain differences. First, the Administrator's determination that an unreasonable risk to health or the environment may be prevented or reduced to a sufficient extent by action taken under a Federal law not administered by the Administrator is not discretionary. Second, if such a determination is made, the Administrator shall submit a report to the agency administering such other law. Such report shall describe such risk and include a specification of the activity or activities which the Administrator has reason to believe caused or contributed to such risk.

Such report shall request such agency to determine whether the risk might be prevented or reduced to a sufficient extent by action taken under such law. Conditioned upon such a determination shall be a request that the agency issue an order declaring whether the activity or activities specified in the Administrator's description caused or significantly contributed to such risk, which determination and order shall be reported to the Administrator.

Like the Senate bill, section 9(b) of the House bill requires the Administrator to coordinate actions taken under this legislation with actions taken under other laws administered in whole or in part by the Administrator; however, the language of the House bill differs regarding the Administrator's authority to regulate a risk to health or the environment associated with a chemical substance or mixture. Unless the Administrator determines that it is in the public interest to protect against such risk by actions taken under this Act, the House amendment requires the Administrator to use the authorities contained in other laws, if such risk could be eliminated or reduced to a sufficient extent.

Sections 9(c) and (d) of the House amendment are identical to the Senate bill. The House amendment contains no provision similar to section 9(e) of the Senate bill.

Conference substitute (section 9)

The conferees have drawn from both the Senate bill and the House amendment to assure that overlapping or duplicative regulation is avoided while attempting to provide for the greatest possible measure of protection to health and the environment.

Section 9(a) establishes the relationship between the Act and Federal laws not administered by the Administrator. If the Administrator has a reasonable basis to conclude that the manufacture, processing, distribution in commerce, use, or disposal of a chemical subtance or mixture presents or will present an unreasonable risk of injury and if the Administrator makes a discretionary determination (which is not subject to judicial review) that the risk may be prevented or reduced to a sufficient extent by action taken under a Federal law not administered by the Administrator, then the Administrator must give the other agency an opportunity to act to protect against the risk before the Administrator uses the authorities in section 6 or 7 to protect against the risk.

If the Administrator determines that another Federal law contains authorities adequate to prevent or reduce the suspected risk to a sufficient extent, the Administrator shall submit to the agency which administers the law a report which describes the risk, including a specification of the activity or combination of activities associated with the substance or mixture which the Administrator believes presents the risk. The report must also include a detailed statement of the information on which it is based. The report shall also request the agency to determine if the risk described in the report may be prevented or sufficiently reduced by action taken under its law and, if such determination is affirmative, to issue an order declaring whether or not the activity specified in the report presents an unreasonable risk.

The agency receiving the request from the Administrator must respond to the Administrator within such time as the Administrator specifies. However, the Administrator must give the other agency at least 90 days.

Section 9(a) prohibits the Administrator from acting under section 6 or 7 with respect to the risk about which the Administrator notified the other agency if the other agency takes one of two alternative courses of action. First, if the other agency issues an order declaring that the activity specified in the Administrator's report does not present the unreasonable risk described in the report, then the Administrator may not take action under section 6 or 7 with respect to such risk. Alternatively, if within 90 days of the publication in the Federal Register of the other agency's response, the other agency initiates action to protect against such risk, then the Administrator is precluded from taking action under section 6 or 7 with respect to such risk. If the other agency does not take one of these actions, then the Administrator is permitted to act under section 6 or 7 to protect against the risk.

The conferees recognize that the other agency may not because of time constraints be able to initiate formal regulatory action to protect against the risk within the specified time period. As long as the other agency has officially initiated an action which will culminate as soon as practicable in effective regulatory action to protect against the unreasonable risk and sets forth a general time schedule of steps for such action, the requirement should be deemed satisfied. However, the requirement that the other agency initiate action to protect against the risk is not satisfied by the mere open-ended possibility of action by the other agency.

Subsection (b) establishes the relationship between this Act and other laws administered in whole or in part by the Administrator. Subsection (b) requires the Administrator to coordinate actions taken under this Act with actions taken under other Federal laws administered by the Administrator.

If the Administrator determines that a risk to health or the environment associated with a substance or mixture could be eliminated or reduced to a sufficient extent by actions taken under the authorities contained in other Federal laws, then the Administrator shall use such other authorities unless the Administrator determines, in the Administrator's discretion, that it is in the public interest to protect against such risk under this Act. While it is clear that the Administrator's determination that it is in the public interest to use this Act, is a completely discretionary decision not subject to judicial review in any manner, it is expected that the Administrator will review the other authorities and present the results of that review at the same time the Administrator takes action under this Act. While the Administrator's decision to use this Act, notwithstanding the other authorities, is unreviewable by any court, a reviewing court is expected to require that the Administrator have examined the other authorities and present the results of that examination when making the finding that it is in the public interest to use this Act. Of course, the requirement to examine other EPA laws and to make determinations applies only when the Administrator takes regulatory action to protect against an unreasonable risk under this Act. It does not apply when the Administrator takes action necessary for the administration or enforcement of the Act, such as issuing recordkeeping requirements.

This provision is not to be construed to relieve the Administrator of any requirement imposed by other Federal laws upon the Administrator, and of course nothing in this Act shall affect any final action taken under other Federal laws administered by the Administrator or in any way affect the extent to which health or the environment is to be protected under such other Federal laws.

SECTION 10. RESEARCH, DEVELOPMENT, COLLECTION, DISSEMINATION AND UTILIZATION OF DATA

Senate bill (section 10)

Section 10 authorizes the Administrator to conduct research and monitoring in cooperation with the Secretary of Health, Education, and Welfare and the heads of other appropriate agencies, as is necessary to carry out the purposes of the Act.

The Administrator shall undertake and support programs of research and monitoring of polychlorinated biphenyls to develop safe methods of disposal. The Administrator shall also establish, administer, and assume responsibility for the activities of an interagency committee to construct within the EPA an efficient system for the collection, dissemination, and use of data submitted to the Administrator under this Act among other Federal agencies. This interagency committee shall also direct its attention to coordinating the regulation of chemical substances among the federal agencies. The Administrator shall design, establish, and coordinate an effective system for the retrieval of toxicological and other scientific data which could be useful to the Administrator in carrying out this Act. This section also authorizes the Administrator to make grants and to enter into contracts in order to carry out his responsibilities under this section.

House amendment (section 10)

The House version of section 10 is substantially similar to the Senate bill. However, the House amendment omits the requirement that the Administrator undertake and support programs of research and monitoring of polychlorinated biphenyls. The House amendment contains additional specific provisions for various research programs such as the development of rapid, reliable and economical screening and monitoring techniques for carcinogenic, mutagenic, teratogenic, and ecological effects of chemical substances and mixtures.

Conference substitute (section 10)

The conference substitute includes provisions found in both the Senate bill and the House amendment, but generally follows the language from the House version. Subsection (a) requires the Administrator to conduct such research, development, and monitoring as is necessary to carry out the purposes of this Act. In doing so, the Administrator must consult and cooperate with the Secretary of Health, Education, and Welfare and with heads of other appropriate departments and agencies. The Administrator may enter into contracts and make grants for the purpose of research and development in this area.

Subsection (b) authorizes the establishment of an interagency committee whose primary responsibility shall be to design an efficient system within the Environmental Protection Agency for the collection of data (submitted to the Administrator under this Act), the dissemination of such data to other instrumentalities of the Feneral Government, and the use of such data.

Subsection (b) specifies that an efficient and effective data retrieval system shall be developed. The conferees emphasize that sufficient data is necessary for successful implementation of this Act, yet they also acknowledge the burden placed on industry by excessive or duplicative reporting. It is essential that toxicological and other relevant scientific data already in the possession of the Federal Government be made available to the Administrator. The efficient exchange of information among Federal agencies and departments will facilitate implementationof this Act, and every effort should be made to achieve this goal and to avoid duplicative requirements in information-gathering.

Subsections (c), (d), (e), (f), and (g) of the conference substitute adopt provisions from the House amendment which concern research and development in the area of data collection. The conferees do not intend that such projects should detract from the primary purposes of the Act, but rather that those purposes should be enhanced by allowing the development of proper tools. Thus the purpose of these subsections is to provide the means to an end. They should in no case detract from the main purposes of the Act nor from other equally important research conducted by the Administrator, but should contribute to the achievement of those purposes where appropriate. Of course, such research and development should not duplicate any research and development already being conducted by other Federal agencies and departments. Thus, careful coordination and consultation with such departments and agencies is required.

INSPECTIONS AND SUBPOENAS

Senate bill (section 11)

The Senate bill authorizes the Administrator or any duly designated representative to inspect any establishment, facility or other premises in which chemical substances or mixtures are manufactured, processed, stored or held before or after distribution in commerce. Inspections are also authorized of conveyances used to transport chemical substances or mixtures in connection with distribution in commerce. Inspections may extend to all things within the premises or conveyances inspected bearing on whether the requirements of the Act have been complied with.

The Senate bill also authorizes the Administrator to issue subpoenas to require the attendance and testimony of witnesses and the production of reports, papers, documents, and answers to questions or other in formation necessary for the Administrator carry out his or her duties under the Act.

House amendment (section 11)

The House amendment contains a similar provision authorizing inspections for the purpose of enforcement of the Act. However, the House amendment provides that no inspection shall extend to financial data, sales data other than shipment data, pricing data, personnel data, or research data (other than research data required by the Act) unless the nature and extent of the data are described with reasonable specificity in the written notice presented to the owner, operator or agent in charge of the premises or conveyance to be inspected. The House amendment contained no subpoena authority.

Conference substitute (section 11)

The conference substitute includes the provision from the Senate bill with the addition of the House provision relating to inspections of financial data, sales data other than shipment data, pricing data, personnel data or research data (other than research data required by the Act or pursuant to any rule issued under the Act).

The conferees recognize that the Administrator will have access to much information under section 5 and section 8 of the Act. Therefore, the conferees expect that the Administrator will use the subpoena authority only when information otherwise available through voluntary means or under other provisions of this Act is inadequate to meet the Administrator's needs under this Act.

It should be noted that the conferees intend that representatives of the Administrator authorized to make inspections should have the opportunity to record the results of such inspections because such records might be required at some later date; therefore, it is intended that persons making the inspection shall be allowed, for example, to photocopy records or photograph premises.

EXPORTS

Senate bill (section 12)

This section outlines the policy for chemical substances and mixtures manufactured, processed, sold, or held for sale solely for export from the United States. Subsection (a) provides that unless the Administrator finds that such substances or mixtures will cause or contribute to an unreasonable risk to the health of persons within the United States or the environment of the United States, such substances are exempt from the Act (other than the reporting requirements of section 8) if proper labeling shows that they are intended for export use only.

However, subsection (b) allows the Administrator to require testing under section 4 to see if such substance or mixture may cause or contribute to a risk of health within the United States or to the environment of the United States. Subsection (b) also requires that any person engaged in export activities shall notify the Administrator if such activities involve chemical substances or mixtures for which data is required under section 4 or 5 or for which a rule has been proposed or promulgated under section 5 or 6 or for which action is pending or relief has been granted under section 7. Should any such circumstance arise, the Administrator shall furnish the appropriate foreign government with relevant information pertaining to the chemical substance subject to the limitations of section 14.

House amendment (section 12)

Except for minor differences in language, the House amendment follows the Senate provision. The House provision also specifically covers articles containing chemical substances.

Conference substitute (section 12)

The conference substitute follows the policy set forth in both the Senate and House provisions to protect the health and environment of persons in the United States and to provide information to foreign governments regarding chemical substances and mixtures, so that such foreign governments can protect their own citizens.

ENTRY INTO CUSTOMS TERRITORY OF THE UNITED STATES

Senate bill (section 13)

The Senate bill instructs the Secretary of the Treasury to refuse entry into the United States of any chemical substance or mixture offered for entry if it fails to conform with any requirement of the Act or any rule in effect under the Act or if it is otherwise prohibited from being distributed in commerce. If a substance or mixture is refused entry, the Secretary of the Treasury is required to notify the consignee of the entry refusal. If the substance or mixture is not exported within 90 days, the Secretary is to cause the disposal or storage of the substance or mixture.

House amendment (section 13)

The House amendment contains a similar provision.

Conference substitute (section 13)

The conference substitute adopts the provision found in both the Senate bill and the House amendment relating to entry into the customs territory of the United States. Although the Secretary of the Treasury is authorized to cause the disposal of substances and mixtures which have been refused entry and are not exported within 90 days, the conferees intend that the Secretary consult with the Administrator before determining the disposal methods for the substance or mixture.

DISCLOSURE OF DATA

Senate bill (section 14)

The Senate bill provides generally that all information obtained by the Administrator under this Act shall be subject to the Freedom of Information Act (5 U.S.C. 552). The Freedom of Information Act makes such information available to the public upon request, unless the information requested falls into one of nine exceptions.

The Senate bill also requires the disclosure of data in certain further specified situations. If officers or employees of the United States request information in connection with their official duties under laws protecting human health or the environment or for specific law enforcement purposes, then the Administrator must disclose the information to them.

Likewise, the Administrator must disclose information to the public whenever the Administrator determines it is necessary to protect human health or the environment. If the Administrator determines that disclosure of information is necessary for a contractor or the contractor's employee to perform official duties satisfactorily under contracts for the United States in connection with this Act, then the Administrator must disclose the information. Finally, the Administrator must disclose information to any duly authorized committee of Congress upon written request.

House amendment (section 14)

The House amendment contains some similarities to, but also some differences from, the Senate bill. Whereas the Senate bill states that the Freedom of Information Act applies except in certain areas where disclosure is mandatory, the House bill statutorily prohibits the disclosure of information which falls into one of the exemption categories (subsection (b)(4) (5 U.S.C. 552(b)(4))) of the Freedom of Information Act.

Subsection (b)(4) of that Act encompasses matters that are trade secrets and commercial or financial information obtained from a person and privileged or confidential. The Administrator may not disclose information under that classification except to officers or employees of the United States in connection with their duties to protect health or the environment or for specified law enforcement purposes or to contractors with the United States or their employees in connection with this Act. Such information may be disclosed when relevant to a proceeding under this Act, but the disclosure must preserve confidentiality as much as possible without impairing the proceeding.

Subsection (b)(1) specifically provides that disclosure of any health and safety study for any chemical substance or mixture which is already being distributed or for which testing is required under section 4 or for which notification is required under section 5, is not prohibited. Data in such a study which discloses a manufacturing process or the proportions of a mixture may not be disclosed if such process or proportions would otherwise be entitled to protection from disclosure.

Subsection (c) authorizes any person who submits data under the bill to designate information he believes is entitled to confidential treatment under section (a). Designated information may not be released for 30 days after notification of release has been received by the person submitting such data.

Conference substitute (section 14)

The conference substitute adopts elements of both the Senate bill and the House amendment. The prohibition against disclosure of information exempt from mandatory disclosure under section (a) of section 552 of title 5, United States Code, by reason of its falling within the exemption under subsection (b)(4) of such section, is included. Section 14 applies to any release of information obtained under the Act.

Mandatory exceptions from this prohibition are also provided. Disclosure of information described in section 552(b)(4) of title 5 is required in the following situations:

(1) To officers or employees of the United States in connection with their official duties to protect health or the environment, and for specific law enforcement purposes.

(2) To contractors with the United States when the Administrator determines it to be necessary for the satisfactory performance of their duties in connection with this Act and under such conditions as necessary to preserve confidentiality as the Administrator may specify.

(3) If the Administrator determines it necessary to protect health or the environment against an unreasonable risk of injury to health or the environment.

In addition, the Administrator may disclose such information when relevant under a proceeding under this Act, except that disclosure under such proceeding shall be made in such a manner as to preserve confidentiality to the extent practicable without impairing the hearing. It is intended that the Administrator exercise due care to prevent the release of confidential information to competitors of persons submitting data merely because the competitors have joined the proceeding.

In any proceeding under section 552(a) of title 5 to obtain information which the Administrator has refused to release on the basis that disclosure is prohibited by section 14(a) of this Act, the Administrator may not rely on section 552(b)(3) of title 5 to sustain the refusal to disclose the information. Thus the Administrator will have to show that the information falls within section 552(b)(4) of title 5. Of course, section 552 of title 5 is the vehicle through which the public can obtain information from the Federal government, and all the provisions of that section will apply to requests for information obtained under this Act.

The conference substitute specifically provides that disclosure of any health and safety study or information from such a study on any substance or mixture which is already being distributed or for which testing is required under section 4 or for which notification is required under section 5, is not prohibited. Data in such a study which discloses manufacturing processes or the proportions of a mixture may not be disclosed if such processes or proportions would otherwise be entitled to protection from disclosure. However, any restriction on the release of such data will not apply to the health and safety study in which it is contained or from which it is derived. To comply with such restriction the Administrator need only to exclude such data when releasing such study.

If a request is made to the Administrator for health and safety study information which is not entitled to protection, the Administrator may not deny a request under section 552 of title 5, United States Code, on the basis that such information is included in the exceptions to mandatory disclosure enumerated in subsection (b)(3) or (b)(4) of such section. It is also intended that the Administrator not use exception (b)(7) of section 552 of title 5, relating to matters under investigation, in an excessive manner as a device for withholding information submitted under this Act. In order to be withheld under that exception, the information must be the subject of an ongoing, active investigation.

In submitting data, a person may designate data which the person believes is entitled to confidential treatment under this Act and submit it separately. If the Administrator proposes to release for inspection designated data, the Administrator must give 30 days notice to the person who submitted the information. Thirty days advance notice need not be given when information is to be released under one of the mandatory exceptions described above or when disclosure is not prohibited because the information is health and safety data. When disclosure is proposed because it is necessary to protect health and the environment from an unreasonable risk, the Administrator shall provide the person submitting the data written notice by certified mail of the proposed release at least 15 days prior to the release. The purpose of this provision is to provide the person submitting the data an opportunity to seek to stop the proposed release if that person disputes the Administrator's determination. The conferees recognize that there may arise emergency situations in which the Administrator determines that earlier release is necessary. In such cases, where the occurrence of the unreasonable risk is imminent, the Administrator need give notice only 24 hours prior to release. The required notice need not be given in writing but may be made by some other means such as telephone or telegraph.

The criminal penalties for wrongful disclosure contained in the House bill have been included in the conference substitute.

PROHIBITED ACTS

Senate bill (section 15)

The Senate bill makes it unlawful for any person to fail or refuse to comply with any rule or order promulgated under section

4, 5, or 6, or any requirement prescribed by section 5 or 6. It also makes it unlawful for any person to use or dispose of a chemical substance or mixture which the person knew or had reason to know was manufactured, processed or distributed in commerce in violation of section 5 or a rule or order under section 6. Failure or refusal to establish or maintain records, submit reports, notices, or other information or to permit access to, or copying of, records is also unlawful. Finally, the Senate amendment makes unlawful the failure or refusal to permit entry or inspection as required by section 11.

House amendment (section 15)

The House amendment makes it unlawful for any person to fail or refuse to comply with any rule or order promulgated under section 4, 5 or 6 or any requirement prescribed by section 5. It also makes it unlawful for any person to use for commercial purposes a chemical substance or mixture which the person using such substance or mixture knew or had reason to know was manufactured, processed or distributed in commerce in violation of section 5, a rule or order under section 5 or 6 or an order issued in an action brought under section 5 or 7. The House provisions respecting maintenance of records, submission of reports, entry, and inspections are identical to the Senate bill.

Conference substitute (section 15)

The conference substitute incorporates the provisions of the House bill with a conforming amendment making violations of the provisions of section 6 relating to polychlorinated biphenyls an unlawful act.

PENALTIES

Senate bill (section 16)

This section outlines the penalties and procedures for assessing penalties against persons who violate section 15. Subsection (a) provides for civil penalties of up to $25,000 per day per violation. Taking relevant factors into account, the Administrator shall assess the amount of such civil penalties in an order made on the record after the opportunity for an adjudicative hearing and proper notification of the person in violation of this Act. Such person may file a petition for judicial review of an order assessing civil penalties in U.S. Court of Appeals within thirty days; however, if a person fails to pay such assessment after it has become a final and unappealable order or after the Court of Appeals has found in favor of the Administrator, then the Attorney General shall recover the amount assessed.

Subsection (b) provides for criminal penalties of up to $25,000 per day or up to one year's imprisonment, or both, per violation for any person who knowingly or willfully violates this Act.

House amendment (section 16)

The House amendment follows subsections (a) and (b) of the Senate provision. In addition, subsection (c) of the House Amendment provides that the Administrator may require a person who has manufactured, processed, or distributed a chemical substance or mixture in violation of regulations issued under paragraphs (1) or (2) of section 6(a) to give notice of the risk associated with that substance to any person who may be exposed to it and to the public at large.

The Administrator may also require such person to either replace or repurchase the substance found to be in violation. The Administrator may choose any or all of the remedies set forth in subsection (c); however, in each case the order must be made on the record with full opportunity for an agency hearing.

Conference substitute (section 16)

The conference substitute adopts the provisions found in both bills concerning civil and criminal penalties for violations of this Act. Under subsection (a), the Administrator shall assess the amount of civil penalties up to $25,000 per day per violation; however, the Administrator must take into account such factors as the gravity and extent of the violation, the ability to pay of the person held in violation, and any prior history of violations under this Act.

Criminal penalties may be imposed on persons who "knowingly or willfully" violate any provision of section 15, which sets forth unlawful acts.

SPECIFIC ENFORCEMENT AND SEIZURE

Senate bill (section 17)

The Senate bill grants the United States district courts jurisdiction, upon application of the Administrator or the Attorney General, to restrain any violation of section 15, to restrain any person from manufacturing or processing a chemical substance before the expiration of the notification period under section 5, and to restrain any person from taking any action prohibited by a requirement prescribed under section 5 or 6 or rules or orders issued under section 5 or 6. In addition, the courts are granted jurisdiction to direct any manufacturer or processor of a chemical substance or mixture who is not in compliance with any order issued under section 5(e) or any rule issued under section 4 or 6 to give notice of such fact to persons within the chain of distribution and to the public, and to either replace or repurchase the substance or mixture. The Senate provision also authorizes the district bution and to the public, and to either replace or repurchase the substance or mixture. The Senate provision also authorizes the district courts to compel the taking of any action required by or under the Act. In addition, the Senate bill provides that any substance or mixture manufactured or processed or distributed in commerce in violation of the Act or any rule or order promulgated under the Act shall be liable to be proceeded against for seizure and condemnation.

House amendment (section 17)

The House amendment grants jurisdiction to the district courts to restrain any violation of section 15, to restrain any person from manufacturing or processing a substance before the expiration of the notification period under section 5, and to restrain any person from taking action prohibited by section 5 or a rule or order under section 5 or 6. Jurisdiction is also granted to compel the taking of any action required by or under this Act. The seizure authority in the House amendment is similar to that found in the Senate bill, except that seizure and condemnation of articles containing chemical substances or mixtures manufactured, processed or distributed in commerce in violation of the Act or any rule or order promulgated under the Act is specifically authorized.

Conference substitute

The conference substitute grants the district courts of the United States jurisdiction to restrain any violation of section 15, to restrain any person from manufacturing or processing a substance before the expiration of the notification period under section 5, and to restrain any person from taking any action prohibited by section 5 or 6 or a rule or order under section 5 or 6. The provision also grants such courts jurisdiction over actions to direct any manufacturer or processor of a chemical substance or mixture manufactured or processed in violation of any order issued under section 5 or any rule or order issued under section 6 to give notice of the risk associated with the substance or mixture to persons in the chain of distribution and to the public. The courts also have jurisdiction to require manufacturers or processors to either replace or repurchase the substance or mixture, whichever the person to whom the requirement is directed elects. The conference substitute also grants jurisdiction to compel the taking of any action required by or under the Act. The seizure authority in the conference substitute is identical to that contained in the House amendment.

PREEMPTION

Senate bill (section 18)

This section outlines the relationship between State authority and the authority under this Act to regulate chemical substances or mixtures. Subsection (a) asserts the State's authority to regulate, with certain limitations. No State may require testing which duplicates testing required by the Administrator under a section 4 testing rule. Further, if the Administrator has prescribed a requirement under section 5 or 6 to protect against a particular risk associated with a chemical substance or mixture, a State may not prescribe any different requirement (other than a total ban) with respect to that risk unless the State obtains permission from the Administrator to do so.

The Administrator may, by rule, grant such permission if the Administrator finds that compliance with the State requirement would not result in a violation of this Act, would result in a significantly higher degree of protection, and would not unduly burden interstate commerce.

House amendment (section 18)

The House amendment is similar to the Senate bill. However, rules promulgated under section 6(a)(5) do not preempt State laws. Moreover, rules promulgated under other Federal authorities such as the Clean Air Act, are not preempted by requirements under this Act.

Like the Senate bill, the House amendment authorizes the Administrator to exempt, by rule, States from prohibitions under subsection (a) in the same manner as the Senate bill.

Conference substitute (section 18)

The conference substitute provides that no State or political subdivision may establish similar requirements for the testing of a substance or mixture after the Administrator has issued a rule under section 4 respecting the substance or mixture. Nor may any State regulate any risk associated with a substance or mixture if the Administrator has prescribed a rule or order under section 5 or 6, which is designed to protect against the risk to health or the environment, unless the rule (A) is identical to that issue under this Act, (B) is adopted under the authority of another Federal law, or (C) prohibits the use of such substance or mixture other than in its use in the manufacture or processing of other chemical substances or mixtures.

In addition to the specific exemptions from the preemption provision, the conference substitute provides a means whereby a State or political subdivision may seek an exemption from the preemptive effects of a Federal requirement in order to provide a higher degree of protection for their citizens than that provided by a requirement under this Act. The Administrator may, by rule, grant an exemption if compliance with the State or local requirement will not cause a violation of the applicable requirement under this Act, if the State or local requirement will provide a significantly higher degree of protection from the risk, and if the State or local requirement will not unduly burden interstate commerce.

JUDICIAL REVIEW

Senate bill (section 19)

The Senate's provision authorized pre-enforcement judicial review of any rule under the Act or an order issued under section 5(e). Any rule promulgated under section 3(b), 5 or 6 shall not be affirmed unless the rule (A) is identical to that issued under this Act, (B) is

House amendment (section 19)

The House provision authorizes pre-enforcement judicial review of rules issued under section 4, 5 or 6(a). Such rules shall not be affirmed unless supported by substantial evidence based on the record taken as a whole.

Conference substitute (section 19)

Section 19 of the conference substitute provides for judicial review in the courts of appeals of the United States for certain rules promulgated under the Act. The jurisdiction for preenforcement review and review of determinations of the Administrator relating to cross-examination is exclusively vested in such courts. Not later than 60 days after the date of promulgation of a rule under section 4(a), 5(a)(2), 5(b)(4), 6(a), 6(e), or 8 any person may file a petition for judicial review of the rule in the appropriate U.S. court of appeals.

The section specifically defines the rulemaking record to include the rule being reviewed (which would include the statement of basis and purpose pursuant to section 553(c) of title 5, United States Code), any transcript required to be made of an oral presentation, any written submission of interested parties, and any other information which the Administrator considers to be relevant to the rule and with respect to which the Administrator published a notice in the Federal Register identifying the information on or before the date of the promulgation of such rule. In addition certain findings and statements required to be made with respect to specific rules must also be included in the rulemaking record. In the case of a rule under section 4(a), the finding required by that section must be included in the record. In the case of a rule under section 5(b)(4), the finding required to be made by that section must be included in that record. In the case of a rule under section 6(a), the finding required by section 5(f) or section 6(a), as the case may be, and the statement required by section 6(c)(1) must be included in the rulemaking record.

The section includes authority for the submission of additional data and oral or written views and for the modification of the rule being reviewed.

Generally section 706 of title 5, United States Code, applies to review of a rule under this section. However, in the case of review of a rule under section 4(a), 5(b)(4), 6(a) or 6(e), the bill provides that the courts shall hold unlawful and set aside such rule if the court finds that the rule is not supported by substantial evidence in the rulemaking record taken as a whole. This provision is in lieu of paragraph 2(E) of section 706 of title 5. It is the intent of the conferees that the traditional presumption of validity of an agency rule is to remain in effect. The conferees recognize that in rulemaking proceedings such as those contained in this bill, which are essentially informal and which involve both determinable facts and policy judgments derived therefrom, the traditional standard for review is that of "arbitrary and capricious". However, the conferees have adopted the "substantial evidence" test because they intend that the reviewing court focus on the rulemaking record to see if the Administrator's action is supported by that record. Of course, the conferees do not intend that the court substitute its judgment for that of the Administrator.

Further, in the case of review of a rule under section 6(a), the court shall set the rule aside if it finds that action by the Administrator in excluding or limiting cross-examination or rebuttal submissions precluded disclosure of disputed issues of material fact necessary for a fair determination of the rulemaking proceeding taken as a whole. Also, in review of such rules, section 706(2)(D) will not apply with respect to review of the Administrators actions respecting limitations or exclusions of cross-examination or rebuttal submissions, and review of such actions can occur only during preenforcement judicial review.

Section 19 also provides that the court may not review the contents and adequacy of any statement required to be made pursuant to section 6(c)(1) or any statement of basis and purpose required by section 553(c) of title 5 of United States Code to be incorporated in the rule except as part of a review of a rulemaking record taken as a whole.

Section 19 provides that in a judicial review proceeding under this section the court may award the costs of suit and reasonable fees for attorneys and expert witnesses if the court determines that such an award is appropriate. In addition, in any review of such an action the Supreme Court may also award such costs of suit and reasonable fees.

The section also provides that the remedies provided in section 19 shall be in addition to, and not in lieu of, any other remedies provided by law. This provision should not be construed, however, to negate the provision in this section which specifically provides that the United States courts of appeals shall have exclusive jurisdiction of any action to obtain judicial review (other than in an enforcement proceeding) if any district court of the United States would have had jurisdiction of such an action but for the provisions of this section.

CITIZEN'S CIVIL ACTIONS

Senate bill (section 20)

Subsection (a) authorizes any person to commence a civil action in specified district courts against (A) any person including the United States or any governmental agency or instrumentality alleged to be in violation of this Act or any rule or order prescribed under sections 4, 5, or 6(a). Such suits may also be brought to compel the Administrator to perform any nondiscretionary act or duty.

Subsections (b), (c), and (d) specify certain procedural provisions. No action may begin until the Administrator and the alleged violator have received proper notice of the alleged violation. If the Administrator has instituted a civil action against an alleged violator to compel compliance, then no action may be brought under this section. However, if the Administrator does not commence such action until after the person bringing the citizen's civil action has notified the alleged violator of intention to sue under this Act, then the person who gave such notification may intervene in the suit brought by the Administrator. The Administrator may intervene in any civil action under this section to which the Administrator is not a party. The court may award the costs of the suit and reasonable fees for attorneys and expert witnesses. The court may also consolidate two or more civil actions involving the same defendant, the same issues, or the same alleged violations when appropriate.

House amendment (section 20)

The House amendment contains the same provision as the Senate bill.

Conference substitute (section 20)

The conference substitute contains the provision included in both the Senate bill and the House amendment with a clarification that citizen's civil actions may also be brought for violations of an order under section 5 or 6.

CITIZENS' PETITIONS

Senate bill (section 21)

Section 21 of the Senate bill authorizes any person to petition the Administrator to issue a rule or order or to take other action for the purpose of protecting against an unreasonable risk of injury to health or the environment. If the petition is denied or not acted upon within 90 days, the petitioner may bring a civil action in a United States district court to compel the Administrator to initiate the requested action. If the petitioner demonstrates by a preponderance of the evidence in a *de novo* proceeding that the action requested in the petition conforms to the applicable requirements of the Act, the court shall order the Administrator to initiate the requested action.

House amendment (section 21)

The House amendment authorizes any person to petition the Administrator to initiate a proceeding for the issuance, amendment, or repeal of a rule under section 4, 5(c), or 6(a). If the petition is denied, the petitioner may file a civil action to compel the Administrator to initiate the rulemaking proceeding. If the petitioner requests the issuance of a rule under section 4, 5(c), or 6(a) (as opposed to the modification or repeal of such a rule) the petitioner has an opportunity for a *de novo* proceeding before the court. If the petitioner makes the requisite showings for the applicable provision, the court must order the Administrator to initiate the requested action unless the court finds that the failure of the Administrator to initiate the requested action was not unreasonable.

Conference substitute (section 21)

The conference substitute authorizes any person to petition the Administrator to initiate a proceeding for the issuance, amendment or repeal of an action under section 4, 5(e), 6, or 8 of the Act. It should be noted that a petition under this section may be used to initiate a proceeding under section 5(f) since a proceeding under that section is for the issuance of a rule under section 6(a). The Administrator must grant or deny any petition under this section within 90 days after it is filed.

The conference substitute thereafter provides for different judicial review of the Administrator's denial of a petition, depending upon whether such petition seeks the issuance of a rule or order or the amendment or repeal of an existing rule or order.

The substitute affords greater rights to a person petitioning for the issuance of a rule or order because in such a situation the Administrator will not previously have addressed the issue by rule or order. If the Administrator denies or fails to respond to a petition for the issuance of a rule or order, the petitioner may commence a civil action in a United States district court to compel the Administrator to take the action requested in the petition. In the court, the petitioner is entitled to a *de novo* proceeding. If the petitioner demonstrates to the court by a preponderance of the evidence that there is an adequate basis for the issuance of the rule or order requested, the court shall order the Administrator to initiate the requested action.

The court may defer requiring the Administrator to take the requested action if it finds that the extent of risk of injury to health or the environment alleged by the petitioner is less than those risks of injury which the Administrator is addressing under this Act and there are insufficient resources to do both. If a deferral is granted, the conferees anticipate that the Administrator may seek extensions as needed.

The conference substitute provides different treatment for review of petitions for amendment or repeal of rules or orders, because the Administrator already will have addressed the general subject matter in an existing rule or order and the Administrator's determination will have been subject to review under section 19 of this Act. Therefore, the conferee's main interest is to make certain that any such petitioner receive timely consideration of such petition. By requiring the Administrator to act on any such petition within 90 days, the conferees will facilitate such a petitioner's right to seek judicial review should the Administrator deny the petition. Otherwise, the Administrator could avoid any judicial review simply by failing to take any action.

The conferees believe that a petition for amendment or repeal of an existing rule or order should contain newly discovered, noncumulative material which was not presented for the Administrator's consideration in promulgating the rule or order. Failure to include such information would be an adequate basis for denying the petition.

At the same time, the conferees do not intend that the Administrator be subjected to constant petitions challenging rules or orders for which adequate judicial review is provided under section 19. Therefore, if the Administrator denies a petition to amend or repeal an action under section 4, 5(e), 6, or 8, the conference substitute permits review of such denial only under the Administrative Procedure Act.

NATIONAL DEFENSE WAIVER

Senate bill (section 22)

The Senate bill directs the Administrator to waive compliance with any provision of this Act upon the request of the Secretary of Defense and a determination by the President that the interest of national defense requires such a waiver. The Administrator shall maintain a written record of the basis for the waiver. In addition, the Administrator shall publish notice of the waiver in the Federal Register, unless the Administrator determines, upon request from the Secretary of Defense, that such publication is contrary to national defense interests, in which case, the Administrator shall notify the Armed Services Committees of the Senate and the House of Representatives.

House amendment (section 22)

The House amendment is similar to the policies and procedures of the Senate bill except that only the President, not the Secretary of Defense, is authorized to request a national defense waiver from the Administrator and to request that publication of the waiver not be placed in the Federal Register for national defense reasons.

Conference substitute (section 22)

The conference substitute includes the provision of the House amendment.

EMPLOYEE PROTECTION

Senate bill (section 23)

Section 23 of the Senate bill provides protection for employees who cooperate with the Administrator in carrying out the Act. The provision prohibits any employer from discharging any employee or otherwise discriminating against the employee with respect to compensation, terms, conditions, or privileges of employment because the employee commenced, caused to be commenced, or is about to commence a proceeding under the Act. Protection is provided for employees who have testified or are about to testify in any proceeding under the Act or who have assisted or participated in a proceeding or any other action to carry out the purposes of the Act. The Secretary of Labor shall conduct investigations of alleged violations and issue orders to require any person who violates the prohibitions to take affirmative action to remedy any such violation. Any person adversely affected by an order of the Secretary may obtain judicial review of the order in the United States court of appeals for the circuit in which the violation allegedly occurred. The Secretary is authorized to enforce the orders in the district court of the United States for the district in which the violation occurred.

House amendment (section 23)

The House amendment contains an identical provision.

Conference substitute (section 23)

The conference substitute adopts the provision found in both the Senate bill and House amendment.

EMPLOYMENT EFFECTS

Senate bill (section 23(f))

The Administrator shall conduct continuing evaluation of the effect on employment of rules or orders under this Act. Any employee who is discharged or whose employment is threatened or who is otherwise discriminated against as a result of any action under this Act may request investigation of the matter by the Administrator. The Administrator shall investigate the matter. If any interested party requests a hearing, the Administrator shall conduct a public hearing in accordance with section 554 of title 5, United States Code, at which the parties are required to present information on any employment effects.

Upon receipt of the investigation report, the Administrator shall make findings of fact as to the employment effects and shall make appropriate recommendations which shall be available to the public.

House amendment (section 24)

The House amendment is similar to the Senate bill. The House provision differs from the Senate's primarily as to whether and how a hearing requested by an interested person shall be conducted.

Upon request, the Administrator must hold a public hearing unless the Administrator determines that there are no reasonable grounds for such hearing. The hearing need not be a formal adjudicative hearing under 5 U.S.C. 554. Provision is made for subpoenas, oaths, and payment of witness fees in connection with any investigation or public hearing conducted under this section.

Conference substitute (section 24)

The conference substitute follows the House amendment with two modifications. First, if the Administrator determines that there are no reasonable grounds for holding a hearing, the Administrator must so find, by order, within 45 days of the date within which time such hearing is requested. Second, if a hearing is held, it shall be in accordance with the requirements of section 6(c) (3) of this Act.

STUDIES

Senate bill (section 24)

The Senate bill requires the General Accounting Office to conduct a study of all Federal laws administered by the Administrator to determine whether and under what conditions, if any, idemnification should be accorded any person as a result of action taken by the Administrator under such laws. The Senate bill also requires the Council on Environmental Quality to coordinate a study of the feasibility of establishing a standard classification system of chemical substances and related substances and a standard method for storing and obtaining rapid access to information respecting such substances.

House amendment (section 25)

The House amendment contains a similar provision except that the indemnification study shall be conducted by the Administrator and reviewed by the General Accounting Office.

Conference substitute (section 25)

The conference substitute includes the House provision.

ADMINISTRATION OF THE ACT

Senate bill (section 26)

Subsection (a) gives authority to each federal department and agency to cooperate with the Administrator, upon request, by sharing services of personnel, facilities, and information in order to carry out the purposes of this Act.

Subsection (b) provides that the Administrator may, by rule, require payment from any person submitting data pursuant to section 4 or 5 to help defray administrative costs, provided that no such fee exceeds $2,500.

Under subsection (c), the Administrator may act with respect to categories of chemical substances or mixtures. For purposes of this section, a category includes chemical substances or mixtures grouped by virtue of similarity of chemical structure, physical, chemical, or biological properties, use or mode of entry into the human body or environment, or some other suitable grouping.

Under subsection (d), any proposed or final rule or order under this Act shall be accompanied by a statement of purpose and justification, which identifies the basis for the action. This statement shall become part of the "record of the proceedings" for purposes of judicial review under section 19(a).

Subsection (e) directs the President to appoint by and with the advice and consent of the Senate, an Assistant Administrator of the Environmental Protection Agency to administer this Act. The Assistant Administrator shall be qualified to direct a program concerning the effects of chemicals on health and the environment by reason of background and experience.

House amendment (section 26)

The House amendment contains provisions similar to the Senate bill concerning cooperation among federal agencies and fees to be paid by persons submitting data under section 4 or 5 to defray administrative costs, except that no small businesses shall be required to pay administrative fees exceeding $100. The House amendment also inincludes a provision on categories similar to that in the Senate bill.

No provision is made in the House amendment for appointment of an Assistant Administrator for Toxic Substances. However, the House amendment establishes an office within the Environmental Protection Agency to provide technical and other nonfinancial assistance to manufacturers and processors of chemical substances and mixtures concerning the requirements and application of this Act.

The House amendment does not contain a specific provision requiring that each proposed or final rule or order be accompanied by a statement of purpose or justification.

The House amendment in section 29 provides that each officer and employee of the Environmental Protection Agency and the Secretary of Health, Education, and Welfare who perform any function or duty under the bill and who has any known financial interest in any person subject to the bill or in any person who applied for or received any financial assistance pursuant to the bill must, beginning February 1, 1977, annually file with the appropriate agency or department a statement concerning all such interests during the preceding calendar year. Such statement must be available to the public.

The House amendment also directs the Administrator and the Secretary, within 90 days after enactment, to define "known financial interest" and to establish methods to monitor, enforce, and review the filing of such statements. They are also directed to report each year to Congress on June 1 regarding such disclosures and actions taken concerning them during the preceding calendar year.

Officers or employees in designated positions of a nonregulatory or nonpolicymaking nature may be exempted, by rule, from the requirements of this section.

The House amendment states that any officer or employer who is subject to, and knowingly violates, this section or any regulations issued thereunder is to be fined not more than $2,500 or imprisoned for not more than one year, or both.

Conference substitute (section 26)

The conference substitute incorporates provisions from both the Senate bill and the House amendment. Subsection (a) gives authority to each federal agency and department to cooperate with the Administrator to carry out the purposes of this Act.

The Administrator is authorized to require, by rule, payment of reasonable fees from any person required to submit data under sections 4 and 5 in order to defray the costs of administering this Act. In no case shall such fees exceed $2,500, or $100 in the case of a small business. In all cases when setting such fees, the Administrator shall take into account the ability to pay of persons submitting data.

The conference substitute includes the House provision concerning categories in subsection (c). The conferees expect that the Administrator will find the authority to categorize especially helpful in promulgating rules under section 5(a)(2) concerning what constitutes significant new use of chemical substances.

The conference substitute adopts the provision from the House amendment which establishes an office within EPA to provide technical and other nonfinancial assistance to manufacturers, processors of chemicals, and others. The purpose of the office is to help manufacturers and processors understand the requirements of the Act in order to assist in its efficient implementation and to avoid unnecessary confusion, which might prove detrimental to the chemical industry and the public interest.

The conference substitute adopts the House provision on financial disclosures for which the Senate bill had no comparable provision.

The procedures and penalties are designed to make sure that persons who perform regulatory functions under this Act divulge any known financial interest such persons may have in any person subject to this Act.

Subsection (f) of the conference substitute modifies the requirement in the Senate amendment that each proposed or final rule or order be accompanied by a statement of basis and purpose to apply only to final orders.

The conference substitute includes the provision found in the Senate bill that the President appoint, with the advice and consent of the Senate, an Assistant Administrator for Toxic Substances who shall direct a program concerning the effects of chemicals on human health and the environment and perform other duties and responsibilities under this Act.

While the Assistant Administrator for Toxic Substances will be assigned responsibilities pursuant to this Act, the Administrator may assign additional duties. Of course this position will be in addition to the existing five assistant administrator positions established by Reorganization Plan No. 3 of 1970.

DEVELOPMENT AND EVALUATION OF TEST METHODS

Senate bill

The Senate bill contains no provision respecting development and evaluation of test methods.

House amendment (section 27)

The House amendment authorizes the Secretary of Health, Education, and Welfare, in consultation with the Administrator and acting through the Office of the Assistant Secretary for Health, to conduct projects for the development and evaluation of inexpensive and efficient methods for determining and evaluating the health and environmental effects of chemical substances and mixtures.

Conference substitute (section 27)

The House provision is included.

STATE PROGRAMS

Senate bill (section 25)

Section 25 authorizes the Administrator to assist up to three states in the establishment of demonstration programs to complement federal efforts under the Act. Subsection (a) describes the functions of such programs. Subsection (b) requires the Administrator to submit annual reports to the Congress on the demonstration programs. Subsection (c) authorizes appropriation of funds to assist the states in funding the demonstration programs. Grants shall not exceed 75 percent of the cost of any demonstration program. Subsection (d) provides that assistance shall be available to those states which can establish a priority need for such assistance. The Senate bill authorizes a maximum appropriation of $2 million for the fiscal year ending September 30, 1977, $2 million for the fiscal year ending September 30, 1978, and $2 million for the fiscal year ending September 30, 1979.

House amendment (section 28)

The House amendment is similar to the Senate bill, but differs in that grants are authorized only to assist states in addressing risks associated with substances and mixtures which the Administrator is unable to address.

The House amendment does not restrict the number of programs which the Administrator may approve. The House amendment authorizes an annual appropriation of $1 million for the fiscal years ending September 30, 1978, September 30, 1979 and September 30, 1980.

Conference substitute (section 28)

The conference substitute generally follows the House amendment with some modification. The Administrator may make grants to States to establish programs to prevent or eliminate unreasonable risks associated with chemical substances or mixtures against which the Administrator is not able or not likely to take action under this Act. The conferees agreed to a compromise on the authorization for such programs of $1.5 million for each of the fiscal years 1977 through 1979.

AUTHORIZATION FOR APPROPRIATIONS

Senate bill (section 27)

Section 27 of the Senate bill authorizes to be appropriated to the Administrator $11,100,000 for the fiscal year ending June 30, 1976, $2,600,000 for the period beginning July 1, 1976 and ending September 30, 1976, and $10,100,000 for the fiscal year ending September 30, 1977. This section prohibits using funds for construction of research laboratories.

Section 27(b) of the Senate bill requires that the Administrator submit concurrently to the Congress any budget requests, supplemental budget estimates, legislative recommendations, prepared testimony for Congressional hearings, or comments on legislation to the President or to the Office of Management and Budget connected with this Act.

House amendment (section 30)

The House amendment authorizes to be appropriated $12,625,000 for the fiscal year ending September 30, 1978, $16,200,000 for the fiscal year ending September 30, 1979, and $17,350,000 for the fiscal year ending September 30, 1980. The House amendment contained no provision relating to simultaneous submissions.

Conference substitute (section 29)

The conference substitute authorizes to be appropriated to carry out the purposes of this Act as follows: $10,100,000 for the fiscal year ending September 30, 1977; $12,625,000 for the fiscal year ending September 30, 1978; and $16,200,000 for the fiscal year ending September 30, 1979.

The conference substitute contains no provision for simultaneous submission of materials to Congress and the Office of Management and Budget.

ANNUAL REPORT

Senate bill (section 28)

The Senate bill requires the Administrator to submit to both the President and the Congress a comprehensive annual report. The report shall include (1) a list of the testing required under section 4 and an estimate of the costs incurred by the person required to perform the tests; (2) the number of notices received under section 5, the number of notices received under section 5 for chemical substances subject to a section 4 rule, and a summary of any action taken during the premarket notification period; (3) a list of rules issued under section 6; (4) a list, with a brief statement of the issues, of completed or pending judicial actions under the bill; (5) a summary of major problems encountered in administration of the bill; and (6) such recommendations for additional legislation as the Administrator deems necessary to carry out the purposes of the bill.

House amendment (section 31)

The House amendment is almost identical to the Senate bill. The only difference occurs with respect to the date that the Administrator must submit the first annual report. The House amendment specifies that the Administrator shall submit the first annual report on or be-before January 1, 1979.

Conference substitute (section 30)

The conference substitute generally follows the provision of the Senate bill. The first submission is due on or before January 1, 1978.

REVIEW

Senate bill

The Senate bill contained no rule review provision.

House amendment (section 32)

Section 32 of the House amendment provides that either House of Congress may veto a rule issued by the Administrator, the Secretary of the Treasury, or the Secretary of Health, Education, and Welfare under this Act, by adopting a resolution of disapproval within 60 days.

Conference substitute

The House recedes.

EFFECTIVE DATE

Senate bill

The Senate bill contained no specific provision specifying an effective date; therefore, the legislation is to take effect upon enactment.

House amendment (section 33)

The House amendment provides that the legislation shall take effect October 1, 1977.

Conference substitute

The conference substitute establishes the effective date as January 1, 1977, except that section 4(f) shall not become effective for two years.

O

Section IV

EPA'S Implementation Plans

THE TOXIC SUBSTANCES CONTROL ACT: WHERE ARE WE GOING?*

Glenn E. Schweitzer
Director, Office of Toxic Substances
U.S. Environmental Protection Agency
Washington, D.C.

When a new law is enacted, hundreds – or in this case thousands – of interested parties immediately begin to direct their attention to those legislative provisions that are to be implemented first. Unfortunately, too few are willing to step back from the heat of recent congressional debates and look ahead a few years to speculate, and indeed to influence, how society can and should be different as a result of new legal ground rules. If we know where we are heading, there is a reasonable chance we will arrive. If we do not know where we are going, the odds of arriving are slim indeed.

In discussing initial implementation activities under the Toxic Substances Control Act, let us not become entangled in a web of administrative and legal jargon that will blur the longer term objectives. Nor should we assume that there is a clear understanding of how this new policy framework will or should impact on government, industry, the scientific community, and the consumer. At the same time, while strategies and philosophies are important contextual ingredients, the sophistication of our foresight and our sensitivity to over-arching societal concerns must permeate specific implementation activities to have a real meaning.

Looking ahead to the 1980s – an era when chemicals should play a more decisive role that ever before in shaping the rate and direction of our economic growth – the Toxic Substances Control Act will probably be a pivotal force influencing the costs of consumer products, the vitality of our research and development efforts, the configuration of our industry, and our international competitive position. It should also be the cornerstone of expanded efforts to understand and to curb harmful chemical pollution.

Surely, the world of the 1980s will be characterized by growing global interdependence, both socially and economically – with an accelerated flow of chemical products and technologies from north to south as well as continued commercial interlocks along the east-west axis. I believe that most of us are optimistic that this economic and technological interdependence will enhance world stability and lead to greater prosperity for all – to more jobs and more products at home and abroad. At the same time, the international specter of ever growing energy problems and depletion of certain natural materials is real, and the international sensitivities to the migration of chemical pollutants across borders will undoubtedly increase.

Domestically, our insatiable appetite for more, better, and cheaper consumer products will continue to tax the ingenuity of our industrial research community, and central to these efforts will be chemistry. Despite the growing reach of government into seemingly every facet of our daily lives, the private enterprise system must continue to be the underlying strength of our country and the small business man must continue to compete side-by-side with our corporate giants.

While chemicals are playing a key role in global and domestic commerce, we will become more aware of their unintended presence in the environment and their effects on man and our ecological resources. Improved analytical techniques and dedicated researchers will undoubtedly show that in more cases than previously expected certain types and levels of chemical concentrations can be harmful. Inevitably, many of the chemical time bombs currently persisting in the environment will have caused serious damage, and despite our best efforts, others will have escaped through the net of assessment and control into the environment.

But within a decade, we should have turned the corner in reducing the frequency and severity of chemical incidents, and in uncovering and controlling many of those latent by-products of the wonders of chemical synthesis.

Within a decade, we should have established sound and reliable procedures and techniques for anticipating the likely consequences of introducing new chemicals into the environment.

And within a decade, we should no longer be caught by surprise by high volume chemicals which are shown to pose grave environmental risks after the damage is done.

How can we measure our progress along the path to control? And how can we assess whether the legal/administrative path we choose is the most direct and efficient route to our goals? In short, how can we implement the Toxic Substances Control Act in a manner that best serves the interests of our society – now and in the future?

[Today,] I will describe the very first implementation steps that are being taken. More importantly, I will be listening for your views on the soundness of this approach into this uncharted sea of chemicals.

Everyone in the Act without Preventing Action

Of all the many substantive and procedural issues related

*The following text is taken from a speech presented as part of the Toxic Substances Law and Regulations Seminar, Government Institutes, Inc., Washington, D.C., December 10, 1976, as provided for publication by Mr. Schweitzer.

TABLE 1

Public Participation in Implementation Planning

Long-term involvement	Interim steps
Policy Advisory Committee Participation in regulation development	Information consultations with 19 organizations (October–November) Public meetings on Toxic Substances Control Act in Washington (December) and elsewhere (January–March) Public meeting on PCBs and chlorofluorocarbons (December–January)

to the new law that have been discussed within EPA [during the past two months,] none have commanded the time or attention that has been devoted to public participation in determining the initial course of implementation – that is, participation before decisions are made. To some seasoned Government officials, this relatively new way of doing business may seem like a threat to their decision-making authority; others foresee a bottleneck to program activities; while others dismiss the approach as a public relations effort. But I am afraid that these people may be living in the past. Fortunately, those officials responsible for this legislation do not share such views. While fully committed to rapid implementation on a variety of fronts, we believe that many interested parties within and outside Government share with us the responsibility for responsible implementation – for fostering greater protection of our environment while insuring the continued viability of probably the most dynamic sector of our industry. Thus, we are determined that to the fullest extent possible everyone will be in the act, but we will still have action. Specifically, what are we doing? This is indicated in Table 1.

In the longer term, we look to a broadly based Policy Advisory Committee to be a key mechanism for bringing the views of many of the interested parties into the mainstream of EPA decision-making. Hopefully, we will soon have the concurrence of the Office of Management and Budget to move ahead and formally establish such a Committee. Also, in the longer term full public participation in the earliest stages of rulemaking should become a way of life.

Meanwhile, beginning with the day after the law was signed, we have been taking a variety of interim steps to involve interested parties in our deliberations. In October and again in November [1976] we met with representatives of 19 diverse organizations vitally concerned with the legislation to seek their views on appropriate consultative mechanisms and approaches for sorting our priorities. [Next week] we will host a public meeting in Washington to hear public views on many of the key strategic policy choices that will undergird the direction of our implementation activities. Similar meetings will be held in other cities [during the next several months].

On the regulatory front, a number of public meetings are already scheduled in connection with the development of regulations concerning polychlorinated biphenyls and chlorofluorocarbons. These meetings have already begun and will extend into next year. Figure 1 represents a map for the national efforts.

A Map for the National Effort

Perhaps the most important order of business facing us in the past [several months] is the development of the initial strategy framework for implementation during the early years. Drawing on our experience in developing strategies for implementation of the Safe Drinking Water Act and the Noise Control Act, efforts are underway to articulate more clearly initial five-year objectives, to develop guiding implementation principles, and to establish program priorities. The types of issues to be addressed during this study are listed below.

Types of Issues

Cross-cutting strategic issues
Cross-cutting technical issues
Section-by-section policy issues
Section-by-section technical issues

The initial product of this activity has been a list of 99 policy issues* – a list that will probably double in length very shortly. Many of these issues cut across the entire Act while others are section specific. These are shown in Table 2. Some of them go to the very heart of the intent of the law:

*While this information was not originally included in Mr. Schweitzer's presentation, it is detailed in the Appendix of the Guidebook.

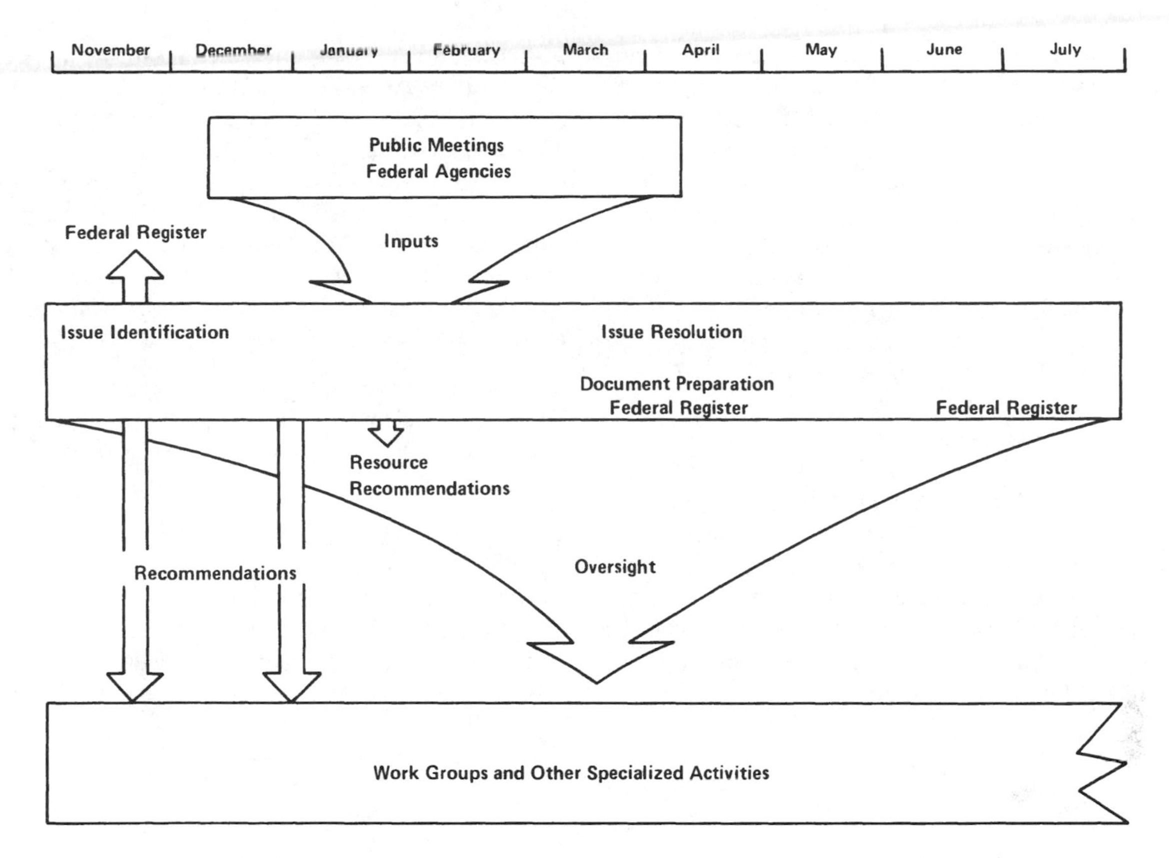

FIGURE 1. Flow chart of proposed national efforts to develop regulations for PCB's and chlorofluorocarbons.

TABLE 2

Cross-cutting Strategic Issues

Emphasis on
- Information acquisition vs. regulatory actions
- Old vs. new chemicals
- Chemical-by-chemical vs. chemical category
- Many chemicals vs. few chemicals
- Acquisition of exposure vs. toxicity data
- Preliminary screening of new chemicals vs. in-depth assessment
- Systematized chemical prioritization vs ad hoc decisions

Interlocking authorities
- Use of reporting for enforcement purposes
- Requiring reporting on chemicals subject to testing
- Reporting new uses of regulated chemicals

Other
- Research reorientation
- Programming flexibility to accommodate petitions and crisis chemicals

Should there be a predetermined emphasis on new or on old chemicals?

Should chemicals be addressed individually or in categories?

Should reporting requirements be linked to testing requirements?

Your views on these and other strategic issues would be most welcome.

A second task is to analyze the law in its totality from several vantage points, including the most effective utilization of its regulatory provisions, its information gathering and dissemination provisions, and its provisions that reinforce the efforts of organizations other than EPA to assess and control toxic chemicals. Then by [Christmas] we should be ready to begin preparing a strategy document. Hopefully, this document will serve as a focal point for debate [during the following four to six months].

At the same time we are developing a strategy, however, it has been necessary to proceed with specific implementation activities due to the urgency of some of the unattended environmental problems, the legislatively imposed deadlines, and the desirability of early procedural guidance.

Starting Down the Road to Control

The first step in the EPA process of developing a major regulation is establishment of a Work Group within EPA. Three such Work Groups have been established and two more are in the process of being established. These are shown in Table 3. The five regulatory activities of immediate concern are directed to (1) polychlorinated biphenyls, (2) selected chlorofluorocarbons, (3) the initial inventory of commercial chemicals and the premarket notification requirements mandated to be effective prior to the end of 1977, (4) initial policies, procedures, and requirements related to testing, and (5) record keeping and health and safety study requirements under Sections 8(c) and 8(d). During the next few months each of these groups will be grappling with a variety of policy issues, and broad input into these deliberations will be actively solicited.

In addition to these major regulatory undertakings a number of internal EPA Task Forces have been established to examine other provisions of the law. These range from determing the rules of procedures governing hearings, to development of research priorities to establishment of data systems (Table 4). There is obviously some urgency to considering the questions associated with the handling of citizen's petitions and notices of substantial risk which may be received shortly after the effective data of the Act. Also, we are examining the backlog of unattended environmental problems to determine additional regulatory actions that should be initiated promptly under this new authority.

Finally, a number of other sections of the law have been identified which also deserve early attention. As a first step, brief staff papers will be prepared which (1) highlight the most important issues, (2) consider preliminary options and recommendations, and (3) prepare timetables for addressing and, hopefully, resolving the issues. These staff papers (as tabulated in Table 5) will simply be the starting point to help focus the long series of deliberations within and outside the Agency that will surely accompany the development of policies in these areas.

A final activity of possible interest related to the Interagency Testing Committee. We anticipate that the Committee will hold its first meeting in early January [1977].

The Fork in the Road

On the one hand, we are vitally concerned with early clarification of the procedural ground rules and of the responsibilities of industry and other affected organizations. These complicated ground rules must be developed

TABLE 3

Work Groups

Topic	Near term milestones	Schedule
PCBs	Proposed regulations	April 1977
Chlorofluorocarbons	Proposed regs	April 1977
Premarket notification/ initial inventory	Draft regs Draft reporting form Initial straw inventory list	February 1977
Testing/categorization	Policy and procedural recommendations	January 1977
Record keeping/health and safety studies	General approach	December 1976

TABLE 4

Recently Initiated Task Forces

Toxic Substances Control Act Section	Topic	Near-term milestones	Schedule
14	Confidentiality/Freedom of Information	General approach	December 1976
26(e)	Financial disclosure	General approach	December 1976
6(c) (4)	Attorney's fees	General approach	December 1976
Several	Hearing procedures	General approach	December 1976
10 & 27	Research priorities	Identification of priority areas	January 1977
10(b)	Data systems	General approach	January 1977
26(d)	Technical Assistance Office	Recommendations	January 1977
6	Additional regulatory actions	Recommendations	January 1977
	Response capabilities	On-line capabilities/Federal Regulation notices	January 1977
21 & 20	Citizen's petitions and suits		
8(e)	Notice of substantial risk		
6(d) & 7	Unreasonable risk and imminent hazard		
	State programs	Receipt of state views	February 1977
28	Grants		
18	Preemption		

TABLE 5

Staff Papers in Preparation

Toxic Substances Control Act Section		Contents
8(a)	Discretionary reporting/	
8(a) (3) (B)	Small business definition	
		Identification of issues
5(a) (2)	Significant new uses	
		Policy and procedural
5(b) (4)	List of risk chemicals	recommendations
25(a)	Indemnity study	Proposed timetable
12 and 13	Exports/imports	
26(b)	Fee schedule	
4(b) (3) & (c)	Testing reimbursement formula	
4(g)	Response to test petitions	

with great care to provide the basis for a long-term program that will command the widest possible support from all sectors of our society. In concert with these developments, research, early warning, and data support programs must be expanded to enable us to better anticipate the problems of the future.

At the same time, we cannot delay actions to address the most serious environmental problems posed by currently uncontrolled entry into the environment of toxic substances. Nor can we delay efforts to acquire timely and relevant production data and data on the toxicity and exposure of commercial chemicals.

This two-pronged effort of immediate action and longer term planning and procedural development is the challenge of the initial months and years of the legislation. As we begin down these two avenues, we must insure that we do not move in diverging directions. This is the only way we will neither exceed the speed limit nor become bogged down in traffic. Our society can ill afford the consequences of either development.

Section V

Summary and Analysis of the Toxic Substances Control Act

SUMMARY AND ANALYSIS OF THE TOXIC SUBSTANCES CONTROL ACT (PUBLIC LAW 94-469)*

TABLE OF CONTENTS

***Prepared for the Synthetic Organic Chemical Manufacturers Association (SOCMA) by Cleary, Gottlieb, Steen & Hamilton, 1250 Connecticut Avenue, N.W., Washington, D.C. 20036, December 6, 1976 and included by virtue of their kind permission, which is hereby most gratefully acknowledged.**

EXECUTIVE SUMMARY

After a number of years of debate, Congress passed a comprehensive law regulating toxic substances in October 1976. The Toxic Substances Control Act (Public Law 94-469) is one of the most important pieces of environmental legislation to be enacted by the Congress. It grants the Administrator of the U.S. Environmental Protection Agency broad regulatory authority over the chemical and allied industries, and it authorizes numerous research projects, state programs, and federal studies. This memorandum will focus on those provisions that are of significant interest to industry.

The Act, which became effective January 1, 1977, gives the Environmental Protection Agency Administrator broad authority to

1. Require chemical manufacturers and processors to conduct extensive testing of chemical substances and mixtures
2. Require premarket notification to EPA of all new chemical substances by December 1977
3. Require premarket notification to EPA of any significant new use of an existing substance
4. Delay manufacture and marketing of a new product if sufficient information to evaluate the substance is not available and if there will be substantial human or environmental exposure
5. Ban or place restrictions on the marketing of existing or new substances that are found to pose an unreasonable risk to health or the environment
6. Impose extensive record keeping and reporting requirements

A. Effective Date

Most provisions of the Act are made effective as of January 1, 1977. However, virtually all provisions of importance to industry, including testing and premarket notification, will not take effect until EPA issues the necessary implementing regulations. An important exception is the reporting requirement discussed on page 5.5. A schedule of important statutory dates and deadlines may be found on page 5.7.

B. Testing

Section 4 of the Act requires EPA to order manufacturers and processors of chemical substances and mixtures for which there are insufficient health and safety data to conduct tests whenever EPA finds that the chemical may present "an unreasonable risk" or there will be substantial human or environmental exposure to the chemical. The Act does not define "unreasonable risk," but the legislative history indicates that EPA should balance the probability, severity, and magnitude of harm against the benefit to society from the chemical substance or mixture.

EPA can require testing of mixtures only if it makes the additional finding that the effect of the mixture on health or environment cannot reasonably and efficiently be determined by testing the substances which comprise the mixture.

Rules will be developed by EPA to establish test protocols and procedures for sharing the expense of testing among manufacturers who are subject to testing requirements for the same product or category of products.

Section 4 also establishes a federal interagency committee to designate up to 50 chemicals that the committee regards as priority candidates for testing. The Administrator must issue a rule ordering testing of these chemicals within a year after they are listed or publish an explanation of his failure to do so.

Flow charts of the actions which EPA and the interagency committee are authorized to take under Section 4 appear in Figures 1 and 2.

C. Premarket Notification for New Substances and Uses

Section 5 requires manufacturers of new chemical substances to notify EPA at least 90 days in advance of manufacture or distribution. EPA will compile an inventory list of all existing substances to enable companies to determine if a chemical substance is a "new" substance. To be eligible for inclusion on the list of existing substances, the substance must have been manufactured, processed, or imported within 3 years of a cutoff date designated by EPA – July 1, 1977 is the most likely date. Mixtures and chemicals manufactured in small quantities for research and product development purposes are not subject to the premarket notification requirement.

The principal purpose of premarket notification is to provide EPA with an opportunity to determine if manufacturing or distribution of a chemical substance should be limited, delayed, or even prohibited. Thus, Section 5 requires the manufacturer of a new chemical substance to submit data on proposed use, employee exposure, any available test data, and health and safety information.

Manufacturers and processors of existing chemicals for a significant new use are subject to the same requirements. EPA will issue rules defining what new uses of existing products are to be considered "significant new uses."

At the end of the 90-day period (which can be extended for an additional 90 days), the chemical can be manufactured and processed, unless EPA takes regulatory action. If EPA determines that there is insufficient information available to evaluate the new chemical substance and that, in the absence of such information, the chemical may present an unreasonable risk it must issue an order proposing to delay or limit manufacture or distribution. If there will be substantial human or environmental exposure to such a chemical, the chemical is, in effect, presumed to present an unreasonable risk. A manufacturer can object to such a proposed order and EPA must then proceed to court

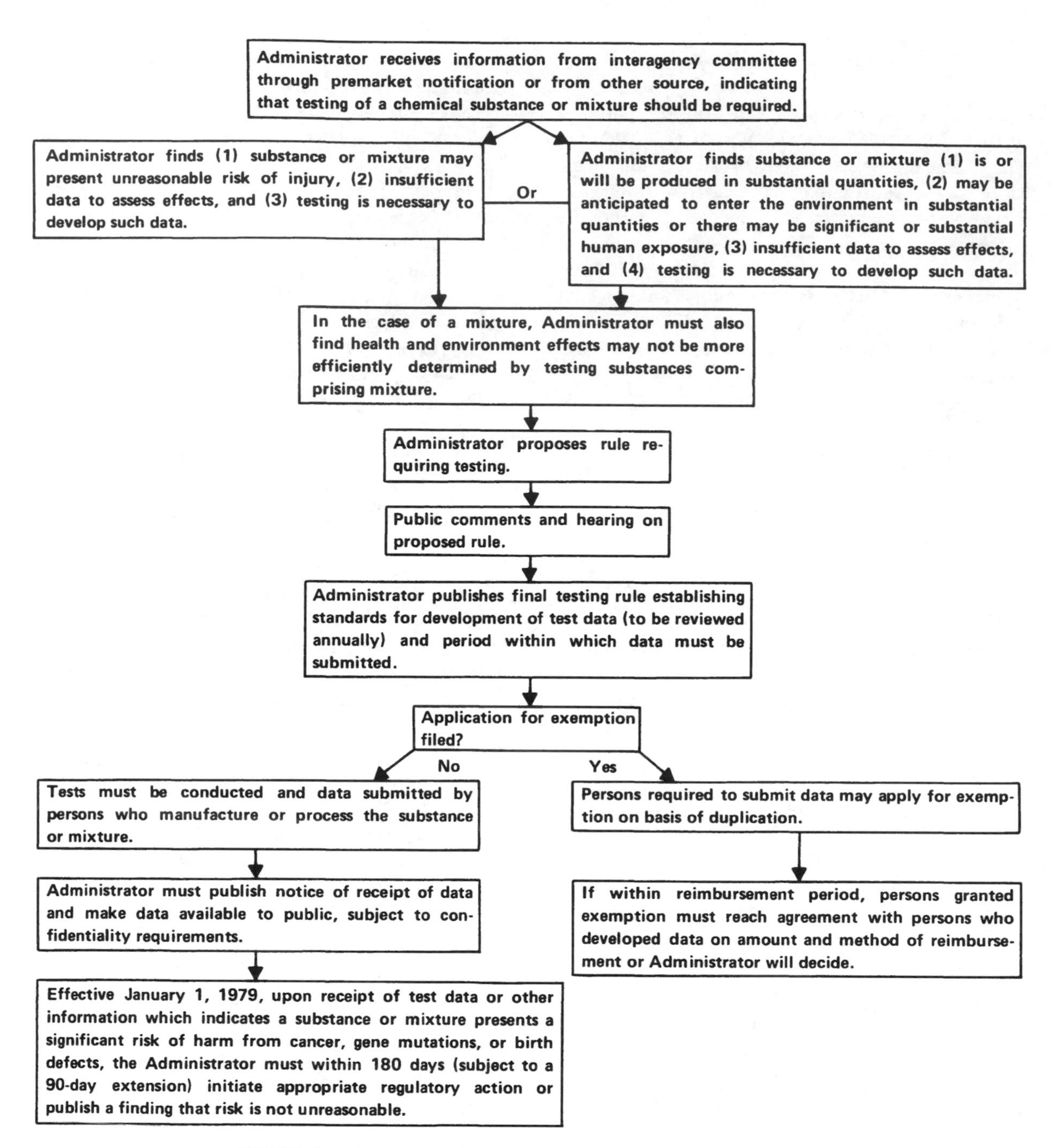

FIGURE 1. Flow chart – Section 4 (of the Act) Testing Requirements.

to block or limit manufacture or distribution. If the court confirms the Administrator's findings, it will issue an injunction.

If EPA receives information during the notice period that suggests there is a reasonable basis to conclude that the new chemical substance may, in fact, present an unreasonable risk, EPA is required to issue an immediately-effective rule restricting manufacture or distribution. The restrictions can range from imposing labeling or record-keeping requirements to limiting manufacture or use. However, total ban on manufacture or distribution can be imposed immediately over the objection of a manufacturer by application to the courts for an injunction.

A flow chart of the premarket notification procedures established by Section 5 may be found in Figure 3.

D. Regulation of Hazardous Chemical Substances and Mixtures

Section 6 requires the Administrator to regulate existing or new chemical substances or mixtures when there is a reasonable basis to conclude that the chemical presents or will present an unreasonable risk of injury to health or the environment. As noted above, the forms of regulation available to the Administrator range from labeling and record-keeping, to limiting use of disposal, or even total prohibition of manufacture. The Act instructs the Admini-

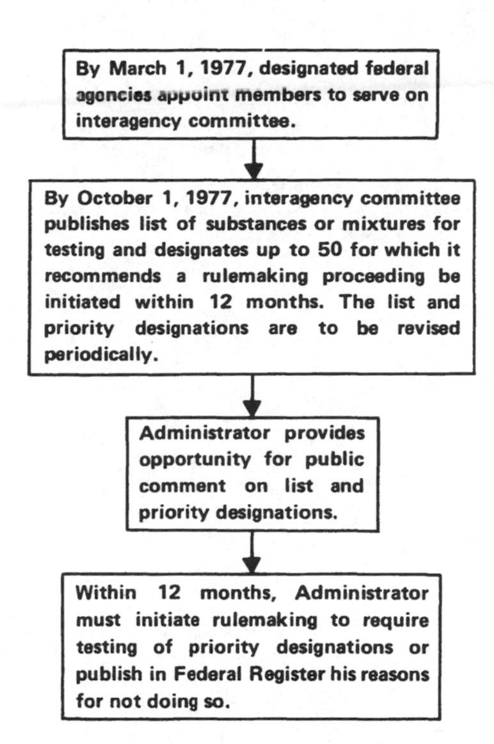

FIGURE 2. Flow chart – Interagency Committee Recommendations on Testing and Priority Designations.

strator to utilize "the least burdensome" measure adequate to protect against the risk and to consider the effect on small business in deciding upon the form of regulation.

While the Administrator ordinarily must proceed by normal rulemaking procedures and allow an opportunity for public comment before regulating manufacture, distribution, use, or disposal under Section 6, EPA may make a rule immediately effective if there is an "unreasonable risk of serious or widespread injury." However, an immediately effective rule prohibiting manufacture must be preceded by a court determination that the substance poses an imminent hazard.

Polychlorinated biphenyls (PCBs) are specifically regulated in the legislation. A ban on nonenclosed PCBs takes effect on January 1, 1978, and a total ban on manufacture or importation of all PCBs takes effect January 1, 1979.

E. Imminent Hazards

The Act provides that when a chemical presents an imminent hazard, the Administrator may commence a court action to obtain an order banning manufacture and distribution. The court can also order the seizure or recall of the hazardous substance (or any article containing the substance), impose limitations on its use or manufacture, or require the manufacturer to replace or repurchase the substance.

F. Reports to EPA

The Act authorizes EPA to require manufacturers and processors of chemical substances to submit to EPA a variety of information including data on categories of use, employee exposure, and health and environmental effects. A manufacturer may be required to submit not only information known to it, but also information that is "reasonably ascertainable" by it.

"Small" chemical companies are exempt from these routine reporting requirements. However, EPA can require small companies to report information about chemicals which are subject to a testing rule or which EPA has previously determined may present an unreasonable risk.

Effective January 1, 1977, the Act imposes the duty on any person who obtains information "which reasonably supports the conclusion" that a substance or mixture manufactured or distributed by that person "presents a substantial risk of injury to health or the environment" inform the Administrator "immediately" of such information, unless that person knows the Administrator has been adequately informed by others.

G. Record Keeping Requirements

The Act also authorizes the Administrator to require manufacturers and processors of chemical substances to keep various records. Manufacturers and processors must maintain records of "significant adverse reactions to health or the environment alleged to have been caused by a substance or mixture," including records of consumer complaints (for 5 years) and employee reactions (for 30 years). EPA is also directed to require the submission of lists of health and safety studies which the manufacturer has conducted or the existence of which is "reasonably ascertainable." The Administrator may require the submission to EPA of any study appearing on such a list.

H. Confidentiality of Data

With some exceptions, the Act prohibits the disclosure of proprietary or trade secret information obtained by EPA under the Act. Data which is designated as confidential generally may not be released by the Administrator without 30 days advance notice to the submitting company. The principal exception to both of these requirements pertains to health and safety studies.

I. Penalties and Enforcement

EPA may inspect any plant or establishment in which chemicals are manufactured or stored and subpoena witnesses, documents, and other information.

The Act sets a civil penalty of up to $25,000/day for violations of the law and a criminal penalty for knowing or willful violations of up to $25,000/day and/or imprisonment for up to 1 year. District courts are authorized to restrain violations of the Act and may order seizure of substances manufactured or distributed in violation of the law.

J. Citizen Suits

Citizens are authorized to bring suit in district courts to restrain violations of the Act or to compel the Administra-

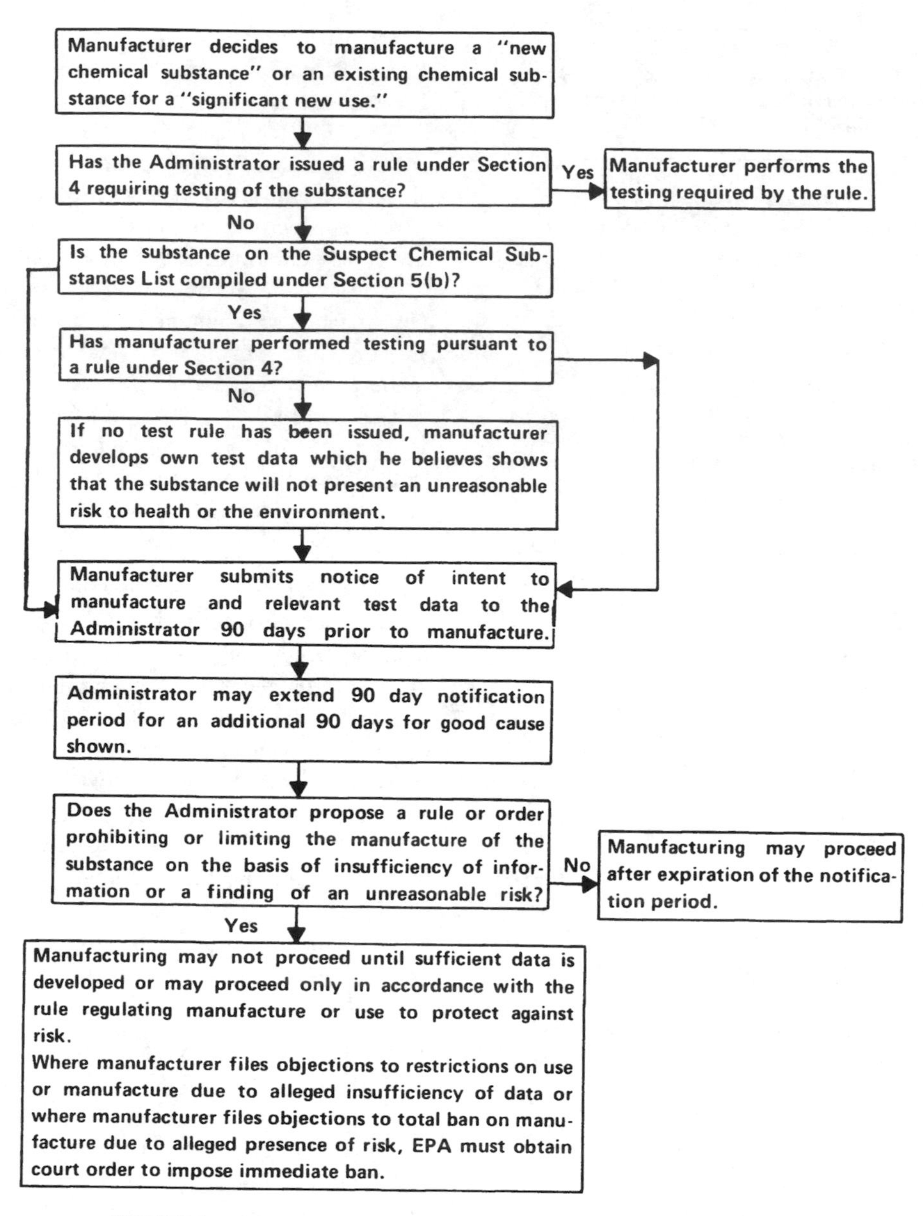

FIGURE 3. Flow chart – Section 5 (of the Act) Premarket Notification.

tor to perform nondiscretionary duties mandated by the Act.

K. Citizens' Petitions

The Act also authorizes citizens to petition the Administrator to initiate proceedings for the issuance, amendment, or repeal of certain orders or rules. If the Administrator fails to act or denies such petitions, the citizen may file a lawsuit to compel initiation of the action.

L. Conflict with Other Federal Laws

If the Administrator determines, in his discretion, that a law administered by another federal agency could be used to prevent or reduce, to a sufficient extent, the risk or injury posed by a substance, the Administrator must give notice to and provide an opportunity for the other federal agency to act before taking action under this Act.

In the case of laws administered by EPA, the Administrator may proceed under this Act rather than other applicable laws if he determines, in his discretion, that it is in the public interest to do so.

In both cases, the discretionary determinations of the Administrator are not subject to judicial review.

M. Conflict with State and Local Laws

The Act removes from the state and local jurisdictions the authority to regulate toxic substances in two major respects. First, no state or locality may establish testing requirements for a purpose similar to that of a testing rule issued by the Administrator. Second, no state or locality may impose a more stringent regulation on the manufacture, processing, or distribution of a chemical substance without EPA approval, except to impose a ban on the end use of the substance within the state or local jurisdictions.

N. Foreign Trade

The Act generally does not apply to exports. Imports must meet the requirements of the Act.

SCHEDULE OF IMPORTANT STATUTORY DATES AND DEADLINES

Effective date of Toxic Substances Control Act* – January 1, 1977

1. Duty to notify EPA of information that a product presents a substantial risk of injury to health or the environment – January 1, 1977
2. Premarket notification
   a. Publication of reporting rules to enable EPA to compile an inventory of existing chemical substances – June 29, 1977**
   b. Publication of inventory of existing chemical substances – November 11, 1977**
   c. Date by which the premarket notification obligation is to be imposed on manufacturers of new chemical substances (assuming above deadlines met) – December 12, 1977**
3. Interagency committee recommendations
   a. Publication by interagency committee of its initial testing recommendations and designating of up to 50 chemical substances and mixtures as priority candidates for testing – October 1, 1977**
   b. Publication by interagency committee of a revised list of the chemical substances and mixtures most in need of testing*** – April 1, 1978**
   c. Deadline for issuance of testing rule and priority designations. The Administrator must publish his reasons for failure to order testing of any chemicals so designated† – October 1, 1978‡
4. First date on which the Administrator is required to take regulatory action under Section 4(f) against chemical substances or mixtures which can reasonably be concluded to present a significant risk of cancer, gene mutation, or birth defects on the basis of information received by him. The Administrator must publish his reasons for failing to act within 180 days of receipt of such information. The 180-day period can be extended for an additional 90 days – January 1, 1979
5. Regulation of PCBs
   a. Publication by Administrator of marking requirements and methods for disposal of PCBs – July 1, 1977‡
   b. Ban on nonenclosed PCBs (except those that EPA concludes do not pose an unreasonable risk) – January 1, 1978
   c. Ban on manufacture and importation of PCBs – January 1, 1979ǂ
   d. Ban on processing and distribution of PCBs except those sold for purposes other than resale prior to this date – July 1, 1979ǂ

III. MEMORANDUM ON THE TOXIC SUBSTANCES CONTROL ACT OF 1976

This memorandum discusses the recently enacted Toxic Substances Control Act (Public Law 94-469). It is intended to provide a basic outline of the provisions applicable to industry and to serve as a reference for interpretation and analysis of the Act's provisions during its implementation.

PART I – PRINCIPAL PROVISIONS OF THE ACT

I. Congressional Policy and Intent

A. The Objectives of the Act

The Toxic Substances Control Act provides that it is the policy of the U.S. to ensure that adequate data be developed on the effect of chemical substances and mixtures on health and the environment, that the development of such data should be the responsibility of industry, and that the federal government should have adequate authority to regulate those substances which present "an unreasonable risk of injury to health or the environment," and to take action against those which are imminent hazards. [Section 2(b)(1), (2)]

Congressmen Broyhill and Eckhardt, the principal managers of the legislation in the House of Representatives, observed that the legislation attempted to strike a balance between environmental concerns and industrial and marketing considerations.◊ The House managers noted that "unreasonable risk" was the trigger for the Administrator of the Environmental Protection Agency (EPA) to take action, because the absence of such a limitation would

*Section 4(f), discussed in point 3, does not take effect until January 1, 1979.

**These are statutory deadlines; the described action can take place prior to these dates. However, past experience suggests that EPA is not likely to act sooner than the relevant deadline, and it may not be able to meet the deadline.

***The Committee must review and/or revise the list every 6 months.

†The Administrator must order testing at the end of each 12-month period following inclusion of a new chemical substance or mixture in the interagency committee's list, or publish a similar statement.

‡This is a statutory deadline; the described action can take place prior to these dates.

ǂThe Administrator has limited authority to grant a temporary exemption from this deadline. [Section 6(e)(3)(B)]

◊122 Congressional Record H8811 (daily edition August 23, 1976) (House Debate – Congressman Broyhill) [hereinafter all citations to Congressional Record are to daily edition]; 122 Congressional Record H8811-2 (August 23, 1976) (House Debate – Congressman Eckhardt).

assume "that a risk free society is attainable, an assumption that Congress does not make."*

Although the primary purpose of the Act is to prevent unreasonable risk or injury to health or the environment, the statement of congressional policy provides that the EPA Administrator authority should not be exercised in such a manner as "to impede unduly or create unnecessary economic barriers to technological innovation." [Section 2(b)(3)] Congressional concern about deterring product development and the disproportionate impact of such a consequence on the smaller chemical company** motivated this provision and the exemption from premarket notification afforded to chemicals used in research and development.***

B. Congressional Intent as to Implementation of the Act

Section 2(c) provides that it is the intent of Congress that the Administrator carry out the Act "in a reasonable and prudent manner" and "consider" the environmental, economic, and social impact of any action. However, this provision does not require EPA to make findings of economic or social impact or to conduct a cost-benefit analysis before acting.† The Senate Report emphasizes that this intent of Congress "should guide each action the Administrator takes under other sections of the bill."‡ The Congressional debate is also replete with cautions that the Administrator should act in a reasonable fashion and consider the economic and societal impact of regulatory actions.‡‡

While certain provisions of the Act provide special treatment for small chemical companies,◇ there is no general statement in the statute requiring the Administrator to consider the effect on small business of each regulatory action taken, except as this obligation may be subsumed under the requirement that the Administrator consider "economic impact." However, Congressman Broyhill stated, "the conferees expect that the Administrator of the EPA will consider the impact which his actions may have on the small businesses regulated under this legislation."♦

II. Key Definitions

The Toxic Substances Control Act imposes testing requirements and other substantial regulatory burdens on persons who "manufacture or process chemical substances and mixtures" which present or may present "an unreasonable risk of injury to health or the environment." Manufacturers of "new chemical substances" must provide premarket notification to the Administrator while manufacturers of "mixtures" need not. Accordingly, the definitions of these terms are critical to an understanding of the Act.

A. Chemical Substances

"Chemical substance" is broadly defined as any organic or inorganic substance of a particular molecular identity, including any combination of such substances thereof occurring as a result of a chemical reaction or occurring in nature or "any element or uncombined radical." [Section 3(2)(A)] Thus, "chemical substance" is defined in such a manner that the Administrator must be provided with advance notice of naturally occurring substances which are produced commercially for the first time as well as of new synthetic chemicals.□

Rules that will be applicable to a chemical substance will include, and not treat separately, "impurities and concomitant products, including incidental reaction products, contaminants, co-products, and trace materials."■ The Conference Report states that the Administrator is expected to develop guidelines to clarify the circumstances under which impurities and concomitant products are to be included within the term "chemical substance" and when they are to be treated separately.♠

*122 Congressional Record 11343 (September 28, 1976) (House Debate on Conference Report – Congressman Broyhill); see also House of Representatives Report No. 94-1341, 94th Congress, 2nd Session 17–18 (July 14, 1976) [hereinafter cited as House Report at]; 122 Congressional Record H8811-8812 (August 23, 1976) (House Debate – Congressman Eckhardt).

**See 122 Congressional Record H8805 (August 23, 1976) (House Debate – Congressman McCollister); 122 Congressional Record H8809 (August 23, 1976) (House Debate – Congressman Collins); 122 Congressional Record H8812 (August 23, 1976) (House Debate – Congressman Slack). 122 Congressional Record H8813 (August 23, 1976) (House Debate – Congressman Ashbrook); 122 Congressional Record H8817 et seq. (August 23, 1976) (House Debate – Congressmen Fuqua and Eckhardt).

***Section 5(h) (3). See page 5.15.

†House Report at 9, 14.

‡Senate Report Number 94-698, 94th Congress, 2nd Session 14 (March 16, 1976) [hereinafter cited as Senate Report at].

‡‡122 Congressional Record H11344 (September 28, 1976) (House Debate on Conference Report – Congressman Broyhill); 122 Congressional Record S16808 (Senate Debate on Conference Report – Senator Pearson); 122 Congressional Record H8804 (August 23, 1976) (House Debate – Congressman McCollister).

◇For example, under Section 6(c), the Administrator must consider the impact of a Section 6(a) rule regulating hazardous substances on small companies. Under Section 8(a), small companies are relieved of reporting requirements under specified circumstances. Under Section 26(b), small companies required to submit test data pay greatly reduced filing fees.

♦122 Congressional Record H11344 (September 28, 1976) (House Debate on Conference Report).

□122 Congressional Record H11346 (September 28, 1976) (House Debate on Conference Report – Congressman Murphy).

■Senate Conference Report No. 94-1302, 94th Cong., 2nd Session 57 (September 24, 1976) [hereinafter cited as Conference Report at]. The Senate Conference Report is identical to the House Conference Report]; 122 Congressional Record 16,803 (September 28, 1976) (Senate Debate on Conference Report – Senator Magnuson).

♠Conference Report at 57. During debate on the Conference Report, Senator Magnuson stated his view that the Conferees intended that the term "chemical substance" be defined broadly to include trace chemicals for purposes of rules restricting or regulating marketing, but interpreted narrowly by EPA when compiling an inventory of existing chemicals to asure that new contaminants are subject to the premarket notification requirements applicable to new chemical substances.

A "new chemical substance" is any chemical substance which is not included in the inventory list of existing substances that EPA is directed to prepare and publish. [Section 3(9)]

"Chemical substance" is defined so as to exclude any commercial pesticide, tobacco, nuclear material, firearm, ammunition, food, food additive, drug, cosmetic, or device, as these substances are (or more appropriately should be) regulated under other statutory authority. [Section 3(2)(B)]

B. Mixtures

Additionally, "chemical substance" is defined so as to exclude "mixtures." [Section 3(2)(B)(i)] "Mixture" is in turn defined as any combination of two or more chemical substances if the combination does not occur in nature and if the substances do not react chemically with each other. Where combinations occur as a result of a chemical reaction, the combination will nevertheless be classified as a mixture if none of the substances comprising the combination are new chemical substances and if the combination could have been manufactured without a chemical reaction occurring at the time of manufacture. [Section 3(8)]* A change in the proportion of inert ingredients creates a new mixture, not a new chemical substance.**

C. Manufacture, Process, and Distribution in Commerce

"Manufacture" is defined to include importing as well as producing or manufacturing. [Section 3(7)] The House Committee Report recognizes that mixtures such as adhesives, paints, inks, and drying oils may incidentally undergo chemical reactions during storage or upon end use. The report states that it was not intended that the person engaged in the end use or storage activity in which the incidental reaction occurs be considered a manufacturer.***

"Process" is defined as the preparation of a chemical substance or mixture after manufacture for distribution in commerce in the same or different form as it was received from the manufacturer or as part of an article containing the chemical substance or mixture. [Section 3(10)]

Only those engaged in processing or manufacturing are subject to the Act's testing and premarket notification requirements. Comparatively few obligations are placed on distributors of chemical products, and those are primarily record keeping and reporting requirements. Distributors are persons who sell, deliver, or store chemical substances, mixtures, or articles containing chemical substances or mixtures "in commerce." [Section 3(4)] "Commerce" is defined so as to include both interstate commerce and intrastate activity affecting interstate commerce.†

D. Unreasonable Risk to Health or the Environment

As noted above, chemical substances and mixtures which may present an "unreasonable risk of injury to health or the environment" are required to be tested, and chemical substances which are in fact found to present an unreasonable risk are subject to regulation.‡

"Unreasonable risk" is not a defined term, but the House Report provides that, in general, such a determination will involve "balancing the probability that harm will occur and the magnitude and severity of that harm against the effect of proposed regulatory action on the availability to society of the benefits of the substance or mixture." EPA must take into account the availability of substitute chemicals and the adverse societal effects of the proposed action.ǂ The Senate Report provides that in determining what is an "unreasonable risk," the Administrator must weigh the effect on small business as one of the factors involved in balancing risk and benefits.◇ This balancing process does not, however, require EPA to make a formal cost-benefit analysis.♦

The Administrator may weigh severity of harm as well as probability of harm in determining whether there is unreasonable risk,□ but may not base this finding on "mere conjecture or speculation."■ Although, as noted above, a finding of "unreasonable risk" is the trigger to regulatory action under several provisions of the Act, the House Report states that "implementation of the standard will of necessity vary depending on the specific regulatory authority which the Administrator seeks to exercise."♠ As an example, the House Report observes that the finding of "unreasonable risk" carries more significant consequences under Section 6 (which authorizes restrictions to be placed on manufacture) than under Section 4 (which relates to testing) or under Section 5(e), which authorizes EPA to delay marketing of certain substances pending the development of information.§

*House Report at 12; Conference Report at 57.

**Conference Report at 72. Thus, where such modifications occur, no new chemical substance results and premarket notification requirements are not applicable.

***House Report at 13; see also Senate Report 94-698 at 19.

†See statement of the General Accounting Office, House Hearings No. 94-41 (1975) at 54–55, urging adoption of this definition.

‡"Health" is not defined but "environment" is stated to include "water, air, and land and the interrelationship which exists among and between water, air, land and all living things." [Section 3(5)]

ǂHouse Report at 14; Senate Report at 13.

◇Senate Report at 12.

♦House Report at 14.

□*Id.*

■*Id.* at 18.

♠*Id.* at 14.

§The House Report also indicates that EPA must present very good reasons for any exercise on its broad Section 6 authority because of the inherent possibility of "far-reaching" societal consequences. *Id.* at 14–15.

III. Testing

Section 4(a) requires the Administrator to order testing of a chemical substance or mixture not only when he finds that a chemical may present an unreasonable risk, but also where he finds that there will be substantial production and substantial exposure and there are insufficient data about the effect of the substance on health or the environment. Testing is required in such cases even though there is no indication that the chemical may be hazardous.*

This provision, which becomes effective January 1, 1977, requires EPA to set standards for testing and to determine, if necessary, a fair allocation of costs among those required to conduct such testing. It also creates a federal interagency advisory committee to prepare a list of the chemicals to which EPA should give priority in issuing testing requirements. The first such list is required to be published no later than October 1, 1977.

A. The Standards for Requiring Testing

Where there are insufficient data upon which the effects of a chemical can reasonably be predicted, the Administrator must publish a rule requiring testing if he finds (1) that the manufacture, processing, distribution, use or disposal of the chemical may present an unreasonable risk, (2) that the chemical will be produced and enter the environment in substantial quantities, or (3) that there is or may be significant or substantial human exposure. [Section 4(a)(1)] Before ordering testing of a mixture, the Administrator must make the further finding that the mixture's effect on the environment may not be more reasonably and efficiently determined by testing the component chemical substances. [Section 4(a)(2)] ** The Administrator must specify whether manufacturers or processors or both are subject to the testing rule.

The Act does not define certain phrases critical to the determination of whether testing will be required, i.e., "substantial quantities" and "significant or substantial human exposure." However, the House Report provides that in making such findings, the Administrator may consider duration of exposure, level or intensity of exposure over various periods of time, the number of people exposed, and the extent of environmental exposure.***

B. Procedures for Issuing a Testing Rule

Rules requiring testing may only be adopted after notice is given in the Federal Register and opportunity is provided for public comment. [Section 4(b)(5)] †

Section 4(b)(1) requires the Administrator to publish a rule specifying standards for the development of test data and a "reasonable" time period for submitting the test results.

In determining the standard and time period, the Administrator must consider the relative costs of the various test protocols and methodologies and the availability of facilities and personnel. [Section 4(b)(1)] Methodologies that may be prescribed include epidemiologic studies, serial or hierarchical tests, in vitro tests, and whole animal tests. Health and environmental effects for which standards for the development of test data may be prescribed include carcinogenesis, mutagenesis, teratogenesis, behavior disorders, cumulative or synergistic effects, and any other effect which may present an unreasonable risk of injury to health or the environment. [Section 4(b)(2)(A)]

Test standards must be reviewed annually to ensure their continuing adequacy. [Section 4(b)(2)(B)] Notice of receipt of test data must be published in the Federal Register within 15 days of receipt. [Section 4(d)]

A company intending to manufacture or process new chemical substances or mixtures as to which no testing rule has been published may request that the Administrator propose standards for the development of test data. The Administrator has 60 days in which to act on the petition. If granted, standards must be prescribed within 75 days thereafter. If denied, the causes for the denial must be published in the Federal Register. [Section 4(g)] This procedure may be useful when the Administrator has found that a particular chemical may pose an unreasonable risk and has placed it on the "risk list" authorized by Section 5(b)(4)(A) or has ordered a delay in marketing the chemical under Section 5(e) because there is insufficient information available to evaluate it. In either case, manufacture and distribution of the chemical may be blocked until test data is submitted to EPA, and it may, therefore, be advantageous to request EPA to specify exactly what data must be submitted.

C. Exemptions from the Testing Requirement

Not everyone subject to a testing rule need actually engage in testing. It is sufficient that the required testing has been done or is being done by one or more companies. If that is the case, all other persons subject to the rule can apply for an exemption. Applications will be granted when the Administrator determines that a chemical substance or mixture subject to a testing rule is equivalent to a chemical substance or mixture for which data have been submitted or are being developed. [Section 4(c)(2)] The Conference Report states that the Administrator should take the

*Conference Report at 61; House Report at 18.

**122 Congressional Record H11346 (September 28, 1976) (House Debate on Conference Report – Congressman Murphy); House Report at 18.

***House Report at 18.

†The Administrator is provided with the authority to issue rules applicable to categories of chemicals as opposed to individual substances under any provision of the Act, including rules requiring testing. Section 26(c)(1); House Report at 61; Conference Report at 102.

presence of any trace contaminants into account in determining whether such a determination can be made.*

If an exemption is granted, cost sharing with any company which has borne the initial testing expense is required for, at the minimum, a 5-year period. [Section 4(c)(3)] ** If the parties are unable to agree upon reimbursement, the Administrator must determine a "fair and equitable" allocation of costs. In determining the fair amount of reimbursement, the Administrator is instructed to consider the competitive and market share positions of the person ordered to make reimbursement and the person to be reimbursed to avoid disproportionate impact on small businesses. A reimbursement order is subject to immediate judicial review. [Section 4(c)(3)(A) and (4)(a)] ***

D. Interagency Advisory Committee Recommendations

Section 4(e) establishes a federal interagency committee to make recommendations concerning testing priorities to the Administrator.† The committee is directed to meet by March 30, 1977 and to publish in the Federal Register by October 1, 1977 a list of those chemical substances or mixtures in the order in which the Committee determines the Administrator should take action, giving priority to those which are suspected of causing cancer, gene mutations, or birth defects. The Committee must set forth the reasons for inclusion of each substance or mixture on this list. The list is to be reviewed and revised every 6 months. The committee is also required to designate up to 50 substances or mixtures on the original or revised list as priority candidates for testing within the 12-month period immediately following designation. [Section 4(e)(1)(A)] The Act provides for public comment on the committee's list and on any revisions of the list. [Section 4(e)(1)(B)]

If the Administrator does not initiate a rulemaking proceeding requiring testing of each chemical separately designated as a priority candidate for testing within a year after such designation (or October 1, 1978 for the first group), he must publish an explanation as to why such action has not been taken in the Federal Register. [Section 4(e)(1)(B)] The Conference Report states that, while the recommendations of the interagency committee are to be afforded "great weight," the decision to require testing rests with the Administrator, and he should not "divert from the regulatory activities of the Agency an inordinate amount of resources to justify the failure to require testing."‡

Effective January 1, 1979, the Administrator must take appropriate regulatory action under Sections 5, 6, or 7 within a stated period after receiving information, which provides the Administrator with "a reasonable basis to conclude that a chemical substance or mixture presents or will present a significant risk of serious or widespread harm to human beings from cancer, gene mutation, or birth defects." [Section 4(f)] The Administrator is given 180 days after receipt of such data to respond (subject to an additional 90-day extension for good cause). A finding by the Administrator that a risk is not unreasonable must be published in the Federal Register and is subject to judicial review.

E. Filing Fees

Where test data must be submitted under Section 4 (or under Section 5), a filing fee of no more than $2500 may be imposed to defray costs of administration. [Section 26(b)] Small companies must pay a filing fee of no more than $100. The Administrator is expected to take ability-to-pay into account in setting the amount of all filing fees.‡‡

IV. The Premarket Notification

The premarket notification provisions of the Act impose an important new regulatory burden on chemical manufacturers and processors. While the requirements themselves are relatively complex, their thrust is simple. EPA must be given advance notice of the proposed manufacture of all new and existing products that are to be put to new uses which may have some health or environmental significance. The notion is that EPA will then be in a position to take action against potentially dangerous chemicals before commercial production begins and that, as a result, the cost of any regulatory action in terms of loss of jobs and capital investment will be minimized.

A. The Premarket Notification Requirement

Section 5 provides that manufacturers of new chemical substances and manufacturers and processors of existing chemical substances for "significant new uses" must notify the Administrator at least 90 days in advance of commercial manufacture. This premarket notification requirement applies to both manufacturers and processors of an existing chemical for a significant new use as determined by the Administrator because the manufacturer may not be aware of the new use.

EPA with an opportunity to determine if the manufacture, processing, distribution, use, or disposal of the substance should be limited, delayed, or prohibited because data are

*Conference Report at 6.

**The reimbursement period may be extended until testing data are developed. [Section 4(c)(3)(B)]

***House Report at 21.

†The committee will be composed of representatives from the following federal agencies: EPA, Occupational Safety and Health Administration, National Institute of Occupational Safety and Health, Council on Environmental Quality, Department of Commerce, the National Science Foundation, the National Cancer Institute, and the National Institute of Environmental Health Sciences. [Section 4(e)(2)(A)]

‡Conference Report at 62.

‡‡Conference Report at 102; 122 Congressional Record H8805 (August 23, 1976) (House Debate – Congressman McCollister).

insufficient or because an unreasonable risk to health or environment is present. [Section 5(a)(1)] * The premarket notification requirement does not apply to mixtures, as noted above, or to new distributions of existing chemicals, unless such distribution constitutes a significant new use.

The notification must specify the name of the chemical, its chemical identity and molecular structure, the categories or proposed categories of use, the amounts manufactured or to be manufactured for each category of use, a description of any by-products, the number of employees exposed, and the manner of disposal. The notification must be accompanied by any existing test data relating to health or environmental effects of the substance and must describe any other data relevant to its health or environmental effects [Section 5(d)] ** Within 5 days of receipt of the notice, a summary version must be published in the Federal Register, and the complete notice must be made available for public inspection subject to the confidentiality provisions of Section 14. (See pages 77–80, *infra.*)

If the chemical substance is subject to a Section 4 testing rule, the manufacturer also is required to submit the required test data with the notification. If an exemption from testing has been granted, the company may not begin manufacture or processing until 90 days after the submission of the test data on which the exemption is based. [Section 5(b)]

The Administrator may, by rule, compile a list of "suspect" substances as to which the Administrator finds that manufacture, processing, distribution, use, or disposal presents or may present an unreasonable risk. [Section 5(b)(4)(A)(i)] If no testing rules have been issued with respect to a substance on this list, then the person required to submit the premarket notification also must submit data demonstrating that the substance or significant new use will not present an unreasonable risk. [Section 5(b)(2)] ***

The Administrator may extend the 90-day premarket notification period up to 180 days for "good cause" [Section 5(c)], but the Conference Report states that the extension from 90 to 180 days should not be automatic if sufficient data to evaluate the product will be available at an earlier point in time.†

B. Determination of Whether a Chemical Substance is a "New" Chemical Substance

To enable the manufacturer to determine whether a chemical substance is a "new" substance, the Administrator is directed to compile and keep current an inventory of existing chemical substances. [Section 8(b)(1)] This inventory list will contain existing chemical substances and new chemical substances reported under Section 5, but will exclude substances not manufactured or processed since July 1, 1974 and research and development chemicals which are manufactured or processed in small quantities.

The Administrator is directed to publish the inventory list by November 11, 1977. Thirty days thereafter, manufacturers of any chemicals not on the list must submit a premarket notification form to EPA and cease any manufacturing commenced prior to December 12, 1977 until the notification period has expired.

The inventory of existing chemical substances may list them by category, rather than individually, "to the extent consistent with the purposes of this Act." [Section 8(b)(2)] The House Report notes that the listing of chemical substances by category will ensure that "minor" modifications in the formulation or structure of a chemical substance which would have "insignificant health or environmental consequences" do not result in that chemical being classified as a new chemical for purposes of the premarket notification requirements.‡ The House Report also recognizes that many small companies can compete only by continually reformulating and making slight changes in existing chemical substances. Accordingly, the Administrator is urged to maintain the inventory list by categories so that "every insignificant change" does not invoke the premarket notification requirement. However,

*122 Congressional Record H11346 (September 28, 1976) (House Debate on Conference Report – Congressman Murphy); 122 Congressional Record H8804 (August 23, 1976) (House Debate – Congressman Clawson).

**See also 122 Congressional Record S16803 (September 28, 1976) (Senate Debate on Conference Report – Senator Magnuson).

***Section 4(g) establishes a procedure under which a company can petition EPA to establish standards for the development of such test data. See page 5.10.

†See Conference Report at 67.

‡House Report at 44. The House Report lists as examples of such minor modifications: polymers or copolymers which vary only in the proportion of starting materials or catalysts used, or in molecular weight, molecular weight distribution, chain structure, or crystallinity; changes with an existing chemical substance in the proportions of colorants, stabilizers, antioxidants, fillers, solvents, carriers, surfactants, plasticizers, and other adjuvants which are themselves reported as existing substances; variations in the proportion of alloyed metals in iron and steel products and other metal alloys; variations in naturally occurring substances or mixtures (such as crude oil, natural gas, minerals, or ores) and the resulting variations in extracts or refined products therefrom; variations in reported reactive mixtures whose commercial or end-use product is electrical energy (batteries); and salts which result from the combination of an existing inorganic anion with an existing inorganic cation.

the Report also notes that minor modifications of existing innocuous compounds may produce highly toxic chemicals. The Report concludes that "the use of categories should be limited to areas where the effects of such minor modifications are well understood to have insignificant health and environmental consequences."*

The Senate Report states that listing by category would be appropriate in compiling the Section 8(b) inventory to assure, for example, that every variation in the distribution of a polymer-chain length did not trigger a premarket notification requirement. However, the Senate Report admonishes that categories should not be used in compiling the Section 8(b) inventory so effectively as to exempt new chemical substances from premarket notification.**

The House Report also notes that certain minor reactions may occur during the mixing or storing of chemicals, which may result in what technically would be considered a "new chemical substance." However, because the "new" substance is not manufactured for commercial purposes, it would not be subject to the premarket notification provisions.***

C. Determination of Whether a New Use Constitutes a Significant New Use

Whether a new use constitutes a "significant new use" will be determined by reference to criteria established by EPA in a rulemaking proceeding.† No such notification is required until a rule has been promulgated. EPA's decision must be based on all relevant factors: including prospective volume of production, the extent to which a use changes the type or form of exposure of human beings or the environment, the extent to which a use increases the magnitude and degree of exposure of human beings or the environment to the substance, and the reasonably anticipated manner of manufacturing, processing, distribution, and disposal. [Section 5(a) (2)] The House Report provides that when the projected volume of a new use is small, a new use normally should not be deemed a "significant new use" unless there is a significant change in the type, form, or duration of human or environmental exposure. Furthermore, a new use generally should not be deemed significant merely because the use is new. However, the Report states that there may be potentially dangerous substances for which any new use may be significant.‡

D. Regulation at the Expiration of the Premarket Notification Period Pending Development of Additional Information

Section 5(e) authorizes the Administrator to issue a proposed order restricting or prohibiting the manufacture, marketing, or use of a new chemical substance or a proposed significant new use of an existing chemical as the premarket notification period draws to a close if there is not enough information to evaluate the health and environmental effects of the new substance or use. [Section 5(e)(1)(B)]

The Administrator may issue such an order if there is insufficient information to permit reasoned evaluation of the health and environmental effects of the substance and in the absence of such information, the manufacture, processing, distribution, use, or disposal of the substance (or any combination of those activities) may present an unreasonable risk to health or the environment or the substance will be produced in substantial quantities and (1) may enter the environment in substantial quantities or (2) there may be significant or substantial human exposure. The prospsosed order must be issued 45 days prior to expiration of the notification period. The manufacturer is entitled to actual written notice: Federal Register publication will not suffice. [Section 5(e)(1)]

The proposed order will take effect on the expiration of the notification period unless the manufacturer files objections "specifying with particularity" the provisions which are objectionable and states the grounds therefor. If such objections are filed, the Administrator must proceed to court to obtain an immediate ban by injunction, unless the objections filed with the Administrator persuade him that an injunction is not necessary [Section 5(e)(2)(A)] ‡‡

This provision was interpreted variously in the Senate and House debate on the Conference Report. Senator Magnuson, the Senate manager of the bill, asserted that the Administrator may require that "reasonable" or "valid" grounds be stated as a condition for recognizing that objections have been filed. Thus, the Administrator would have some "flexibility" in determining whether the objections filed rendered his order ineffective. A manufacturer who disagreed with the Administrator's determination that the grounds for objection were not reasonable would be entitled to judicial review.◊

*House Report at 44; 122 Congressional Record H8805-6 (August 23, 1976) (House Debate – Congressman Eckhardt).

**Senate Report at 31.

***House Report at 31.

†The Administrator is provided with the authority to issue rules applicable to categories of chemicals as opposed to individual substances under any provision of the Act, including rules defining what constitutes a "significant new use." [Section 26(c)(1)] House Report at 61; Conference Report at 102. During Senate debate on the Conference Report, Senator Magnuson stated that the Administrator "is expected to promulgate rules concerning significant new uses by categories in order to avoid a multiplicity of rulemaking proceedings." 122 Congressional Record S16803 (September 28, 1976).

‡House Report at 24; see 122 Congressional Record S16803 (September 28, 1976) (Senate Debate on Conference Report – Senator Magnuson).

‡‡This procedure is borrowed from Section 701(e) of the Federal Food, Drug, and Cosmetic Act. Conference Report at 68–69.

◊122 Congressional Record S16803 (September 28, 1976) (Senate Debate on Conference Report – Senator Magnuson); see also *id.* at S16807 (statement of Senator Tunney).

In contrast, Congressman Broyhill interpreted the provision as simply a notice requirement. Once objections are filed, the Administrator must go to court to enforce the order even though he may regard the manufacturer's objections as unmeritorious or not of sufficient specificity.* This dichotomy in approach appears to mirror the consistent difference in the two chambers' approach to premarket notification and to reflect the concern of the House that EPA will not be able to ban the marketing of a chemical without an administrative hearing, unless it first obtains the concurrence of a neutral court.** Case law interpreting similar provisions in the Food, Drug, and Cosmetic Act suggests that the Administrator will be required to proceed to court unless the manufacturer's objections are "frivolous" or inconsequential.***

To obtain an injunction in such a proceeding, the Administrator must demonstrate to the court the presence of the same factors which led the Administrator to propose the ban or limitation on manufacture, processing, distribution, use, or disposal. [Section 5(e)(2)(B)] The Conference Report states that this requirement is intended to entirely supplant the traditional requirements for obtaining an injunction (e.g., likelihood of success on the merits and irreparable injury).†

Moreover, the court is authorized to issue a temporary restraining order or preliminary injunction to preserve the status quo if the court finds that the notification period may expire before action on the injunction can be completed. [Section 5(e)(2)(C)]

The injunction would be dissolved after submission of adequate test data,‡ but only if a proceeding has not been initiated to regulate the chemical under Section 6. If such a rulemaking proceeding had begun, the injunction remains in effect until the publication of a rule under Section 6(a) or the termination of the Section 6 proceeding. [Section 5(e)(2)(D)]

E. Regulation of Hazardous Substances at the End of the Premarket Notification Period

Section 5(f) requires the Administrator to take action to limit or ban a new chemical substance or a proposed significant new use of an existing substance when there is a reasonable basis to conclude that the substance presents or will present an unreasonable risk of injury to health or the environment before a rule under Section 6 can be adopted and become effective.

While the Administrator has the authority to issue immediately effective rules under Section 6(d), that provision applies only when there is imminent, unreasonable risk of "serious or widespread injury." A Section 5(f) rule is authorized, with respect to new chemicals and significant new uses, when there is an imminent, unreasonable risk of injury, regardless of whether it is "serious or widespread." [Section 5(f)(1)] ǂ A rule is not authorized once the notification period has expired.

Once the Administrator has made the required finding, i.e., that manufacture, processing, distribution, use, or disposal of a chemical substance will present an unreasonable risk before a rule to protect against that risk can be promulgated under Section 6, the Administrator may propose an immediately effective rule. Such a rule could ban or limit production for a particular use, regulate use or disposal, impose labeling or record-keeping requirements on the manufacturer, or otherwise regulate its production, distribution, and disposal. [Section 5(f)(2)] ◇ Immediately thereafter, the Administrator would be required to initiate expedited rulemaking in accordance with Section 6(d).

In those instances in which a rule under Section 5(f) is equivalent to a total prohibition of the manufacture, processing, or distribution of a chemical substance due to the relative paucity of uses, the Administrator must issue a proposed rule 45 days prior to the expiration of the notification period and follow the procedure outlined above where there is an insufficient amount of data. Thus, if objections are filed to the proposed rule, EPA must seek a court injunction to obtain an immediate total prohibition on manufacture, processing, or distribution. [Section 5(f)(3)(C)] ♦

The traditional standards for obtaining an injunction would not apply to such a court proceeding. The Administrator need only show that there is a reasonable basis to conclude that manufacture, distribution, use, or disposal of a chemical, or any combination of those activities, presents or will present an unreasonable risk to health or the environment before a rule can be promulgated under Section 6. [Section 5(f)(3)(B)]

If the Administrator declines to take regulatory action at the close of the notification period, the Administrator must publish a statement explaining his failure to act in the Federal Register with respect to (1) significant new uses of existing chemicals, (2) new chemical substances which are subject to a testing rule, or (3) chemicals which have been listed as "suspect" in accordance with Section 5(b)(4). [Section 5(g)] New products falling in these three categories are given this special treatment because EPA has

*122 Congressional Record H11344 (September 28, 1976) (House Debate on Conference Report – Congressman Broyhill).

**122 Congressional Record H8815 (August 23, 1976) (House Debate – Congressman Eckhardt).

***See *Dyestuffs & Chemicals, Inc.* vs. *Fleming,* 271 F.2d 281 (8th Cir. 1959), *cert. den.,* 362 U.S. 911 (1960); *American Home Products Corp.* vs. *Finch,* 303 F. Supp. 448 (D.Del. 1969).

†Conference Report at 69.

‡Section 4(g) establishes a procedure under which EPA can be requested to prescribe standards for the development of such test data. See page 5.10.

ǂConference Report at 70.

◇Section 5(f)(2) authorizes the Administrator to exercise the regulatory authority provided under Section 6(a), described at pages 5.16 and 5.17.

♦See also Conference Report at 70–71, 122 Congressional Record S16802 (September 28, 1976) (Senator Magnuson).

previously singled them out as justifying close scrutiny. In the course of Senate debate on the Conference Report, Senator Magnuson stated that the Administrator's response in a case where he had not required testing should be that an unreasonable risk does not exist or that there is no need for testing.* The Administrator's failure to perform this duty provides no statutory authority to delay, and the Conference Report states that the Administrator's failure to perform this duty should not be permitted to delay manufacturing or processing of the new chemical substances.**

F. Exemptions from the Premarket Notification Requirements

Section 5(h) provides for exemptions from the requirement to submit premarket notification and test data. Section 5(h)(1) authorizes an exemption for test marketing under the restrictions specified by the Administrator. The Administrator, upon a satisfactory showing determines that test marketing will not present an unreasonable risk.

Section 5(h)(2) exempts (after application) manufacturers of chemical substances from premarket submission of test data if the data would be duplicative of data already submitted to the Administrator. Cost reimbursement of the person who previously submitted the data is required, as under Section 4.

Section 5(h)(3) automatically exempts from the premarket notification and test data requirements of Section 5(a) and (b) chemicals produced "in small quantities" which are utilized solely for research and development, provided that those persons exposed to the chemicals are notified by the manufacturer or processor of any risk to health. The purpose of this exemption is to ensure that research and product development are not impeded.*** Research and developmental chemicals are eligible for this exemption whether the product is developed "in house" or is sold by the manufacturer to another company for research and evaluation by the customers.† The term "small quantities" is to be defined by rule, and its meaning may vary according to the chemical substance's intended use. The House Report notes that laboratory reagents may be tested in terms of grams, while textile fibers or paper processing materials may have to be manufactured in larger quantities to be adequately evaluated. Although exempt from premarket notification, these chemicals would still be subject to regulations under Section 6. The research and development exemption also does not permit test marketing of a product; the exemption for testing marketing is contained solely in Section 5(h)(1).‡

Section 5(h)(4) authorizes the Administrator to exempt new substances or significant new uses of existing substances from any Section 5 requirement if the Administrator determines, upon application and by rule, that the chemical substance will not present an unreasonable risk of injury to health or the environment. The rule must be promulgated in accordance with Section 6(c)(2) and (3) (see page 5.16 below).‡

Finally, the Administrator is authorized to exempt intermediates (which may temporarily exist as a result of a chemical reaction in the manufacture or processing of a mixture or another chemical substance) from premarket notification and test submission requirements, provided that there is no human or environmental exposure. [Section 5(h)(5)]◊

V. Regulation of Hazardous Chemical Substances and Mixtures

Section 6 requires the Administrator to prohibit, limit, or otherwise regulate existing or new chemical substances or mixtures where there is "a reasonable basis" to conclude that the chemical "presents or will present an unreasonable risk" of injury to health or the environment. [Section 6(a)]

A. Alternative Forms of Regulation Under Section 6

A wide range of alternative and cumulative forms of regulation of dangerous chemicals is authorized under Section 6. The Administrator's arsenal of remedies includes total prohibition of manufacture, processing, or distribution of the product; prohibition of a specific use; imposition of labeling requirements on the substance or on articles containing the substance; prescribing maximum concentration levels; testing; record keeping; regulation of methods of disposing of the substance or articles containing the substance; a requirement that manufacturers provide notice of risk to customers, persons in possession of the product, and the public; and a requirement that the manufacturer replace or repurchase the product. [Section 6(a)]

Section 6(a) provides that the Administrator must utilize the "least burdensome" requirement necessary to adequately protect against the risk, as developed by evidence

*122 Congressional Record S16802 (September 28, 1976) (Senate Debate on Conference Report – Senator Magnuson).

**Conference Report at 71.

***122 Congressional Record H8805 (August 23, 1976) (House Debate – Congressman McCollister); 122 Congressional Record H8819 (August 23, 1976) (House Debate – Congressman Murphy).

†House Report at 30; 122 Congressional Record H8805 (August 23, 1976) (House Debate – Congressman McCollister); 122 Congressional Record H8826 (August 23, 1976) (House Debate – Congressman Murphy).

‡House Report at 29–30; 122 Congressional Record H8818 (August 23, 1976) (House Debate – Congressman Eckhardt).

‡House Report at 30.

◊See also Senate Report at 19.

presented in the rulemaking proceeding.* Where limits on production are imposed, they must be applied equally to all companies in an equitable manner.** Additionally, when quantitative limitations are proposed, the Conference Report states that the Administrator should consult with the Federal Trade Commission (FTC) and the Attorney General to avoid anticompetitive consequences.***

The Administrator also can order a manufacturer or processor to revise its quality-control procedures if the Administrator determines, after a hearing, that they are inadequate to prevent a chemical substance or mixture from inadvertently presenting an unreasonable risk. If the Administrator determines that the use of such inadequate quality-control procedures has resulted in the distribution of chemicals which present an unreasonable risk, he may order notice of that fact to be given to purchasers and the public, and he may require the manufacturer to replace or repurchase (at the option of the manufacturer) any such substance or mixture if "necessary to adequately protect health or the environment."[Section 6(b)]

B. Procedures Necessary for Issuance of Section 6(a) Rules†

1. Findings to be Made by the Administrator

At the time of promulgating a rule under Section 6(a), the Administrator must publish a concise statement in the Federal Register concerning the effects of the chemical substance or mixture on health and the environment and the magnitude of human and environmental exposure; the benefits of the chemical and the availability of substitutes;‡ and the reasonably ascertainable economic consequences of the rule, considering the effect on the national economy, small business,ǂ technological innovation, the environment, and the public health.[Section 6(c)(1)] The House Report states that industry should come forward with relevant data on the economic consequences of the rule for industry, including any negative impact on manufacturers or on the development of substitutes. However, the economic saving to society from removal of hazardous substances must also be considered by the Administrator. The House Report states that a chemical which causes an unreasonable risk should not be permitted to be marketed solely because of economic costs to producers if not permitted to be sold.◇ The Administrator's statement of findings is part of the record subject to judicial review.♦

The Administrator is instructed to proceed under other EPA-administered laws rather than under Section 6(a) if he determines, in his discretion, that these laws are adequate to protect against risk of injury, considering relative efficiency of acting under the other laws and costs of compliance.[Section 6(c)(1)] Such a discretionary determination is not subject to judicial review.□

2. Rulemaking and Hearing Procedures

Section 6(a) rules must be published in accordance with the rulemaking procedures of the Administrative Procedure Act, and in addition, opportunity is provided for a public hearing with cross-examination.[Section 6(c)(2),(3)]■ Where there are disputed issues of material fact, cross-examination may be conducted at the rule-making hearing by any interested person. The statute also provides that EPA may conduct the cross-examination or designate a single representative for a particular group of participants.[Section 6(c)(3)]♠

Attorney's and expert witnesses' fees may be awarded where any person represents an interest which will "substantially contribute to a fair determination" of the issues, and the economic interest of the participant is small in comparison to the cost of effective participation in the proceedings or who is without sufficient financial resources. Financial burden on the participant generally will be considered in determining whether to allow attorney fees and the amount of compensation [Section 6(c)(4)] §

*See also Conference Report at 75; House Report at 34; 122 Congressional Record H11344 (September 28, 1976) (House Debate on Conference Report – Congressman Broyhill); 122 Congressional Record S16808 (September 28, 1976) (Senate Debate on Conference Report – Senator Pearson); 122 Congressional Record H8805 (August 23, 1976) (House Debate – Congressman McCollister); 122 Congressional Record S4411 (March 26, 1976) (Senate Debate – Senators Cannon and Tunney).

**Conference Report at 75; 122 Congressional Record H8828-29 (August 23, 1976) (House Debate – Exchange of Congressman Murphy and Congressman Hagedorn).

***Conference Report at 75.

†Section 6(b) orders relating to inadequate quality control are subject to the normal APA provision relating to adjudicatory hearings. See 5 U.S.C. § 554.

‡However, the Administrator should not rule on the efficacy of the chemical for its intended use. 122 Congressional Record H8828 (August 23, 1976) (House Debate – Exchange between Congressmen Murphy and Eckhardt and Congressman Hagedorn).

ǂThe impact on small business is to be given careful consideration. See 122 Congressional Record H11346 (September 28, 1976) (House Debate on Congressional Report – Congressman Murphy). 122 Congressional Record H8805 (August 23, 1976) (House Debate – Congressman McCollister).

◇House Report at 35.

♦Conference Report at 76.

□Conference Report at 76.

■These procedures are borrowed from the Magnuson-Moss Warranty – Federal Trade Commission Improvements Act. Conference Report at 76.

♠However, where the Administrator determines that substantial and relevant issues divide an interest group, individual members will be allowed to conduct cross-examination.

§House Report at 37–38; 122 Congressional Record S16804-5 (September 28, 1976) (Senate Debate on Conference Report – Senator Magnuson).

3. The Effective Date of Section 6(a) Rules

While proposed, Section 6(a) rules ordinarily will be issued with delayed effective dates to allow opportunity for comment. Section 6(d)(1) provides that such rules should be made effective "as soon as feasible" after the date when the proposed rule is published in the Federal Register. Additionally, where there is an "unreasonable risk of serious or widespread injury," the rule may be made immediately effective if the Administrator determines that it is in the public interest to do so. [Section 6(d)(2)] There must be "a real need to protect the public safety" to justify deviation from customary rule-making procedures.* Such an immediately-effective rule is not final agency action for purposes of judicial review. [Section 6(d)(2)] **

After an immediately-effective rule has been promulgated, a full rulemaking must thereafter be initiated on an expedited basis. Where an immediately-effective rule has the effect of totally prohibiting manufacturing, processing, or distribution, a court injunction must first be obtained pursuant to the imminent hazard provisions of Section 7. [Section 6(d)(2)(A)(ii)]

C. Regulation of PCBs

Section 6(e) specifically provides for the phase out of the manufacture of polychlorinated biphenyls. By July 1, 1977, the Administrator is to publish regulations governing labeling and disposal of PCBs. After January 1, 1978, PCBs may not be manufactured, processed, or distributed other than in a totally-enclosed manner (except those PCBs that EPA concludes do not pose an unreasonable risk). [Section 6(e)(2)(A)] *** "Totally enclosed manner" is to be defined by rule to ensure that there will be no significant human exposure.

Manufacture† of PCBs is totally prohibited after January 1, 1979. Processing or distribution is prohibited 6 months later (July 1, 1979).

The Administrator is authorized to grant a temporary exemption from the ban on manufacture or distribution if he finds that an unreasonable risk of injury to health or the environment will not result and that good-faith efforts have been made to develop a substitute chemical. [Section 6(e)(3)(B)] ‡ An exemption is also provided for the distribution of PCBs sold for purposes other than resale prior to July 1, 1979 to permit the reuse of existing PCBs.

VI. Imminent Hazards

Section 7 authorizes the Administrator to initiate a court proceeding against any person who manufactures, processes, distributes, uses, or disposes of an imminently hazardous chemical substance or mixture. [Section 7(a)(1)] This is a mandatory requirement where the Administrator has failed to issue an immediately-effective order under Section 6(d). [Section 7(a)(2)]

A. The Definition of "Imminent Hazard"

"Imminent hazard" is defined as any chemical substance or mixture which "presents an imminent and unreasonable risk of serious or widespread injury to health or the environment." [Section 7(f)] ⧧ The risk to health or the environment is deemed "imminent" if the chemical substance or mixture is likely to result in injury before a final rule can be published under Section 6. [Section 7(f)]

While the risk of injury must be imminent, manifestation of physical injury is not necessary to establish that an imminent hazard exists.◇ The Conference Report expresses the intent that the Administrator take action early enough to prevent final injury from materializing.◆ An injury is "widespread" if a substantial number of people are affected, although the geographical area may be limited. If the risk to health or environment is "serious," it does not need to be shown that it will be widespread.

B. Procedures for Taking Action Against Imminent Hazards Under Section 7

The Administrator may file suit in a federal district court for seizure of the hazardous chemical or for a range of alternative forms of injunctive relief. Authorized relief includes recall of the chemical (or article containing the chemical), notification to purchasers, replacement or repurchase, as well as seizure of the chemical (or article containing the chemical). [Section 7(b)(2)]

The availability of the additional remedies, seizure, and recall may influence the Administrator to proceed to court under Section 7 rather than issue an immediately effective rule under Section 6(d). However, Section 6(a) rulemaking must be initiated concurrently with the filing of the court action or "as soon as practicable" thereafter. [Section 7(d)]

VII. Reporting and Record Keeping Requirements

Section 8(a) requires manufacturers and processors to

*122 Congressional Record S4411 (March 26, 1976) (Senate Debate – Senators Cannon and Hartke).

**Although not final, such agency action may nevertheless be subject to judicial review under certain circumstances. See L. Jaffe, *Judicial Control of Administrative Action*, 358, 418 (1965).

***122 Congressional Record S16803-4 (September 28, 1976) (Senate Debate on Conference Report – Senator Magnuson); 122 Congressional Record S4408 (March 26, 1976) (Senate Debate – Senator Nelson).

†Manufacture is defined to include importation. [Section 3(7)]

‡122 Congressional Record H8831 (August 23, 1976) (House Debate – Congressman McCollister).

⧧122 Congressional Record S16804 (September 28, 1976) (Senate Debate on Conference Report – Senator Magnuson).

◇Conference Report at 79; House Report at 40.

◆Conference Report at 79.

□Conference Report at 79.

submit reports to EPA, keep records, and otherwise provide the Administrator with a variety of information concerning chemical substances and mixtures and their effect on health and the environment.

A. Reports Required to be Submitted by Manufacturers and Processors

EPA is authorized to issue rules requiring manufacturers and processors of chemical substances and mixtures to maintain records and submit reports specifying the name of the chemical, molecular structure, categories of use, amounts manufactured, by-products, existing data concerning health and environmental effects, number of employees exposed, and manner of disposal. [Section 8(a)(2)]

The manufacturer must provide all such information which is known to him or "reasonably ascertainable." The Conference Report states that "reasonably ascertainable" is to be defined by referral to an objective standard, i.e., "information of which a reasonable person similarly situated might be expected to have knowledge."* The same phrase was interpreted by the House Report as that information which a manufacturer or processor could obtain "without incurring unreasonable costs or burdens."**

In the case of mixtures and research and development chemicals in small amounts, reporting and record-keeping requirements may only be imposed to the extent necessary "for the effective enforcement of the Act." [Section 8(a)(1)(B)] The Administrator is directed to avoid requiring unnecessary and duplicative reporting to the extent possible. [Section 8(a)(2)] For the same reason, the House Report urges EPA to utilize information in the possession of other federal agencies.***

B. Small Business Reporting Requirements

Small business is exempt from these reporting requirements, except for information the Administrator may require in order to compile the first inventory list and information which may be necessary to determine whether a rule or order should be promulgated or to enforce a rule or order. [Section 8(a)(3)]† The House Report provides that "reporting and record-keeping by small companies is to be held to a minimum.‡

The administrator is authorized to obtain reports from small manufacturers if a substance is the subject of a testing rule under Section 4(a); if it is listed as a "suspect" substance in the Section 5(b) list; if it is subject to an order or rule issued at the end of the premarket notification period under Sections 5(e) or 5(f); if it is subject to a rule regulating manufacture or use under Section 6; or if it is subject to a court order obtained under Sections 5 to 7. The Administrator may only require reports by small companies after he has established a reporting requirement by rulemaking procedures. [Section 8(a)(3)(A)]

The term "small business" is to be defined by EPA following consultation with the Small Business Administration. [Section 8(a)(3)(B)] The House Report indicates that several factors should be taken into account in determining what is a small business: not only the number of employees, but also the resources available for reporting and the amount of production.ǂ

C. Record Keeping Requirements

By rules, the Administration will require the maintenance of records of chemicals which have produced a "significant adverse reaction" on health or the environment must be retained for 5 years by any person who manufactures, processes, or distributes chemical substances or mixtures. [Section 8(c)] In addition, reports of significant adverse reactions of employees must be retained for 30 years. These records are meant to include consumer allegations as well as, for example, reports of occupational disease and environmental damage and are subject to inspection by EPA representatives.

The act does not define "significant adverse reaction," but the Conference Report provides that seriousness, duration, and frequency of reactions are to be taken into account in determining whether a reaction is "significant."◇ The Conference Report notes that trifling symptoms may be significant if experienced by several persons, because minor symptoms may indicate serious risks to health. In general, the Conference Report indicates that the requirement to retain records should be interpreted to "err on the side of safety."◆

D. Submission of Health and Safety Studies

Section 8(d) directs the Administrator to publish rules requiring submission of lists of health and safety studies by companies which manufacture, process, or distribute chemical substances or mixtures. "Health and safety study" is defined to include any study of any effect of a chemical substance or mixture on health or the environment, including underlying data and epidemiological studies, occupational exposure studies, toxicological, clinical, and ecological studies, and any test performed pursuant to the Act. [Section 3(6)].

Companies may be required to submit not only a list of

*Conference Report at 80.
**House Report at 42.
***House Report at 42; 122 Congressional Record H8805 (August 23, 1976) (House Debate – Congressman Murphy).
†See also Conference Report at 80.
‡ House Report at 43; 122 Congressional Record H8812 (August 23, 1976) (House Debate – Congressman Murphy and Congressman Slack).
ǂHouse Report at 43.
◇Conference Report at 81.
◆Conference Report at 81.

studies which they have conducted or commissioned, but also studies whose existence is known to them or "reasonably ascertainable" by them. EPA can request a copy of any study on such a list. [Section 8(d)] These studies are exempt from the protections afforded under the confidentiality provisions, Section 14(b).*

The Conference Report states that this reporting requirement should not be administered in a manner which is unduly burdensome.** Thus, the Administrator is authorized to exclude categories of studies which are "unnecessary to carry out the purposes of the Act." [Section 8(d)(1)] However, the Conference Report urges that "too much information should be required rather than too little."***

E. Submission of "Substantial Risk" Information

Section 8(e) requires any company which manufactures, processes, or distributes a chemical substance or mixture to notify the Administrator "immediately" when the company obtains information which indicates that the chemical presents a "substantial risk of injury to health or the environment." Notification is required unless the company has "actual knowledge" that the Adminstrator has already been informed of the risk.

The term "substantial risk" is not defined in the statute.† However, the provision was clearly intended to be a limited reporting requirement; it does not require the submission of all information concerning any adverse effect of a chemical substance or mixture on health or the environment. In general, if the possible harm to health or the environment is great, a product should be deemed to pose a "substantial risk," even if the probability of harm is quite small. Thus, information that a chemical substance or mixture poses some risk of (1) serious or widespread injury to the environment or (2) an increase in mortality or serious irreversible or incapacitating reversible illness, should be treated as "substantial risk" information. On the other hand, information that a product may have a slight adverse impact on health or the environment would constitute "substantial risk" information only if the risk of such impact was fairly high.

The reporting requirement only applies to information received on or after January 1, 1977. It does not require a company to submit on January 1, 1977 information which may have been obtained prior to that date. Of course, in assessing whether information obtained on or after January 1, 1977 is "substantial risk" information, a company will have to consider all health and safety information in its possession concerning the product in question.

PART II – ADMINISTRATIVE PROVISIONS OF THE ACT

I. Resolution of Conflict or Overlap with Other Agencies and Laws

The Act attempts to provide some guidance to the Administrator on how to avoid overlapping or duplicative federal regulation. Section 9 of the Act, the vehicle for this effort, contains two separate provisions, one for laws not administered by EPA (e.g., OSHA) and one for laws administered by EPA (e.g., the Federal Water Pollution Control Act, the Clean Air Act, the Safe Drinking Water Act, and the Solid Waste Disposal Act, as amended by the Resource Conservation and Recovery Act of 1976).

A. Laws Not Administered by EPA

Section 9(a) deals with the actions that the Administrator must take when he determines, in his discretion, that a federal law administered by another agency could be used to prevent or adequately reduce an unreasonable risk of injury to health or the environment posed by a chemical substance or mixture. If the Administrator makes the discretionary determination that there is no such law, a determination that is not subject to judicial review,‡ then this Act applies.

If he determines that there is such a law, the Administrator must submit a report to the other agency describing the risk and requesting that the other agency determine whether such a risk may be prevented or reduced by the other agency and if the answer is affirmative, issue an order declaring whether or not the activity presents an unreasonable risk. [Section 9(a) (1)] The other agency must respond within the time requested by the Administrator, which may not be less than 90 days from the date of request.

The Administrator may not take regulatory action or seek judicial relief if the agency to which the request was directed either issues an order declaring there is no unreasonable risk or, within a stated period, initiates action under the law which it administers. If the other agency takes neither of these actions, the Administrator may proceed under this Act. The legislative history makes clear that formal regulatory action may not be possible within the statutory time constraints. Accordingly, the requirement that the other agency take action is satisfied if (1) the other agency officially initiates an action, (2) the action will culminate as soon as practicable in effective regulatory action, and (3) the other agency establishes a general time schedule for such action.‡‡

*See below, page 79.
**Conference Report at 81.
****Id.*
†The legislative history suggests that it was intended to be a risk that was greater than an "unreasonable risk." See 122 Congressional Record E2915 (May 26, 1976) (Congressman Eckhardt).
‡Conference Report at 84.
‡‡Conference Report at 84–85.

B. Laws Administered by EPA

Section 9(b) provides guidance to the Administrator on coordinating this Act with other laws administered by EPA. If the Administrator determines that a risk to the health or the environment could be eliminated or reduced to a sufficient extent under other laws administered by EPA, he is instructed to use those laws unless he determines, in his discretion, that it is in the public interest to protect against such risks under this Act. The Administrator's determination that it is in the public interest to use this Act instead of other Acts administered by EPA is not subject to judicial review.* However, the Conference Report states that the Administrator should review the other authorities and publish the results of that review when he takes action under this Act. Furthermore, a reviewing court is authorized to determine whether the Administrator, in fact, examined the other authorities at his disposal and published the results of that examination when making his finding that it is in the public interest to use the Act, rather than those other authorities.** Thus, while the ultimate determination is intended to be nonreviewable, the Conference Report indicates that Congress intended the courts to be able to review and, where appropriate, set aside the Administrator's actions if he has failed to make the required examination of other authorities and publish the results thereof.

The requirement to examine other laws administered by the Administrator applies only when the Administrator takes regulatory action to protect against unreasonable risks under the Act. It does not apply, for example, when the Administrator takes action necessary for administration or enforcement of the Act, such as promulgating record-keeping requirements.***

C. Coordination with Other Federal and State Agencies

1. Occupation Safety and Health Act (OSHA)

The Act makes clear that the exercise of authority by the Administrator, under this Act, does not constitute a limitation on the authority of OSHA to prescribe or enforce standards or regulations affecting occupational safety and health. [Section 9(c)]

2. State Agencies

Section 18 outlines the relationship between state and EPA's authority under this Act. The Act removes the authority to regulate toxic substances in two major respects from the states.

First, no state or political subdivision may establish requirements for the testing of a substance or mixture for purposes similar to the purposes of testing required by the Administration under Section 4. [Section 18(a)(2)]

Second, no state may regulate any chemical substance, mixture, or article containing such substance or mixture, which is being regulated by the Administrator under Sections 5 and 6,† unless the requirement imposed is (1) identical to that issued under the Act, (2) adopted pursuant to another federal law (e.g., the Clean Air Act), or (3) prohibits the end use of such substance or mixture in such a state or political subdivision.‡

The Act also provides that a state or locality may apply to the Administrator for an exemption from the preemptive effect of a federal requirement. [Section 18(b)]

II. Review of Administrative Actions

A. Judicial Review

Agency actions, including actions of the Administrator, are generally subject to the judicial-review provisions of the Administrative Procedure Act (APA) and are generally reviewable in district courts. The Act, however, provides for judicial review of important rulemaking actions of the Administrator in the courts of appeal.ǂ Generally, the record for review in such rulemaking proceedings is the entire record before the Administrator, and the standard review is whether the Administrator's actions are supported by "substantial evidence on the record as a whole."

1. Petition for Review of a Rule

A petition for judicial review of certain specified rules promulgated by the Administrator◊ may be filed by any person within 60 days of promulgation of the rule in the appropriate U.S. court of appeals, whose court has exclusive jurisdiction over the matter. [Section 19(a)(1)(A)]

*Conference Report at 85.

***Id.*

****Id.*

†However, the state is not preempted from more stringent regulation of substances or mixtures being regulated by the Administrator under Section 6(a)(6) (rules governing disposal).

‡The purpose of the provision is to "enable States to totally ban a specific end use of a substance or mixture while not interfering with the manufacture or processing and distribution of such substances or mixtures. For instance, a State could totally prohibit the use within its boundaries of a detergent containing a particular chemical substance. However, the state could not prohibit the manufacture or processing within the State of either the substance or the detergent, nor could it prohibit or limit interstate distribution of the substance." 122 Congressional Record H11346 (September 28, 1976) (House debate on Conference Report – Congressman Murphy).

ǂThe legislative history indicates that certain discretionary actions of the Administrator were intended to be nonreviewable. See discussion at pages 5.19 and 5.20.

◊These rules are rules promulgated under Section 4(a) (testing requirements), 5(a)(2) (significant new use for premarket notice), 5(b)(4) (list of suspect chemical substances for premarket notice), 6(a) (regulation), 6(e) (PCB), or 8 (reporting and record keeping).

2. Record on Review of Rules

The rulemaking record upon which the reviewing court will base its decision is comprised of the rule, any findings or statements which have been made, the transcript of any oral presentation, any written submission of interested parties, and any other information the Administrator considered relevant and with respect to which the Administrator published a notice in the Federal Register. [Section 19(a)(3)] *

The Act authorizes a petitioner or the Administrator to apply to a reviewing court for permission to make additional submissions or presentations respecting the rule in question. If the court is satisfied that the submissions would be material and that there were reasonable grounds for the failure to come forward at the time of the administrative proceeding, the court may order the Administrator to provide an additional opportunity to make submissions and presentations. The Administrator may modify or set aside the rule being reviewed and must file the new rule, which is then to be reviewed by the court. [Section 19(b)]

3. Scope of Review

Upon the filing of a petition under Section 19(a)(1), the Act grants courts of appeal jurisdiction to review rules promulgated under various sections of the Act and to grant appropriate relief, including interim relief. [Section 19(c)] In reviewing a rule issued under provisions listed in Section 19(c)(1)(B)(i), the court must hold the rule unlawful if it finds the rule is not supported by substantial evidence in the rulemaking record taken as a whole. [Section 19(c)(1)(B)] Thus, rules issued under Section 4(a) (testing), 5(b)(4) (suspect chemical substances), and Section 6(a) or (e) (regulation of hazardous chemical substances) will be reviewed under the substantial evidence standard.

However, the legislative history indicates that, insofar as premarket notification is concerned, the substantial evidence test does not require the Administrator to look beyond the notification documents required by Section 5. Additionally, since the testing requirements of Section 4(a) frequently will be based on insufficiency of data, the Administrator ordinarily need not develop substantial evidence of that insufficiency.**

In reviewing a rule promulgated under Section 6(a) (regulation), the court must set aside the rule if it finds that the actions of the Administrator in excluding or limiting cross-examination, rebuttal, or oral presentations precluded disclosure of disputed material facts necessary for a fair determination of the rulemaking proceeding by the Administrator. [Section 19(c)(1)(B)(ii)]

Review of rules and orders issued under sections of the Act not specified in Section 19(c)(1)(B)(i) (see page 61) are subject to the narrower "arbitrary and capricious" standard of review found in 5 U.S.C. §706. Under that standard, agency action will be overturned only when found by a court to be "arbitrary, capricious, an abuse of discretion, or otherwise not in accordance with law."

B. Citizens' Civil Actions

The Act authorizes any person to commence a civil action in the appropriate district court to restrain a violation of the Act by any person, including any governmental agency. [Section 20(a)(1)] Suits may also be initiated to compel the Administrator to perform any nondiscretionary act or duty under the Act. [Section 20(a)(2)] The Act does not, however, authorize the collection of money damages in citizens' actions.

No citizens' action may be brought under this section to restrain a violation of the Act until 60 days after the Administrator and the alleged violator have received notice of the alleged violation. If the Administrator is diligently prosecuting a proceeding to obtain compliance with the Act or if the Attorney General has initiated a civil action against an alleged violator to compel compliance, no citizens' action may be brought. However, if the governmental action or proceeding is not commenced until after the citizen has notified the alleged violator of his intention to sue under this Act, then the citizen has the right to participate in the action or proceeding. [Section 20(b)(1)]

C. Citizens' Petitions

Citizens may petition the Administrator to initiate a proceeding for the issuance, amendment, or repeal of certain orders or rules. If the Administrator fails to act or denies the petition, citizens may bring a lawsuit to force initiation of the action.

Any person may petition the Administrator to initiate a proceeding for the issuance, amendment, or repeal of certain rules or orders. [Section 21(a)] *** The Administrator must grant or deny the petition within 90 days. If he denies the petition, he must publish his reasons for doing so in the Federal Register. If the Administrator denies or fails to respond to a petition, the petitioner may bring a civil action in the district court to compel the Administrator to initiate the requested action. [Section 21(b)(4)(A)] †

*The Administrator need only publish a list of the information. He need not summarize it. However, the Administrator must keep current and make available throughout the proceeding a listing of information to be included in the record. House Report at 55. Furthermore, the Administrator is not required to document "each and every widely accepted scientific principle or fact which may support the rule or order." Senate Report at 27.

**122 Congressional Record S16804 (Senate debate on Conference Report – Senator Magnuson).

***The petitions may be directed to rules under Section 4 (testing), 6 (regulation), or 8 (reporting and record keeping) or orders under Section 5(e) (order prohibiting or limiting activities) or 6(b)(2) (quality control).

†Although Section 21(b)(4)(A) only refers to rule making, the legislative history indicates the section is also applicable to orders. Conference Report at 98–99; 122 Congressional Record S16807 (September 28, 1976) (Senate debate on Conference Report – Senator Tunney).

In an action in which a petition requesting the issuance of a rule or order is at issue (as opposed to a petition for amendment or repeal of a rule or order), the Act recognizes the absence of an adequate record for the reviewing court and accordingly provides the petitioner with an opportunity to have the petition considered by the court in a *de novo* proceeding, i.e., evidence can be introduced in a hearing before the court. [Section 21(b)(4)(B)]

In an action for issuance of a rule or order, the court must order the Administrator to take the requested action if it finds by a preponderence of the evidence, that the Administrator should have taken the requested action.* However, the court may defer requiring the Administrator to take the requested action if it finds that the extent of risk of injury to health or environment alleged by the petitioner is less than those risks of injury which the Administrator is addressing under the Act and that there are insufficient resources to do both. [Section 21(b)(4)(B)]

The Act provides for different treatment for review of citizens' petitions for amendment or repeal of existing rules or orders (as opposed to issuance of new rules or orders) because the Administrator will have addressed the general subject previously, and his decision will already have been subject to the judicial review provisions of the Act. Accordingly, the Act provides that if the Administrator denies a petition to amend or repeal a rule or order, a review of that denial may be had only under Administrative Procedures Act (APA) (5 U.S.C. §700 et seq.), and there will be no opportunity for a *de novo* court hearing under Section 21. Furthermore, the legislative history makes clear that the failure to present newly discovered, noncumulative material in such a petition is an adequate basis for denying the petition.**

D. Attorneys' Fees and Other Costs

The judicial review, citizens' suits, and citizens' petitions provisions of the Act generally permit the award of costs of suit and reasonable fees for attorneys and expert witnesses. Fees and costs may be awarded to any litigant, plaintiff, or defendant, whether successful or unsuccessful, and against any party, including the U.S., where appropriate.*** These provisions, however, may not be of benefit to parties who stand to gain significant economic benefit from participating in such proceedings or who could otherwise afford to participate.†

III. Penalties and Enforcement

A. Penalties

Civil penalties of up to $25,000/day are provided for persons who violate the Act. [Section 16(a)] A person charged with violating the Act may request an administrative hearing. Subsequent to the hearing, he may file a petition in the appropriate U.S. court of appeals, within 30 days from the date of the administrative order for review of an order assessing the penalty.

Criminal penalties of up to $25,000/day, in addition to or in lieu of a civil penalty, and/or imprisonment for up to 1 year, are provided for persons who knowingly or willfully violate the Act. [Section 16(b)]

B. Enforcement

The Act grants U.S. district courts jurisdiction to (1) restrain any violation of the Act, (2) restrain any person from taking any action prohibited by Section 5 (premarket notification) or 6 (regulation) or by a rule or order issued thereunder, and (3) compel the taking of any action required by the Act. Any manufacturer or processor of a chemical substance or mixture manufactured and distributed in violation of Section 5 or 6 (or any rule or order issued thereunder) can also be required to give notice to distributors and (to the extent reasonably ascertainable) to others exposed to the substance or mixture. Additionally, the manufacturer or processor can be ordered to replace or repurchase the substance or mixture at the election of the manufacturer or processor. [Section 17(a)(1)]

Any chemical substance or mixture distributed in violation of the Act or any rule promulgated thereunder is liable to seizure and condemnation by any district court in whose district the substance, mixture, or article is found. [Section 17(b)]

IV. Inspections, Subpoenas, and Confidentiality of Data

A. Inspections

For purposes of administering the Act, the Administrator may inspect without a search warrant any premises in which chemical substances or mixtures are manufactured, processed, or stored and any conveyance used for transport of substances or mixtures. Each inspection may be made only upon presentation of proper credentials and written notice to the owner or person in charge of the premises or conveyance. Inspections must be started and completed with reasonable promptness and must be conducted at reasonable times, within reasonable limits, and in a reasonable manner. [Section 11(a)] Inspections may not extend to financial data, sales data (other than shipment data), pricing data, personnel data, or research data (other than research required by the Act), unless the nature and extent of the data to be inspected are described with reasonable specificity in the written notice of inspection. [Section 11(b)]

*Although a court may order the Administrator to initiate a rule making, it may not determine the content of the rule. Senate Report at 12; House Report at 59.

**Conference Report at 98–99; House Report at 57–58.

***122 Congressional Record S16804 (September 28, 1976) (Senate debate on Conference Report – Statement of Senator Tunney.)

†122 Congressional Record S16805 (September 28, 1976) (Senate debate on Conference Report – Senator Magnuson).

B. Subpoenas

Section 11(c) authorizes the Administrator to issue subpoenas to require the attendance and testimony of witnesses and the production of documents. The legislative history makes clear that the subpoena power should be used only when information otherwise available through voluntary means or other provisions of the Act (e.g., the premarket notification or record-keeping section) is inadequate to meet the needs of the Administrator.*

C. Confidentiality of Data

Cognizant that some information which the Administrator may obtain from companies may be of considerable competitive and commercial value, the Act contains specific prohibitions against the release of certain information, primarily of a trade secret nature. Companies filing confidential data may designate it as such, and in these instances, the Administrator generally must give advance notice of his intention to release such data. [Section 14]

1. Data Subject to Confidential Treatment

Except for certain authorized disclosures discussed below, Section 14(a) prohibits the Administrator or his representatives from disclosing any information reported or otherwise obtained under the Act which is exempt from disclosure under the Freedom of Information Act (FOIA) by reason of exemption 4 to the FOIA, 5 U.S.C. §552(b)(4). Exemption 4 provides that the mandatory disclosure provisions of the FOIA do not apply to "trade secrets and commercial or financial information obtained from a person and privileged or confidential."

Information which falls under exemption 4 to the FOIA is, therefore, generally exempt from disclosure under the Act. However, such information can be disclosed to

1. Officers or employees of the U.S. in connection with their official duties to protect health or the environment or for specific law enforcement purposes
2. Contractors of the U.S. and their employees when the Administrator determines it to be necessary for the satisfactory performance of their duties in connection with the Act
3. The public if the Administrator determines it necessary to protect health or the environment against an unreasonable risk. [Section 14(a)]

Furthermore, such information may be disclosed in any proceeding under the Act provided disclosure is accomplished in a manner that preserves confidentiality, to the extent practicable, without impairing the proceeding. [Section 14(a)] **

Disclosure of any health and safety study or information from such a study (with the exception of information on processes used in manufacturing or processing or on proportions of mixtures)*** is not prohibited. [Section 14(b)] †

2. Procedures for Protection and Release of Data

Any person submitting data to the Administrator may designate and separately file data which he believes is entitled to confidential treatment. [Section 14(c)(1)] If the Administrator proposes to disclose such data, he must give written notice by certified mail to the person who submitted the data, and the data may not be released until the expiration of 30 days after the person who submitted the data has received the required notice. [Section 14(c)(2)(A)]

The notice requirement embodied in Section 14(c)(2)(A) does not, however, apply to the release of information to Federal officials and government contractors or during the course of a proceeding under the Act. If the Administrator proposes to release data on the grounds that it is necessary to do so to protect health or environment against an unreasonable risk, he must notify the person submitting the data at least 15 days before release. When the Administrator determines that immediate release is necessary to protect against an imminent, unreasonable risk, 24-hr advance notice will suffice, which may be provided by telephone or telegram. [Section 14(c)(2)(b)(i)] ‡ The statutory notice requirement does not apply to the release of information described in Section 14(b) (data from health and safety studies), unless that information includes trade-secret data on processes used in manufacturing or processing or on proportions of mixtures. [Section 14(b)] ‡‡

Criminal penalties are provided for officers, employees, and contractors of the U.S. who knowingly or willfully disclose information protected from disclosure. [Section 14(d)] All information reported to or otherwise obtainable by the Administrator under the Act is available upon written request to any duly-authorized committee of the Congress. [Section 14(e)]

V. Foreign Trade

A. Exports

With some exceptions, the Act does not apply to any chemical substance or mixture that (1) is being manufactured, processed, or distributed solely for export from

*Conference Report at 87.

**The legislative history indicates that in such proceedings the Administrator should exercise due care to prevent the release of confidential information to competitors of persons submitting data merely because they have joined in the proceeding. Conference Report at 90, Senate Report at 25.

***The Administrator, however, may disclose chemical substances in mixtures by their order of quantity. House Report at 51.

†122 Congressional Record S 16805 (Senate Debate on Conference Report – Senator Magnuson).

‡Conference Report at 91.

‡‡*Id.*

the United States,* and (2) is labeled to show that it is intended for export. [Section 12(a)(1)] **

The exemption does not, however, apply to any substance, mixture, or article that the Administrator finds will present an unreasonable risk of injury to health or the environment within the United States. [Section 12(a)(2)] *** For example, this exception to the export exemption gives the Administrator control of substances exported to Canada, which may impact the Great Lakes or substances to be disposed of by ocean dumping. The Administrator may also require testing, under Section 4, of an otherwise exempt substance or mixture to determine whether it presents an unreasonable risk to health or environment in the United States. [Section 12(a)(2)]

Finally, a person who exports or intends to export a chemical substance must notify the Administrator if

1. The submission of data is required under Section 4 (testing) or 5(b) (premarket notification):
2. An order has been issued under Section 5 (premarket notification):
3. A rule has been proposed or promulgated under Section 5 or 6 (regulation): or
4. An action is pending or relief has been granted under Section 5 or 7 (imminent hazard).

The Administrator must, in turn, furnish timely notice of the availability of the relevant information to the government of the country of destination.

B. Imports

Imports are subject to the Act. The Act requires the Secretary of the Treasury to refuse entry into the U.S. of any chemical substance or mixture which (1) fails to comply with any rule in effect under the Act or (2) is offered for entry in violation of Section 5 (premarket notification) or 6 (regulation) or an order issued in an action brought under Section 7 (imminent hazard) or an order limiting the processing or manufacture of a substance pending development of information under Section 5(e). [Section 13(a)]

VI. Special Protection for Employees

A. Employment Effects

The Act requires the Administrator to continually evaluate the potential effect on employment of requirements imposed under the Act. [Section 24(a)] Adversely-affected employees (or their respresentatives) may request the Administrator to investigate an actual or threatened layoff or an adverse or threatened adverse effect on a worker's employment. Upon receipt of the request, the Administrator must conduct the investigation and, if requested by any interested person, hold public hearings thereon, informal in nature, unless he determines that there are no reasonable grounds for holding such hearings. [Section 24(b)(2)(A)] Upon completion of the investigation, the Administrator must make public findings of fact and recommendations. Although Congress carefully noted that this procedure was not to be construed to require the Administrator to amend or repeal a rule or order [Section 24(b)(4)], such a proceeding, of course, may persuade him to do so.

B. Employee Protection

The Act prohibits an employer from discharging or otherwise discriminating against an employee because the employee has caused a proceeding to be commenced under this Act, has testified or is about to testify in such a proceeding, or has assisted or participated in a proceeding "or in any other action to carry out the purpose of this Act." [Section 23(a)] This provision is inapplicable to any employee who, without directions from his employer, deliberately violates any requirement of the Act. [Section 23(e)]

*The exemption is inapplicable, however, where the substance, mixture, or article was, in fact, manufactured, processed, or distributed for use in the U.S. For example, a person confronted with an enforcement action for manufacturing a substance (for domestic use) in violation of the Act could not claim the benefits of the exemption in order to export the noncomplying substance. House Report at 48–49.

**The provisions of Section 8 relating to record keeping and reporting remain applicable.

***See also House Report at 49.

Section VI

Industries' Assessment of the Toxic Substances Control Act

THE CHEMICAL INDUSTRY'S ASSESSMENT OF THE TOXIC SUBSTANCES CONTROL ACT

Fred D. Hoerger
Manager, Regulatory and Legislative Affairs
Health and Environmental Research
Dow Chemical Co.
Midland, Michigan

Marvin E. Winquist
Manager, Special Chemicals Business
Designed Products Department
Dow Chemical Co.
Midland, Michigan

Etcyl H. Blair
Director, Health and Environmental Research
Dow Chemical Co.
Midland, Michigan

TABLE OF CONTENTS

I. THE MANDATE: A MORE EFFICIENT REGULATORY PROGRAM AND IMPROVED CONTROL OF UNREASONABLE RISKS

An associate recently made a few comments about the Federal Water Pollution Control Act of 1972. He indicated that the law set up a number of procedures and regulations requiring permits and reports on water quality for every discharge pipe in the U.S. – procedures for dealing with a spill of one pound of chlorine in the Mississippi identical to those for the sinking of an entire barge of chlorine, and a whole host of other strategies applied universally across the waters of the U.S. Despite all this, we have today the problems of PCBs, Kepone, and taconite tailings in Lake Superior. This example typifies the problems of our current regulatory agencies that deal with health and environmental risks – frequently vast regulatory programs are applied somewhat uniformly across the industries, without developing solutions to some of the more significant health and environmental concerns.

It seems that Congress, in enacting the Toxic Substances Control Act, has tried to provide a prescription for improving regulatory efficiency. Several provisions of the law are somewhat pioneering and unique in regulatory procedures:

1. A mechanism provides for prioritizing chemicals for consideration of testing.
2. Regulation is to be selective. The concern for health and environmental risk is to be balanced with the findings on practical factors, such as social and economic benefit, and the reasonable availability of testing facilities.
3. EPA authority under the law, although very broad, is intended as a backstop after use of authorities under other laws has been reasonably exhausted.
4. Finally, EPA is to coordinate with other agencies and bodies on certain research, data gathering, and data filing activities to minimize duplication and to provide greater relevancy.

In implementing the new law, EPA must find ways to apply the new regulatory procedures and to establish public credibility. Public expectations for the Toxic Substances Control Act are very high. For example, President Ford, EPA, and the Council on Environmental Quality have hailed the Toxic Substances Control Act as landmark legislation. EPA credits the law with providing federal authority to

1. Gather more and better information about the health and ecological effects of chemicals
2. Deal with problems after they "have occurred as well as identifying and avoiding problems before they occur"
3. Balance the various costs, risks, and benefits in deciding appropriate regulatory actions
4. Help coordinate all federal approaches in matters involving toxic substances

Various public interests and industries generally concur with the overall objectives of the Act – to control unreasonable risks while not unduly inhibiting innovation. However, viewpoints differ on the degree of change to be brought about in dealing with risk evaluation, chemical problems, and the degree of responsibility to be assumed by the private sector, compared to that of the federal agencies. Hundreds and thousands of subissues will ultimately shape the public policies which bring about change in the bureaucracies, in the structure of the many and varied regulated industries, and in the variety and quality of goods and services becoming available to the public.

II. INDUSTRY ATTITUDE

Generally, the chemical industry accepts the mandates of more testing, of careful evaluation of risk potentials for new products, and of minimizing chemical incidents and unreasonable risks of injury to health and environment. The larger chemical companies recognize and emphasize, the very significant potential impact of the new law on their operations. This is underscored by their very active participation in trade association formulation of recommendations for implementation, numerous discussions with suppliers and customers, development of additional experts on regulatory and legal procedures, and by an increasing emphasis on health and safety testing.

The toxic substances law is one of the most complex of all the numerous laws regulating the industry. Control strategies available to EPA are far-reaching. Furthermore, the provisions of the law and the EPA implementation policy provide for widespread public participation, ranging from the philosophical and general to specific issues in citizen petitions. Thus, attitudes of the larger companies reflect adjustments by creating staff and practices to deal with complex compliance procedures, greater uncertainties, and involvement in the formulation of public policy.

The attitude of Dow Chemical Co. toward the new law, expressed in the December 1976 quarterly stockholders newsletter, is probably typical of the attitudes of most larger companies.

> We believe that the well-known Dow practices regarding products and manufacturing operations have been highly consistent with the purposes of the Toxic Substances Control Act. Although disappointed with the immediacy of still another bureaucratic law, Dow will certainly comply with the legislation – adequate control of health and environmental risks – as well as the regulations.
>
> Dow's concern for health and environment precede 1934 when a toxicological laboratory was initiated. Over the years toxicological and environmental testing, engineering and investment for reduction

of discharges and emissions, industrial hygiene and medical surveillance programs, analytical chemistry, and customer service have been welded into an effective force, product stewardship. This experienced staff will form the basis for continued and balanced effort consistent with the legislation and the needs of chemical processors and fabricators.

We are concerned about the very considerable Federal regulatory power and bureaucratic red tape created by this bill. Increased costs and a slowing in new job opportunities in the productive sector are real possibilities.

We hope that the Environmental Protection Agency will be judicious and reasonable in administering this new program and base its actions on scientific fact.

Smaller companies appear to have strong anxieties because of the potential for a single regulation to impact a major portion of a company's business. The many processors, fabricators, and other downstream users of chemicals, appear to have varying degrees of awareness of the law and varying degrees of sophistication in dealing with it. Thus, we observe that, whether large or small, chemical companies are greatly concerned about the quality of regulatory programs that will be utilized in implementing this new law.

III. INDUSTRY RECOMMENDATIONS FOR EPA POLICY IN IMPLEMENTATION

A. Priorities

EPA conducted a public meeting to solicit input on implementing the law in December 1976. The Manufacturing Chemists Association (MCA), broadly representing 180+ chemical companies, had several general recommendations for EPA:

1. Learn from the history of recent regulatory programs:
   - Statutory deadlines must be met.
   - Lack of well conceived early-on priorities eventually lead to wasted resources and slowed regulatory action.
   - Citizens' petitions have consumed significant human effort and budget dollars in the agencies, the regulated industries, and the courts.
   - Uniform standards and testing requirements, in contrast to selective programs, contribute to large costs and slowed innovation.
2. Establish priorities for implementation which:
   - Emphasize those subsections with statutory deadlines
   - Place minimal and/or "pilot effort" on those subsections with discretionary timing
3. Achieve adequate staffing and procedures before the end of 1977 to:
   - Consider testing rules for the list of priority chemicals recommended by the interagency committee
   - Evaluate premarket notifications
4. Emphasize correct and comprehensive compilation of the "Inventory List" of existing chemicals to avoid disruption of goods and services in late 1977. (Chemicals omitted from the list cannot be marketed after December 1977.)
5. Approach the testing of chemicals from the standpoint of specifics rather than classes.
6. Selectively implement the "significant new use" clause on a chemical-by-chemical basis.
7. In research authorized under the law, emphasize development of simple and concise test methods.

Ford Motor Co., a major consumer of chemicals, participated in the December public meeting in Washington and emphasized the importance of lead time needed for compliance with a, let us say, hypothetical regulation which might curtail use of a chemical substance in automobiles. Probably few in the audience of 400 had previously appreciated the fact that dozens of chemical substances used in automobiles have to be specified – at the design state – 1 to 2 years prior to model introduction dates.

B. The Interagency Priority Committee

Comments from several sectors at the EPA public meeting indicated widespread interest in the activities of the Interagency Committee in nominating chemicals for EPA's priority attention for testing. This interagency committee includes representatives from the Environmental Protection Agency (EPA), the Occupation Safety and Health Administration (OSHA), the National Institute of Occupational Safety and Health (NIOSH), the National Cancer Institute (NCI), the National Institute of Environmental Health Science (NIEHS), the National Science Foundation (NSF), the Council on Environmental Quality (CEQ), and the Department of Commerce.

In creating this committee, it seems apparent that Congress was attempting to assure that testing of existing chemicals would occur and to delineate a selective approach to testing. The committee represents scientific research programs on health and environment and programs imposing regulations on chemical risks. The committee must publish its reasons for prioritizing a chemical. After the list is published, and there is much speculation that the list will include at least 50 chemicals, EPA has 1 year in which to initiate rulemaking for testing for up to 50 chemicals on the list specially designated by the committee.

Public participation can occur after the priority list is published and during the rulemaking for testing. Presumably, literature surveys or supplemental surveys will occur in this postlisting stage prior to the initiation of testing. The public-hearing portion of the rulemaking for a chemical will undoubtedly encompass spirited debate on what effects really need further investigation. The law prescribes the practical reality that testing rules reflect the availability of resources.

There is much uncertainty to this process, but we at Dow Chemical believe that the system of checks and balances will suffice if all sectors work positively toward implementation.

In December 1976, prior to the effective date of the law, the National Resources Defense Council filed a citizens petition requesting the testing of the thermal-oxidative decomposition of a chlorofluorocarbon. This action forces the question of priorities and where emphasis in implementation will occur. EPA might consider referring petitions for testing to the Interagency Committee to establish a degree of factual and risk comparison with other priority concerns.

C. Testing of Chemicals

Policies toward testing of chemicals are among the most complex areas to be faced by EPA. For many years, Dow has produced and used many of the so-called toxic chemicals with minimal health and environmental problems. We attribute a part of our performance to testing of chemicals on a selective basis and to careful translation of this data to operating and customer conditions. The general principles used in priority in our toxicological work and designing experiments have been a screen of physical and chemical properties, a screen of several simple acute mammalian and environmental tests, and literature background. This is followed by selective implementation of subacute, chronic, and special tests. We have evolved a testing guidance, but not testing requirements.

We recommend that EPA emphasize selectivity in setting priorities for testing regulations – both on which chemicals are to be tested and for which effects. Expert interdisciplinary and administrative skill in EPA, along with cooperation with interdisciplinary scientists in industry and academia, will be necessary if the mandated thrust toward selectivity is to be achieved.

D. Small Business

The small business segment of the chemical industry stands to be the most vulnerable to the impact of implementation of various facets of the Toxic Substances Control Act. Small companies, in general, exist and thrive because they contribute something unique and needed. They do not and cannot compete head-on in the commodity-size products of the large companies. Rather, they complement large companies by supplying the needs of U.S. industry for unique products, often used as minor components, e.g., catalysts, thickeners, polymerization initiators, and stabilizers. As a result, they deal in small volumes. They also provide unique know-how in specialized end uses and/or an efficient service-sales program tailored to the needs of a particular industry or technology. Yet, because of size and competitiveness, they do not have the resources in people or money to do much more than they are now doing.

In many cases, they will not be able to ride on the coattails of the toxicology testing done by larger companies and simply pay their allocated share. If they have a unique product, they must pay the full cost, comparable to that required for a commodity chemical. In contrast, many commodity-chemical costs can be shared among large chemical companies.

Another major impact to be dealt with by the small companies is the need for monitoring the Federal Register, a new people-demanding task, to anticipate potential threats or requirements on their products or services. Record keeping, although given some relief in the form of special consideration of the problem by the EPA Administrator [Section 8(1)(A)(B)], is still a potential burden for those many small producers and processors of unique chemical products or mixtures [Section 8(3)(A)].

Large companies will be concerned about the viability of the smaller chemical companies because they often are key suppliers of unique low-volume products to the majors and, on the other hand, represent significant customers for the products of the major producers. The smaller companies often are the key links in transferring technology (and products) of the large companies to the ultimate users. Even if they do not chemically modify a product that they buy from larger companies, they still can be held responsible for reporting significant uses, new and old, which only they are aware of as part of their corporate know-how.

The inability of the small companies to deal with the pressure of TSCA, on top of the other major health and environmental laws they are already struggling with, could cause them to abandon their business. This then could force larger companies to acquire their business if it is vital to the larger companies. As a result, competition would be reduced and technological innovation slowed in the many low-volume chemical-consuming industries so well covered now by these specialist companies.

A number of the minor chemical-consuming industries are not necessarily minor industries, e.g., metal processing, automotive, and aircraft manufacture. Furthermore, the newer industries (electronics, communication and word processing, solar energy, and improved fossil fuel processing) all require small volume, but highly essential chemicals for their current base, as well as for continuing technological advances.

The EPA Administrator will have to pay particular attention to this problem of small business vitality for its impacts on the competitiveness of U.S. chemical industry, both in the U.S. and in foreign industry. It seems highly desirable that EPA begin now to study the role of small chemical businesses in providing many benefits to our society.

E. Product Choices

The natural tendency resulting from the Act will be to make toxicity of a product a much higher priority factor in judgments on manufacture or use of a chemical. These choices will be made not because toxic products cannot be handled safely, but rather because of the uncertainty of government action on suspect or potentially-suspect products.

The effect of this tendency will be to keep everything status quo for as long as possible, except where there will be obvious attack on a product being currently used.

Products that have an obviously lower hazard or risk in

toxic properties will be sought and used wherever possible. Companies that have the capability of generating these kinds of products and safety data on them faster, more quickly and thoroughly than others will have a competitive advantage. This is now the case in the pesticide products.

On the other hand, the law provides a range of control strategies besides product bans (product substitutes). Limitations on uses, quality control procedures, and information dissemination on hazardous properties (labels) are strategies designed to improve the "management of risk." Few would propose that we give priority to a search for a safer dynamite or for a fuel less combustible than gasoline. Even so, because of the publicity given to movements to ban DDT, certain aerosol propellants, PCBs, and the bottle to name only a few, much of the public currently perceives banning of materials as the only approach to chemical problems. EPA leadership will be needed to explain the reasons and common sense bases for strategies aimed at management of risk rather than bans. As with pesticides, there could evolve a trend of industry abandoning activities that have minor economic impact on their total product mix and concentrating on "big ticket" items. This could cause a serious upset in a broad range of American industry.

The discouraging regulatory history affecting the health of the drug and pesticide businesses should teach both government and industry to design a much better set of regulations and rules in these product fields than have previously been developed. Much more of the economy is at stake because this Act affects the whole chemical industry. The potential hazards to health are much lower in general chemicals than in pesticides and drugs. The intent of the Act [Section 2(2)] is aimed at unreasonable risk situations, implying that only a limited number of situations or products will be significantly affected.

IV. DIRECTIONS FOR CORPORATE POLICY AND PRACTICES

A. Customer-Supplier Relationships

The chemical industry will be increasing communication in the near future between suppliers and customers of chemicals to assure that their raw materials and markets are not restricted by failure to have products on the inventory list, as required in Section 8(b). In general, the Act will tend to increase the flow of information between customers and suppliers. More product knowledge, especially in the toxicology area, will be exchanged. Those suppliers with the best programs for information gathering and transferring of their products to their customers will be rewarded for their efforts.

New product introductions, covered by Section 5, will be the most difficult part of the Act, with respect to customer-supplier relationships. The rulings promulgated in this area of the Act will have significant effect on the number of new product introductions and the rate at which they can evolve from research projects to commercial status. The transition from research to commercial is a difficult line to define in the abstract or the generalization. Many products need to be tested in large quantities to measure and define their performance and potential for success as a new commercial product. For example, polyols used for flexible polyurethane foams for automotive molded seats or for slabstock production often require tank-truck quantities for a production run to determine processing and finished foam properties.

A greater level of business risk or uncertainty on any product at the research stage must be borne by the producer and user than previously experienced. Therefore, a greater degree of planning, discussion, and trust is necessary between these parties to conduct these kinds of experiments. There must be a clear definition between the parties as to whether a product is experimental or commercial and when the transition is to be made.

There will be a conflict of interest in this regard. The supplier will want to run trials, as close to commercial size as possible, as early as possible, to define whether the cost of further development is justified. The user, risking time, equipment, and money in testing these products, will want assurance that there is a reasonable chance that his investment is going to be recovered.

Another difficulty is the probable decline in the speed of commercialization of new products because of the time required for additional testing and EPA approvals of the test data to allow commercial manufacture.

The definition of significant new uses raises a similar set of potential conflicts of interest between customer and supplier. Many consumers of chemicals for competitive reasons do not reveal the uses of the products to their supplier. It is not yet clear whether they will be forced to under Section 5(a)(1)(B) of the Act. The definitions of significant and new use in Section 5(a)(1)(B)(2) and 5(s) will have a lot to do with how well this section of the law can be executed.

If, during the life of a chemical, one of its major uses becomes severely restricted under the Toxic Substances Control Act, a serious imbalance in the market place could develop. In the case of major commodity chemicals, the producers usually do not equally or proportionally share in each end use. One supplier usually dominates one market while being a minor supplier in another. The loss of a major segment of one of these uses could put one company in severe economic straits while barely affecting the others.

In products like polyols for urethane foams (epoxy resin or polyester resin intermediates), it would be even more difficult for those affected by a single-use restriction, as many would lack the technology to supply products acceptable to remaining markets or uses with which they had no previous experience. This facet of the Toxic Substances Control Act will have to be developed very carefully to avoid severe economic stress on the industry.

B. Maintaining Innovation

The chemical industry's hallmark is and has been innovation. Its strength, vitality, and great contributions to industry and society as a whole have depended on its innovative skills. One of the great challenges under the Toxic Substances Control Act will be to keep the optimism and commitment within the chemical industry to continue to create and develop new products and processes to provide a better life for our country.

Historically, it has taken 5 to 10 years for a new product to go from an idea to a money-making commercial venture. The yield on research projects that achieve successful commercial product status is probably 1:100. Management has always tried, so far in vain, to increase the yield, decrease the time to commercialization, or both.

TSCA will have the effect of adding one more major hurdle to the already exasperating obstacle course or gauntlet every research product must go through. However, on the positive side, winners will be winners for a longer period of time.

Under a carefully administered Toxic Substances Control Act, it is possible to achieve improved safety for our public with a minimum impact on constructive innovation. But the scope of the Administrator's powers are so wide that the threat of overkill is great. It is imperative that EPA, the public, and industry concentrate on carefully selected priorities to maximize the effective utilization of our scarce toxicological testing skills and to monitor the impacts of rulings on the selected priority products so that more effective and efficient rules and priorities can be evolved through experience as the programs move to cover additional products.

Improperly developed rules and regulations in administering the Toxic Substances Control Act could have the effect of turning the large research and development operations of the chemical industry to a purely defensive mode, focusing primarily on major mature product lines. The results of this could be (1) lost innovation of new technology – leaving it to our powerful foreign competitors or (2) general reduction of research and development investment because of its low productivity and very questionable benefits in terms of improved safety to our citizens. The precedent for these possibilities already exists in drug and pesticide research.

Innovation in the research and development phase will take on some new slants and twists. New innovation in project selection will become significantly more rewarding because it will be so much more expensive to advance research and development projects toward the commercialization state than before. Innovation in the development protocol, which will minimize the time required to commercialize a project, will be a fruitful effort. These two examples apply to research management, but even the bench chemist can begin redefining his new product ideas with the ultimate advantages of unrestricted sales under the Toxic Substances Control Act as one of the criteria by which he mentally screens his alternative projects.

Product and process improvements aimed at safer conditions for the manufacture and use of products will have strong appeal. The investment in toxicological screening capability, either in-house, contract, or both, will challenge the innovative skills of research management and, if done positively and effectively, will give a competitive edge over those companies that react only defensively. More and more frequently the chemist or engineer in research will team up with his counterpart expert in toxicology, ecology, and industrial health in research efforts. Not the least important area of innovation payoff is the developing of techniques to work smoothly and efficiently with the EPA-Office of Toxic Substances to minimize delays, misunderstandings, and reversals of positions.

Another area to consider is the need to develop the ability to communicate what is necessary from a legal point of view, while minimizing the loss of proprietary company information about technology, markets, opportunities of their products, and processes to U.S. and foreign competitors.

Finally, of considerable importance to business is the leakage of information to foreign competitors of new technology under development in the U.S. by monitoring the Federal Register and public documents. With both chemical structure and uses being defined, the main keys to chemical technology are bared. Industry and EPA need to resolve this problem as the definitions of chemicals and uses are finalized. Failure to achieve a workable solution to this dilemma could have a negative impact on new-product research in the U.S. As a result, international companies may be inclined to conduct more of their research activities overseas.

C. Corporate Policy, Organization, and Staffing

The Toxic Substances Control Act is, of course, only one of the numerous health and environmental laws that have an impact on the industry. It appears that all companies at all levels of management are striving to comply with these laws. Most of these laws have a flexibility of reach – as more problems occur more regulations follow, and at times, the relationship appears exponential. If regulation is to be reasonable, we believe that industry must place more emphasis on avoiding significant problems in the manufacture, use, and disposal of chemicals, whether or not there is an applicable regulation.

In other words, a corporate policy that merely states "we will comply with all regulations" is incomplete. Policy, attitude, staffing, and day-to-day decisions must visibly reflect a component of good citizenship toward health and environment. Dow is striving toward this approach by evolving a practice called product stewardship.

Product stewardship is an active program of identifying and solving problems (current or anticipated) which are related to the toxicity, safe handling, and environmental considerations of all products during manufacture, shipment, use, and ultimate disposal.

All companies must be concerned with crisis incidents, regulations, and compliance. However, we believe the industry need not be unduly preoccupied with such matters if each company is continually evaluating and developing its health and environmental program on the basis of its internal perceptions of what should be done – value judgments based on a sense of responsibility to employees, customers, the public, and the environment, i.e., internal citizenship or product stewardship.

We recommend several priorities for the chemical industry and the chemical-processing industries. First, appropriate staffing for environmental, public, and occupational health problems must receive top management priority. Although line management bears the ultimate responsibility, we must obtain or train active practitioners in industrial hygiene, occupational medicine, epidemiology, environmental sciences, and toxicology. Specialists in the collection and interpretation of hazard information are also needed. Furthermore, we should not be waiting for EPA, OSHA, or labor to force our staffing. Our decisions should be based more on a need factor determined by the nature and size of our operations and less on the imminency of an OSHA standard, a proposal by a public interest group, or a Section 6 regulation of a chemical.

Dow, while quite large in the aggregate, consists of many small units similar to small companies. Consequently, the industry has perhaps over emphasized the differences between large and small operations. From the standpoint of generating new data on chemicals, the large company may have some advantage but the smaller companies must learn how to tap greater resources such as the Chemical Industry Institute of Toxicology, MCA-sponsored studies, or the expertise of their large suppliers.

In terms of environmental and occupational health programs, all companies must recognize themselves as high-technology industry. Consequently, all units, large or small, must adequately staff occupational health and environmental programs. Processes, product mixes, and support personnel vary, but here are some reference points for evaluating staffing:

1. Industrial hygiene – 1 full time hygienist per 1000 employees (some operations may be advised to base their staffing on the number of physical and chemical stresses in the workplace, i.e., 1 hygienist per 250 stresses)

2. Medicine – 1 practicing physician per 1200 employees

3. Collection and interpretation of existing hazard data – 1 specialist per 5000 employees

These are guidelines that smaller companies may find useful in contracting for necessary services.

Another important priority must be an integrated-management approach to organization and communications. The organization must stimulate a management style that actively involves the health professionals and environmental scientists in day-to-day operations, including many middle and top management deliberations. Nonlinear communication networks, ad hoc problem-solving teams, overview or audit committees, and policy committees are all approaches worthy of consideration. The mode of organization must be selected to fit overall corporate organization, but the goal is to get frank, open, critical, constructive, and timely communication.

We live in an age when environmental contamination and occupational health concerns may be reported or dramatized in the *Wall Street Journal, The Washington Post* or *Rolling Stone* magazine. Some of these stories may appear before our scientists, managers, and physicians are aware of them. An article in the local press can stimulate employee questions that plant managers and physicians must answer and clarify within hours. Management decisions, for example, on customer relations or product continuance must often be made in the same time frame. The communications channels must be clear and effective.

A third priority involves obtaining data and information to minimize the uncertainties of chemical risks. Again, this is a moving-target area which requires interdisciplinary effort. Carcinogenicity, mutagenicity, birth defects, environmental impact, and epidemiology are today's buzz words. However, fundamental considerations such as acute toxicity, industrial hygiene, monitoring records, and general chronic effects still need strong and increased emphasis.

Every company must frequently reassess its capability to evaluate the need for testing; to set priorities for testing; to determine meaningful tests; and to conduct, evaluate, and report both the results and the significance of the study.

These three priorities must be addressed constructively, and with a transcendental sense of good citizenship or conscience aggressively, by each chemical producer, regardless of size.

V. THE CHALLENGE TO THE CHEMICAL INDUSTRY – PARTICIPATION IN THE REGULATORY PROCESS

It is a fact that industry has a Toxic Substance Control Act to live with. The degree of constraint or impedance caused by the Act is still to be defined. There is no reason that the basic purpose of the Act [Section 2(b)(2)], removing unreasonable risk of injury to health or the environment due to chemical substances, cannot be achieved without causing undue restriction on the chemical industry. Another intent expressed in Section 2(b)(3) that "the authority given EPA under this law to regulate the chemical industry should be exercised in such a manner as not to impede unduly or create unnecessary economic barriers to technological innovation" supports this contention.

However, the only ones who can and who have the will to present the true industrial point of view are industries themselves. They are the only ones to blame should the evolving policies and regulations be unreasonable and restrictive. Historically, industry has remained outside the sphere of the political process. It has been considered unbusinesslike and unproductive of industrial manpower to involve any significant part of the line organization in the political process. The Toxic Substances Control Act has built into it an opportunity and, in fact, a need for active industrial participation in the evolution of the Act into a workable set of regulations.

Every company in the chemical industry needs to reassess its priorities, organizational structure, expertise, and job assignments in the light of this Act. There must be a serious, meaningful commitment to involve the proper resources and people in creating and presenting the industrial perspective to the regulations' development. This can be done through trade association and direct single-company inputs. In either case, the source of the positions and participants in the dialogue must be chemical company personnel, experts in and committed to this new phase of chemical business priorities.

One can be sure that views opposite to those of industry will be developed and presented. The opposition is made up of highly intelligent, motiviated people with a mission. Industry must be willing to match with its top quality, highly motivated people. Without an industrial position of at least equal weight, the Toxic Substances Control Act could develop into an unbearable burden for the U.S. chemical business.

VI. THE CHALLENGE TO THE TRADE ASSOCIATIONS – A NEW RELEVANCY

Numerous trade associations provide forums for individual companies and practitioners to establish views on regulations. Generally, by such actions they are effectively energizing industry to get involved in the regulatory process. Trade association recommendations appear to be effective in "tinkering" with regulatory proposals to achieve a degree of practicality. Based on current activity, it appears that a number of trade associations, i.e., Manufacturing Chemists Association, the National Paint and Coatings Association, Soap and Detergent Association, the Chamber of Commerce, the Synthetic Organic Chemical Manufacturer's Association, and numerous others will provide viewpoints to EPA.

However, the regulatory activities of the past few years under OSHA, the air law, and the water law indicate new needs for the industry. Facts sometimes must be collected on an industry-wide basis. The citizen suit and court challenge have potential for defining the limits of agency authority. Benefits and essentiality of chemicals and chemical services need credible dramatization – to Congress, the press, and numerous publics. Priorities need articulation. Regulatory actions on a specific chemical, produced by only one or two manufacturers, may set precedents affecting many chemicals involving all members of the trade association.

Trade associations, like all voluntary organizations, must continually update their programs toward the future. Those that can become effective and cooperative, yet firm, advocates in the public arena are the ones which will remain relevant and will be supported. Trade associations must provide facts and viewpoints beyond that which can be generated by the individual member companies. Trade associations must provide vision and leadership.

THE TOXIC SUBSTANCES CONTROL ACT OF 1976: THE FORMULATOR'S VIEWPOINT

R. E. Copland
Chemical Specialties Manufacturers Association, Inc.
Washington, D.C.

There is no doubt that the passage of the Toxic Substances Control Act affects the entire chemical industry. Because its impact on basic manufacturers is direct, TSCA's impact on formulators/processors is more subtle, thereby demanding careful attention and astute planning. This conclusion primarily grows out of the proposition that what affects the manufacturer will probably affect the formulator, except that it may be difficult for the formulator to grasp that fact in a timely and practical manner. That is, TSCA purports to establish a mechanism to regulate and control the use of toxic chemical substances; as formulators may utilize toxic substances in their mixtures, so they are also vulnerable to regulation. Whatever may affect their supply of chemical substances affects their ability to produce mixtures.

The formulator and basic manufacturer are defined by what they produce. A formulator produces mixtures. The definition of a mixture, then, describes the identity of formulators and must be understood before a basic comprehension of TSCA's impact for formulators can be ascertained. A mixture is:

> Any combination of two or more chemical substances if the combination does not occur in nature and is not, in whole or in part, the result of a chemical reaction; except that such terms does [sic] include any combination which occurs in whole or in part, as a result of a chemical reaction if none of the chemical substances comprising the combination is a new chemical substance and if the combination could have been manufactured for commercial purposes without a chemical reaction at the time the chemical substances comprising the combination were combined.

This means that a mixture can involve chemical reactions if the mixture could be produced without a reaction at the time of combination for commercial distribution.

A simple illustration will demonstrate this point:

I

Substance A + Substance B ⟶ Substance E
(reacts to form)

Substance E
+
Substance D
+
<u>Substance C</u>
Product X

II

Substance C
+
Substance D
+
<u>Substance E</u>
Product X

Both products in these two examples are mixtures. Although example I involves a reaction, the fact remains that product X could have been produced, as in example II, without a reaction at the time of combination. However, if product X is then reacted with substance Y, creating product Z, and that reaction is indispensable to the production of Z, then product Z is a chemical substance for purposes of the Act.

It should be noted that the mixture definition appears to contain a prohibition against the use of new chemical substances as a component of the combination where a reaction is involved. As a general rule, this apparent prohibition should not present a problem to formulators.

Under the Act, new chemical substances are those not appearing on the §8(b) inventory list of old substances. Under Section 5, new chemical substances are subject to premarket notification. Ostensibly, once a new substance is permitted by the Administrator to be marketed, it will be included on the inventory list and become an old substance. As Section 8(b) states: "In the case of a chemical substance for which a notice is submitted in accordance with Section 5, such chemical substance shall be included in such (inventory) list as of the earliest date (as determined by the Administrator) on which such substance was manufactured or processed in the United States."

Therefore, the formulator will be purchasing from his supplier what is, in effect, an old substance because of its inclusion on the inventory list. An exception occurs if the formulator wishes to test market a mixture containing a newly developed chemical substance for which no §5 notification has been submitted. This situation will be discussed below.

The more important sections of TSCA – 4, 5, 6, and 8 – and their relationship to the formulator should be thoroughly digested in order to fully grasp the potential pitfalls of the Act.

§8 – Reporting and Retention of Information

Section 8 pertains to reporting and record-keeping requirements and the compilation of an inventory list. Reporting and record keeping for mixtures will be determined by rulemaking with the admonition that such rules may only be promulgated if necessary for the effective enforcement of the Act. Thus, reporting and record keeping for mixtures will likely relate to a specific informational need of the Administrator, almost certainly obviating a shotgun approach.

At this time it appears likely that the Administrator will

require the submission of information concerning use categories, concentrations, and volumes of selected chemical substances employed in mixtures. The need for such information arises because, in some cases, basic manufacturers do not know how their substances will ultimately be used. The Environmental Protection Agency (EPA) is most interested in learning how chemicals are used in terms of the magnitude and manner of exposure to man and the environment. Therefore, information bearing on such exposure, such as use categories, will undoubtedly be necessary. However, if the Administrator considers the end use of chemical substances to be "reasonably ascertainable" [§8(a)(2)] by the basic manufacturer, there would be no need to require such information from the formulator. Section 8(e) requires immediate submittal of a notice to the Administrator of information obtained indicating that a substance or mixture "presents a substantial risk of injury to health or the environment." This reporting requirement is now effective and no implementing rule is needed.

"Use category" is, at this preliminary stage of implementation, a nebulous term. Basically, EPA is seeking to classify use patterns in a systematic way to avoid a confusion of terminology and obtain a handle on the ultimate location of toxic substances. In fact, the Agency is considering requesting information on a plant-by-plant rather than a corporate-wide basis in order to accurately assess the population-at-risk.

One proposed system would break down uses into 100 major functions (e.g., solvent, surfactant, etc.), 20 major applications (e.g., paint, detergent, etc.), and 10 application environments (e.g., institutions, manufacturing plants, etc.), with further subdivision of function and application. The system is based on a 10-digit decimal system so that, theoretically at least, the appropriate numbers would be inserted into the appropriate slots. Thus, a chemical used in car wash detergents may be described by function (grease cutter), application (car wash), and application environment (outdoors and service establishment). It remains to be seen whether such a system is practical or, indeed, does not produce extraneous information that will itself pose an administrative burden to the Administrator.

It is clearly impractical to require the submission of reports for the literally millions of mixtures that are in commercial distribution. EPA surely recognizes this fact and will take, out of necessity, a selective approach. In requiring reports from formulators, the Administrator's concern primarily relates to the effects of component chemical substances. It is logical to expect that mixture reporting will be confined to mixtures containing those chemical substances which merit special attention or are expected to be used in a manner which constitutes a significant new use. This concept will be explained below.

Therefore, a practical requirement suggests itself: available evidence in the hands of EPA should indicate health or environmental problems with a substance before reporting from a formulator is required. It may be that vehicles exist in TSCA as "triggering" devices for reporting, such as the §5(b)(4) "risk" list, the compilation of which is discretionary upon the Administrator. The §5(b)(4) list would contain those substances which the Administrator finds present or may present an unreasonable risk of injury to health or the environment.

Records of adverse health or environmental effects (including consumer complaints) allegedly caused by mixtures must be maintained by formulators in a manner to be determined by rule. These records are available on request of the Agency, so that formulators should maintain accurate and complete complaint files. In addition, the Administrator must promulgate rules requiring the submission of health and safety studies initiated by the formulator (known to him or reasonably ascertainable by him). It is likely that the basic manufacturer, to whom this prescription also applies, would be in a better position to have conducted or know of such studies, so that the brunt of the requirement will probably fall on him.

Compilation of the §8(b) inventory list will determine what chemicals have to undergo the rigors of the premarket notification process and, thus, is of crucial importance. The formulator must be certain that the components of his mixtures are included on the list so as to avoid a subsequent uncertainty of supply. The list must be published by November 11, 1977 with a 30 day grace period after which no new chemical substance may be manufactured without prior notification. Thus, every effort should be made to ensure that all formulations, currently marketed or on the drawing board for 1977 introduction, have their components listed.

EPA has published a preliminary inventory list consisting of substances gleaned from various government files and other sources. This list should be scrutinized by formulators who may either supplement it if their component substances are missing or arrange for their suppliers to do so. Nomenclature may present problems to the formulator since chemical terminology can differ widely. Currently, EPA is employing the Chemical Abstract System and hopefully will provide a cross-index of common names.

Statutory consideration is afforded small formulators relative to reporting and record keeping, but the practical benefit of this consideration may not be consequential. Small businesses will be defined by rule, but the bulk of the chemical formulation industry will no doubt fall into this category. While TSCA limits the reporting and record-keeping requirements that can be imposed on small formulators, the requirements that will be imposed on any formulator should not substantially surpass these limits, with the possible exception of adverse effects on health and safety records.

The Administrator may require "small" formulators of mixtures to submit information pertaining to substances or mixtures subject to:

1. a testing rule under §4
2. the "risk" list of §5(b)(4)
3. a regulation under §6
4. an effective order under §5(e)
5. for which relief has been granted pursuant to a civil action brought under §§ 5 or 7

It seems extremely unlikely that formulators as a whole would be asked to supply information on substances or mixtures falling outside this list; in fact, one could justifiably argue that the list as constituted is too inclusive and that rules promulgated under §4 ought not to incur a need for other than basic test data. If the results of the testing suggests a need for more information, then the Administrator has the tools at his disposal for accumulating it.

§4 – Testing of Chemical Substances and Mixtures

The testing of chemical substances and mixtures should not prove burdensome to formulators in as much as TSCA clearly contemplates that manufacturers have the primary responsibility of conducting prescribed tests. §4(a)(2) is of key importance to the formulator. It states:

> In the case of a mixture, the effects which the mixture's manufacture, distribution in commerce, processing, use, or disposal or any combination of such activities may have on health or the environment may not be reasonably and more efficiently determined or predicted by testing the chemical substances which comprise the mixture.

The meaning of this somewhat cryptic sentence is spelled out more fully in the report by the House Committee on Interstate and Foreign Commerce:

> The assessment of safety of a mixture may well be based upon the toxicity of particular components, and tests of the entire mixture with its varying component ratios may be unnecessary or unrewarding. At the same time, the Committee recognizes that there may be instances in which a particular combination must be tested to reasonably evaluate the effects of the mixture. For instance, the effect of two chemicals, when combined, may be greater than the sum of the effects of the components taken independently. The Committee bill does not prohibit the Administrator from requiring testing of the mixture in such instances.

It appears from this passage that some evidence of a synergistic effect must be indicated before the Administrator can require testing of mixtures per se.

On the other hand, situations may arise where the formulator wants to conduct appropriate testing. For example, the formulator may have a large economic stake in a particular mixture, whereas the component chemical substance suppliers may see a scant profit. In such cases, it may be worthwhile to the formulator to conduct testing of either the component substances, if required, or the mixture itself in order to protect the continued marketing of his formulation. It is possible that a formulator, whose mixture contains a suspect component, may wish to conduct tests on the mixture *in toto* in order to demonstrate that the low concentration or particular chemical properties of the mixture renders the suspect component innocuous.

Under §4(g), a formulator can petition the Administrator to prescribe test standards for the development of data pertaining to a particular substance or substances. While the Act is silent as to test standard petitions for mixtures, it is probable that EPA would be cooperative if presented with a sound argument as to the need for such standards.

§4(e) provides for the establishment of a list of chemical substances which will be given priority consideration for the promulgation of testing rules. The list is compiled by an interagency task force, and it is entirely possible that this group may designate, in some cases, classes of chemical rather than individual substances.

The formulator should examine this priority list (which will be published in the *Federal Register*) along with the stated reasons for inclusion. It is probable that some of these substances will eventually fall within a §6 regulation. As is more fully detailed below, §6 rules can eliminate a mixture from the marketplace. Thus, if a substance on the priority list is a mixture component, it is imperative that the formulator follow its regulatory path very closely. As a general rule, test results will be available for public examination. Under §4(d), the Administrator must publish a notice of receipt; formulators will therefore have ample opportunity to track substances subject to testing rules.

Inclusion on the list should alert the formulator as to potential problems, but by itself, however, should not precipitate formulation changes. It is by no means clear that the background and expertise of the interagency committee will actually result in the selection by them of substances that merit "priority" status. Sociopolitical pressure and conflicting judgments by EPA personnel may render the priority list something less than the name would imply.

An important impact of TSCA implementation pertains not merely to enforcement of the Act itself: the accumulation of data which may lead to new regulations promulgated under other statutes such as the Occupational Safety and Health Act. EPA itself has posed the question: "How important is strict enforcement of specific TSCA regulations in the context of the broader impacts that the legislation will undoubtedly have?" These "broader impacts" likely pertain, to a substantial degree, to the gathering and disseminating of information on toxic substances in order to support regulation under other statutory authorities. In the future, it seems apparent that formulators will be required to increase their corporate awareness of other statutes and agencies that can affect their business operations. The Federal government will undoubtedly tighten its regulation of the chemical industry on all fronts – workplace safety, product safety, environmental pollution, and waste disposal treatment.

§5 – Manufacturing and Processing Notices

As §5 seems to pertain exclusively to chemical substances, many formulators appear to have the mistaken notion that its provisions do not concern them. While it is true that new mixtures are not per se subject to premarket notification, circumstances will exist under which mixtures may be subject to notification requirements.

§5(a)(B) states: "No person may manufacture or process any chemical substance for a use which the Administrator has determined. . . is a significant new use" without notifying the Agency at least 90 days before processing (formulation).

The drafters of the legislation noted their concern that the hazards of chemical substances may be more acute when placed in a particular environment; some uses for a particular substance may be perfectly innocuous while other uses would present an unreasonable risk. Therefore, clearance of a substance for manufacture always contains one important qualification: if the substance is used in a different fashion than previously indicated, a reevaluation of such substance may be necessary.

While the definition of a "significant new use" is subject to rulemaking, TSCA lists several criteria which must be considered:

1. projected volume (to determine if an increase in volume will affect anticipated exposure to humans and the environment
2. different type or form of human or environmental exposure
3. increased magnitude and duration of exposure
4. anticipated manner and method of processing, distribution, and disposal

The formulator, then, when planning to market a new mixture must consider whether a component chemical substance or substances would be used in a manner comprising a significant new use. Clearly, it is not within the purview of the formulator to decide when the marketing of his mixture constitutes a significant new use for a chemical substance. Nor is it possible that the Administrator can promulgate a rule that describes for general application what use will constitute a significant new use. Rather, it is probable that established and approved uses, concentrations, and/or volume for selected substances or chemical classes outside a prescribed range will constitute a significant new use actuating a notification requirement.

Notification of a significant new use may pose confidentiality problems. §5(d)(2) specifies that the Administrator shall publish in the *Federal Register* a notice which identifies the chemical substance and lists the intended uses of the substance. Since notification must take place 90 days before processing, and the *Federal Register* notice is published 5 days after receipt of notification, it seems that sensitive marketing plans may be publicly exposed.

While the name of the company submitting the notification need not be included in the public notice, the fact that a chemical substance will be put to a particular use may compromise valuable lead time. EPA personnel are considering restricting the public notice to information concerning potential types of exposure as indicated by the particular form in which the chemical would be used, its concentration, and manner of its general distribution. If this laudable concern can be reduced to a functional system, it would seem that the spirit of the Act can be met without compromising sensitive information.

Notification of a significant new use must be given at least 90 days before manufacture or processing. However, it appears that no action will be taken with regard to significant new uses for at least 2 years. Under §5(e), should the Administrator determine that insufficient information is available to evaluate the potential health and environmental effects of the chemical substance when used in the anticipated manner, he may propose an order that would prohibit or limit the processing of a chemical substance for a significant new use until such time as data is submitted that, in effect, either clears the substance or leads to a §6 rule.

The procedures surrounding a §5 regulation pending development of information are somewhat complex. They should be understood by the formulator since court action may be involved that could entail considerable legal expense.

If it seems likely that the processing of a chemical substance for a significant new use will raise safety questions, data should be developed as early as possible that will support utilization of the substance for the proposed new use. Such advanced planning will unquestionably benefit the formulator by negating subsequent marketing delays imposed by §5 regulation. It may be desirable to direct the supplier to discreetly run appropriate tests to ensure that §5 roadblocks can be overcome. If safety testing for a significant new use is not favorable, marketing plans, in all likelihood, may need reassessment. Certainly, once notification of a significant new use is submitted, the formulator has a vested interest in marketing his mixture as soon thereafter as possible to avoid losing the obvious advantage of being first in the field.

It may even be advisable to ask the Agency to describe what test standards would be appropriate for clearance purposes. This may be accomplished by petition under §4(g). The Administrator must either grant or deny the petition within 60 days of receipt. However, if he should decide not to prescribe test standards, he must publish his reasons for such denial in the *Federal Register.* While it is not readily apparent that such publication would constitute a breach of confidential marketing plans, under some circumstances announcement that a manufacturer or processor of a particular chemical substance seeks test standards may have the same practical effect. The formulator must balance these considerations and plan accordingly.

Many formulators manufacture intermediates that they employ in the formulating process. Furthermore, a formulator may wish to test market a mixture that may involve a significant new use of a chemical substance. The question arises, then, as to whether the manufacture of "new" intermediates and test marketing of mixtures comprising significant new uses are vulnerable to the requirements of §5. The answer is a qualified yes.

Intermediates will be treated as chemical substances so that development of a "new" intermediate will impose a notification requirement prior to utilization in the formulation process. However, intermediates that exist only technically, which cannot be practically isolated and do not involve human or environmental exposure will be exempt from notification requirements upon application [§5(h)(5)].

If a formulator intends to test market a "new" chemical substance in a mixture, or a mixture which comprises a significant new use for a component substance, he may apply for an exemption from premarket notification. The term "test market" contemplates the distribution of the substance for an assessment of marketability rather than performance evaluation that is a part of product development. Actually, if the "new" substance is developed by a supplier for possible use in the formulator's mixture, either the manufacturer or supplier could probably apply for the exemption. §5(h)(1) states that the Administrator may grant an exemption for test marketing purposes "upon a showing. . .satisfactory to the Administrator" that no unreasonable risk will be presented. The Administrator has authority to impose such restrictions on the activity as he deems appropriate.

Receipt of exemption applications for both intermediates and test marketing are published in the *Federal Register*; interested persons may comment, and the application must be approved or denied within 45 days after receipt. The decision to grant or deny the application must also be published. Again, lack of confidentiality concerning sensitive marketing plans may present a problem. In fact, it is not clear at this time that an exemption from premarket notification will provide any benefit whatsoever to the formulator, since the data that must be submitted to obtain an exemption will probably be similar to data that must be submitted with a premarket notification.

§6 – Regulation of Hazardous Chemical Substances and Mixtures

The impact of §6 is straightforward: the Administrator has authority to ban, or otherwise restrict, the manufacturing of any chemical substance or mixture if he finds "a reasonable basis to conclude" that the substance or mixture presents or will present an unreasonable risk to health or the environment. A range of regulatory options are available to the Administrator, such as banning for a particular use or requiring special labeling. All regulations must be promulgated by rule; an immediately effective rule can be issued without prior hearing if the Administrator determines that serious or widespread harm is imminent and in the cases of an order of total prohibition, a court has preliminarily agreed with him.

In addition, the Administrator is empowered to require formulators to submit a description of quality control procedures if he "has a reasonable basis to conclude" that the formulation process itself is causing health or environmental problems. Should the Administrator determine that quality control procedures are inadequate to prevent risk of injury, he may order an appropriate revision. If, in his opinion, the use of inadequate procedures has resulted in the distribution of hazardous products, the Administrator may order the formulator to give notice of risk to distributors or any other person who reasonably might be expected to have possession of the product, to give public notice, and to provide replacement or repurchase. Obviously, the public notice provision could prove troublesome to formulators marketing consumer products. Before either ordering revision of quality control procedures or ordering notice and replacement or repurchase, the Administrator must provide an opportunity for a hearing.

Section 6, along with §5, comprises the real teeth of TSCA. However, formulators conscious of the safety of their component chemical substances should not be caught napping by a §6 regulation. For the near term, it is likely that chemical substances subject to §6 regulation will be those with a strongly documented history of health or environmental problems. As previously mentioned, the priority list, §4 testing rules, and the §5(b)(4) "risk" list must be monitored closely in order to recognize what substances or classes of substances may be restricted in the future. In addition, quality control procedures should be reviewed so as to avoid contamination or possible reaction after packaging.

The preceding paragraphs have by no means exhausted the subject. Specific interpretations by Agency personnel can be anticipated but not predicted; time and practical experience will more fully reveal the precise nature of TSCA's impact on formulators. Moreover, TSCA contains other, more exotic sections that can also potentially affect the formulator, e.g., citizens' petitions and employee protection. Nevertheless, a sound comprehension of §§ 4, 5, 6, and 8 and their attendant obligations and responsibilities, if communicated effectively to appropriate personnel, should aid significantly in creating a functional post-TSCA environment.

To achieve corporate preparedness, formulators must review all component chemicals contained in their mixtures and, if questions arise, search the literature for health and environmental effects. If questions persist, the formulator is well advised to obtain appropriate assurances from his supplier and conscientiously monitor the *Federal Register* for salient mention of the suspect chemical. One of the major purposes of TSCA is to achieve a heightened industry awareness of toxic substances.

In writing legislation, Congress established within EPA "an identifiable office to provide technical and other nonfinancial assistance to manufacturers and processors respecting the requirements of this Act. . .and the means and methods by which such manufacturers and processors may comply with such requirements." Formulators, particularly those of small or medium size, are advised to make full use of this office. This Industry Assistance Office is intended to provide practical help; formulators should be able to obtain answers from it that will not be available elsewhere.

TSCA will affect most formulators, but for the great majority, its impact will affect the conduct of business operations rather than products or their development. Certainly, some products will be affected, but careful monitoring and perhaps occasional advocacy, should greatly ameliorate those effects. In sum, prudent administration by the government, along with the active participation of industry, should produce an orderly transition as well as meet the intent of the Act.

Section VII

Corporate Preparedness

CORPORATE PREPAREDNESS FOR THE TOXIC SUBSTANCES CONTROL ACT*

J. D. Behun
Mobil Chemical Company
Edison, New Jersey

INTRODUCTION

The Toxic Substances Control Act is the latest and most encompassing federal law concerning toxicology and safety to which industry must comply. For 40 years there was only one law in this area, the Food, Drug, and Cosmetic Act. As shown in Figure 1 however, the number and rate of enactment of major laws has increased greatly. There are now twelve significant laws on toxicity and safety, the bulk of which were enacted since 1970. The National Environmental Policy Act of 1969 set the stage for much of the ensuing laws.

The historical background of the development of the Toxic Substances Control Act (TSCA) is discussed in Section 1 of this Guidebook. However, to put the TSCA in perspective, one can see from Figure 1 that there are already laws to protect workers, consumers, and to regulate the quality of water, air, food, as well as control transportation and storage of chemicals. As shown in the diagram in Figure 2, the TSCA is intended to close gaps in the current array of environmental laws and provide broad comprehensive protection.

With the signing of the bill on Columbus Day, October 11, 1976, the Environmental Protection Agency (EPA) has been granted a charter to search for, find, and control "unreasonable risks to health and the environment" caused by existing as well as new chemicals. Industry, as responsible citizens, is committed to comply with the law. The Office of Toxic Substances (OTS), who must implement this law, recognizes that it is embarking on a journey into vast, unmapped areas. Dr. Schweitzer, Director of OTS, has indicated that they will seek assistance from industry in determining how they can best administer the law to achieve its intent. Both industry and EPA have limited resources, and each understands that regulations must be reasonable, so that neither will be unnecessarily overburdened and ultimately create greater penalties to consumers and taxpayers. Although a government regulating agency and the group it regulates may on occasion have opposing views, they must work together towards a common goal. It is in this spirit that industry must accept EPA's invitation to participate in, and contribute to, the development of meaningful and reasonable regulations under the TSCA.

DISCUSSION

Our earth is composed of a dynamic array of chemical substances. Man, animals, plants, and all living matter are composed of a highly ordered combination of chemicals. Living things depend on certain chemicals for nourishment, but will react adversely to others – the biologically disruptive materials we generally class as toxic substances. The fact that the chemical industry has produced and processed millions of chemicals in multibillion pound quantities in such a delicately balanced system while man has grown bigger, stronger, with a longer life expectancy is to some extent evidence of the voluntary toxic substances control program that industry has practiced for years. The relatively few examples of serious problems are atypical, yet the need for a Toxic Substances Control Law to regulate a growing industry in a growing society cannot be denied.

It must be recognized that the difference between a toxic and nontoxic behavior of any chemical is in the level of exposure of the living organism to this chemical. We all know that too much table salt or even water can be toxic. However, chemicals do differ in the level at which they will produce an adverse response. Putting it another way, some are more toxic than others. Most chemicals are safe at relatively high levels, but a few are dangerous even in very small doses. As population increases and as industry grows, it is probable that more people will become exposed to higher amounts of more toxic substances. Thus, we are in need of a Toxic Substances Control Act. After all, it wasn't until there were sufficient cars on the road that it was necessary to put up stop lights and stop signs. But just as we do not need a stop light or stop sign on every corner to slow down the few habitual violators, not every chemical needs to be regulated as a dangerous poison. The purpose of the law is to define the risk associated with some existing chemicals and to detect potentially hazardous new chemicals before they enter commerce. It will be the responsibility of the Office of Toxic Substances of EPA to distinguish those substances that can present an unreasonable risk.

Industry must await guidance from EPA in order to establish a detailed action program for conforming to the

*Presented in part before the Toxic Substances Law and Regulations Seminar, sponsored by Government Institutes Inc., Washington, D.C. December 9–10, 1976.

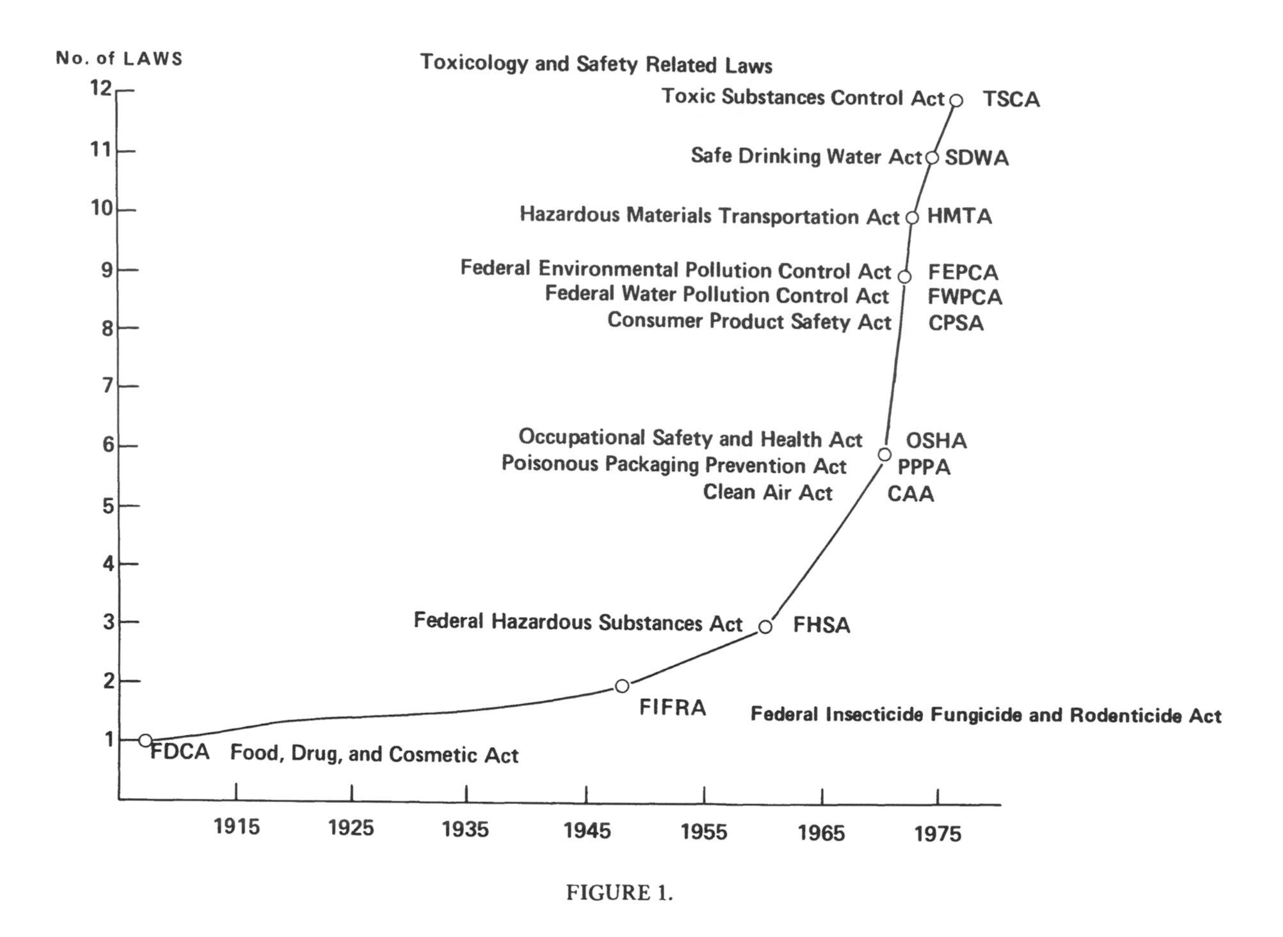

FIGURE 1.

TSCA. The impact upon different companies and their resources for complying will vary, but certain general and specific activities will be undertaken by all. We will consider some of the major features of industries' implementation program that are apparent at this point in time.

Organizing and Staffing

No two companies are likely to organize and staff in the same way to meet the requirements of the TSCA. However, there are certain common professional skills that will be needed by all in order to perform the activities that this law will demand (Table 1). Taking them alphabetically, first, industries' implementation program will require administrative personnel to carry out the coordinating and liaison functions including developing a program; information gathering and dissemination; arranging for record keeping; reviewing company status; maintaining contact with trade associations; and reporting to EPA. Chemical expertise and familiarity with the composition of raw materials, intermediates, and products will be needed, as well as knowledge of processes and by-products to assure that proper records are kept, operations are monitored, and changes are not overlooked. Engineering efforts can provide the necessary quality assurance, effluent controls, and minimize worker contacts with chemicals. Legal assistance will be called for not only to help interpret the law but also to assess potential liabilities and to deal with possible citizens' civil actions. A public relations function must inform employees of potential or real risks that are detected and the protective actions to be taken by the individual and company. It must also assure that appropriate communications are maintained with unions, the local, and national public, with particular attention towards distinguishing between real and imaginary or alleged risks. Last, but not least, professional toxicology will be required to determine the appropriate tests to be performed and to interpret the results of the tests.

TABLE 1

Organizing and Staffing

Administrative
Chemical
Engineering
Legal
Public affairs
Toxicology

By this listing of skills and activities it is not implied that each company must hire at least one more of each of the professionals named in order to comply with the TSCA. Many companies can probably redistribute existing professional skills to satisfy these new needs. In some cases,

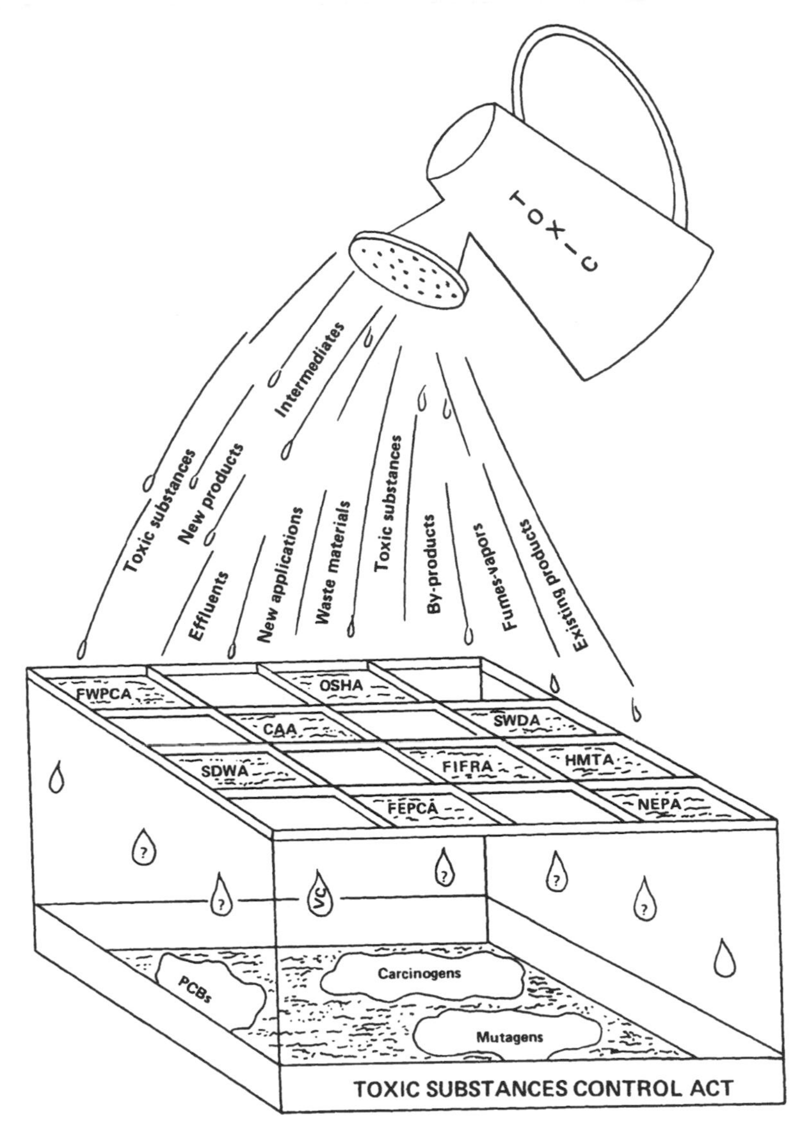

FIGURE 2.

a combination of several of these skills may reside in one or more individuals who can perform most of these functions. The approach of smaller companies will, most assuredly, be different than that of larger companies. The latter will have to decide whether it suits their organizational structure and their requirements for complying with the law to go the centralized or decentralized organizational route. I will have more to say later about integration with other corporate activities.

Familiarization and Information

In order to cope with the law, each company should develop a thorough understanding of it. Without rehashing what has already been covered in earlier sections, the key elements of the TSCA (Table 2) are:

1. Testing of new and existing substances or mixtures deemed potentially harmful to health and the environment.
2. Requirement for premarket notification before manufacture of any new chemical substance or an existing substance for a significant new use.
3. EPA's broad discretionary authority for regulating substances that present unreasonable risk. This includes possible banning of the chemical, limiting its use, requiring labeling statements, or prescribing quality control procedures.
4. Manufacturers and processors may be required to keep and report detailed records on products, processes, uses, human exposure, as well as adverse health and environmental effects.

These four key Sections only define the broad scope of

TABLE 2

Familiarization and Information

The Law	
Testing	Established
Premarket notification	New
EPA authority	
Record keeping	
Information sources	
Seminars	
Brochures	
Manuals	
Manufacturing Chemist Association	
Environmental Protection Agency/ Office of Toxic Substances	
Who should know	
Management	
Supervision	
Staff	
Employees	

TABLE 3

Inventory and Records

Inventory
- Raw materials
- Intermediates
- Products
- Imports

Records
- Identity
- Volume
- Uses
- By-products
- Human contact
- Disposal

Adverse effects
- Health
- Environment

the law. There are 31 Sections and many subsections of the law with which manufacturers, processors, distributors, users, and disposers of chemical substances should become familiar. But an understanding of the law is only the beginning. As OTS develops the specific regulations, companies will have to keep up to date and familiarize themselves with these as they evolve.

The question then is, how does one become informed? Various sources of information are and will become available. Both the Manufacturing Chemists Association (MCA) and OTS have scheduled information seminars. There will also be descriptive literature available. OTS has issued a ten-page summary of the law. The MCA has issued a summary, including some typical questions and answers, an outline of activities for compliance, flow diagrams for testing and premarket notification, and a chronology of major event deadlines. Telephone or written inquiries can be made to MCA or to EPA's Office of Toxic Substances for clarification or additional information. The *Federal Register* is an important source of current developments.

Once having obtained information, it should be disseminated throughout the company to individuals who will be directly or indirectly affected. This includes just about everyone in a chemical company: management, supervision, staff, and employees. The means by which information is distributed can be any combination of formal presentations, internal seminars, memos, bulletins, or one-to-one communication.

Inventory and Records

By November 1977, EPA must publish a final inventory list that will define those substances and mixtures which are established materials and do not require premarket notification. OTS has indicated that a preliminary list will be published by mid-1977 in the *Federal Register* to allow time for corrections. Every company will have to compile its own inventory list to compare with EPA's preliminary list. It would be well to include all raw materials, intermediates, products, and imports (Table 3). Since EPA will have the option of listing by categories instead of specific chemicals, care must be taken to see that no item is overlooked to the extent consistent with the purposes of the Act. Clarification with EPA will be needed to determine whether a specific substance qualifies for inclusion in one of EPA's defined categories.

Under Section 8 of the Act dealing with reporting and retention of information, a manufacturer or processor must maintain records and submit to the Administrator such reports as he may reasonably require. We assume, at this time, that this will apply to selected chemicals. The information that may be required will include: identity, volume, uses, by-products, extent of human exposure, and methods of disposal. Reports also may be requested on health and safety studies performed in the past. All who manufacture, process, or distribute chemicals are required to keep records on any alleged adverse health or environmental effects. Records on employees must be kept for 30 years and on other effects for 5 years.

The enormous burden placed on industry for gathering, storage, and retrieval of information may require a systematic (perhaps computerized) method for data handling.

Toxicology Testing

Toxicology data will provide the basis for EPA to determine whether a substance presents an unreasonable risk, and the data will help them judge to what extent that substance should be controlled. Some of the data already exist but more will be called for by EPA. Industry will have to develop the bulk of this information.

A few of the larger chemical companies, such as Carbide,

TABLE 4

Toxicology Testing

- In-house
- Contract
- Multi-sponsor
  - Manufacturing Chemists Association
  - American Petroleum Institute
  - Chemical Industry Institute of Toxicology
- Reference sources

Dow, duPont, and Eastman Kodak have their own in-house toxicology testing facilities and have been conducting tests for many years. DuPont's Haskel Laboratory goes back 40 years. To meet the anticipated demands of the new law, as well as to provide more long-term exposure tests (the need for which has become apparent in recent years), many of the in-house testing laboratories have already consented to more than doubling their present capacity. Other companies such as Monsanto have decided to build their own testing laboratories, while still others are considering similar moves.

What about companies that don't have their own toxicology testing laboratory and have no plans for doing their own testing? Many independent testing laboratories exist who will do testing under contract. The Society of Toxicology issued a survey of toxicology laboratories in March 1976 (see Section X, page 10.1). This survey names the known laboratories, describes the types of tests with which each laboratory has experience, their facilities and services, number and type of professional staff, and indicates whether they do contract work. In addition to individual contracts, many companies participate in multi-sponsored testing programs of commodity chemicals. The tests are generally long-term exposure tests that are coordinated by trade associations such as Manufacturing Chemists Association, American Petroleum Institute, and International Institute of Synthetic Rubber, to name a few. Recently, a unique organization known as the Chemical Industry Institute of Toxicology has been formed, headed by Dr. Leon Goldberg. The Institute, which now has 27 members, has initiated contract testing on two large-volume commodity chemicals, ethylene and toluene, and is establishing its own laboratory to conduct further testing, devise new tests, and perform toxicology research.

In assessing individual testing requirements, it will be important to know what toxicology data already exist on established chemicals. A number of reference books are available and there are some computerized data banks established and being set up. One such existing service is called Tox-Line. It should also be noted that the National Library of Medicine Office of Toxicology performs a service of listing reported long-term tests being undertaken or under way in a publication entitled Tox-Tips.

The big question in toxicology testing is what tests should be performed and how far should the testing be carried for a specific chemical. EPA will have to provide the guidance here. The language of the law offers some clues – generally, the greater the degree of expected human exposure, the greater the amount of data that may be required. Chemicals which have a close structural and property relationship to known carcinogens will need sufficient testing to determine their risk potential.

TABLE 5

Tests for Health and Environmental Effects

Carcinogenesis	Persistence
Mutagenesis	Acute toxicity
Teratogenesis	Subacute toxicity
Behavior disorders	Chronic toxicity
Cumulative effects	
Synergistic effects	

Methodologies

- Epidemiology
- Hierarchical
- In vitro
- Whole Animal

The law gives the Administrator authority to require testing for health and environmental effects such as carcinogenesis, mutagenesis, teratogenesis, behavioral disorders, cumulative effects, and synergistic effects. EPA can prescribe testing protocols or standards to test for persistence, acute toxicity, subacute toxicity, chronic toxicity, and any other characteristic which presents an unreasonable risk. Methodologies that may be prescribed in the standards include epidemiology studies, serial or hierarchical tests, in vitro tests, and whole animal tests (see Table 5).

To sum it all up, very extensive toxicology information may be demanded. Whether a company has in-house testing capability or not, the best insurance for seeing that appropriate tests are performed and that the results are properly interpreted is a capable industrial toxicologist. Some contract testing laboratories can provide a limited amount of guidance and interpretation.

Identify-Assess Risks

Even though some of the major provisions of the TSCA are not scheduled to become effective until late in 1977 (e.g., issuance of the Priority List, Inventory List, and Premarket Notification), consideration must be given right now to some areas of immediate vulnerability that will exist once the law becomes enacted.

TABLE 6

Identify-Assess Risks

Substantial risk	Premarket notification
Imminent Hazard	
Citizens actions and petitions	Risk/benefit
Priority List	Decisions

As of January 1, 1977, upon receipt of any information that "reasonably supports the conclusion that a substance or a mixture presents a substantial risk to health or the environment," such information must be reported immediately to EPA by any person who manufactures, processes, or distributes such material. The term substantial risk has not yet been defined by EPA. However, each company will be obliged to carefully examine significant adverse effects shown in any new test data, worker or customer exposure data, long-term medical records, or environmental data, and may wish to check with EPA to be certain not to violate the law.

As of January 1, 1977, both EPA and private citizens will have legal basis for bringing civil action against anyone contributing to an unreasonable risk. Under Section 7, Imminent Hazard, EPA has the authority to seize substances or mixtures and initiate civil or punitive action. Under Section 20, private citizens will have the right to bring civil action against a manufacturer or processor for violation of any rule or against EPA for failure to exercise its nondiscretionary authority. Under Section 21, citizens may also petition EPA to propose, modify, or amend a regulation. All of these sections of the law represent possible provisions under which any company could be faced with a toxic substances control issue starting January 1, 1977.

Coming back to portions of the TSCA that are scheduled to take effect later, companies cannot afford the luxury of not preparing now. According to the law, the Interagency Priority List may not issue until October 1977, after which OTS has up to 12 months to propose rules. All of this sounds like it is a long way off, but one must keep in mind that it may take 2 or 3 years of toxicology testing before a definitive indication of unreasonable risk is obtained. In the meantime, the fate of the product is in the balance. If a company thinks one or more of its major products may appear on EPA's Priority List, it would do well to get a head start on assessing its risk. Conceivably, data might be obtained that could help convince EPA that the product needs only a minor amount or perhaps no further testing. If on the other hand, early testing raises a warning flag, would it not be better to know sooner to appropriately plan for extensive testing or curtailment if necessary.

It would appear from a cursory look at the TSCA that industry has a period of about 11 months to prepare before it is subject to the Notification Provisions in Section 5. However, on closer examination, it is apparent that anyone contemplating introducing a new product after December 13, 1977 will need a great deal more lead time than in the past. Although the law states that EPA must be notified only 3 months before manufacturing a new product, EPA can extend this time period another 3 months. In addition, time must be allocated for testing to assure oneself and EPA that there is no unreasonable risk. It should be noted that under the law, EPA can be petitioned to define what tests are required. The time for action with respect to TSCA is now on any new product that will be initiated 1 year or so from this time.

Human nature being what it is, some may wait until EPA offers explicit guidelines or specific regulations that will affect them directly before they start testing. As pointed out frequently in current trade literature, the number of toxicologists and the number of well-equipped and staffed contract testing laboratories are limited. These factors could further delay introduction of new products and new uses or could cause EPA to take action on established materials which are suspect because of lack of sufficient data. This is another reason for testing now.

As companies test substances, either on their own initiative or because of EPA rules, some unpleasant surprises are bound to occur. When a potential unreasonable risk is identified, a risk/benefit analysis must be performed to determine to what extent long-term exposure testing should be undertaken, whether a new product should continue to be developed, and whether an established product should be curtailed or abandoned. Each company must devise its own method for making such assessments and ultimately decide on the best course of action.

Information and Communication

Earlier reference was made to the needs for informing employees about the law, the inventory list requirement, and toxicology data storage and retrieval. The information and communication requirements engendered by the TSCA will be considerably broader than just these items and will in some cases require additional organization and staffing (see Table 7). Section 8 of the Law on Reporting and Retention of Records describes the scope of information that may be called for; and it allows one to infer what staffing and communication networks should be instituted.

TABLE 7

Information and Communication

Monitoring
Recording
Coordinating
Reviewing
Reporting

Although it was said earlier that a broad data bank will be needed on all materials handled by each company, including, raw materials, intermediates, and products, we cannot overlook the fact that this information must be monitored and kept up to date. In addition to materials, data must be kept on processing, distribution, uses, and disposal; and these data must be kept current. A company concerned with a multiplicity of products and/or processes will undoubtedly want to consider a computerized system of data storage and retrieval. Professional staff in different departments of the company, e.g., purchasing, engineering, manufacturing, marketing, transportation, and environmental control, must provide the basic input data and must have the necessary clerical help. Periodic reviews by a coordinator will be essential to see that monitoring and updating are being accomplished. Operational line management and executive management must be kept informed and alerted to any unusual situations that may be potential unreasonable risks. Finally, each company must decide on how it will interact with trade associations and with EPA. A system must be established for making inquiries, petitioning, notifying, and reporting to the OTS as needed.

Planning and Impact

Because the specific regulations will not be available for some time, quantitative planning is not possible at this time. However, qualitative consideration can be given to planning based on the general provisions of the law and anticipation of more specific guidance forthcoming from EPA.

TABLE 8

Planning for Impacts

Testing
Research and development costs
Decreased innovation
Operating expenses
Capital costs
Delays
Penalties
Curtailment
Seizure
Bans

New product planning will have to take into consideration the premarket notification requirement, in terms of added lead time and costs for toxicology testing, as well as process integrity and product quality assurance. If a company wishes to maintain a constant level of innovation, added research and development costs necessary to offset the above penalties must be accommodated in planning research and development for future years. There will be increased costs for all testing, established as well as new products. To this must also be added administrative and record-keeping requirements. In manufacturing, higher operating expenses and increased capital costs will be associated with closer controls. As the law is more fully implemented, there may be additional time and financial penalties associated with new product introduction, civil actions, as well as curtailments, seizures, and conceivably, bans.

Current Awareness

Each company will want to maintain an active current awareness program to allow for orderly planning and minimum adverse impact (Table 9). Through trade associations, direct contact with EPA and other government agencies, and reading the *Federal Register* we can all follow the new developing regulations. Newspapers and trade publications can help up keep abreast of legal actions and petitions. Technical journals, meetings, and seminars will allow us to stay abreast of recent advances in toxicology, testing, epidemiology, and new analytical techniques. Additionally, consultants and contract studies can help us answer specific questions.

TABLE 9

Current Awareness

Trade associations
EPA and other agencies
Federal Register
Media
Journals
Meetings
Seminars
Consultants
Contract studies

Integration

The TSCA recognizes that there may be conflict and overlap with other federal laws and agencies in Section 9(d), dealing with EPA's coordination with other federal agencies. Since industries' organizations have been developed to meet the requirement of these laws and agencies as they evolved, it is natural that within many companies, similar conflicts and overlaps may exist. Since toxic substances control relates to health and the environment, it follows that TSCA matters will require interaction with the related activities of environmental control, industrial hygiene, and medical, safety, and regulatory affairs groups within any company (Table 10). This will be accomodated in various ways.

TABLE 10

Integration

Toxic substances
Environmental
Industrial hygiene
Safety
Medical
Regulatory affairs

Some companies have reorganized to put all of these functions under one health and environmental group. Others have appointed high-level advisory councils to determine policy and recommend action on major issues. The organizational direction that each company takes, centralized vs. decentralized or staff vs. line emphasis, is less important than assuring that lines of communication exist among the above named functions and that they remain open at all times. Information or the actions of one group will affect that of the other. Furthermore, the interests of the company and the compliance with not only TSCA but also all laws and regulations can best be served by a cohesive company approach.

Before closing, it seems appropriate to take note of the fact that small companies are particularly vulnerable to the new law as they are to any form of regulation. By virtue of their size and limited resources, they will be unable to respond as fully to all of the above mentioned features of the law as will the medium-size and larger companies. The law does grant some relief to smaller companies in record keeping, reporting, and the fee schedule for review of required data. Small companies would be well advised to pay particular attention to current awareness and to continue to exercise the resourcefulness that has characterized their maintenance of a significant position in our society.

CONCLUSIONS

This section attempts to provide a broad overview of some of the key elements of industries' implementation program for the TSCA to the extent they are definable at this time. How much emphasis each company will have to place on each element will depend on many factors. Not the least of these is the regulations to be developed by the OTS. With the large measure of discretionary authority afforded the Administrator by the TSCA, he will have a great deal of latitude in choosing how severely he regulates. We can only hope he will place emphasis on the first word of the key phrase "unreasonable risk," which sums up what the law is designed to control and not on imaginary or speculative risks. Neither the industry nor the public need any additional unnecessary financial burden at this time. It had been estimated by McGraw Hill that the chemical industry may spend $760 million, 11% of its total capital outlay in 1976, to meet environmental standards. The cost of complying with TSCA will be on top of this already large and growing expenditure. Industry is, of course, concerned about escalating costs of doing business. Nevertheless, it will respond to the TSCA diligently and responsibly. When EPA submits its annual report on the performance of the TSCA each year, there will be little outcry from industry if the intent of the law is being carried out.

OTS, in the introductory portion of its ten-page summary of the law, has publicly expressed its understanding of the social and economic contributions afforded by the chemical industry. We hope these words, showing an understanding of benefits, are not forgotten when regulations are being written. We never have or will live in a totally risk-free society, and therefore, emphasis must be placed on controlling the unreasonable risks, which means taking into consideration and balancing the risks against the benefits. The many benefits afforded by the chemical industry in the past and potentially in the future were pointed out in a September 1976 address by the President of the American Chemical Society, Dr. Glenn T. Seaborg. He cited the many contributions of our food, shelter, health, transportation, and leisure, and the potential that lies ahead for even greater benefits. Although a recent Harris Poll indicates that 78% of the people believe in government regulation, eight out of ten also think that goverment should create a good climate for industry. If we are to maintain a healthy free enterprise system that can provide the potential future benefits while protecting from unreasonable risks, government and industry must cooperatively seek the common goal. It is in this spirit and toward this end that industry will comply with the Toxic Substances Control Act.

Section VIII

Toxological and Environmental Effects Testing

TOXICOLOGICAL EFFECTS TESTING

Theodore Ellison
Mobil Chemical Co.
Edison, New Jersey

Under Section 4 of the Toxic Substances Control Act, Testing of Chemical Substances and Mixtures, a manufacturer of chemicals which are to be introduced into commerce may be required to furnish information concerning the manufacture, distribution in commerce, use, or disposal of such substances. Furthermore, information on the effect of such activities on the health or environment may have to be provided. If there are insufficient data for the Administrator to make a responsible decision, this information must be supplied by the manufacturer. One of the requirements of the law is that

> the health and environmental effects for which standards or the development of test data may be prescribed include carcinogenesis, mutagenesis, teratogenesis, behavioral disorders, cumulative or synergistic effects, and any other effect which may present an unreasonable risk of injury to health or the environment. The characteristics of chemical substances and mixtures for which such standards may be prescribed include persistence, acute toxicity, subacute toxicity, chronic toxicity and any other characteristic which may present such a risk.

Four progressive levels of investigational effort, depending upon extent, frequency, and nature of chemical use, are identified:

1. Single or infrequent exposure – acute and irritation tests
2. Occasional low level exposure – short term repeated dose, sensitization, fish and bird studies
3. Frequent low level, occasional high level exposures – teratology, mutagenicity, metabolism, biodegradation, environmental transport studies
4. Frequent high level, general consumer, unavoidable exposures – reproduction, lifetime, carcinogenic, environmental fate, food chain studies

Level I exposure – This originally occurs when the chemical is still in laboratory or pilot plant production where a limited number of persons may be exposed. Later, this level also includes the single or infrequent chemical contact through accidental splash, spill, or container rupture. Trivial exposures such as might be experienced by laboratory personnel developing potential end uses are considered here. For this level, the studies usually conducted are

1. Single-dose studies
   - Oral LD_{50}
   - Dermal LD_{50}
   - Inhalation LC_{50}
2. Irritation studies
   - Eye, mucous membrane
   - Dermal
   - Corrosiveness

Level II exposure – As more persons become involved in the production or industrial use of the chemical where the exposure rates are low, high concentrations can be avoided, or precautionary procedures can be employed, toxicological investigations are expanded to include:

1. 10 to 90-day studies
   - Oral
   - Inhalation
   - Dermal routes
2. Environmental behavior
   - 96-hr fish LC_{50}
   - 5 to 7-day bird toxicity
   - Biological or chemical oxygen demand
   - Biodegradation
3. Sensitization studies
   - Dermal – guinea pig
   - Dermal – man
4. Mutagenicity
   - Ames test (Salmonella)
   - Tier I

Level IV exposure – This level is usually defined as occurring when use has become so widespread that nearly everyone is exposed to the chemical or use, so that exposure to it in small amounts becomes unavoidable. Other types falling within the category are frequent high-level or general consumer exposures. Additional studies conducted for this level are

1. Mutagenicity – Tiers II and III
2. Teratology
3. Reproduction
4. Carcinogenic or lifetime in two species
5. Environmental

6. Disposition
   Sewage treatment
   Incineration
   Landfill
7. Food chain accumulation
8. Long-term aquatic/wildlife

Because of existing Federal regulations regarding the transportation and labeling of chemicals, nearly every new chemical entity produced in commercial quantities will undergo applicable Level I testing procedures. As the chemical enters a wider range of distribution, a beginning should be made concerning Level II exposure studies, including 90-day studies by the oral, inhalation, and dermal routes. Some environmental studies will be beneficial to have on hand should the Administrator ask for additional information.

If the product is widely distributed in the consumer market and a specific problem is anticipated, further testing beyond Level II should not be initiated unless fully discussed with the Administrator of the Act.

I. ACUTE TESTING

A. Toxicity Tests

1. Acute Oral Median Lethal Dose (LD_{50})

Young albino rats of both sexes will be used as test animals. All animals are kept under observation for 5 days prior to experimental use, during which they are checked for general health and suitability as test animals. The animals are housed in stock cages and are permitted a standard laboratory diet plus water ad libitum, except during the 16-hr period immediately prior to oral intubation when food is withheld.

Initial screening is conducted in order to determine the general level of toxicity of the test material. The experiment will consist of ten rats per dose level (five male and five female) using six levels plus a control. All doses are administered directly into the stomach of the rats using a hypodermic syringe equipped with a ball-tipped intubating needle.

After oral administration of the test material, the rats are individually housed in suspended, wire-mesh cages and observed for the following 14 days. Initial and final body weights, mortalities, and reactions are recorded. A necropsy examination is conducted on all animals.

At the end of the observation period, LD_{50} of the test material is calculated, if possible, using the techniques of Litchfield and Wilcoxon.[1] The test material is then assigned a classification in accordance with Harold C. Hodge.[2] The classification system is presented in Table 1.

2. Acute Dermal Median Lethal Dose (LD_{50})

New Zealand strain young adult albino rabbits of both sexes with a weight range of 2.5 to 3.0 kg are used as test animals. All rabbits were maintained under observation in the laboratory for at least 7 days prior to testing. During the pretest period, the animals are examined with respect to their general health and suitability as test animals. The rabbits are housed individually in suspended, wire-bottomed cages and maintained on a standard laboratory ration. Food and water are offered ad libitum.

Twenty-four hours prior to the dermal applications, the backs of the rabbits are shaved free of hair with electric clippers. The shaved area on each animal constitutes about 30% of the total body surface area. The animals are then returned to their cages to await testing on the following day. The 24-hr waiting period allows recovery of the stratum corneum from the disturbance which accompanied the close-clipping procedure and permitted healing of any microscopic abrasions possibly produced during the process. Half of the animals are further prepared by making epidermal abrasions every 2 or 3 cm longitudinally over the area of exposure. The abrasions are deep enough to penetrate the stratum corneum (horny layer of the epidermis), but not to disturb the derma. Prior to determination of the LD_{50}, single rabbits are dosed at various levels to determine a range within which the LD_{50} can be determined. Four rabbits are used per dose level for a minimum of three dose levels to determine the dermal LD_{50}. The test material is applied at the highest reasonable dose level. The test site is covered by wrapping the trunk of the animal

TABLE 1

Classification of Test Materials Based on Acute Oral LD_{50}

Acute oral LD_{50} (range of values, mg/kg)	Classification	Probable LD for a 70-kg man in commonly used measures
<5	Extremely toxic	A taste (less than 7 drops)
5–50	Highly toxic	Between 7 drops and 1 t
50–500	Moderately toxic	Between 1 t and 1 oz
500–5,000	Slightly toxic	Between 1 oz and 1 pt or 1 lb
5,000–15,000	Practically nontoxic	Between 1 pt and 1 qt
>15,000	Relatively harmless	More than 1 qt

TABLE 2

Classification of Test Materials Based on Acute Dermal LD_{50}

Acute dermal LD_{50} (range of values mg/kg)	Classification	Probable LD for a 70-kg man in commonly used measures
<20	Extremely toxic	Approximately 30 drops
20–200	Highly toxic	Between 30 drops and 4 t
200–500	Moderately toxic	Between 4 t and 1 oz
500–3,000	Slightly toxic	Between 1 oz and 1 pt or 1 lb
3,000–10,000	Practically nontoxic	Between 1 pt and 1 qt
>10,000	Relatively harmless	More than 1 qt

with impervious plastic sheeting which was securely taped in place. This plastic wrap insured close contact of the epidermis and test material.

The test material remains in contact with the skin for 24 hr. At the end of this period, the plastic sheeting and all residual test material are removed. The test sites are examined for local skin reactions and the animals are returned to their cages. Observations for mortality, local skin reactions, and behavioral abnormalities are continued for a total of 14 days following the skin applications. Initial, 7-, and 14-day body weights are recorded. A necropsy examination is conducted on all animals.

In the case of significant mortality following the initial study, additional experiments are conducted at lower dose levels in order to obtain data sufficient to calculate the acute dermal median lethal dose (LD_{50}), if possible, using the techniques of Litchfield and Wilcoxon.[3] The test material is then assigned a classification. The classification system is presented in Table 2.

3. Acute Inhalation – Single Level and Acute Inhalation Median Lethal Concentration ([LC_{50}])

The only available guidelines for tests of inhalation toxicity in small laboratory animals are those set forth in the Federal Hazardous Substances Act. No standardized procedure for the testing of chemicals has yet been recommended.

Young adult albino rats are employed as test animals. The rats are selected after having been under observation for at least 5 days to insure their general health and suitability for testing. The animals are housed in stainless steel cages and permitted a standard laboratory diet plus water ad libitum, except during inhalation exposure. During the exposure period, observations are made with respect to incidence of mortality and reactions displayed. At the end of the exposure period, the rats are returned to their cages for observation. A body weight is determined for each animal prior to inhalation exposure and for each surviving animal at the end of the observation period. The data are recorded as an index to growth. Gross pathologic examinations are scheduled to be conducted upon all animals which might succumb during the test period and upon those sacrificed at the end of the observation period.

Test animals (groups of ten, five male and five female for each exposure level) are exposed in a specially constructed inhalation chamber for a period of 1 hr. The chamber is designed so that the animals could be introduced into the test atmosphere after 99% of the maximum vapor concentration is established. The average nominal vapor concentration is calculated by dividing the weight loss of the chemical by the total volume of air used during the test.

Variables

1. Whole body vs. snout-only exposure
2. Dust vs. vapor vs. mist vs. heated vapor
3. Sampling of test material concentration – actual chemical analyses vs. average analytic vs. weight difference method
4. Duration of exposure – 1 vs. 4 hr
5. Number of dosages
6. Species and number

The classification system is presented in Table 3.

B. Irritation Tests

1. Primary Dermal Irritation

Six male albino New Zealand rabbits with a weight range of between 2.0 and 2.5 kg are used in this experiment. The animals are individually housed and maintained in accordance with standard laboratory procedures. The trunks are clipped free of hair and 1 $in.^2$ abrasions are made through the stratum corneum with care being taken not to abrade deep enough to cause bleeding and disturb the derma. The abraded areas are rotated throughout the six rabbits. A 0.5-ml portion of the test material is introduced under a 1 $in.^2$ gauze patch to the abraded and nonabraded skin and held in place with Dermicel tape. The trunks of the animals are wrapped with rubberized cloth during the

TABLE 3

Proposed Criteria for Inhalation Hazard Toxicity Classification System

Nontoxic	A substance that has a 1 hr inhalation LC_{50} of more than 200 mg/l of mist, fume, or dust, or more than 20,000 ppm of gas or vapor.
Slightly toxic	A substance that has a 1 hr inhalation LC_{50} of more than 100 mg/l of mist, fume, or dust, or more than 10,000 ppm of a gas or vapor but less than or equal to 200 mg/l or 20,000 ppm.
Moderately toxic	A substance that has a 1 hr inhalation LC_{50} of more than 20 mg/l of mist, fume, or dust, or more than 2,000 ppm of a gas or vapor but less than or equal to 100 mg/l or 10,000 ppm.
Toxic	A substance that has a 1 hr inhalation LC_{50} of more than 2 mg/l of mist, fume, or dust, or more than 200 ppm of a gas or vapor but less than or equal to 20 mg/l or 2,000 ppm.
Highly toxic	A substance that has a 1 hr inhalation LC_{50} of 2 mg/l or less of mist, fume, or dust, or 200 ppm or less of a gas or vapor.

TABLE 4

Primary Skin Irritation Test – Albino Rabbits Scoring Criteria for Skin Reactions

Reactions	Description	Score[a]
Erythema	Barely perceptible (edges of area not defined)	1
	Pale red in color and area definable	2
	Definite red in color and area well defined	3
	Beet or crimson red in color	4
Edema	Barely perceptible (edges of area not defined)	1
	Area definable but not raised more than 1 mm	2
	Area well defined and raised approximately 1 mm	3
	Area raised more than 1 mm	4
Injury in depth	Escharosis, necrosis	8

[a]Maximum primary irritation score = 8.

24-hr exposure period, and the animals are immobilized in an animal holder.

The animals are observed closely during the first 8 hr for discomfort and other adverse physical signs. Twenty-four and 72 hr after exposure, the treated areas are observed and evaluated for erythema, eschar, and edema formation in a range of 0 to 4 as described in the method previously cited. The classification system is presented in Table 4.

2. Primary Dermal Corrosion (DOT)

A group of six albino New Zealand rabbits with a weight range of 1.8 to 2.4 kg are used in this study. The test method is essentially that of Draize et al.[4] It consists of application of 0.5 ml (0.5 g) of the test material to clipped areas of intact skin. Applications are made under occlusive patches (1 in. × 1 in. gauze, covered by adhesive tape). Following application of the test material, the entire trunk of each animal is covered with an impermeable occlusive wrapping. The animals are then immobilized. The wrapping and test material are removed 4 hr following application. The sites are individually examined and scored separately for erythema and edema at 4 and 48 hr. The mean scores for 4- and 48-hr gradings are averaged to determine final irritation indices. Corrosiveness seen at 4 and/or 48 hr alone indicates a corrosive material. Tissue destruction (corrosiveness) does not merely include sloughing of the epidermis, or erythema, edema, or fissuring.

3. Primary Eye Irritation (FHSA Method)

This method is described in the *Federal Register* (September 27, 1973, 1500:42). Six albino New Zealand rabbits are used in this experiment. The animals are housed and fed individually and maintained in accordance with standard laboratory procedure. Water is available at all times. The test material is placed in one eye of each animal by gently pulling the lower lid away from the eyeball to form a cup into which the test material is dropped. The lids are then gently held together for 1 sec and the animal is released. The other eye, remaining untreated, serves as a control. For testing liquids, 0.1 ml is used. For solids or pastes, 100 mg of the test substance is used except for substances in flake, granule, powder, or other particulate form. The amount that has a volume of 0.1 ml (after compacting as much as possible without crushing or altering the individual particles, such as by tapping the measuring container) shall be used whenever this volume weighs less than 100 mg. In such a case, the weight of the 0.1-ml test dose should be recorded. The eyes are not washed following instillation of the test material except at the optional test following the 24-hr observation.

The eyes are examined and the grade of ocular reaction is recorded at 1, 24, 48, and 72 hr and at 4 and 7 days. After the recording of observation at 24 hr, any or all eyes may be further examined after applying fluorescein. For this optional test, one drop of fluorescein sodium ophthalmic solution (USP) or equivalent is dropped directly on the cornea. After flushing out the excess fluorescein with sodium chloride solution (USP) or equivalent, injured areas of the cornea appear yellow.

At each scoring interval, the cornea, iris, and peripheral conjunctiva are examined and graded for irritation and injury according to a standard scoring system.[4] The maximum possible score at any one examination and scoring period is 110 points, which indicates maximal irritation and damage to all three ocular tissues. Zero score indicates no irritation. The scoring system is presented in Table 5. In this scoring system, special emphasis is placed upon irritation or damage to the cornea, while less emphasis is placed upon damage to the iris and conjunctiva.

After the completion of the test, the scores are analyzed, and a descriptive eye irritation rating is assigned to the test material. The criteria used for assignment of the descriptive rating are the frequency, extent, and persistence of irritation or damage which occur to the three ocular tissues.

4. Primary Eye Irritation (Draize Method)

Nine albino New Zealand rabbits without ocular defects are used. The procedure followed is a modification of that suggested by Dr. J. H. Draize in Appraisal of the Safety of Chemicals in *Foods, Drugs and Cosmetics,* compiled by the staff of the Division of Pharmacology, Food and Drug Administration, Department of Health, Education and Welfare.

In the technique of determining toxicity of substances to eye mucosa, observation of injuries are made on the cornea, the iris and the bulbar, and palpebral conjunctivae. Numerical scores are assigned to lesions observed according to a standard scoring system. In this system of scoring, the injuries to the cornea and iris account for approximately 80% of the score; these structures are purposely weighted because of their vital role in vision. Healthy New Zealand rabbits are used for this test. One tenth of a milliliter of the test substance is instilled in the right eye; the left eye, remaining untreated, serves as a control. The treated eyes of three rabbits remain unwashed. Since washing the eye may or may not alleviate symptoms of injury, the six remaining animals are divided into two equal groups. In the first of these two groups, the eyes are treated with test substance and washed with 20 ml of lukewarm water (at approximately 37°C) 4 sec after instillation of test substance. In the second group, the treated eyes are washed 30 sec after instillation of test substance. The washing procedure is regarded as significant, since it is important to know the effect of such a procedure, i.e., whether it is beneficial or detrimental, and if beneficial, the extent of the benefit. Readings facilitated by hand-held lenses are made 1, 2, 3, 4, and 7 days after treatment. The scale for interpretation of eye scores is the same as used for the FHSA method (Table 5).

II. SUBACUTE TESTING

A. Subacute Oral Studies

1. 10-Day Toxicity Study (Rat)

Ten albino rats (five males and five females) weighing 200 to 300 g are dosed by oral gavage to a concentration approximately one fifth that of the LD_{50} for once per day, 5 days/week, for 2 weeks (ten exposures). A control group of animals is dosed, simultaneously, under the same conditions except for administration of test material. One half of the animals from each group (test and control) are sacrificed after the last dose, while the remaining animals are sacrificed after a 14-day recovery period for gross and histopathologic examination of major organs. Acute toxicity (LD_{50}) is a prerequisite.

2. 90-Day Feeding Study (Rat)

This study is to determine any pharmacologic or toxic effects resulting from the administration of the test material to rats for 90 consecutive days.

Materials and methods – Eighty albino rats, weighing between 200 and 225 g, are individually housed in wire mesh cages in a temperature and humidity-controlled environment having an artificial-light cycle of 12 hr. Prior to initiation of the study, the animals are acclimated to laboratory conditions for a minimum period of 7 days. The rats are divided into four groups with 20 animals in each of the three treatment groups and 20 animals in the control group. There are an equal number of males and females with approximate even weight distribution in each group. Administration of the test material may either be by incorporation of the compound in the diet or by gavage, depending upon the client's preference and the nature of the compound. The rats are individually weighed at the initiation of the study, weekly thereafter (on the same day of the week), and on the final day of the study. Food consumption rates are recorded weekly. Observations of any pharmacologic or toxic effects evident in any animal are recorded when they occur.

Hematology – Hematological evaluations will be determined initially and at 30, 60, and 90 days. The animals are fasted overnight and blood is then drawn the following day from the orbital sinus. The hematologic parameters along with the methods used are listed below:

1. Erythrocyte count
2. Total leucocyte count
3. Hemoglobin
4. Hematocrit
5. Differential leucocyte count

TABLE 5

Eye Irritation Test – Albino Rabbits Scale of Weighted Scores for Grading the Severity of Ocular Lesions

Ocular tissues	Description	Grading
Cornea	Opacity (D) – degree of density (area which is most dense is taken for reading)	
	Scattered or diffuse area, details of iris visible	1
	Easily descernible translucent areas, details of iris slightly obscured	2
	Opalescent areas, no details of iris visible, size of pupil barely descernible	3
	Opaque, iris invisible	4
	Area of cornea involved (A)	
	One quarter (or less) but not zero	1
	Greater than one quarter but less than one half	2
	Greater than one half but less than three quarters	3
	Greater than three-quarters, up to whole area	4
	Score equals D × A × 5, total maximum = 80	
Iris	Values (V)	
	Folds above normal, congestion, swelling, circumcorneal injection (any or all of these or a combination of any thereof), iris still reacting to light (sluggish reaction is positive)	1
	No reaction to light, hemorrhage, gross destruction (any or all of these)	2
	Score equals V × 5, total maximum = 10	
Conjunctiva	Redness (R) – (refers to palpebral conjunctiva only)	
	Vessels definitely injected above normal	1
	More diffuse, deeper crimson red, individual vessels not easily discernible	2
	Diffuse beefy red	3
	Chemosis (S)	
	Any swelling above normal (includes nictitating membrane)	1
	Obvious swelling with partial eversion of the lids	2
	Swelling with lids about half-closed	3
	Swelling with lids about half-closed to completely closed	4
	Discharge (D)	
	Any amount different from normal (does not include small amount observed in inner canthus of normal animals)	1
	Discharge with moistening of the lids and hairs just adjacent to the lids	2
	Discharge with moistening of the lids and hairs and considerable area around eye	3

Score equals (R + S + D) × 2, total maximum = 20

Clinical chemistry – Clinical chemistry evaluations will be determined initially and at 30, 60, and 90 days. The animals fast overnight and blood is drawn the following day from the orbital sinus. The clinical chemistry parameters along with the methods used are listed below:

1. Serum glutamic pyruvic transaminase
2. Serum glutamic oxyalacetic transaminase
3. Serum alkaline phosphatase
4. Blood urea nitrogen
5. Total cholesterol
6. Uric acid
7. Glucose

Urinalysis – Urologic evaluations will be conducted initially and at 30, 60, and 90 days. The urologic parameters along with the methods used are listed below:

1. pH
2. Specific gravity
3. Protein
4. Glucose
5. Color
6. Appearance

Animal sacrifice – Upon completion of the 90-day study, all animals are anesthetized and sacrificed by exsanguination. A complete gross pathological examination is done on each animal and any observed abnormality is recorded. A similar examination is carried out on any animal that dies during the course of the experiment. The following organs of each animal are weighed, recorded, and reported as individual organ weights and percent of body weight:

1. Heart
2. Liver
3. Adrenals
4. Kidneys
5. Spleen
6. Gonads

The following organs are taken from each animal and preserved in 10% neutral buffered formalin:

1. Heart
2. Liver
3. Lungs
4. Spleen
5. Kidneys
6. Stomach
7. Intestine
8. Pancreas
9. Thyroid
10. Adrenals
11. Gonads
12. Brain
13. Bone marrow
14. Any abnormal lesions

Histopathological examinations are done on the above tissues of all animals in the high-dose and control groups. Tissues from the mid- and low-dose groups should be examined if a no-effect level must be determined.

3. 90-Day Feeding Study (Dogs)

This study is to determine any pharmacologic or toxic effects resulting from the administration of the test material to dogs for 90 consecutive days.

Materials and methods – Eighty purebred beagle dogs are individually housed and maintained according to standard laboratory procedures. Before initiation of the study, the animals are acclimated to laboratory conditions for a minimum of 14 days. The dogs are divided into four groups with 20 animals in each of the three test groups and 20 animals in the control group. There are an equal number of males and females with approximate even weight distribution in each group. Administration of the test material may be by repelleting, encapsulation, or by injection, depending upon the client's preference and the nature of the compound. The dogs are individually weighed at the initiation of the study, weekly thereafter (on the same day of the week), and on the final day of the study. Food consumption rates are recorded weekly. Observations of any pharmacologic or toxic effects evident in any animal are recorded when they occur.

Hematology – Initially and at 30, 60, and 90 days, hematologic measurements will be made and will consist of the following:

1. Hemoglobin
2. Hematocrit
3. Erythrocyte count
4. Total and differential leukocyte count
5. Packed cell volume
6. Erythrocyte sedimentation rate
7. Platelet count
8. Reticulocyte count
9. Prothrombin time

Clinical chemistry – Initially and at 30, 60, and 90 days, blood chemistry will be performed and will include the following:

1. Serum glutamic pyruvic transaminase
2. Serum glutamic oxalacetic transaminase
3. Serum alkaline phosphatase

4. Bilirubin
5. Blood urea nitrogen
6. Lactic acid dehydrogenase
7. Creatinine phosphokinase
8. Electrophoretic examination of protein
9. Reducing substances
10. Serum electrolytes
11. Total cholesterol
12. Total lipid and uric acid

Urinalysis – Urinalysis, including pH determination, specific gravity, protein, glucose, ketone bodies, and bile pigments, along with microscopic examination of deposits, will be done initially and at 30, 60, and 90 days.

Animal sacrifice – On completion of the 90-day study, all animals are anesthetized and sacrificed by exsanguination. A complete gross pathological examination is done on each animal with any observed abnormalities recorded. A similar examination is carried out on any animal that dies during the course of the experiment. The following organs of each animal are weighed fresh: heart, kidneys, liver, and spleen. The adrenals and ovaries or testes are weighed after formalin fixation. The weights are recorded and reported as individual organ weights as well as percent of body weight. Organs taken from each animal and preserved in 10% buffered formalin are as follows:

1. Heart
2. Liver
3. Lungs
4. Spleen
5. Kidneys
6. Stomach
7. Intestine
8. Pancreas
9. Thyroid
10. Adrenals
11. Ovaries or testes
12. Brain
13. Bone marrow
14. Any abnormal lesions

Histopathological examinations are done on the above tissues on all dogs in the control and high-dose groups. Tissues from the mid- and low-dose groups should be examined if a no-effect level must be determined.

B. Subacute Dermal Studies

1. 10-Day Toxicity Studies (Rat)

Ten albino rats (five males and five females) weighing 200 to 300 g are selected for the administration of the test material. The hair is removed from the back and flanks of each rat with an electric veterinary clipper. Each rat is administered one fifth of the LD_{50} to the shaved surface once per day, 5 days/week for 2 weeks. A control group of animals consisting of the same number of animals is treated, simultaneously, without administration of the test material. One half of the animals from each group (test and control) are sacrificed after the last application of the test material. The remaining animals are sacrificed after a 14-day recovery period for gross and histopathological examination of major organs. Acute toxicity (LD_{50}) is a prerequisite.

2. 21-Day Dermal Toxicity (Rabbit)

This experiment is designed to evaluate the safety of a test material on the abraded and nonabraded skin sites of New Zealand albino rabbits when exposed daily, 6 hr/day for 21 consecutive days.

Procedure – Twenty-four adult New Zealand albino rabbits, weighing between 2.5 and 3.5 kg, will be selected at random and placed into three groups (control, high-dose level, and low-dose level) consisting of four males and four females per group. The animals are fed, housed, and maintained in accordance with standard laboratory procedures. Water is available at all times with the exception of the 6-hr exposure period. The animals are immobilized in an animal holder and the trunks clipped free of hair with an Oster animal clipper. One half of the animals are further prepared by making epidermal abrasions every 2 or 3 cm longitudinally over the area of exposure, with fresh abrasions made at time intervals as needed. The abrasions are made deep enough to penetrate the stratum corneum, but not deep enough to disturb the derma. The test material is introduced over approximately 10% of the animal's body surface, remaining in place for the 6-hr exposure period each day. The test material will be removed by washing gently with warm water and a soft cloth at the end of the exposure period. Dose levels for the study will be selected on the basis of pilot dermal irritation levels performed. Observations:

1. The animals are observed closely during the first 6 hr for signs of discomfort or other adverse physical signs and daily thereafter for a total of 21 days.
2. Individual body weights and food consumption will be recorded 1 week before the initiation of the study and weekly thereafter.
3. Necropsies will be performed on any animal that dies during the experiment.
4. Hematologic studies, clinical blood studies, and urine analyses will be performed.
5. Blood samples will be withdrawn from the orbital sinus of each rabbit at 0 and 21 days.
6. Urine samples will be collected from two males and two females in each group at 2 and 21 days. The animals will be placed in metabolism cages for periods of 24 hr.

The following parameters will be investigated:

1. Hematologic studies
   - Total leukocyte count
   - Differential leukocyte count
   - Erythrocyte count
   - Hemoglobin concentration
   - Hematocrit value
2. Clinical Blood Chemistry Determinations
   - Blood urea nitrogen concentration
   - Serum alkaline phosphatase activity
   - Serum glutamic-pyruvic transaminase activity
   - Serum glutamic-oxaloacetic transaminase activity
   - Cholinesterase activity in plasma and erythrocyte
3. Urine analyses
   - Reducing substances
   - Albumin
   - Microscopic elements
   - pH
4. Gross and microscopic pathologic studies

Arrangements will be made to subject any animal which might die during the test to a gross autopsy. Also, in those instances where post-mortem changes are not advanced, sections of representative tissues and organs will be scheduled to be taken for histopathologic study.

At the conclusion of the investigational period, all surviving rabbits in each group will be sacrificed and subjected to a gross pathologic examination. At this time, the brain, liver, kidneys, heart, spleen, gonads, adrenal glands, and thyroid gland will be removed, trimmed, and weighed. The organ weights will then be tabulated and expressed as the percentage of the total body weight.

In order to ascertain the presence or absence of histopathologic change as a result of the repeated dermal applications, the following tissues and organs were removed from each animal and examined microscopically:

Adrenal glands
Aorta
Brain
Caecum
Colon
Esophagus
Gall bladder
Gonads
Heart
Kidneys
Liver
Lungs
Lymph nodes (mediastinal and mesenteric)
Pancreas
Parathyroid glands
Peripheral nerves (sciatic and femoral)
Pituitary gland
Prostate glands
Salivary glands
Urinary bladder
Seminal vesicle
Skeletal muscle (thigh)
Skin from the application site
Small intestine (duodenum, jejunum, ileum)
Spleen
Sternum
Stomach
Thyroid gland
Trachea
Uteri

Skin sections were fixed in Bouin's fixing fluid and all other tissues in 10% buffered formalin. All sections were stained with hematoxylin and eosin.

C. Subacute Inhalation Studies

1. 10-Day Toxicity Studies (Rat)

Ten albino rats (five males and five females) weighing 200 to 300 g are exposed to a concentration approximately one fifth that of the LC_{50} for 1 hr/day, 5 days/week for 2 weeks (ten exposures). A control group of animals is exposed, simultaneously, under the same conditions except for administration of test material. One half of the animals from each group (test and control) are sacrificed after the last exposure and the remaining animals sacrificed after a 14-day recovery period for gross and histopathologic examination of major organs. Acute toxicity (LC_{50}) is a prerequisite.

2. 30-Day Aerosol Toxicity (Rat)

This experiment is designed to evaluate the safety of a test material to aerosol inhalation in albino rats when exposed 6 hr/day for 5 days/week. This involves 40 calendar days or a total of 30 exposures.

Procedure – Forty young adult albino rats weighing 200 to 300 g will be placed in three groups plus one control group (five male and five female per group) at dosages to be determined from the acute inhalation data.

Exposure chambers – Each group of ten rats will be exposed in a Plexiglas® inhalation chamber having a volume of 40 l. The chamber is designed so that the animals can be introduced to the test atmosphere after the desired aerosol concentration has been established. Each animal will be caged separately during exposure to prevent crowding and minimize filtration of inspired air by animal fur.

Aerosol of the test material will be generated with an Ohio Ball-Jet Nebulizer.® A stream of clean dry air (–40°C dewpoint) will be passed through the nebulizer. The resulting air-aerosol mixture will be combined with additional clean dry air, if necessary, to achieve the desired final aerosol concentration. The resulting test atmosphere will then be passed into the exposure chamber at the top center, dispersed by a baffle plate, and exhausted at the bottom of the chamber. Air flow rates through the system will be measured with calibrated rotameters connected in the air supply lines upstream of aerosol contamination. Temperature and absolute pressure of the test atmosphere will also be measured.

Analyses of Chamber Air – The average nominal aerosol concentration, determined by dividing the weight loss of the nebulizer by the total volume of air used, will be calculated daily. The actual concentration of nonvolatile material will be determined by sampling the test atmosphere in the breathing zone of the animals being exposed. The weight gain of a glass fiber filter will be divided by the total volume of air drawn through the filter during the sampling period.

Animal Parameters
- Body weights and weight gains
  - All animals will be weighed before exposure and weekly thereafter.
- Mortality and reactions
  - Mortality and untoward behavioral reactions will be recorded daily.
- Hematology, clinical blood chemistry studies, and urine analyses
  - The following studies will be conducted on three male and three female rats from each group before exposure and after 4 weeks of testing:
  - Hematology
    - Erythrocyte count
    - Hematocrit value
    - Hemoglobin concentration
    - Total and differential leukocyte counts
    - Mean corpuscular hemoglobin (MCH)
    - Mean corpuscular hemoglobin concentration (MCHC)
    - Mean corpuscular volume (MCV)
  - Clinical blood chemistry
    - Serum alkaline phosphatase activity (SAP)
    - Blood urea nitrogen concentration (BUN)
    - Serum glutamic-pyruvic transaminase activity (SGPT)
    - Fasted blood glucose
  - Urine analyses
    - Albumin
    - Glucose
    - Crystals
    - pH

D. Pathology

Any animals that die during the test will be autopsied. At the end of the 4-week exposure period, all surviving animals will be sacrificed and subjected to a complete gross pathologic examination. The weight of the liver, kidneys, spleen, lungs, gonads, heart, and brain of each animal will be recorded. Organ to body weight and organ to brain weight ratios will be calculated. A complete set of organs and other tissues (see below) will be removed from each animal and preserved in 10% formalin solution. The lungs will be fixed inflated and *in toto.*

Adrenal gland
Aorta (thoracic)
Bone
Bone marrow
Brain (cerebrum, cerebellum, and pons)
Caecum
Colon
Esophagus
Eye
Gonad
Heart
Kidney
Liver
Lung
Lymph Node (cervical, peribronchial, and mesenteric)
Muscle
Optic nerve
Pancreas
Parathyroid
Peripheral nerve (sciatic)
Pituitary gland
Prostate
Salivary gland (submaxillary)
Small intestine (duodenum, jejunum, and ileum)
Spinal cord
Spleen
Sternum
Stomach (cardia, fundus, pylorus)
Trachea
Thyroid gland
Urinary bladder
Uterus and skin

Microscopic examinations of the following tissues will be conducted:

Heart
Kidney
Liver
Lung
Lymph nodes
Bronchi
Pituitary gland
Spleen
Trachea

Urinary bladder
Gonads
Brain

E. Delayed Neurotoxicity Studies

1. Single dose (LD_{50}) method (Leghorn hens)

The purpose of this study will be to establish whether or not the compound will produce delayed locomotor ataxia (demyelination) in white Leghorn hens.

The study will be conducted in two phases. The acute LD_{50} will be determined in the first phase, while the possible delayed neurotoxic effects of the approximate LD_{50} dose will be studied in a separate group of animals in the second phase. Both phases will employ white Leghorn hens, deep litter raised and over 9 months of age, as test animals. The birds will be housed in deep litter pens and will be maintained on Purina Euggena CF® and water ad libitum.

Three groups of ten hens each will be utilized in the neurotoxicity study. One group will be used as an untreated control, a second positive control group will be dosed with 500 mg/kg triortho cresyl phosphate (TOCP), and the test group will receive the test material at the calculated LD_{50}. Birds surviving this dose will be observed for 21 days for delayed locomotor atoxia. If the response is positive, the birds will be sacrificed at 21 days and the spinal cord (fixed *in situ*), sciatic nerve, and brain will be examined histologically for demyelination.

If the response is negative, the surviving birds will be redosed after 21 days with the LD_{50} and observed for an additional 21 days. At the end of this period, the hens will be sacrificed, and samples of spinal cord, sciatic nerve, and brain will be taken for possible future histological examination.

The chickens will be observed every 2 to 3 days during the 21-day periods to determine abnormalities in gait, activity, and agility. Any acute cholinorgic effects that interfere with locomotor activity usually subside within 72 hr. Positive clinical neutotix effects would be histopathologically confirmed to eliminate the possibility that the locomotor ataxia did not arise from intercurrent disease such as neurolymphomatosis.

2. Multiple dose (MTD) method (Leghorn hens)

While procedures for neurotoxicity testing recommend dosing hens at or near the single oral LD_{50}, with many compounds the hens survive after administration of the largest single oral dose feasible. The present technique employs a maximum-tolerated multiple dosing (MTD). In general, dosing is carried out in a group of ten white Leghorns hens, at least 9 months of age, twice daily, for 3 consecutive days. Each dose is administered by gavage and corresponds to the maximum feasible dose. A second group of 10 hens are utilized as the untreated controls with TOCP used in a third group of 10 hens as the positive control.

Birds are observed for gross or overt signs of neurotoxicity for 21 days. If no abnormal signs are observed, the same dose schedule was repeated. On day 42, all remaining birds are sacrificed and subjected to gross and microscopic pathologic examination.

The birds displaying positive clinical reactions are examined by fixing the brain and spinal cord with anterior horn cells for 48 hr in 10% formalin. The sections of the cervical, thoracic, and lumbosacral areas are stained by the Marchi method with a yellow-orange color reaction indicating demyelination.

III. SENSITIZATION TESTING

A. Guinea Pig Sensitization Test

Ten albino guinea pigs (five male and five female) weighing 300 to 500 g will be used to evaluate the skin sensitizing properties of the test material. Prior to the test, the hair will be clipped from the backs and flanks of the animals. Each animal will be insulted every other day with single closed patches containing a nonirritating concentration of the test material for a total of ten times. Two weeks after the last patch, the animals will be challenged with duplicate patches. Each exposure period will last for a period of 6 hr. Since a nonirritating concentration of the test material will be used, any reaction at challenge will be considered positive.

Closed patches will be applied to the guinea pigs in the following manner. A Webril® pad (7/8 in. × 1 in.) wetted with 0.5 ml of the test material will be applied near the midline of the shaved back of the guinea pig. The Webril® pad will be occluded with a standard size Elastoplast® coverlet (1 1/2 in. × 2 in.) and the entire trunk of the animal will be wrapped with impervious plastic sheeting. The animals will then be placed in a restrainer and adjustments will be made so that the guinea pig will be restrained but not immobilized. At the end of the 6-hr exposure period the animals will be removed from the restrainer, unwrapped, and returned to their stock cages. The animals will be scored for irritation 24 hr after each exposure.

A positive control group consisting of 10 albino guinea pigs will also be employed according to the above procedure. A 0.05% (wt/vol) (w/v) dilution of chlorodinitrobenzene in ethanol will be used as the positive control material.

The scoring criteria for erythema and edema are presented in Table 6.

B. Schwartz-Peck Patch Test (Human)

Human subjects (at least 50) of different age, sex, and race are used in this test.

Phase 1 – to detect primary irritation – A 10% aqueous solution of the substance under test is applied directly to the skin by means of a patch, secured and made air-tight

TABLE 6

Scoring Criteria for Skin Reactions

Reaction	Description	Score[a]
Erythema	Barely perceptible (edges of area not defined)	1
	Pale red in color and area definable	2
	Definite red in color and area well-defined	3
	Beet or crimson red in color and/or injury in depth (necrosis, escharosis)	4
Edema	Barely perceptible (edges of area not defined)	1
	Area definable but not raised more than 1 mm	2
	Area well-defined and raised approximately 1 mm	3
	Area raised more than 1 mm	4

[a]Maximum primary irritation score = 8.

TABLE 7

Scoring Criteria for Skin Reactions

Erythema and edema formation	
Very slight erythema (barely perceptible)	1
Well-defined erythema	2
Moderate to severe erythema	3
Severe erythema (beet redness) to slight eschar formation (injuries in depth)	4
Total possible erythema score = 4	
Edema formation	
Very slight edema (barely perceptible)	1
Slight edema (edges of area well-defined by definite raising)	2
Moderate edema (area raised approximately 1 mm)	3
Severe edema (raised more than 1 mm and extending beyond area of exposure)	4
Total possible edema score = 4	
Total possible primary irritation score = 8, sensitization score = 8	

Primary irritation index	Sensitization
2 or less – mild irritant	2 or less – mild sensitizer
2–5 – moderate irritant	2–5 – moderate sensitizer
6 or above – severe irritant	6 or above – severe sensitizer

with plastic tape. This application remained in contact with the skin for 48 hr. The patch is then removed and the reaction of the underlying skin is graded. Follow-up evaluations are continued until the skin returns to normal.

Phase 2 – to detect sensitization – After the skin is allowed to recuperate for a 2 week period, the substance is reapplied at a 10% aqueous solution to the skin of the same subjects for another 48-hr period. This second patch is removed and the reaction of the underlying skin is again graded. Follow-up evaluations are continued until the skin returns to normal. Distilled water is used as a control for this patch. A primary irritant will cause harm to the skin during the first application. A sensitizing agent will cause harm to the skin during the second or retest application. Primary irritants are able to exceed the protective resistance threshold of the skin on one application, while a sensitizing agent is detected when a reaction occurs upon final application. The scoring criteria for skin reactions are given in Table 7.

C. Repeated Insult Patch Test (Human)

Human subjects (at least 50) of different age, sex, and race are used in this test. The repeated insult technique is used in the patch test. It calls for a series of nine induction patches of each test material to be placed on eacn of the subjects. The series is followed 12 days later by a single challenge patch of each test material to detect skin sensitization. The upper arm (outer surface) of the male subjects is utilized for patching, while the female subjects are patched on the upper back, shoulder area.

The series of nine induction patches of each test material is applied according to the following schedule. Patches are applied on Monday, Wednesday, and Thursday and allowed to contact the skin for 24 hr, after which time they are removed and the skin sites graded for irritation. Thursday's patches are placed immediately after removal and grading of Wednesday's applications. After the ninth induction patches have been graded, a nonpatching period of 12 days elapses before the challenge patch of each test material is applied to detect sensitization reactions. For these 24-hr patches new skin sites are used. These sites are invariably chosen adjacent to the induction sites, i.e., where repeated applications have been made during the series of nine patches. These sites are observed at patch removal for sensitization reactions and again 24 and 48 hr after patch removal to detect possible delayed reactions.

Skin applications of the test materials were made using Readi-Band® clear plastic patches 1 and 1/2 in.^{2} with nonwoven Webril centers. Approximately 0.05 ml of each test material was placed on the Webril center of the patch just prior to application.

Variables

1. Number of subjects
2. Frequency of patching
3. Test concentration
4. Duration of patching

IV. ENVIRONMENTAL TESTING

A. 96-Hour Static Fresh Water Fish Toxicity (Bluegill Sunfish and Rainbow Trout) LC_{50}

This experiment is designed to determine the acute toxicity of a given compound to bluegill sunfish (*Lepomis macrochirus*) and rainbow trout (*Salmo gairdneri*) under static conditions.

Bluegill sunfish (200) and rainbow trout (200) are used in this experiment. All fish are between 35 and 75 mm in length and 0.5 to 3.0 g in weight. Care is taken when selecting the fish in each group to insure that the largest fish used will never be more than one and one-half times the size of the smallest fish in the group.

Stock fish are acclimatized for 10 days prior to bioassay. Water temperature is kept at 19 ± 2°C and 13 ± 2°C for sunfish and trout, respectively. The fish receive a standard commercial fish food daily until 2 days prior to testing, at which time feeding is discontinued. Only fish from stock groups with mortality rates of less than 10%, 48 hr prior to testing, are chosen for testing.

Thirty fish of each test species, per concentration level, are utilized for each of the five concentrations of test compound and one negative control group. The maximum allowable load for each bioassay vessel is 1 g of fish per liter of water. All aeration will be withheld during the 96-hr exposure period.

The test is observed for 96 hr with particular attention being given to general behavior, partial or total loss of equilibrium, and mortality. Mortality rates, dissolved oxygen, and pH are recorded at 6, 24, 48, 72, and 96-hr time intervals.

B. Subacute Wildlife Feeding Study (Bobwhite Quail and Mallard Ducks)

This test is designed to determine the effects of a test material upon bobwhite quail and mallard ducks when administered on a daily basis as a dietary additive.

Test subjects are obtained from stock which has been reared and maintained at all times on standard game bird diets (Purina Game Bird Chow). Each test and control group consists of ten birds which are housed in battery brookers equipped with thermostatically controlled heating units. Food and water are supplied on an ad libitum basis.

Each test involves six negative control groups (fed the basal diet at all times), six positive control groups (fed ppm Dieldrin at logarithmic levels appropriate for determing an LC_{50} value), and six test groups to be fed the test material at dosage levels appropriate for determining the LC_{50} value. Birds are randomized into treatment and control groups and placed on a basal ration diet mixed with the appropriate vehicle for at least 2 days prior to being placed on the test diet. Test diets are administered to the appropriate groups for the first 5 days of the 8-day testing period. Basal diets are fed during the final 3 days of observation.

The test compounds and the dieldrin used in the positive control diets are dissolved in the appropriate vehicle prior to incorporation in the rations. Concentration of the material in solution is adjusted so that the addition of 2 parts (by weight) of this solution to 98 parts of the standard game bird ration will result in the desired concentration of test material in the finished diet. The basal diet, which is fed to the negative control groups at all times and to the positive control and test groups during the pretreatment (acclimation) and posttreatment (observation) periods, consists of 2 parts vehicle to 98 parts of the standard ration. The following records are kept for each test:

1. Body weights of all birds at start of pretreatment period; birds are weighed as a cage group
2. Food consumption (gram/bird/day) based upon the total consumption of the group
3. Time (day of test) of death of each bird
4. Signs of systemic or toxic effects

The following parameters will be determined:

1. LC_{50} – 8-day, defined as the concentration (ppm) in the diet computed to produce 50% mortality at the end of the 8-day test and observation period; computed in accordance with the method of Miller and Tainter[5]
2. Toxicity of the test material relative to dieldrin and the concurrently determined LC_{50} values

C. Acute Toxicity on Estuarine Organisms (Marine Crabs, Shrimp, and Oysters)

Estuarine organisms will be exposed to various concentrations of the chemical in water to determine LC_{50} values of the test organisms. A range finding experiment will first be carried out to determine the appropriate lethal range for the organisms. After the range has been determined, a series of concentrations, within this range, will be used to determine the LC_1, LC_{50}, and LC_{95} values upon the test organisms.

The following physical conditions will be used to establish an artificial seawater media for the testing of all the marine organisms under consideration. Artificial seawater, as proposed by Zaroegins (1969), will be utilized giving a salinity of 22% and a pH adjusted to within a range of 7.6 to 8.0 using $NaHCO_3$. The temperature will be maintained between 20 and 25°C throughout the entire length of the experiment. Aeration of the water will not be done and an artificial substrate will not be used.

Grass shrimp *(Palaemonetes vulgaris)* – Juvenile grass shrimp will be used in both the range-finding experiment and the dose-level experiment. Twenty test organisms will be used at each concentration level. A control test and duplicate tests will be run at each concentration. The shrimp will be placed in a 10-l container along with the chemical at the appropriate concentration. Toxicity will be determined by loss of equilibrium and death by loss of motility. Symptoms will be recorded after 24, 48, 72, and 96 hr, respectively. The organisms will not be fed during this time period. The number of organisms may be increased at the dose-level experiment to accommodate histological or metabolic analysis.

Blue crabs (*Callinectes sapidus*) – Juvenile blue crabs will be used in both a range-finding experiment and dose-level experiment, similar to the grass shrimp procedure. Twenty organisms will be used per concentration level with a control and duplicate test being run at each concentration. The crabs will be placed in a 10-l container along with the chemical at the appropriate concentration. The death of the organisms, as determined through loss of motility, will be recorded at 24, 48, 72, and 96-hr intervals, respectively. The organisms will not be fed during the 96-hr test period. The number of test organisms may be increased to accommodate histological and metabolic studies.

Amercian oyster (*Crassostrea virginica*) – This study will utilize 48-hr juvenile oysters as the test organisms. Approximately 2000 will be placed in a one-liter container of media along with the chemical for both the range-finding and dose-level experiments. The experiment will be conducted for 14 days with the motality rate checked every 2 days. Death will be determined by counting the number of shells in which the edible meat has decayed. The test media will be changed every 2 days to rid the system of biological waste products and the oysters will be fed a suitable artificial food. The system will not be aerated. A control and duplicate test will be run at each concentration for the dose-level experiment.

V. MUTAGENICITY TESTING

A. In Vitro Assays

1. Basic Plate Test Mutagenicity Screen (Ames Salmonella/ Microsome Assay)

This assay evaluates the test chemical in a rapid sensitive screen utilizing the hepatic 9000 × g supernatant fractions from Aroclor® 1254 induced rat liver as the activation component. Noninduced tissues may be selected. Although any of the indicator strains listed may be used, TA-1535, TA-1537, TA-1538, TA-98, TA-100 (*Salmonella* strains), and D4 (Saccharomyces) are recommended for this assay because of their sensitivity and range of responses. The D4 strain is added to provide an eucaryotic (nonbacterial) cell type to the screen, since it has been found that some chemicals may be preferentially active in eucaryotic cells. All tests will be run at four dose levels both with and without activation, and the results will be reported as numbers of revertans per plate. Solids, liquids, and gases can be evaluated.

The primary advantage of the plate test is in situations where large numbers of compounds are to be screened or as a prescreen to help define compounds that should be examined with more definitive tests. The basic nonactivation and activation systems are designed to be completely flexible and can be expanded by the addition of other tissues or mammalian species as desired.

Plate tests – Plate tests that use bacteria strains as indicators provide a rapid and an inexpensive indication of mutagenic activity. They are most applicable in obtaining a positive or a negative response with certain limitations imposed on the results. The limitations are 1. No information can be obtained regarding the degree of mutation compared with the exact total surviving population and 2. Statistical evaluation of data is difficult to perform.

Suspension tests – Quantitative suspension tests that use bacteria and yeast indicators provide a precise picture of the mutagenic and/or recombinogenic activities of a chemical by generating mutation and recombination frequencies and calculated dose-response curves. These tests supplement the results from plate tests and make possible an accurate assessment of the genetic activity of the chemical for procaryotic and eucaryotic indicator organisms.

The following reaction mixture is employed in the activation tests:

Component	Final concentration/ml
TPN (sodium salt)	4 μmol
Glucose-6-phosphate	5 μmol
Sodium phosphate dibasic (pH 7.4)	100 μmol
$MgCl_2$	8 μmol
KCl	33 μmol
Tissue Homogenate (S-9)	0.1–0.15 ml

Table 8 lists chemicals that are used as positive chemical controls in nonactivation and activation assays.

Tissue homogenates used in activation tests – Either crude tissue homogenates or 9000 X g supernatant (S-9) fractions are added to the reaction mixture as activation sources. Preparations from liver, lung, kidney, or testes are normally employed although almost any organ or tissue can be utilized. Tissues from random bred mice and Sprague-Dawley rats pretreated with microsomal enzyme inducers, such as Aroclor 1254,® are routinely used in screening assays.

Interpretation of results from plate tests and suspension tests – Multiple testing modes are offered because each yields a different type and reliability of information. Plate tests are an inexpensive and a rapid means of examining large groups of chemicals for presumptive mutagenic activity. Chemicals that are capable of inducing mutations at or greater than ten times background spontaneous levels can easily be detected in plate tests. Dose-response curves will increase the reliability of plate tests. Chemicals with weak mutagenic activity that induce only fourfold or fivefold increases in mutation rates may not be detected because they fall within the reliable confidence limits of the control test. Quantitative suspension tests, however, can be constructed so that increases of four or five times background can be confidently detected. Therefore, the selection of testing modes should reflect the anticipated level of reliability required.

Dose levels and their meaning to in vivo situations – Selection of a testing mode (qualitative or quantitative) in which more than a single dose level is used will often provide additional confidence in an evaluation of the mutagenic activity of a chemical. The selection of dose levels for microbial systems is based upon the toxicity of the test compound to the indicator cells and may not have any relationship to in vivo activity of the agent. In fact, the degree of mutagenic effect by the test compound in the indicator cells may have no relationship to the level of activity in vivo. Thus, it cannot be stated that a chemical that induces high levels of mutagenic activity in a bacterial or yeast strain would be potentially more active in vivo than a compound exhibiting only weak mutagenic activity in the indicator organisms. Therefore, these facts should be taken into consideration when attempting to determine the potential effect of chemicals in mammals based on in vitro microbial screens.

2. L5178Y Mouse Lymphoma Forward Mutation Assay

This assay evaluates the mutagenic potential of chemicals in a specific locus mutation assay using mammlian cells. In mammalian cells, this test can be used to confirm presumptive mutagenicity demonstrated in microbial screens.

TK ± BUdR-sensitive L5178Y lymphoma cells are used as indicator organisms and ethylmethanesulfonate is used as the positive control. EMS is considered to induce predominantly base-pair substitutions and not to require activation. Dimethylnitrosamine at 7 μmol/ml is used as the positive control compound in activation assays.

Dosage determinations will be made from preliminary toxicity curves established from treatment with a standard

TABLE 8

Positive Controls Used in Nonactivation and Activation Assays

Assay	Chemical	Solvent	Probable mutagenic specificity[a]
Nonactivation	Ethylmethanesulfonate (EMS)	Water or saline	BPS
	Methylnitrosoguanidine (MNNG)	Water or saline	BPS
	2-Nitrofluorene (NF)	Dimethylsulfoxide[b]	FS
	Quinacrine mustard (QM)	Water or saline	FS
Activation	2-Anthramine (ANTH)	Dimethylsulfoxide[b]	BPS
	2-Acetylaminofluorene	Dimethylsulfoxide[b]	FS
	8-Aminoquinoline (AMQ)	Dimethylsulfoxide[b]	FS
	Dimethylnitrosamine (DMNA)	Saline	BPS

[a]BPS, base-pair substitution; FS, frameshift.
[b]Previously shown to be nonmutagenic.

set of concentrations of the test chemical. Four dose levels of the test chemical will be used in the evaluation, plus positive and negative control tests.

Since many chemicals have been shown to require metabolic biotransformation before demonstrating mutagenic activity, L5178Y tests that incorporate microsomal enzymes will be conducted. These assays are essentially performed as described previously with the exception that 9000 X g postmitochondrial supernatant tissue fractions plus required cofactors are added during the chemical treatment phase of the assay.

B. In vivo Assays

1. Host Mediated Assay (Modification of Microbial Assays)

The host mediated assay employs microbial indicator cells as described previously, but uses the intact animal as the source of chemical biotransformation, detoxification, and distribution.

Microbial cells are injected intraperitoneally into mice (flow, ICR random bred male mice, 25 to 30 g). The test chemical is then administered to the animal either by oral intubation or by intramuscular injection. Following a suitable incubation period, the mice are killed, and the microbial indicator cells are aseptically removed from the peritoneal cavity. The recovered cells are then assayed in vitro for evidence of mutation induction.

The following microbial strains can be used in this modification:

Salmonella typhimurium	*Escherichia coli*	*Saccharomyces cerevisiae*
G-46	WP_2 urvA	D3
TA-1535*		D4*
TA-1537*		
TA-1538*	WP_2	
TA-98		
TA-100		

*Stains routinely used.

A test chemical is evaluated at three dose levels plus positive and negative controls. The dose levels may be either client prescribed or selected by an LD_5 determination. If the latter method is used, the suggested dose levels are the LD_5, 1/10 LD_5, and 1/100 LD_5. Chemicals will be evaluated by administering the dose levels of the chemical over a 5-day period. On the last day of dosing, the indicator organisms will be injected and the assay conducted.

2. Dominant Lethal Assay

The dominant lethal assay is an in vivo test to determine the germ-cell risk from a suspected mutagen once its genetic activity has been clearly demonstrated using in vitro or other suitable mutagenicity screens.

The following protocols describe the type of dominant lethal assay available for germ-cell risk assessment using either rats or mice. Assays are performed at three dose levels (supplied by the sponsor or established from toxicologic LD_{50} determinations). Positive (triethylenemelamine [TEM]) and negative controls are also conducted.

The standard procedures for the assays with rats and mice are similar and consist of administering the test compound for 5 days to male animals. This is followed by sequential mating of the dosed males to new sets of female animals each week for a sufficient number of weeks to cover the total spermatogenic cycle. This assay is generally referred to as the Standard Dominant Lethal Assay and is described in detail in the following protocols.

A modified dominant lethal assay has been developed and is currently being used in a number of laboratories, including the FDA. The only significant change in the protocol is that instead of a single dosing followed by sequential mating of the male animals, the males are dosed continuously over the spermatogenic cycle and mated in duplicate at the end of the dosing period. The overall results are similar with both procedures. Each has some minor advantages over the other but, in general, either is acceptable for germ cell analysis.

Subchronic dominant lethal assay in rats – In this test, male and female random bred rats (250 to 300 g) from a closed colony are employed. These animals are 10 to 12 weeks old at the time of use. Ten male rats are assigned to each of five groups; three dose levels are selected with a positive control (TEM) and a negative control (solvent only). The positive control is administerd intraperitoneally. Administration of the test compound is normally oral by intubation in this subchronic study (one dose per day for 5 days). Following treatment, the males are sequentially mated to females per week for 7 weeks. Two virgin female rats are then removed and housed in a cage until killed. The male is rested on Saturday and Sunday and two new females are introduced to the cage on Monday. It has been our experience that conception has taken place in more than 90% of the females by Friday and that the 2-day rest is beneficial to the male in regard to subsequent weekly matings. Females are killed using CO_2 at 14 days after the midweek of caging, and at necropsy, the uteri are examined for fetal deaths and total implantations.

Sufficient animals are provided in our experimental design to accomodate any reduction in the number of conceptions. Each male is mated with two females per week, and this provides for an adequate number of implantations per group per week for negative controls even if there is a fourfold reduction in fertility of implantations. Corpora lutea, fetal deaths, and total implantations per uterine horn are recorded.

Each animal is dosed by gastic intubation or other methods as appropriate once a day for 5 days. Doses are spread 24 hr apart. Three dosages, including a high, medium, and low dose, are normally employed. The test compound is administered in undiluted corn oil or another appropriate vehicle. Positive controls consist of animals that are given one dose of the known mutagen TEM administered intraperitoneally at a level of 0.30 mg/kg in 9.85% saline. Two negative control groups are employed: undiluted corn oil is normally used as a solvent control for the test compound and 0.85% saline is normally used as a solvent control for the positive control compound TEM.

Subchronic dominant lethal assay in mice – In this test, male and female random bred mice (25 to 30 g) from a closed colony are employed. These animals are 7 to 8 weeks old at the time of use. Ten male mice are assigned to each of five groups; three dose levels are selected with a positive control (TEM) and a negative control (solvent only). The positive control is administered intraperitoneally. Administration of the test compound is normally oral by intubation in this subchronic study (one dose per day for 5 days). Following treatment, the males are sequentially mated to two females per week for 8 weeks. Two virgin female mice are housed with a male for 5 days (Monday through Friday). These two females are then removed and housed in a cage until killed. The male is rested on Saturday and Sunday and two new females are introduced to the cage on Monday. Conception usually has taken place in more than 85% of the females by Friday, and the 2-day rest is beneficial to the male in regard to subsequent weekly matings. Females are killed using CO_2 at 12 days after the last day of mating, and at necropsy, the uteri are examined for viable and dead implants.

Sufficient animals are provided in our experimental design to accomodate any reduction in the number of conceptions. Each male is mated with two females per week, and this provides for an adequate number of implantations per group per week for negative controls even if there is a fourfold reduction in fertility of implantations. Fetal deaths and total implantations per uterine horn are recorded.

The results of the 8-week dominant lethal subchronic study are evaluated statistically. In this study, each of ten male mice are mated with two different females each week for 8 weeks at each dose level. The total number of implantations and dead implantations are counted for each pregnant female. From this base data, the following six parameters are calculated and evaluated by the following methods:

1. Fertility index – The fertility index was computed as the number of pregnant females/number of mated females and evaluated by a chi-square test to compare each treatment group and the positive control to the negative control. Armitage's trend for linear proportions was used to test whether the fertility index was linearly related to arithmetic or log dose. Results are summarized in tabular form.

2. Total number of implantations – The total number of implantations was evaluated by a t-test to determine whether the average number of implantations per pregnant female for each treatment group and the positive control group differed significantly from the negative control group. Results are summarized in tabular form.

3. Dead implantations – Dead implantations are computed for each female by subtracting number of dead implants from the number of total implants. A Freeman-Tukey transformation was performed on the number of dead implants for each female prior to being evaluated by t-test to compare each treatment group and the positive control to the negative control. Regression analysis was used to determine whether the average number of dead implants per female was related to the arithmetic or the log dose. Results are summarized in tabular form.

4. Proportion of females with one or more dead implantations – The proportion of females with one or more dead implants was computed as the number of females with dead implants/number of pregnant females. The quotient was evaluated by the statistical method used for deriving the fertility index, with the addition of a probit regression analysis to determine whether the probit of the proportions was related to log dose.

5. Proportion of females with two or more dead implantations – The proportion of females with two or more dead implantations was computed as the number of females with two or more dead implants/number of pregnant females. The data were evaluated by the same method used for determining the proportion of females with two or more dead implants.

6. Dead implants/ total implants – Dead implants/ total implants were computed for each female and then transformed by way of Freeman-Tukey arc-sine transformation prior to being evaluated by t-test to compare each treatment group and positive control to negative control.

3. Heritable Translocation Assay (Mice)

The heritable translocation assay (HTA) in mice consists of determining the ability of a chemical to induce reciprocal translocation in the germ-line cells of treated male mice. The presence of induced translocations can be detected by mating the male F_1 progeny of the treated males with unrelated females and scoring for a reduction in the number of viable fetuses. This semisterility has been directly attributed to the F_1 males being translocation herterozygotes, that is, having inherited one member of a pair of reciprocal translocation between nonhomologous chromosomes, thus producing duplication/deletion gametes at meiosis. These gametes are usually nonviable and/or lethal. The presence of these reciprocal translocations can be verified cytogentically by the presence of translocation figures among the double tetrads at meiosis. The reciprocal translocation assay provides a sensitive method for estimating heritable genetic damage resulting from chromosome breakage and also has relevance to the human experience. It is more specific than the dominant lethal test that may measure other aberrations in addition to chromosome breakage. The primary advantage that the HTA has over other in vivo tests, such as the dominant lethal assay or rat bone marrow cytogenetic assay, is that the HTA sensitivity measures a transmissible lesion. Genetic alterations which can pass through meiotic stages and can be transmitted to subsequent generations are potentially more damaging to the gene pool than dominant lethal and other nontransmissible chromosome alterations.

The test will require approximately 30 weeks to complete and will utilize a minimum of 500 male and 1000 female mice. One hundred males will be evenly distributed among four groups. Each group will receive a specified, predetermined dosage level of the test compound or the positive or negative control compound for 7 weeks. At the termination of the dosing interval, each of these males will be mated with two females. One hundred healthy F_1 males will be selected from among the progeny in each group. When they reach reproductive maturity, each of these F_1 males will be mated to two virgin females for 1 week. Two weeks after a copulation plug is detected in these females, they will be killed with CO_2, surgically opened, and the number of living fetuses and resorbed embryos present in their uteri determined. Semisterility will be determined from this information on the basis of a specific set of criteria. Basically, the criteria will be the total number of implantations both living and dead. The F_0 males will be housed five animals to a cage for dosing. Males will be mated with two females as stated in the protocol. Females will be assigned the number of their mate and housed together until just prior to the parturition date, at which time they will be isolated or killed. After weaning, F_1 males will be housed five to a cage until mated. Mated F_1 males will be retained until a decision is made concerning their semisterility. At this time, they may be killed, or if the data are equivocal, they may be either remated or killed and their gonadal cells analyzed for cytogenetic evidence of translocations.

About 6 weeks are required for the progeny of a stem cell to traverse the entire periods of the spermatogenic and spermiogenic cycles and emerge as mature sperm. Theoretically, since genetic damage can be induced in premeiotic as well as postmeiotic phases of these cycles, one should, in the absence of specific information to the contrary, examine all parts of the gametogenic cycle for possible reciprocal translocations or at least ensure that cells in each stage have been exposed to the compound. This latter goal can be achieved with surety if the males are dosed for 7 weeks prior to mating them.

The F_1 animals that show presumptive semisterility will be killed and their gonadal cells analyzed for evidence of translocations. The basic method will be that of Ford and Evans,[6] in which the testes will be teased apart with fine forceps and then subjected to a mild trypsinization. The released cells will be washed, processed, and then air-dried onto slides. Stained slides will be examined for the presence of translocation figures indicative of translocation heterozygosity among them.

VI. CHRONIC TESTING (TOXICITY AND/OR CARCINOGENICITY)

A. Twenty-four Month Skin Painting (Mice)

The purpose of this study is to evaluate the carcinogenic potential of test compounds when painted on the skin of 100 C_3H male mice (Jackson Laboratories, Bar Harbor, Maine) three times per week for 24 months.

The test compounds are applied with a small paint brush, syringe, or other device (approximately 20 mg) to the backs of each mouse denuded with clippers. Hair removal is routinely done once a week. If the viscosity of the test material inhibits direct application, dilution can be with mineral oil, and the volumes administered can be adjusted. Groups of 100 mice are used for untreated controls, negative controls if a suitable diluent is used, and a positive control.

Animals are observed daily for general physical appearance, observable masses, and mortality. Records should be kept of the following:

1. Number of animals with observable masses
2. Number of observable masses on each animal
3. Time of appearance of observable masses in each group
4. Mortality and condition of carcass

Gross autopsies are performed on any animal that succumbs during the course of these experiments, all animals sacrificed *in extremis,* and all animals at termination. Microsopic examination will be conducted on tissues and organs from the animals sacrificed *in extremis* and at termination, and on ten males from each group of the surviving animals.

B. Three-Generation Reproduction Study (Rat)

Reproductive studies in rats are designed to evaluate the effects on reproductive performance of parents and developmental processes in their offsprings. These studies are normally performed in conjunction with a long-term feeding study.

Thirty weanling Fischer 344 rats per group will serve as breeding stock (F_0) for F_1 offspring. The second mating of the F_0 stock will produce the F_1b breeders. The cycle of breeding the F_1b and F_2b offspring will continue until the F_3b generation has been produced. The rats will be housed individually, except during mating trials, under standard laboratory conditions of temperature (72 ± 2°C) and humidity (50 ± 5%) with water and the appropriate diets as described below available ad libitum.

Group	Treatment	F_0 breeding stock (number of rats) Female	Male
1	High dosage	20	10
2	Moderate dosage	20	10
3	Low dosage	20	10
4	Control	20	10

Dosage levels will be based on data from preliminary feeding studies. Each new diet preparation (animal diet mash plus chemical) will be assayed for agricultural chemical content and will be used only if it is within ±10% of theoretical.

1. Breeding

The P_1 generation will be individually housed and fed the appropriate assigned diet until the animals are approximately 100 days of age or for 60 days prior to breeding. At the end of this period, one male and two females from groups receiving the same diet will be placed in a double-screen, bottom-breeding cage for mating. Brother-sister matings will be avoided. The presence of sperm in a vaginal smear will be considered evidence of positive mating, and the males will be removed and returned to single cages. If after 1 week, no positive mating is observed, the male will be removed and a different male from the same diet group will be placed with the females. Females will be considered barren if no positive mating is observed after 2 weeks. Such females will be removed and a different female from the same diet group will be substituted. The day sperm is found in a vaginal smear will be considered day zero of gestation. After mating, all animals will be returned to their individual cages.

2. Parturition

Approximately 5 days prior to parturition, females will be placed in individual plastic shoe-box nesting cages equipped with a stainless steel top, feed jar, and water bottle. Bedding will be changed three times weekly, or more often if needed, through weaning. Females will be allowed to deliver and care for their own young with a minimum of disturbance. The first day pups are observed will be considered day 1 for that litter.

3. Nursing

The pups will be allowed to nurse for a period of 21 days after which they will be weaned. On day 4, pups will be culled to ten if possible, five males and five females.

4. Rebreeding

Following weaning of the F_1a litters, animals will be allowed a 1-week rest period, then mated a second time, as described above, for the production of a second litter (F_1b). A different male from the same diet group will be used as sire for the second litter. Males may be sacrificed following the second mating.

5. Reproduction Cycle

At weaning of the F_1b litter, 10 males and 20 females from each diet group will be selected at random for the second or P_2 generation. They will be individually housed and maintained for approximately 100 days on the same diet as their parents. They then will be mated in the same manner described previously producing the F_2a and F_2b litters. At weaning of the F_2b litter, 10 males and 20 females from each diet group will be selected at random for the third or P_3 generation. They will be individually housed and maintained for approximately 100 days on the same diet as their parents, at which time they also will be mated in the manner described above producing the F_3a and F_3b litters.

6. Observations and Data Collection

Parental data:

1. All animals will be observed daily for general physical appearance, behavior, and pharmacotoxic signs.
2. Weekly records will be maintained for gross signs of systemic toxicity.
3. Records of individual body weights, food consumption, and survival of all male and female parents (P_1, P_2, P_3), initially (weaning weight), at 4 and 9 weeks, and immediately before and after mating will be maintained.
4. Weight of females on days 1, 4, and 21 following parturition will be maintained.
5. Necropsies will not be performed on any of the parents (P_1, P_2, P_3). Tissues will not be preserved or examined unless above observations so indicate.

Breeding data:

1. Identification of female
2. Identification of all males placed with female, the last listed being considered the sire
3. The number of days required to mate
4. Duration of gestation
5. Fertility, gestation, live birth, and lactation indices

Litter data:

1. Identification of parents
2. Number of pups born: alive, dead, total, number of each sex
3. Individual pup weight and average weight on day 1
4. Number of survivors on day 4 and percent survival of live births
5. Number of each sex, individual pup weight, and average weight on day 4

6. Number of survivors on day 21 and percent survival of live pups by sex

7. Weight and sex of pups on day 21 (weaning)

8. Gross necropsies will be performed on approximately one third of the pups in litters F_1b, F_2a, F_2b and F_3a at weaning and on approximately one half of the F_3b litters at weaning. The entire animal will be preserved for possible future gross and pathological examination.

9. The following tissues from 10 male and 10 female weanling pups, chosen at random from the F_3a litters of the control and each treatment group, will be preserved and held for possible future examination:

Adrenal	Lung
Bone	Ovary/testis
Bone marrow	Pancreas
Brain	Pituitary
Eye	Spleen
Heart	Stomach
Intestine (large)	Thyroid
Intestine (small)	Urinary bladder
Kidney	Unusual lesions and tumors
Liver	

7. Statistical Evaluation

1. Pregnancy ratio
2. Gestation index
3. Viability index
4. Lactation index

C. Twenty-seven Month Toxicity Study (Rat)

This study permits the administration of a compound in dosed-diet to F_1 rats for at least 27 months or until 50% of the initial population remains to evaluate the chronic toxicity and oncogenicity potential of the experimental compound.

Thirty weanling Fischer 344 rats (20 female/10 male per group) will serve as breeding stock (F_0) for F_1 offspring to continue on the study for at least 27 months. The F_0 rats will be housed individually under standard laboratory conditions of temperature (72 ± 2°C) and humidity (50 ± 5%) with water and appropriate diets as described below available ad libitum.

		F_0 breeding stock (number of rats)		F_1 experimental animals (number of rats)	
Group	Treatment	F	M	F	M
1	High dosage	20	10	60	60
2	Moderate dosage	20	10	60	60
3	Low dosage	20	10	60	60
4	Control	20	10	60	60

Dosage levels are to be determined from a preliminary study. Each new diet (laboratory animal diet mash plus chemical) will be assayed for agricultural chemical content and will be used only if it is within ±10% of theoretical. Following 8 weeks on the diet, the females will be examined by vaginal smear and exposed to males on evidence of estrus. One male will be mated to only two females within experimental groups. Progeny from each treatment group will be selected so that there is no more than 4 days difference in age. A maximum of three males and three females will be selected from each litter. A total of 60 rats per sex per dosage will be assigned to each treatment group.

Study Parameters

1. Appearance and behavior of all rats will be recorded daily throughout the study.

2. Individual body weights (to the nearest gram) will be determined prior to the start of the study and thereafter at weekly intervals at a uniform time of day and uniform day of the week throughout the study.

3. Food consumption (to the nearest gram) will be recorded weekly throughout the study.

4. Clinical studies will be performed on ten animals of each sex from each of the experimental and control groups after 52 weeks and on all animals at 116 weeks of study and will include:

a. Hematology – hematocrit, hemoglobin, erythrocyte count, total leukocyte count, and differential leukocyte count

b. Clinical biochemistry – fasting blood sugar, blood urea nitrogen, gamma-glutamyl transpeptidase, serum glutamin-pyruvic transaminase, and serum glutamic-oxalacetic transaminase

c. Urine analysis – pH, specific gravity, glucose, ketones, total protein, bilirubin, microscopic examination of sediment

d. Terminal studies on animals sacrificed at 52 weeks including organ weights, tissue fixation, and histopathologic examination

e. Animals which succumb as well as all animals in which sacrifice is necessitated for a moribund condition will undergo a gross necropsy; providing lesions are demonstrable, selected sections will be taken from target organs and submitted for histopathologic examination

5. Terminal studies at 27 months (116 weeks) will include sacrifice of all survivors and the performance of gross necropsies.

a. Organ weights for each animal sacrificed will include the thyroid, heart, liver, spleen, kidney, adrenal, testes with epididymis or ovary, and uterus.

b. Tissue fixation by preservation in 10% neutral buffered formalin will be undertaken on the following 30 tissues and organs:

Brain
Pituitary
Thymus
Thyroid
Salivary gland
Lung
Heart
Liver
Spleen

Adrenal
Kidney
Bladder
Stomach
Pancreas
Duodenum
Jejunum
Ileum
Caecum
Colon
Sciatic nerve
Spinal cord
Skin
Bone marrow (long bone, sternum)
Eye
Testes
Seminal vesicles
Prostate
Ovaries
Uterus
Mammary gland

c. Histopathologic examination of all the aforementioned tissues will be performed on all male and female rats of the high dosage and control groups. In addition, selected tissues or target organs from remaining test groups will be examined on the basis of pathology seen in the high-dosage group. Any tumor or other tissue from any animal which appears abnormal at the time of necropsy will also be examined.

d. Statistical comparison of means using the student "t" test will be used for evaluation of weekly body weight data, organ/body weight ratios, and several of the clinical studies. Significance at $P < 0.05$ and $P < 0.01$ will be noted.

D. Twenty-four Month Toxicity Study (Mouse)

This study permits the administration of a compound in dose-diet to F_1 mice for at least 24 months or until 50% of the initial population remains to evaluate the chronic toxicity and oncogenicity potential of the experimental compound.

Thirty weanling B6C3F1 mice (20 female/10 male per group) will serve as breeding stock (F_0) for F_1 offspring to continue on the study for at least 24 months. The F_0 mice will be housed individually under standard laboratory conditions of temperature (72 ± 2°C) and humidity (50 ± 5%) with water and appropriate diets as described below available ad libitum.

		F_0 breeding stock (number of mice)		F_1 experimental animals (number of mice)	
Group	Treatment	Female	Male	Female	Male
1	High dosage	20	10	50	50
2	Moderate dosage	20	10	50	50
3	Low dosage	20	10	50	50
4	Control	20	10	50	50

Dose levels are to be determined from preliminary studies. Each new diet preparation (laboratory animal diet mash plus chemical) will be assayed for agricultural chemical content and will be used only if it is within ±10% of theoretical.

Following 8 weeks on the diet, the females will be examined by vaginal smear and exposed to males on evidence of estrus. One male will be mated to only two females within experimental groups. Progeny from each treatment group will be selected so that there is no more than 4 days difference in age. A maximum of three males and three females will be selected from each litter. A total of 60 mice per sex per dosage will be assigned to each treatment group.

Study Parameters

1. Appearance and behavior of all mice will be recorded daily throughout the study.
2. Individual body weights (to the nearest gram) will be determined prior to the start of the study and thereafter at weekly intervals at a uniform time of day and uniform day of the week throughout the study.
3. Food consumption (to the nearest gram) will be recorded weekly throughout the study.
4. Complete gross pathologic examination for tumor formation will be conducted upon all post-mortem animals, all animals sacrificed when found moribund, and upon all mice surviving 24 months of study.
5. Terminal studies at 24 months (104 weeks) will include sacrifice of all survivors and the performance of gross necropsies.

a. Tissue fixation by preservation in 10% neutral buffered formalin will be undertaken on the following tissues and organs:

Brain
Pituitary
Thymus
Thyroid
Salivary gland
Lung
Heart
Liver
Spleen
Adrenal
Kidney
Bladder
Stomach
Pancreas
Duodenum
Jejunum
Ileum
Caecum
Colon
Sciatic nerve
Spinal cord
Skin
Bone marrow (long bone, sternum)
Eye
Testes
Seminal vesicles
Prostate
Ovaries
Uterus
Mammary gland

b. Histopathologic examination of the aforementioned tissues will be performed on all male and female mice of the high-dosage and control groups. Selected tissues or target organs from remaining test groups will be examined on the basis of pathology seen in the high-dosage group. Any tumor or other tissue from any animal which appears abnormal at the time of necropsy will be submitted for histopathological examination and classification.

ACKNOWLEDGMENT

The author gratefully acknowledges the use of protocols prepared by the following testing laboratories: Cannon Laboratories, Reading, Pennsylvania; Consumer Product Testing, Fairfield, New Jersey; Food and Drug Research Laboratories, East Orange, New Jersey; Gulf South Research Institute, Baton Rouge, Louisiana; Industrial Bio-Test Labs., Northbrook, Illinois; and Litton Bionetics, Kensington, Maryland.

REFERENCES

1. **Litchfield, J. T., Jr. and Wilcoxon, F.,** A simplified method of evaluating dose-effect experiments, *J. Pharmacol. Exp. Ther.*, 96, 99, 1949.
2. **Hodge, H. C.,** The LD_{50} and its value, *Am. Perfum. Cosmet.*, 80, 57, 1965.
3. **Litchfield, J. T., Jr. and Wilcoxon, F.,** A simplified method of evaluating dose-effect experiments, *J. Pharmacol. Exp. Ther.*, 96, 99, 1949.
4. **Draize, J. H., Woodard, G., and Calvery, H. O.,** Methods for the study of irritation and toxicity of substances applied topically to the skin and mucous membranes, *J. Pharmacol. Exp. Ther.*, 82, 377, 1944.
5. **Miller, L. C., and Tainter, M. L.,** *Exp. Biol. Med.*, 57, 261, 1944.
6. **Ford, V. M. and Evans, G. L.,** *Comparative Mammalian Cytogenetics*, Springer-Verlag, New York, 1969, 461.

ENVIRONMENTAL EFFECTS MONITORING

C. Ganz

TABLE OF CONTENTS

I. INTRODUCTION

The Toxic Substances Control Act (TSCA) stipulates that new products should not cause an unreasonable risk of injury to health or the environment. As a consequence of this stipulation, originators of new products must be able to assess the potential environmental and health effects of their products prior to the introduction of these materials into the marketplace. In this section, potential areas of environmental concern for the chemicals producer will be focused upon. In so doing, possible ways in which discharged products can intrude upon and interact with the environment will be discussed. Techniques and methods currently being used to determine whether a product may cause adverse environmental effects will also be described.

Although this section should be of general interest to those involved in dealing with TSCA, the discussion has been especially designed to give the neophyte a basic awareness of the multidimensional nature of environmental effects. Armed with this information, those concerned should then be able to formulate many of the important questions which must be answered in assessing the environmental safety of particular products. For further details as to how these questions may be answered, references to more comprehensive discussions are annotated throughout.

For the purposes of establishing a working definition of environmental risk, an *ecosystem* must first be defined, as the complex set of interrelationships existing between various life forms and the physical-chemical factors influencing the maintenance of their life processess. A chemical product can thus be considered a potential environmental risk if it can alter physical, chemical, or biological conditions in an ecosystem in a way which adversely affects life processess.

For those making decisions on the acceptability of a product, the degree of risk must be weighed against the relative benefits of the product. Obviously, some materials will present unacceptable risks no matter what the potential benefits. The producer should bear in mind that, according to TSCA, the burden of proof of environmental safety will rest with the chemicals manufacturer. Consequently, from an economic standpoint, the market potential of the product must be weighed against the cost of accumulating sufficient data to demonstrate that the product presents an acceptable environmental risk.

II. DISPOSAL AND DISTRIBUTION OF CHEMICAL PRODUCTS IN THE ENVIRONMENT

Clearly, the relative environmental risk of a chemical product will be strongly influenced by its potential for entering the environment. Consequently, a review of the routes of disposal and distribution of waste chemical products is presented first. Figure 1 illustrates some of the many ways in which man-made wastes may be disposed of and subsequently distributed in the environment. The primary sources of man-made wastes are shown in the upper part of the diagram. These are comprised mainly of stationary industrial and domestic point sources. Manufacturers, processors, and commercial users of chemicals account for most industrial chemical wastes while ultimate consumers of products made from these chemicals (paper, detergents, plastics, etc.) comprise the domestic dischargers. Runoff of chemicals used in agriculture, automobile exhaust emissions, and accidental chemical spills are some examples of major nonpoint sources.

Industrial wastes may consist of solids, airborne emissions, or liquids, while domestic wastes are usually in liquid or solid form. A number of primary modes of disposal exist. Solid wastes, for example, may be deposited in open dumps, buried in landfill sites, or burned. Solids may also be discharged with liquid wastes either in a dissolved or suspended form. Until recently, disposal of solid wastes into offshore ocean dumping sites was widely used. However, this practice has been severely curtailed because of potential adverse effects upon sea life in the dumping areas and because of suspected transport of the wastes back to inhabited shoreline areas.

Airborne wastes, which may be in the form of vapors, particulates or aerosol droplets, are discharged directly into the atmosphere through process stacks, workplace air exhaust systems, and incinerators. Liquid effluents are most commonly discharged either directly into surface waters or alternatively into waste water treatment plants (WWTP) or septic systems.

Because the WWTP is the collection point for enormous quantities of industrial and domestic wastes, its importance in the waste disposal and distribution chain cannot be overemphasized. The basic operations of WWTPs must be understood before a realistic environmental assessment of most products can be made.

The processes in a WWTP are designed to utilize many processes which nature uses for purification but in a more concentrated and controlled manner. A WWTP may be characterized as primary, secondary, or tertiary depending upon the extent to which it is equipped to treat incoming wastes. The main WWTP processes are illustrated schematically in Figure 2. In the primary treatment step, the aqueous effluent enters a semiquiescent chamber where suspended solids are allowed to settle out. The settled solids are then removed, while the liquid phase is either disinfected and discharged or diverted to a secondary treatment process.

The two secondary procedures most often used are the activated sludge process and the trickling filter process. In the activated sludge process, an influent containing dissolved organic matter (various hydrocarbons and to a lesser extent organic nitrogen, sulfur, and phosphorus compounds) is continuously mixed with an aerated solution of

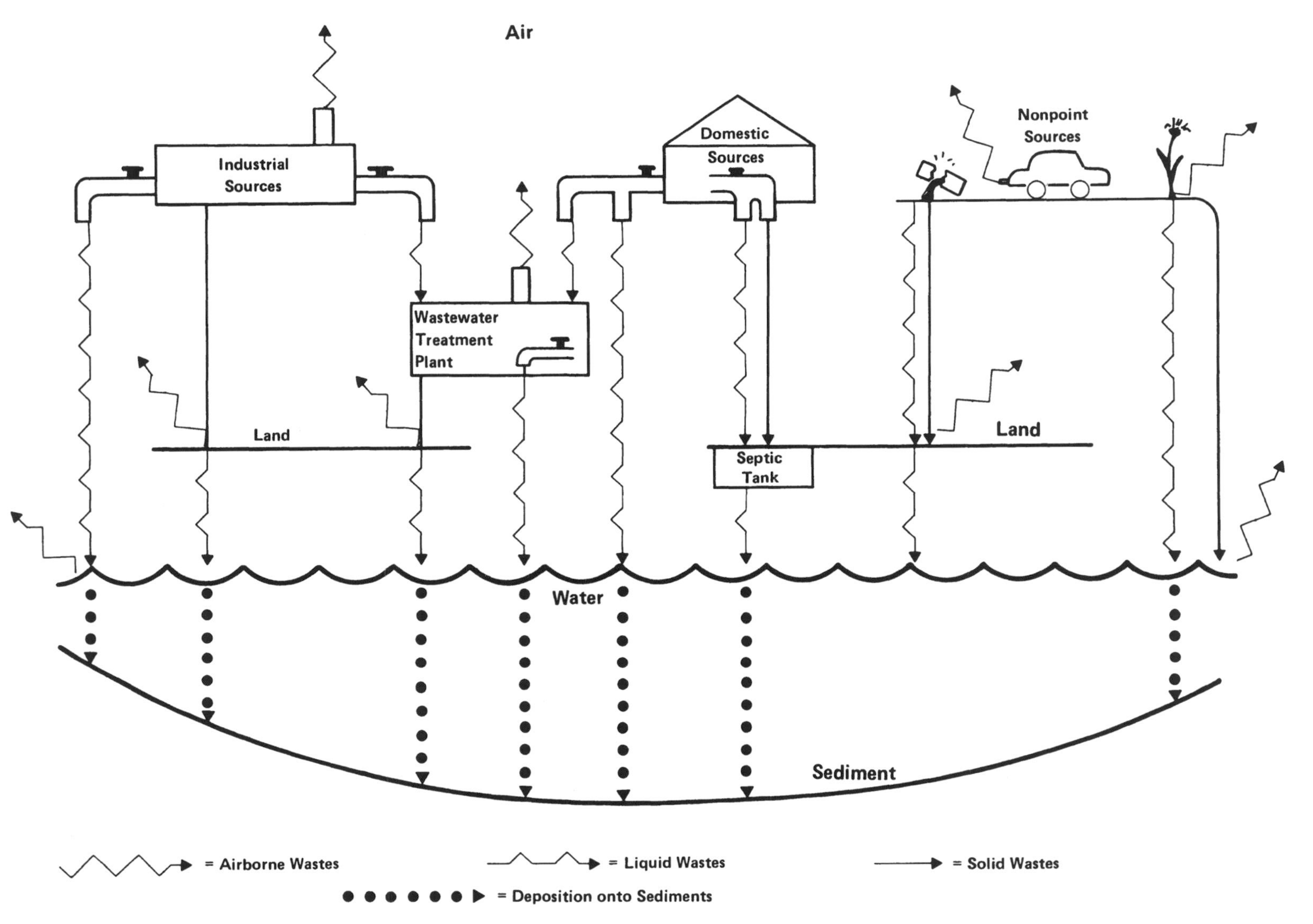

FIGURE 1. Modes of disposal and distribution of man-made wastes in the air, land, and water environments.

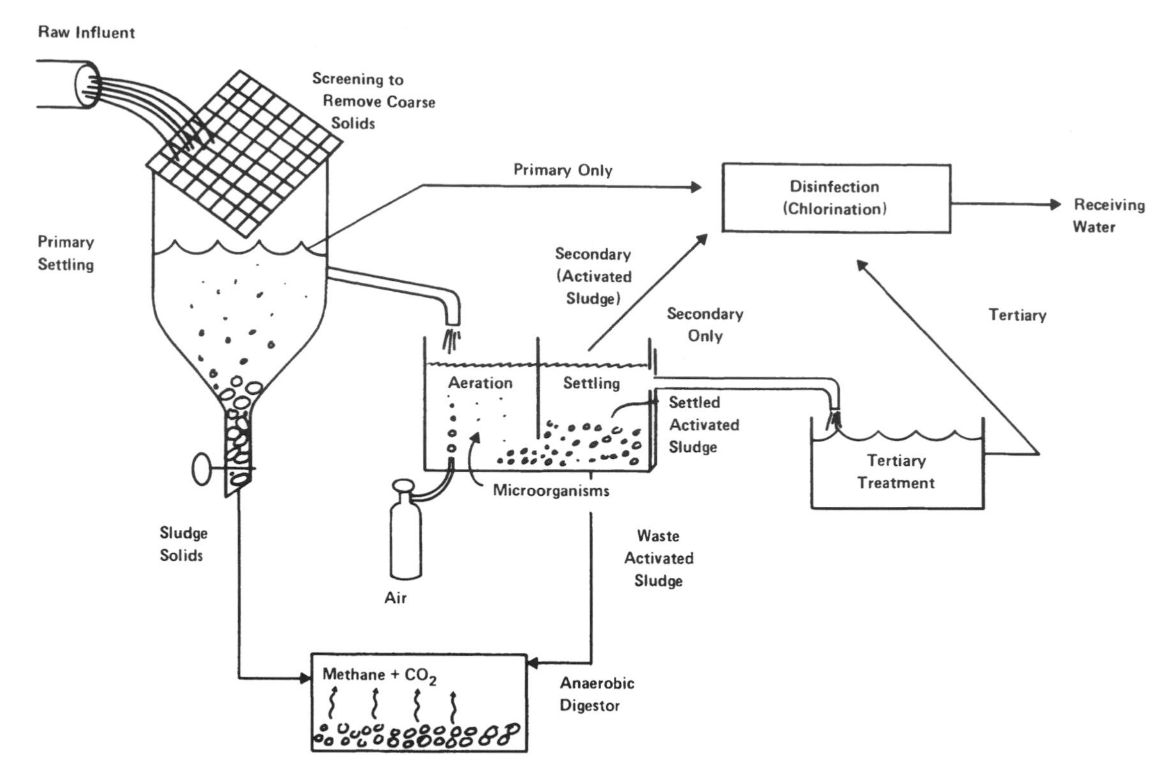

FIGURE 2. Schematic diagram of the activated sludge waste treatment process.

microorganisms. The organisms, with the aid of dissolved trace metals, digest the organic materials and simultaneously use the dissolved oxygen from aeration for respiration. The end results of this process are the degradation of dissolved organic compounds to biologically stable fragments (ultimately to carbon dioxide) and the concommitant creation of an increased mass of microorganisms known as activated sludge. The activated sludge thus serves as a continuously regenerable digestion medium for dissolved organics. In addition, the activated sludge can physically adsorb undigested or partially digested residues so that they are in effect physically removed from the waste stream. The trickling filter also makes use of the ability of microorganisms to digest dissolved organic materials. Here, the influent is sprayed over a pit filled with stones or a similar solid material onto which a slime of microorganisms has formed. As effluent trickles down through the layers of slime-covered support, the dissolved organic matter is taken up and digested.

Oxidation lagoons or ponds, which are shallow basins usually large enough to permit detention times of at least several days, are another form of biological treatment which can be especially effective for certain slowly degradable materials. The lagoon normally contains an upper aerobic (oxygen rich) section and a lower anaerobic (oxygen poor) section where various types of bacteria digest wastes. Dissolved oxygen is supplied either by growth and respiration of microscopic algae plants or by mechanical aeration. Lagoons may be used both as final treatment systems and as pretreatment systems. In the latter, the lagoon may also serve as an equalization basin where effluents of widely varying concentration can be mixed to yield an effluent of some median concentration level. Equalization reduces the possibility of "slug loads" (effluents with high concentrations of chemical wastes) being injected into WWTPs or receiving waters.

WWTP microorganisms are extraordinary in that they can adapt to a wide range of organic materials. They are not, however, capable of utilizing all organic matter. For example, branched chain alkylbenzene sulfonate detergents and certain chlorinated hydrocarbons are likely to pass through secondary treatment systems essentially unchanged. As living organisms, microbes are susceptible both to sudden changes in their environment and to the presence of toxic or inhibitory agents. For instance, sudden shifts in temperature or pH, an accidental spill of a toxic chemical, or a reduction in available dissolved oxygen due to an unusually high waste load may all result in substantial reductions in the plant's purification efficiency.

Following secondary treatment, the effluent water may be subjected to any one of a number of advanced treatment processess. These so-called tertiary treatment processes are used to remove undesirable effluent components which may remain after secondary treatment. The method chosen is largely dependent upon the particular problem at hand. For example, activated carbon adsorption is used to remove low concentrations of unwanted organics. Alum, ferric chloride, and polyelectrolytes have been used to clarify turbid effluents and remove residual color. Sodium bisulfite and ion exchange systems have been used to cleanse metal containing wastes. Electrolysis and reverse osmosis have also been used in specific instances while ozone treatment appears to show promise in both disinfection and the conversion of certain refractory chemicals to more environmentally acceptable forms.

Figure 1 demonstrates that wastes from many sources are funneled into WWTPs. Consequently, the treatment facilities themselves become potential primary sources of liquid, solid, and airborne pollutants. More specifically, liquid effluents from WWTPs, although in a much purified state, may still contain objectionable products not removed during the treatment process. Similarly, air pollutants may result from volatiles and aerosols produced during aeration processes and from volatiles and particulates created during incineration and digestion of solid wastes.

Many of the purification steps in a WWTP involve conversion of some of the dissolved or suspended materials into a solid residue. Some WWTPs treat their solid waste products by using digestion with anaerobic bacteria, i.e., bacteria which operate in the absence of oxygen. In these systems, specific bacteria first convert the organic matter into volatile organic acids. The organic acids are then acted upon by methane-forming bacteria to yield as final products methane, carbon dioxide, and additional microorganisms. In a properly operating digestor, the waste gases may be of sufficient BTU content to be used for energy production. Unfortunately, anaerobic digestors are based upon delicate biochemical equilibria and may be easily upset by a number of factors including the presence of toxic chemicals. Once the balance is disturbed, it can be quite difficult to reestablish optimum operating conditions.

Solid waste disposal has become an increasingly difficult problem as regulatory agencies reduce the available disposal alternatives. As resource and energy recovery from solid wastes become more economically attractive, it is probable that new solutions to many solid waste problems will materialize.

Figure 1 also demonstrates that the primary modes of disposal are only the beginning of a complex distribution network. Following the diagram down from the primary sources, one can see that many opportunities exist for interchange of pollutants between air, land, and water environments. For example, wastes that were initially airborne may subsequently be transported back to land and water depositories by settling or by combining with atmospheric water droplets that eventually condense to yield various forms of precipitation. Similarly, solid wastes which were disposed of on or beneath land surfaces can be leached into ground water or washed into surface waters by rain. Volatilization of specific solid waste components may also occur particularly if they are present as thin films on the substrate material. Poorly soluble components may also be redistributed from surface waters into the air by evaporation from or coevaporation with water. Lake and river sediments often become repositories for many waterborne chemicals. Factors such as flooding and bacterial activity may subsequently result in the transport of such chemicals from sediments into water and land environments.

The processess of interchange will be more or less continuous until the chemical is degraded and becomes a constituent of the pool of natural materials (natural organic matter, CO_2, H_2O, NO_3^-, $SO_4^=$, $PO_4^{\equiv}$, etc.). The extent and time frame of these conversions will depend upon the physicochemical properties of the material and its reactivity in the environment.

Before an environmental assessment program is formulated, the manufacturing processes, the expected commercial uses, and the anticipated end uses of the product should be thoroughly studied. From such a study, it will usually be possible to chart the most likely primary pathways by which the chemical product will be discharged. Some knowledge about the chemical properties and biological degradability of the product and its components will also help in predicting how the various constituents may be transported following the primary discharge. With knowledge of this type, the environmental assessment can be designed to concentrate on those interactions deemed most likely to occur.

III. ENVIRONMENTAL INTERACTIONS OF DISTRIBUTED CHEMICAL WASTES

Now that the distribution routes for disposed chemicals have been enumerated, let us turn to some of the interactions which occur as a chemical product makes its way along the various environmental pathways. Table 1 summarizes some of the interactions which must be examined when considering environmental effects.

A. Atmospheric Discharges

If the uses and properties of a product suggest that it will be discharged in significant quantities directly into the atmosphere, an assessment of the dispersion characteristics of the product and its reactivity in the atmosphere must be made.

Familiarity with dispersion properties will permit the extent of transport and the ground level concentrations of the chemical to be estimated. If these estimates are combined with knowledge about potentially adverse health effects of the product (irritability, odor, respiratory effects, acute and chronic toxicity, carcinogenic potential, etc.) and its potential for adverse environmental effects (damage to vegetation, animal life, man-made structures, etc.) one can make a partial evaluation of the potential atmospheric risk of the product.

Dispersion is usually estimated from mathematical models.[1,2] Important variables taken into account in these models include: type of source (point source, e.g., a single stack; or area source, i.e., a number of closely grouped

TABLE 1

Interactions Between Chemical Products and Environmental Systems

Air	Land	WWTP	Water
Effects on air and land creatures	Degradation by soil microorganisms	Treatability	Biodegradation
Objectionable odor	Adsorption to soil	Precipitation	Oxygen depletion
Photochemical reactions	Leaching through soil	Degradation by WWTP microorganisms	pH effects
Smog formation	Uptake by plants	Adsorption to suspended solids and to bacterial residues	Effects on aquatic nutrient supplies
Transport	Inhibition or toxicity to soil microorganisms	Chemical degradation (hydrolysis, oxidation etc.)	Toxicity to aquatic plants and animals
Effects on vegetation	Photochemical reactions	Inhibition or toxicity to WWTP microorganisms	Accumulation in fish
Effects on man-made articles and structures			Transmission through the food chain
			Accumulation and reaction in bottom sediments
			Complex formation
			Chemical reactions
			Photochemical reactions

sources), height of the source, wind direction and velocity, distance from the source, and emission rate.

A dispersion equation has been developed[2] to provide a general guide for estimating maximum ground level concentrations for emissions from small stacks (10 m height). Assuming a set of average meteorological and emission conditions, the form of the equation is

$$\overline{C} = 3.8 \times 10^{-5}\ Q$$

where: $\overline{C}$ = 30 day average concentration (g/m^3) at a distance 150 m downwind from the source, and Q = emission rate (g/s).

Similarly, an estimated dispersion equation has been developed[2] for groups of urban area sources. This expression takes a form:

$$\overline{C} = 35\overline{Q}$$

where: $\overline{C}$ = 30 day average concentration of material over an urban area, and $\overline{Q}$ = average emission (g/m^{2}s).

These equations may be used as first approximations to estimate dispersed concentrations of the product. Obviously, the way in which the chemical is to be used will influence important variables in the equation and must be given ample consideration in assessing air pollution potential. For example, a high temperature process such as steel making will have relatively large emission rates and will probably utilize tall stacks. On the other hand, a textile dyeing and finishing plant, where moderately high temperatures may be used, might utilize smaller stacks and have a concomitantly lower emission rate. A spray painting process, which is usually performed at ambient temperature, may be serviced simply by a ventilation hood and thus would probably have even smaller stacks and lower emission rates.

How information on dispersion may be used to estimate potential risk is still a subject of much debate and study. One suggestion[3] has been to tie the human safety levels to threshold limit values (TLVs) established by the American Conference of Government Industrial Hygienists (ACGIH) for human exposures to various chemicals in the workplace.[4] The proposed atmospheric environmental goal (EG) would be expressed as

$$EG = TLV\ (8/24) \times \text{Safety factor}$$

where: TLV = the airborne concentration of a substance to which a human can be repeatedly exposed in the workplace (8-hr day/40-hr week) without adverse health effects; 8/24 = factor assuming that the safe exposure level over 24 hr is 1/3 of the safe level for 8 hr, and Safety factor = factor required to extrapolate worker safety limits to the general population.

As will be discussed below, this approach is an oversimplification since it neglects interactions which may serve to remove the chemical from the atmosphere. However, it may serve to estimate concentrations which would be safe under worse conditions. Table 2 shows the present national primary air quality standards for major man-made pollutants.

The physical form of an emitted chemical will have a significant influence on the fate of the material in the atmosphere. Chemicals may be emitted as aerosols (microdroplets capable of remaining suspended in air), vapors, or particulates.

The process in which a chemical is used will be among the major factors determining the form in which it is discharged.[5] Spraying of herbicides and pesticides for example may produce both particulate and aerosol pollutants. In addition, the resultant crop residues may volatilize and become vaporous pollutants. Inorganic particulates

TABLE 2

National Primary Ambient Air Quality Standards

Pollutant	Maximum permissible concentration ($\mu g/m^3$) Annual mean	Short-term exposure	Applicable time interval for short-term exposure (hr)
Sulfur dioxide	80	365	24
Particulate matter	75	260	24
Carbon monoxide	10,000[a]	40,000	1
Photochemical oxidant	–	160	1
Hydrocarbons (corrected for methane)	–	160	3
Nitrogen dioxide	100	–	–

[a]Maximum 8-hr exposure. No annual mean is specified.

Taken from Reference 6.

may result from blasting and quarrying operations as well as from ducts venting process air from operations involving asbestos fibers and the like. Combustion sources will, in all probability, produce copious amounts of both particulate and vapor emissions. Vapors can also be expected in varying concentrations from virtually all operations in which volatile chemicals are used. Some examples include:

1. Organic chemicals production (solvents, volatile intermediates and products)
2. Textile dyeing and finishing (formaldehyde, dye carriers)
3. Paper manufacturing (mercaptans, hydrogen sulfide)
4. Plastic resin manufacture (monomers, solvents)
5. Metals processing (mercury)

Automobile exhaust also accounts for large amounts of aerosol, particulate, and vapor emissions. Among these are lead particulates, hydrocarbons from unburned and partially burned gasoline and oil, particulate combustion products, carbon monoxide, nitrogen oxides, and sulfur oxides.

Once airborne, a chemical may undergo or mediate a number of secondary interactions in the atmosphere. Of major interest are the mechanisms by which potential pollutants are removed from the atmosphere. Most pollutants are stripped from the air by physical processes. Large particulates may be dense enough to settle simply as a result of gravitational forces. The smaller particulates which remain airborne for longer periods may eventually be washed down, removed by collision with physical barriers (trees, hills, other particles), or adsorbed to condensed droplets of moisture. Actually, particulates play an important part in the moisture precipitation process by serving as condensation nuclei, i.e., surfaces onto which water vapor can readily condense to form droplets. These condensation nuclei cause precipitation to occur much more rapidly than would be the case in their absence.

Aerosols behave similarly to particulates with the major portion being removed by precipitation processes. A substantial portion of gaseous emissions are also washed out of the atmosphere by precipitation. The formation of so-called "acid rain" is a result of concentrated sulfur dioxide emissions being transformed to sulfuric acid by atmospheric moisture and subsequently precipitated with rain.

Gaseous pollutants may also undergo a number of chemical reactions in the air. The chemistry of atmospheric reactions is extremely complex and usually involves high energy species formed by photochemical reactions.[2] Sunlight provides an ample source of photons for these reactions. Ozone in the lower layers of the atmosphere provides a filter for the sun's rays allowing very little UV light below 290 nm to reach the earth's surface. At higher elevations where less shielding is available, a somewhat greater portion of shorter wavelength UV irradiation may be available.

Atmospheric reactions can be beneficial or harmful. The positive reactions are those which can degrade potential toxicants to less toxic analogues. Much of the emphasis, in research has been placed on reactions which cause negative effects. A prime example of the latter is the formation of photochemical smog. This phenomenon is particularly troublesome in areas such as Los Angeles, California where frequent temperature inversions occur and pollutant emissions are heavy. Temperature inversions result when heating and cooling patterns cause a layer of warm air to be sandwiched between two layers of colder air, one at the earth's surface and the second above the warm air layer. The warm air layer tends to act like a lid on a pot, keeping the air pollutants trapped for extended periods of time.

Photochemical smog is thought to result from a series of reactions involving nitrogen oxides, hydrocarbons (particularly olefins), atmospheric oxygen and sunlight.[2,7-9] The products of these reactions include ozone, aldehydes, peroxides, and peroxyacyl nitrates (PAN). The latter are known to be potent plant-damaging agents while the general class of compounds formed are a probable cause of increased eye irritation and respiratory ailments during episodes of photochemical smog.

Other types of smog may result from photochemical reactions which convert industrial chemicals and natural organic vapors into polymeric particulates. The mists characteristic of densely forested areas such as the Smoky Mountains are caused in part by reactions of this type.

Still another type of smog (London Type) is characterized by the build up of sulfur dioxide in stagnant air and is especially prevalent in areas where fossil fuel is burned. The formation of sulfuric acid from the oxidation of SO_2 to SO_3 and subsequent reaction of the latter with moisture constitutes a health hazard to both human and plant life. Indeed, mortalities during several air pollution disasters have been traced to inordinately high SO_2 levels.

As a final example of the atmospheric effects of man-made chemicals, recent investigations of the possible depletion of stratospheric ozone by chlorofluorocarbons (Freons) are cited.[10-12] The stratosphere is a relatively stable layer located above the air layer in which most weather conditions form. Atmospheric components may reside in the stratosphere for relatively long periods of time. The stratosphere also contains a store of ozone which acts as a filter for the sun's potentially damaging, short wavelength UV rays. Extensive theoretical studies made with the advent of the supersonic transport plane focused increased attention upon the fragile nature of the stratospheric ozone layer. Investigators have now found in laboratory experiments that chlorofluorocarbons can be photochemically degraded to chlorine by UV light of the shorter wavelengths found in the stratosphere. The chlorine thus formed can react with ozone resulting in a net depletion of the latter. Although tests to verify this phenomenon are still in progress, it is clear that in future environmental safety evaluations stratospheric effects will have to be taken into account. Furthermore, considering the present limited knowledge of atmospheric reactions, those concerned with product safety evaluations must remain cognizant of the fact that new dimensions will continue to be added to assessments of the air pollution potential of new products.

B. Interactions of Land Disposed Wastes

Wastes have too often been applied to land surfaces or buried in landfill sites with the mistaken notion that the wastes will be immobilized there. In actuality, land surfaces are dynamic systems which are strongly influenced by such factors as topography, soil characteristics, climate, depth of the water table, and microbiological activity. As with air pollutants, our two primary concerns must be, first, the possibility of chemical wastes being transported from the original disposal site to places where they can adversely affect environmental and human health and, second, what reaction processes are available to degrade the chemical to either more or less environmentally acceptable forms.

Wastes may be applied to land in solid, semisolid (sludge) or liquid forms. The liquids and sludges will normally penetrate the soil to some extent upon application. The solids will usually remain intact until water from rain or melting snow carries them into the soil. The extent of penetration through the soil is primarily a function of soil type and aqueous solubility of the waste material. Sandy soils tend to have poor retentive qualities so that liquid can readily percolate down into lower soil layers. Although sand does act as a filtration medium, its surface adsorption properties are not nearly as pronounced as those of other types of soil. Loams contain large quantities of decaying organic matter and can be quite effective at adsorbing and retaining organic chemical residues. Clay soils are also excellent adsorbents but are much less porous than either sand or loam.

A major concern in land disposal is the transport of disposed wastes into ground and surface water supplies. Surface runoff carries significant quantities of soluble, suspended, and soil-adsorbed chemicals directly into lakes, streams, and oceans. Agricultural runoff and untreated runoff from urban areas, usually carried through storm sewers, are major contributors. Penetration of contaminants through soils into ground water supplies which often feed wells, streams and lakes, is a second process which must be carefully evaluated before using land disposal alternatives. In this case, the depth of the water table is of extreme importance since a deeper water table will generally permit more opportunity for purification mechanisms to operate. Our laboratory has investigated several cases in which a high water table was responsible for transport of land discharged chemicals into surface waters. These cases were characterized by the appearance of periodic stream contamination during and after periods of heavy rainfall. The episodes could not be related to inadvertant spills or clandestine dumping of the chemicals. Upon careful analysis, it was discovered that the ground close to the suspected sources was nearly saturated with chemical. In one case, the saturation was due to a leaking underground transfer line while, in a second case, land disposal of the chemical was implicated. In both instances, portions of a water insoluble organic chemical phase were forced up through the ground by elevation of the water table during heavy rainfalls. The result was periodic injections of the contaminants into the stream.

The reactivity of chemical products and wastes applied to land is mainly the result of microbiological action. The breakdown or chemical modification of complex molecules by living organisms is often referred to as metabolism. Soils contain a variety of organisms capable of decomposing a large number of chemical substrates.[13] Aerobic and anaerobic bacteria, nitrogen-fixing bacteria, cellulose bacteria, fungi, yeast, mold, and protozoa all perform important metabolic functions in the soil.

Some of the major soil metabolism steps include:[14]

1. Decomposition of the more degradable organic components (amino acids, simple proteins)
2. Degradation of more resistant materials, such as cellulose, and formation of humic acid, a complex end product which undergoes only slow deterioration
3. Mineralization of humus (conversion of humic acids to inorganic substances)

In general, aerobic organisms will convert organic materials to CO_2, H_2O, NH_3, and energy (heat or growth of new organisms). Anaerobic organisms, on the other hand, will generate methane and H_2S in addition to CO_2, H_2O, and energy.

Among the reactions occurring enroute to the ultimate degradation products are oxidations, reductions, methylations, hydroxylations, and carbon-carbon bond scissions. Specialized soil bacteria also mediate the systems which cycle various forms of nitrogen, phosphorus, and sulfur through the environment. Because of their key role in the earth's natural recycling processes, care must be exercised to ensure that disposal or use of chemical products will not disrupt the chain of organisms so critical to these processes. Consequently, a test of toxicity to soil microorganisms might be in order if widespread land disposal or land use of the product is anticipated.

Another important mechanism for degradation of land disposed chemicals is photochemical reaction.[13] Of course, for this process to be operative, the chemical substance must be exposed to sunlight. Therefore, this mechanism is limited to those chemicals which tend to remain on land surfaces. Photochemical reactions are capable of driving molecules to significantly higher energy states than are normal thermal and chemical processes. This factor makes photoreactions particularly useful in the decomposition of resistant compounds.

Chemicals which are resistant to microbial, photochemical, and other available degradative processes are said to be persistent materials. Some persistent materials such as bricks, glass bottles, and aluminum cans are merely nuisances. Others such as DDT and polychlorinated biphenyls (PCBs) remain intact for sufficiently long periods to be transported through virtually all segments of the environment.[15,16] In a recent study,[17] for example, a number of chlorinated pesticides were found at ppb levels in human breast milk. It is clear that the appearance of such chemicals in systems other than those for which they were designed may pose a serious, long-term threat to human and environmental health.

A final area of interest in land disposal of solid wastes is the use of WWTP sludges as soil conditioners for agricultural applications. Many chemical components of wastes including trace metals are concentrated in WWTP sludges. Studies[18,19] have shown that some of these metals tend to accumulate in soils and may be translocated into the crops grown in their presence. These studies have also shown that accumulated trace metals may also inhibit plant growth. Hence, transport of waste chemicals into crops and other flora is another aspect which must be taken into account when formulating environmental assessment programs.

C. Interactions of Chemicals in the Aquatic Environment

1. The WWTP

This phase of our discussion will begin with the WWTP since a large fraction of industrial and domestic wastewater passes through these facilities. Many of the processes of concern in the WWTP are also of interest in receiving waters. Some of the important questions which must be answered in assessing the fate and effect of a chemical in a WWTP can be summarized as follows:

1. Will the chemical be removed during the waste treatment process?
2. By what processes will the chemical be removed?
3. How much of a load will the product place on the capacity of the WWTP?
4. What effect will the product have upon the physical processes taking place in the plant?
5. What effect will the chemical have on the biological processes occuring in the plant?
6. How may the chemical product be altered during the treatment process?

As mentioned in the previous discussion on WWTPs, several mechanisms for removal of influent chemicals exist. Among the more important ones are precipitation, volatilization, adsorption to solids and bacterial matter, biological degradation, and chemical reactions.

Particle size and aggregation potential are important variables in determining precipitation characteristics. Product formulations often contain additives specifically designed to stabilize suspensions or emulsify immiscible liquids. Household detergents can perform similar functions which may interfere with settling processes. Sometimes the presence of other solids in the waste mixture can, by aggregation and coprecipitation, aid in settling out chemicals which would otherwise remain suspended. The addition of coagulants such as lime, alum, polyelectrolytes and ferric chloride are also often used to assist in the settling process. Adjustment of pH can sometimes aid in particle aggregation or in converting soluble ionic species to insoluble neutral species. In the latter case, dissolved solids (metals or organics) are converted to suspended solids for removal in the sludge.

Organics with poor water solubilities and relatively high vapor pressures (e.g., solvents) will be readily purged from water during aeration processes. Aeration may sometimes lead to foaming if large quantities of surfactant materials are present. Controlled foaming is sometimes used to float materials to the surface where they can be skimmed off. Under proper conditions, inorganics such as cyanide ion and mercury salts can also be converted to volatile materials. Adsorption to solids suspended in the sewage is an important removal mechanism for some chemicals. A good example of this form of removal is that of certain dyes and optical brightening agents. The latter are added to detergents in small quantities to enhance the brightness of laundered fabrics. Many of these materials are designed so they will be readily adsorbed on cotton, a cellulose derivative, during the wash cycle. The abundance of cellulosic materials in sewage (paper, fecal material, vegetable trash, etc.) provides a large number of binding sites for the cotton optical brighteners. Studies[20] have actually shown that this is an important mechanism for removal of brighteners from sewage. In a similar manner, poorly soluble organic chemicals such as chlorinated hydrocarbons can readily bind to organic and inorganic solids. An investigation of polychlorinated biphenyls (PCBs) in treatment plant effluents[21] revealed high concentrations of PCBs in digestor and primary sludges indicating that adsorption was an important removal mechanism. The cell walls of bacteria and other microorganisms present in the WWTP also function as organic binding sites for a variety of chemical entities.

The heart of many WWTPs lies in the biological treatment segment. Here the dissolved organics are removed by physical adsorption, volatilization and biochemical degradation. The latter process is performed by bacteria which use the organics as food and convert the food to biologically stable compounds and to CO_2. The biological processes are the most fragile in the waste treatment process and must be protected accordingly. Some of the conditions needed to maintain a biological treatment system in good operating condition include:

1. Sufficient dissolved oxygen to maintain bacterial growth
2. A sufficient quantity of food for the bacteria
3. A sufficient concentration of noncarbon bacterial nutrient factors such as nitrogen and certain trace metals (iron, magnesium, etc.)
4. Relatively uniform concentration loading
5. Relatively uniform waste characteristics (i.e., types of materials to be treated)
6. Little or no toxic chemicals

To expand on these requirements, the bacteria in a biological treatment process require oxygen in order to perform their purification function. If the oxygen is depleted or in scarce supply, the aerobic bacteria will die and a septic condition in which anaerobic bacteria thrive will be created. The anaerobic bacteria take their oxygen not from dissolved molecular oxygen but from oxygen-containing chemical entities such as sulfate. This change results in the generation of foul odors (usually related to hydrogen sulfide formation) and the incomplete decomposition of organics.

Organic chemicals can be characterized by the quantity of oxygen which will be required for the complete conversion of the chemical to inorganic substances (carbon → carbon dioxide, hydrogen and oxygen → water, nitrogen → nitrate, and sulfur → sulfate). This quantity is called the theoretical oxygen demand (ThOD). The ThOD for sulfanilic acid ($C_6H_7NO_3S$), for example, can be calculated as follows:

C_6 requires 12 O $\longrightarrow$ 6 CO_2

H_7 requires 3.5 O $\longrightarrow$ 3.5 H_2O

N requires 3 O $\longrightarrow$ 1 NO_3

O supplies 3 O $\longrightarrow$ 3 O

S requires 4 O $\longrightarrow$ $SO_4^=$

Net oxygen requirement = 12 + 3.5 + 3 − 3 + 4 = 19.5; therefore

$C_6H_7NO_3S + 19.5\ O = 6\ CO_2 + 3.5\ H_2O + SO_4$

(173.8 g) (312 g)

therefore, 312 g of oxygen is needed for complete degradation of 1 mol of sulfanilic acid. Each gram of sulfanilic acid will then require 1.80 g oxygen for complete degradation (or 1 lb of sulfanilic acid will require 1.80 lb oxygen).

A limiting factor in aerobic digestion is the poor solubility of oxygen in water. The amount of oxygen available to bacteria at a given time is limited. At 35°F, (2°C), oxygen saturated water will contain 13.8 mg O/l. At 50°F (10°C), only 11.3 mg/l will saturate the water, while at 78°F (25°C), the water will be able to hold only 8.4 mg/l. Consequently, at summer water temperatures, 1 million gal of water can hold only about 70 lb of oxygen at saturation. We saw that 1.80 lb oxygen would be needed to completely degrade just 1 lb of sulfanilic acid. Theoretically then, less than 8 mg/l(ppm) of sulfanilic acid in water would be sufficient to completely utilize the dissolved oxygen at 78°F.

In actual practice, the bacteria in a waste treatment plant do not completely degrade all the organic chemical wastes. Certain materials are only slowly degraded or partially degraded while others are not degraded at all. The biochemical oxygen demand (BOD) is used as a measure of the relative ability of bacteria to utilize specific chemical compounds or mixtures as food. In the BOD test, bacteria and the test material are incubated in oxygen saturated water for a preset period of time (usually 5 days at 20°C). The bacteria utilize the test compound as a carbon source and take up a proportional amount of oxygen for respiration. At the end of the test, oxygen depletion is measured and the quantity (g) of oxygen utilized per gram of test material is calculated.

The conditions in the BOD test are more similar to those in a receiving water than to those in a biological treatment process. For instance, the concentration of bacteria is much lower in the BOD test than in the WWTP. Also, in the BOD test, the test material is the only carbon source present while in the typical treatment system many other sources of organic carbon are readily available. In spite of these and other differences, the BOD is a useful indicator of biodegradability. Table 3 presents BOD values for a number of common chemicals as well as the BOD values as a percent of the ThOD. Other methods for assessing the potential oxygen demand of chemical products and wastes will be described in the section on test methods.

A low BOD can indicate several things other than poor utilization of the material by bacteria. Figure 3 shows a set of hypothetical curves which may result from BOD measurements. The curves labeled A show BOD determinations of a material at three different concentrations. At the lowest concentration, the BOD has reached a high level within 5 days. This curve is typical of a readily biodegradable material. The curves at the higher concentrations however yield lower BOD levels. This result indicates that at higher concentrations the chemical may be inhibiting bacterial activity or may be toxic to the bacteria. Curve B results when bacteria are initially unable to utilize the test material but eventually acclimate to it. Once acclimation occurs, the rate of biodegradation usually increases substantially. Curve C may represent a material which is very poorly degradable or is toxic or inhibitory even at low concentrations.

Caution must be exercised in drawing conclusions from BOD data since apparent decreases in BOD with increasing concentration may be due to such factors as solubility limitations and pH changes (extreme pHs are toxic to bacteria).

TABLE 3

Five-day Biochemical Oxygen Demand of Some Common Chemicals[a]

Chemical compound	BOD (5 day): g oxygen/g compound	% of ThOD (BOD × 100 ÷ ThOD)
Benzoic acid	1.65	85
Cellulose	0.15	13
Formaldehyde	0.37	34
Ethyl alcohol	1.25	60
Glucose	0.71	71
Polyvinyl alcohol	0.01	1
Naphthalene	0.00	0
Detergent (LAS)	0.80	34
Detergent (ABS)	0.03	1
Perchloroethylene	0.06	15

[a]Values were obtained from Reference 22.

The usefulness of chemical coagulants and precipitants in wastewater treatment has already been mentioned. A number of chemical agents are used particularly in tertiary treatment processes to further purify or to disinfect the effluent. Again, it is important to be certain that toxicants are not generated during these chemical reactions. A good example of just such an occurrence is the case of chlorination of effluents and drinking water. Chlorination has been widely used during the past 40 years to disinfect water. Much emphasis has been placed on the disinfecting qualities of chlorination while relatively little has been placed on the chemical interactions. Only recently has it become known that the chlorinating agents will react with many trace contaminants in water to yield chlorinated hydrocarbons, many of which are suspected of being toxic or carcinogenic.[23-27]

From the foregoing discussion, it is apparent that a chemical waste can have a substantial effect on the waste treatment process. High concentrations of high BOD waste may place a severe strain on the bacterial processes resulting in less than adequate removal. Similarly low BOD materials are likely to pass through the biological treatment process relatively unchanged. Sufficient concentrations of toxic materials such as acids, heavy metals, and a number of industrial chemicals may lead to serious upsets in the biological treatment process by killing the bacteria which are carrying out the purification. The possibility that the waste treatment process may convert a chemical to a more toxic form should not be overlooked. All of these factors should be given serious consideration in assessing product environmental safety.

2. Surface Water Interactions

Since the WWTP is an attempt to copy and improve upon natural purification mechanisms, many similarities exist in the way chemicals can affect both systems. Several important differences, however, can also be cited. First, receiving waters usually contain a much more diversified biota (group of life forms) than a WWTP. These range from microscopic organisms and plants to fish and wildlife. In fact, the diversity of life forms is often used as an indicator of surface water quality. Secondly, concentrations of life forms and of contaminants in natural waters are generally more dilute than in WWTPs. Thirdly, runoff contributes a great deal of natural organic and inorganic matter to receiving waters while much of the influent into WWTPs is comprised of man-made wastes. Finally, natural waters contain no mechanical aerators and therefore must depend upon natural processes to maintain dissolved oxygen concentrations adequate to sustain life.

A natural water body is a balanced ecosystem so long as it is not impinged upon by abnormal external inputs. Figure 4 diagrams a simplified version of the aquatic life cycle. The

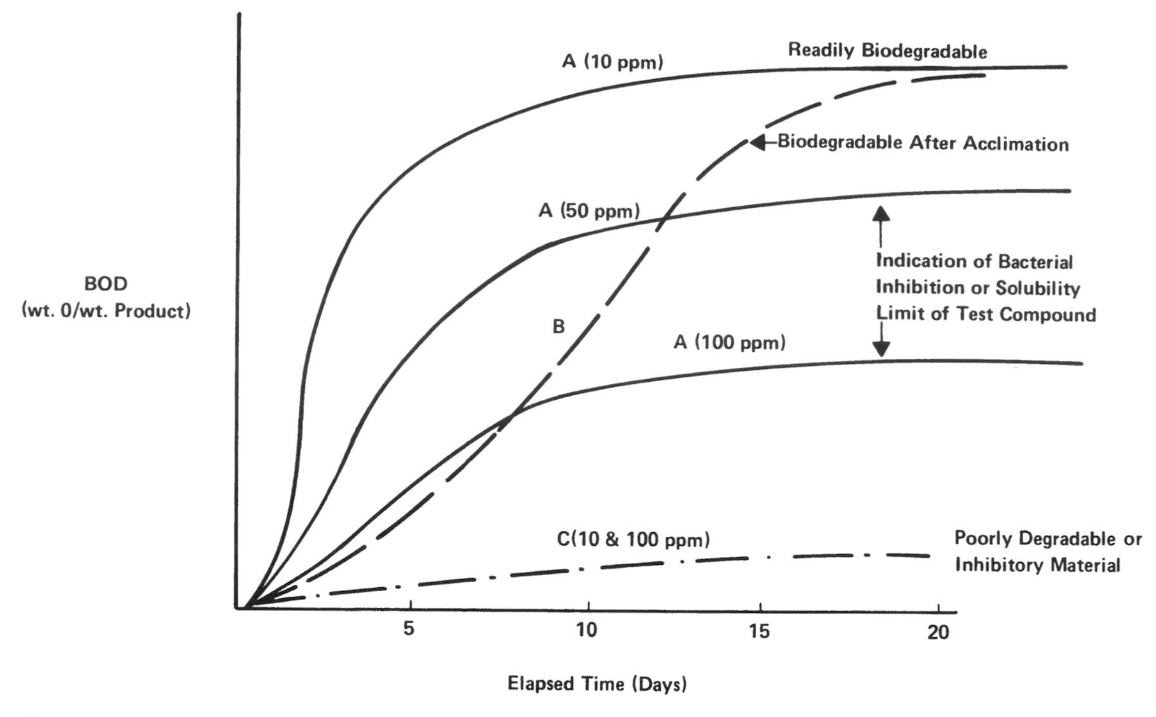

FIGURE 3. Schematic representation of BOD curves demonstrating biodegradation behavior of various types of chemicals.

driving forces behind the life cycle are derived from inputs of natural organic and inorganic nutrients, and sun energy.[28] Aquatic plants (mainly algae) use the nutrients and energy for growth and generate food and oxygen for aquatic animals. The latter are in turn consumed by larger aquatic or land animals. Microorganisms complete the cycle by decomposing accumulated wastes and dead plant and animal life thereby regenerating stored nutrients for reuse. Wildlife, which feeds on the aquatic life forms, removes some of the biomass partially balancing the influx of nutrients from external sources.

Disruptions of aquatic ecosystems by the introduction of man-made wastes can be traced to two major causes. The first is the addition of excessive loads of oxygen-demanding wastes and nutrient materials. Bacterial decomposition of the oxygen-demanding wastes depletes the dissolved oxygen supply and inhibits aquatic life. Due to this so-called oxygen sag, regions downstream from major effluent outfalls are often found to contain much reduced populations of aquatic biota.

Phosphate pollution offers an excellent illustration of how excess nutrients may upset the biological equilibria in an aquatic ecosystem. High concentrations of phosphorus originating from phosphate detergents have been implicated as a prime cause of abnormal algae blooms in surface waters. Bacterial degradation of dying portions of these blooms saps oxygen from the water and results in anaerobic conditions (eutrophication), stagnancy and death to members of the water community. In the case of phosphates, no substitute of equivalent performance and equal general safety has been found.[29] However, the problem has, for the most part, been mitigated by a combination of reduced phosphate levels in detergents and removal of the phosphate by specially designed waste treatment systems.

Much attention is now being focused on the important problem of toxic chemicals discharged into the aquatic environment. On the one hand, toxic chemicals may act by direct toxic effects both of an acute (immediate) and chronic (long-term) nature. In addition, due to the cyclic nature of the aquatic food chain, chemicals which are absorbed by lower life forms can easily be transmitted to higher life forms. Often, the concentration of chemical increases as the material is transmitted up through the food chain. This process is commonly referred to as biomagnification. Certain members of the food chain also tend to accumulate or concentrate ingested chemicals. The fatty tissue of fish, for example, is a particularly good repository for many chemical entities. Usually a 10-fold increase in tissue concentration above that present in the water is considered significant bioaccumulation.[30] Accumulated chemicals can also be transmitted to nonaquatic fish consumers such as birds and wildlife. The pesticide DDT is a prime example of the chemical which has been transmitted through the aquatic and wildlife food chain.[15]

D. Toxic Pollutants

On June 7, 1976 the U.S. District Court for the District of Columbia approved a settlement between several environmental groups and the Environmental Protection

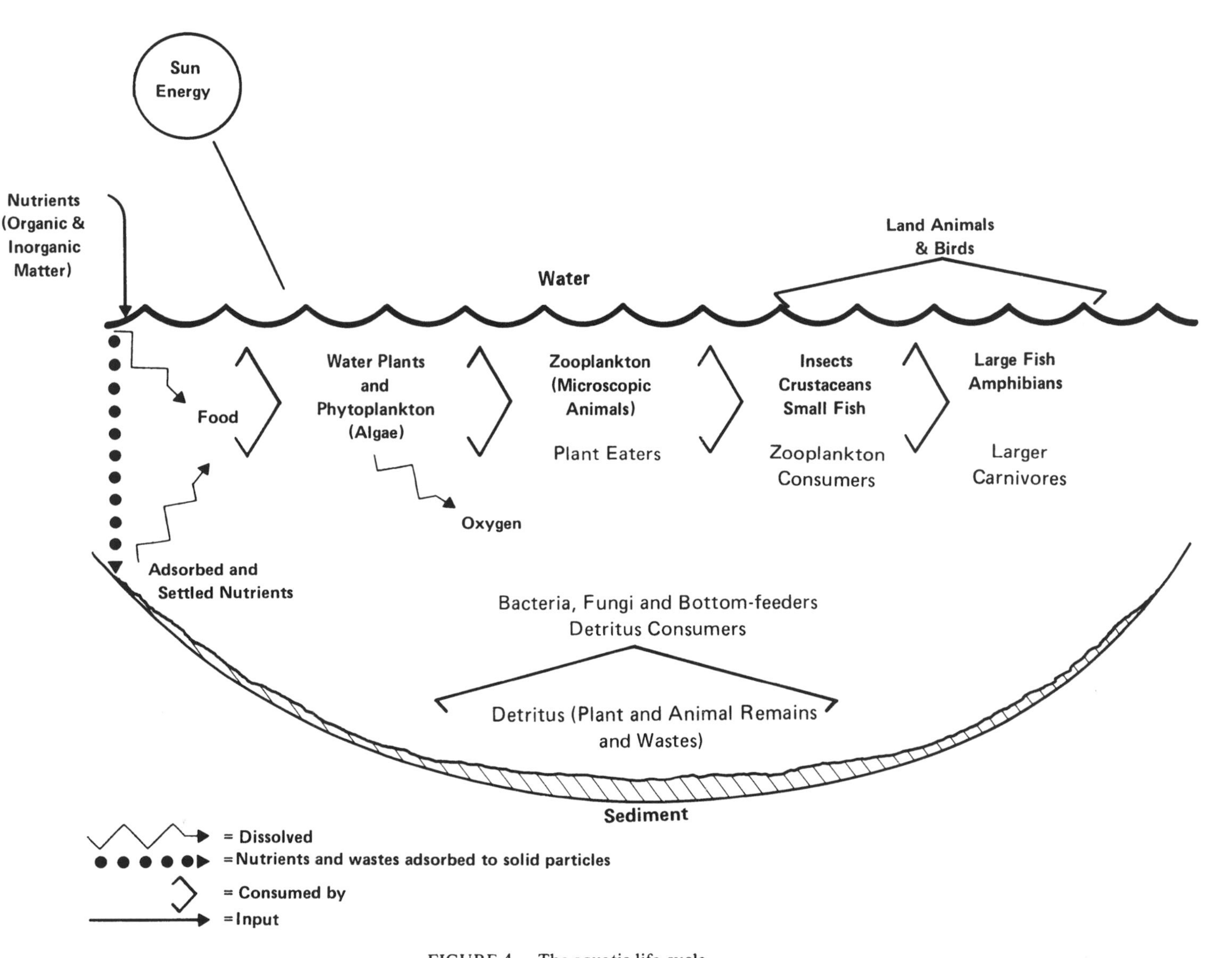

FIGURE 4. The aquatic life cycle.

Agency regarding reconsideration of effluent guidelines for various discharge sources. A key element in this settlement was the requirement to consider a list of toxic chemicals in establishing effluent limitations which dischargers must meet by 1983. The original list which included 65 chemicals and chemical classes has been expanded and at present includes 129 specific chemicals. Table 4, which lists these chemicals, will serve as an excellent guide to the types of materials presently considered as toxic pollutants.

In viewing the aquatic environment, one must bear in mind that it is a biochemically active system. Sediments in particular contain large populations of microorganisms capable of converting relatively inert chemicals into more active species. An example of such a process is the bacterial conversion of insoluble mercury metal to highly toxic organic mercury compounds (e.g., methyl mercury).[31] The organic mercury compounds are more easily taken up by various aquatic animals and are thus transported through the food chain.

Natural waters also have assimilative capacity which permits them to accept a certain level of contaminants and toxic substances without sustaining serious damage. Bacteriological activity provides a detoxification mechanism by degrading certain toxicants. Other important detoxifying mechanisms include inactivation of toxic metals by complexation, photochemical reactions, and immobilization of toxicants by adsorption to suspended particles and bottom sediments.[32]

IV. ENVIRONMENTAL EFFECTS TESTING

Having discussed, in broad terms, the multiplicity of transport mechanisms and interactions by which discharged chemicals may influence the environment, we will now attempt to review methods available for testing chemical products for potential adverse environmental effects. As mentioned in several instances above, it is wise to devise an initial testing program based upon the projected uses and discharge modes of the product. In some cases, an approach of this type may reveal, even before any testing is performed, potential problem areas which can be solved by appropriate modifications in methods of preparation, formulation or modes of usage of the product. For the most part, we will provide brief discussions of the methods along with references to more detailed discussions or examples.

A. Atmospheric Effects Testing

Developments of predictive laboratory tests for monitoring atmospheric effects of chemicals have been hampered both by the complexity and relative inaccessability of the atmosphere, as well as by the low concentrations and transient nature of the species responsible for many atmospheric reactions.

To a large extent, present work involves computer

TABLE 4

EPA Recommended List of Priority Pollutants

Acenaphthene[a]
Acrolein[a]
Acrylonitrile[a]
Benzene[a]
Benzidine[a]
Carbon tetrachloride (tetrachloromethane)[a]
Chlorinated benzenes (other than dichlorobenzenes)[a]
  Chlorobenzene
  1,2,4-trichlorobenzene
  Hexachlorobenzene
Chlorinated ethanes (including 1,2-dichloroethane, 1,1,1-trichloro-ethane and hexachloroethane)[a]
  1,2-dichloroethane
  1,1,1-trichloroethane
  Hexachloroethane
  1,1-dichloroethane
  1,1,2-trichloroethane
  1,1,2,2-tetrachloroethane
  Chloroethane
Chloroalkyl ethers (chloromethyl, chloroethyl and mixed ethers)[a]
  bis(chloromethyl) ether
  bis(2-chloroethyl) ether
  2-chloroethyl vinyl ether (mixed)
Chlorinated naphthalene[a]
  2-chloronaphthalene
Chlorinated phenols (other than those listed elsewhere; includes trichlorophenols and chlorinated cresols)[a]
  2,4,6-trichlorophenol
  Parachlorometa cresol
Chloroform (trichloromethane)[a]
2-chlorophenol[a]
Dichlorobenzenes[a]
  1,2-dichlorobenzene
  1,3-dichlorobenzene
  1,4-dichlorobenzene
Dichlorobenzidine[a]
  3,3′-dichlorobenzidine
Dichloroethylenes (1,1-dichloroethylene and 1,2-dichloroethylene)[a]
  1,1-dichloroethylene
  1,2-trans-dichloroethylene
2,4-dichlorophenol[a]
Dichloropropane and dichloropropene[a]
  1,2-dichloropropane
  1,2-dichloropropylene (1,3-dichloropropene)
2,4-dimethylphenol[a]
Dinitrotoluene[a]
  2,4-dinitrotoluene
  2,6-dinitrotoluene
1,2-diphenylhydrazine[a]
Ethylbenzene[a]
Fluoranthene[a]
Haloethers (other than those listed elsewhere)[a]
  4-chlorophenyl phenyl ether
  4-bromophenyl phenyl ether
  bis(2-chloroisopropyl) ether
  bis(2-chloroethoxy) methane

[a]Specific compounds and chemical classes as listed in the consent degree.

TABLE 4 (continued)

EPA Recommended List of Priority Pollutants

Halomethanes (other than those listed elsewhere)[a]
- Methylene chloride (dichloromethane)
- Methyl chloride (chloromethane)
- Methyl bromide (bromomethane)
- Bromoform (tribromomethane)
- Dichlorobromomethane
- Trichlorofluoromethane
- Dichlorodifluoromethane
- Chlorodibromomethane

Hexachlorobutadiene[a]
Hexachlorocyclopentadiene[a]
Isophorone[a]
Naphthalene[a]
Nitrobenzene[a]
Nitrophenols (including 2,4-dinitrophenol and dinitrocresol)[a]
- 2-nitrophenol
- 4-nitrophenol
- 2,4-dinitrophenol[a]
- 4,6-dinitro-*o*-cresol

Nitrosamines[a]
- N-nitrosodimethylamine
- N-nitrosodiphenylamine
- N-nitrosodi-*n*-propylamine

Pentachlorophenol[a]
Phenol[a]
Phthalate esters[a]
- bis(2-ethylhexyl)phthalate
- Butyl benzyl phthalate
- di-*n*-butyl phthalate
- di-*n*-octyl phthalate
- Diethyl phthalate
- Dimethyl phthalate

Polynuclear aromatic hydracarbons[a]
- Benzo(a)anthracene (1,2-benzanthracene)
- Benzo(a)pyrene (3,4-benzopyrene)
- 3,4-benzofluoranthene
- Benzo(k)fluoranthane (11,12-benzofluoranthene)
- Chrysene
- Acenaphthylene
- Anthracene
- Benzo(ghi)perylene (1,12-benzoperylene)
- Fluorene
- Phenanthrene
- Dibenzo(a,h)anthracene (1,2,5,6-dibenzanthracene)
- Indeno(1,2,3-cd)pyrene (2,3-*o*-phenylenepyrene)
- Pyrene

Tetrachloroethylene[a]
Toluene[a]
Trichoroethylene[a]
Vinyl chloride (chloroethylene)[a]
Pesticides and metabolites
- Aldrin[a]
- Dieldrin[a]
- Chlordane (technical mixture and metabolites)[a]

DDT and metabolites[a]
- 4,4'-DDT
- 4,4'-DDE (p,p'-DDX)
- 4,4'-DDD (p,p'-TDE)

Endosulfan and metabolites[a]
- a-endosulfan-alpha
- b-endosulfan-beta
  - Endosulfan sulfate

TABLE 4 (continued)

EPA Recommended List of Priority Pollutants

Endrin and metabolites[a]
- Endrin
- Endrin aldehyde

Heptachlor and metabolites[a]
- Heptachlor
- Heptachlor epoxide

Hexachlorocyclohexane (all isomers)[a]
- a-BHC-alpha
- b-BHC-beta
- γ-BHC (lindane)-gamma
- g-BHC-delta

Polychlorinated biphenyls (PCBs)[a]
- PCB-1242 (Arochlor 1242)
- PCB-1254 (Arochlor 1254)
- PCB-1221 (Arochlor 1221)
- PCB-1232 (Arochlor 1232)
- PCB-1248 (Arochlor 1248)
- PCB-1260 (Arochlor 1260)
- PCB-1016 (Arochlor 1016)

Toxaphene[a]
Antimony (Total)[a]
Arsenic (Total)[a]
Asbestos (fibrous)[a]
Beryllium (total)[a]
Cadmium (total)[a]
Chromium (total)[a]
Copper (total)[a]
Cyanide (total)[a]
Lead (total)[a]
Mercury (total)[a]
Nickel (total)[a]
Selenium (total)[a]
Silver (total)[a]
Thallium (total)[a]
Zinc (total)[a]
2,3,7,8-tetrachlorodibenzo-*p*-dioxin (TCDD)[a]

Note: Recommended list as of April 29, 1977. Further additions or deletions may occur.

From U.S. Environmental Protection Agency, Selected Programs Contracting Section, Washington, D.C., 1977.

dispersion modeling based upon anticipated processes, emission rates, and plant locations. These models allow estimates of ground level concentrations around major sources to be developed.[2,33] A fair amount of study on phenomena such as photochemical smog has been performed in laboratory smog chambers. Recently developed analytical techniques such as tunable lasers[34,35] and Fourier transform infrared spectroscopy[35,36] are permitting elucidation of reaction mechanisms in the atmosphere involving low concentrations of reactive species.

An assessment of emissions during pilot plant runs or in laboratory scale process simulations may be useful in determining the type and approximate quantity of various chemical entities being emitted. Such studies can serve not only to determine substances which may be of environmental concern but also to anticipate which air pollution control devices may be required. The investigator should be aware that in sampling emission point sources such as stacks and ducts the form of the emission must be known to obtain valid information about emission rates. Vapor components of stack emissions are assumed to move at equal rates when diverted into a sampler. The representativeness of a vapor sample will therefore be essentially independent of the sampling rate. Aerosols and particulates, on the other hand, have varying size distributions and therefore contain fractions with differing aerodynamic properties. In this case, if sampling is performed at flow rates different from that present within the source, a bias is created due to the differing inertias of large and small particles.[37] Consequently, sampling of particulate and aerosol emissions should be performed isokinetically, i.e., the flow rate into the sampling device should match the flow rate in the stack or duct. In addition, since flow velocity and concentrations across the stack usually vary, sampling should be performed at a number of points across the source. These points are located so that each point represents an equal sampling area. In a circular stack, the sampling sections are generated by mapping a set of concentric circles which come closer together as they move farther out from the center. The circles can thus be spaced in such a way that the area between each pair of neighboring circles is the same.

Trapping and analysis of the components of interest can be achieved by a number of methods. General sampling methods for various components[38] include:

1. Filtration
2. Impaction (stopping particles or aerosols by physical collision with a solid or liquid barrier)
3. Bubbling through solvents or reactive liquids
4. Cold (cryogenic) trapping
5. Solvent trapping
6. Adsorption onto solid polymers or activated carbon

The latter method is now widely used to trap and concentrate trace concentrations of organic compounds present in ambient air.[39–41] The trapped organics are heat desorbed for analysis usually by gas chromatography possibly coupled with mass spectroscopy.

Table 5 summarizes EPA recommended methods of analysis for major ambient air pollutants. A number of other analytical procedures may be used if equivalency to the recommended method can be demonstrated.[42] Preferably, methods which do not require that discrete samples be taken should be used when possible. Several instruments which can continuously draw test air directly into the analyzer are now available and permit continuous monitoring without the need to obtain and store samples.[43] Stevens and Herget[44] have edited an excellent book which reviews modern analytical approaches to specific air pollution problems. See also References 1 and 37.

B. Testing the Effects of Land Disposed Chemicals

1. Transport These can be performed either on actual field sites or in glass tubes in the laboratory. In both cases, a known quantity of the chemical is applied as a solution, slurry, or solid to the top of a column of soil. It is advisable to use various soil types for testing unless projected use suggests that only a single soil type will be important. Measured quantities of water are subsequently added at various times after application. In field situations,

TABLE 5

Recommended Analytical Methods for Ambient Air Pollutants[6,42]

Pollutant	Sampling method	Analytical method	Ref.
SO_2	Trap by bubbling through a solution of potassium tetrachloromercurate	Pararosaniline, colorimetric	6
Particulate matter	Trap by filtering air through a high volume sampler	Gravimetric	6
Carbon monoxide	Draw air continuously through an infrared cell	Nondispersive, infrared	6
Photochemical oxidant	Draw air sample and ethylene simultaneously through a sample cell	Chemiluminescence reaction	6
Hydrocarbons corrected for methane	Collect air sample in a suitable container	Inject aliquot into gas chromatograph equipped with a flame ionization detector	6
NO_2[a]	Draw air over hot catalyst to convert NO_2 to NO	Chemiluminescence reaction of NO with ozone	33

[a]This method is being recommended to replace the Jacobs-Hochheiser diazotization-colorimetric method which has been withdrawn.

rainfall is monitored as a measure of applied water. Core samples from the field are taken at selected time intervals after application. Crosswise slices of each core are then analyzed by a method specific for the applied chemical. The analytical results show the quantity leached and the depth of leaching as a function of time and rainfall. In the laboratory, the water which penetrates the soil column can be conveniently collected. In this way, the quantity of chemical actually carried through the soil by the water can be measured. In addition, the soil plug can be removed from the column and analyzed in the same manner as the field samples. See Reference 45 for a good example of a laboratory study of leaching behavior.

Plant uptake – Plant uptake and effects on growth may be studied by growing the plant in soils containing various concentrations of the test material. The plants can be harvested at various times and the different parts of the plant analyzed for the compound of interest and its metabolites. Higher concentrations will usually be found in the plant during the early stages of growth when the ratio of plant tissue to nutrient uptake from the soil is at a maximum. Studies of organic chemicals are greatly facilitated by the use of test compounds labeled with a radioactive carbon (^{14}C) tracer. Analyses of the labeled test compound and its metabolites at low concentrations can be accomplished with a minimum of sample preparation. In addition, the location of the test chemical in the plant can be mapped by placing the plant parts on X-ray film. The film will develop only in those areas which contain the radioactive tracer. The result is a negative outlining the transport pathways of the labeled chemical. References 18, 19, 46, and 47 are examples of studies involving translocation of chemicals into plants.

Runoff – Prediction of runoff requires a complex model encompassing many sources and pathways. Reference 48 affords a conceptual model which can be applied to agricultural sources.

2. Reactivity

Microbiological activity – Both biochemical reactivity and toxic effects on soil microorganisms should be studied. Biochemical reactivity is often tested with the aid of radioactively labeled compounds. Using essentially closed systems which provide for adequate oxygen and moisture input, products resulting from microbiological assimilation of the test material such as volatile organic fragments and carbon dioxide can be readily trapped and analyzed. Experiments may be arranged so that sample systems can be withdrawn at preselected times and the soil analyzed for starting compound and nonvolatile metabolites. Toxicity to soil microorganisms can be simultaneously tested using the same experimental arrangements. Since aerobic decomposition generates carbon dioxide, a reduction in carbon dioxide generation in treated soil versus untreated soil would signal a reduction in aerobic biochemical activity. At the same time, soil microbiological plate counts in specific media can yield information about which strains are being affected. Reference 13 contains a number of excellent discussions of results obtained using this type of methodology.

Photochemical reactions – Photochemical reactivity will be dependent upon the wavelength of incident radiation. As discussed above, the UV irradiation at and near the earth's surface is comprised only of wavelengths greater than 290 nm. Laboratory experiments must therefore use irradiation conditions which preferably simulate the solar spectrum at the earth's surface but which in any case produce only UV light above 290 nm.[13,49] Field experiments may be carried out using sunlight. In either case, some measure of the incident energy (photons) should be made. Photochemical reactivity will also be influenced by the physical form of the substance. Chemicals in solution may react somewhat differently than chemicals adsorbed to soil particles. Also other components can absorb light energy and transfer the energy to an otherwise unreactive molecule. Thus merely irradiating a chemical applied to a glass plate, as is sometimes done, may yield photoreactions which are somewhat different from those which occur on a moist soil particle where a solvent and possibly photosensitizers may be present. See References 13 and 49 for further discussion.

C. Testing Effects of Chemicals on the Aquatic Environment

A large number of tests for assessing effects of chemicals on the aquatic environment have been developed. EPA has published a listing of accepted procedures for measuring many water quality parameters.[50] Many of the recommended procedures are described in the APHA publication "Standard Methods for the Examination of Water and Wastes."[51]

1. Oxygen Demand Tests

As explained in the text,[22,52] bacterial utilization of products having a high oxygen demand can have negative effects on receiving waters and may be important in determining anticipated loads on WWTPs. Conversely, inability of bacteria to utilize a chemical compound may indicate that the chemical will be persistent to the environment. Although oxygen demand tests are primarily designed for testing wastewater and receiving waters they can also be useful in testing dilute solutions of raw products. In addition to the BOD test, several chemical and instrumental methods have been developed for the purpose of obtaining rapid indications of the expected oxygen demand of waste waters. It should be kept in mind that except in rare instances the BOD will be less than the ThOD. Consequently experimental correlations must be made between the BOD and the nonbiological tests to use the latter for predictive purposes.

Biochemical oxygen demand (BOD)[51] – This test is performed by mixing a sample with oxygen saturated water containing certain trace metals and bacteria from a WWTP, river, or other appropriate source. The dissolved oxygen (DO) is measured with a DO membrane probe both initially and after incubation for 5 days at 20°C (longer incubation periods may provide additional information about slowly degradable compounds). The oxygen depletion, corrected for blank respiration by the bacteria, is calculated and the resultant value ratioed to the concentration of product. In this way the BOD is determined on weight of oxygen/weight of product basis. Possible methods of shortening the time required for the BOD test have been reviewed.[53]

The standard BOD test affords a picture of the total oxygen consumption over 5 days. As shown in Figure 3, it is sometimes of interest to obtain a better picture of the rate of bacterial activity during the test. To accomplish this, tests which measure the evolution of carbon dioxide have been developed.[52,54,55] These tests are normally performed in a closed system fitted with a manometer. In the Warburg apparatus, the bacteria and test material are incubated in a special flask fitted with a carbon dioxide trap. During the test, aerobic bacterial respiration consumes dissolved oxygen and evolves carbon dioxide. The carbon dioxide is adsorbed by potassium hydroxide contained in the trap. Since oxygen is being utilized from the space above the test solution while the evolved carbon dioxide is being adsorbed, there is a net pressure reduction which can be measured with the manometer. The manometric readings reflect bacterial utilization of the test compound and produce curves similar to those shown in Figure 3.

Short-term oxygen demand test[56] – In this test, the rate of dissolved oxygen utilization by an activated sludge sample is compared before and after introduction of a test chemical. An increased rate of DO consumption after the chemical is added is an indication that the test material is readily utilized by the sludge microorganisms. A decreased rate indicates that the test compound is toxic or inhibitory. This test is particularly useful for assessing the effect of "slug loads" of the product on activated sludge.

The chemical oxygen demand test (COD)[56] – This test utilizes digestion by a strong chemical oxidant, potassium dichromate in sulfuric acid, to determine ultimate oxygen demand. After the digestion, unused dichromate is determined by titration with ferrous ammonium sulfate. The quantity of dichromate consumed is used to calculate the COD. Some organics such as aromatic and straight chain aliphatic hydrocarbons are not appreciably oxidized by this procedure. If complete oxidation of the sample does take place, then the COD will equal the ThOD.

Total organic carbon analysis (TOC)[51] – This method gives a measure of the total carbon ultimately available for utilization by bacteria. The carbon in the sample is combusted to CO_2 and the CO_2 measured by infrared techniques or by conversion to methane and measurement by flame ionization detection.

Total oxygen demand (TOD)[22] – This method yields the total amount of oxygen utilized by the material during combustion. A nitrogen gas stream containing a known fraction of oxygen is passed over the heated sample and the decrease in oxygen in the exit gases measured. The TOD should equal the ThOD.

2. Biodegradability Tests

Although the BOD test[57] gives some measure of biodegradability, particularly if the Warburg respirometer is used, several other tests have been developed which better simulate conditions in a WWTP or a receiving water.

The semi-continuous activated sludge test[58] – This test was adopted by the Soap and Detergent Association as the method to test biodegradability of synthetic detergents in activated sludge plants. Activated sludge, a nutrient solution and test material are aerated for 24-hr periods. A sample is removed daily and replaced with fresh nutrient and test material. The daily change in concentration of test material in the supernatant liquor after the sludge has been allowed to settle, is a measure of removal of the compound by the sludge. Further information about the mode of removal (degradation, adsorption, volatilization) and the effects of the test chemical on the activated sludge can be determined by examining the sludge solids, by trapping the volatiles and by performing bacterial plate counts.

The continuous activated sludge test[55,59,60] – This test is performed in an apparatus which contains, in addition to an aeration chamber like the semicontinuous apparatus, a settling and effluent overflow chamber similar in design to those present in many WWTPs. Test solution mixed with nutrient is continuously pumped into an aeration chamber charged with activated sludge. Supernatant from the settling chamber continuously overflows and is collected. Influent and effluent concentrations are compared after an acclimation period and percent removal by the activated sludge is calculated.

The river die-away test[55,61,62] – This test is used to assess biodegradability in actual samples of river water. The test material is added to the water at various concentrations in dark bottles to minimize algae growth. Samples removed at various time intervals are tested for the presence of starting material.

Other biodegradability tests include the shake flask test[57,58] in which test material is shaken with bacterial seed and monitored and the CO_2 generation test[55] in which CO_2 generated by bacterial utilization of a test material is trapped and monitored as a measure of rate of biodegradation.

3. Aquatic Bioassays

One of the quickest and best indicators of a waste material's effect on the aquatic community is the bioassay. In this test, either a single specie or group of interrelated species are exposed to varying test concentrations and toxicological, accumulation and/or biomagnification effects monitored. The tests for fish will be described as an illustration of the general approach.

Fish toxicity bioassays[51,52,63] – These may be designed to test either acute or chronic toxicity. In the acute test, fish are exposed usually for 96 hr to at least five concentrations of chemical equally spaced on a logarithmic scale (e.g., 10, 18, 32, 56, or 100 mg/ml). At the end of the test, total mortalities at the various concentrations are tallied, and by statistical treatment of the data, a concentration lethal to 50% of the test animals (variously designated LC_{50}, TLM, or TL_{50}) is calculated. The test can be performed in a static system (no water changes during 96 hr) or in a flow-through system.[64] In the latter case, water streams containing each concentration are continuously fed into the appropriate test tanks. Although more expensive and troublesome, the flow-through system is preferred since it appears to better simulate actual conditions in a surface water. Treated wastewater is often diluted with receiving water in a flow-through system and passed through fish tanks as a continuous indicator of toxicity. When the fish show stress, the waste flow can be diverted to a holding basin until the cause is discovered and corrected. The specie of fish chosen for the test should be common to the region where the discharge is to be made. For chemicals expected to be used in widely distributed areas, it may be wise to perform the test on a warm-water specie (e.g., fathead minnow, *Pimephales promelas*, or bluegill sunfish, *Lepomis macrochirus*), a cold water specie (e.g., rainbow trout, *Salmo gairdneri*), and a bottom feeder (e.g., channel catfish, *Ictalarus punctatus*). The chronic test is performed under conditions similar to the flow-through test but for longer periods of time and usually at concentrations below the acute toxic levels.

Accumulation studies[65] – These are performed in a similar manner to the flow-through toxicity tests but concentrations of the test material are usually close to those expected in surface waters. Fish are removed and assayed for the test material at intervals or until no significant further accumulation is noted. The fish are then transferred to clean water and tissue concentrations monitored to determine if accumulated chemical will be released from the tissue.

Food chain biomagnification – This is determined by designing model ecosystems containing species which represent various segments of the food chain (see Figure 4). Physiological effects on the various species as well as biomagnification of the chemical as it is transmitted up the food chain is followed. Reference 66 is a good example of such a study.

Life cycle bioassays – These are designed to study the effects of chemicals on test species through different stages of life. These studies are often carried through several generations to determine reproductive effects. Detailed discussions of bioassay protocols for a number of aquatic life forms are described in Reference 51, section 800.

4. Analysis of Toxic Chemicals in the Aquatic Environment

Toxic chemicals can be divided for analytical purposes into three categories: metals, volatile organics, and nonvolatile organics.

Metals – The standard technique for analyzing trace metals in water, wastewater, and other aquatic samples is atomic absorption (AA) spectroscopy.[51] The sample is usually treated with strong acids to digest organic components and to convert certain metals to forms amenable to atomization in a flame. In the AA technique, a lamp containing a heated source fashioned from the metal of interest directs a beam at wavelengths characteristic of that metal through a flame. The digested sample is aspirated into the flame to convert the element of interest to atoms. These atoms absorb energy from the beam in proportion to the quantity aspirated into the flame. The net reduction in the beam intensity is detected by a photomultiplier tube to give a quantitative estimate of the amount of metal present. Atomic absorption is both specific and sensitive. Modifications permitting the metals to be atomized without a flame (flameless atomic absorption) have reduced detection limits to μg/l (part per billion) levels and below.

Volatile organic compounds – Compounds with boiling points below 150°C which do not have high water solubilities can be conveniently concentrated and separated from many interferences in the water sample by purging the volatiles from the solution with an inert gas. The organics in the exit gases are collected on a porous polymer trapping tube.[67] The organics are then heat desorbed from the tube directly onto a gas chromatography (GC) column. The latter separates the components by selective partitioning between a flowing gas phase and a liquid or solid phase contained in the column. Detectors specific for various elements or classes of compounds detect the components as they elute from the gas chromatograph. These detectors include:

1. Flame ionization – hydrocarbons
2. Electron capture – halogenated compounds, nitro compounds
3. Electrolytic conductivity – halogenated, nitrogenous or sulfur-containing compounds
4. Thermal electron – N-nitroso compounds
5. Flame photometric – phosphorus or sulfur containing compounds
6. Alkali-flame – phosphorus- and nitrogen-containing compounds

Probably the most powerful detector to be coupled to a gas chromatograph is the mass-spectrometer (MS).[68] As the components elute from the GC into the MS, they are bombarded with electrons and break up into characteristic fragments. The fragmentation patterns permit unequivocal identification of most organic compounds. In addition, by directing the MS to "look" only at characteristic fragments, the gas chromatogram can be greatly simplified since most interfering components are not detected.

Nonvolatile organic compounds – These compounds require concentration and isolation from interferences in order to be analyzed. A widely used technique involves extracting the organics from the water with an immiscible solvent such as methylene chloride. This is followed by further partitioning into other solvents and/or purification of the extract with solid adsorbents.[50,69,70] Adsorption of trace organics from water onto macroreticular resins[71] appears to be useful in concentrating and purifying organics. If the compounds are sufficiently volatile and heat stable, gas chromatography is an analytical method of choice for the purified higher boiling organics. For many compounds not amenable to GC, high pressure liquid chromatography (HPLC) offers an excellent quantitative analysis alternative.[72] In HPLC, the components are separated by partitioning between a moving liquid and a stationary solid phase. Recent advances in HPLC have made it an extremely versatile separation and analytical tool. Detectors used in HPLC include UV absorption, fluorescence and more recently mass spectrometers.[73]

SUMMARY

This chapter has attempted to provide an overview of the factors which must be considered when evaluating the environmental risk of a product. Although the treatment was by no means exhaustive, it is hoped that sufficient information has been provided to allow those concerned to develop a realistic and intelligent approach to assessing environmental effects.

REFERENCES

1. **Ledbetter, J. O.,** *Air Pollution, Part A: Analysis,* Marcel Dekker, New York, 1972.
2. National Academy of Sciences, *Principles for Evaluating Chemicals in the Environment,* National Academy of Sciences, Washington, D.C., 1975, Appendix F.
3. **Denny, D. A.,** Estimating Future Air Pollution Control Requirements in the Textile Industry, paper presented at the Textile Technology/Ecology Interface – 1977, American Association of Textile Chemists and Colorists, Atlanta, Georgia, March 29 to 30, 1977.
4. Threshold Limit Values for Chemical Substances in Workroom Air, American Conference of Governmental Industrial Hygienists, Cincinnati, Ohio, 1974.
5. **Ledbetter, J. O.,** *Air Pollution, Part A: Analysis,* Marcel Dekker, New York, 1972, chapter 2.
6. *Fed. Regist.,* 6(84), April 30, 1971.
7. **Manahan, S. E.,** *Environmental Chemistry,* Willard Grant Press, Boston, Mass., 1972, chapter 13.
8. **Fox, D. L., Kamens, R., and Jeffries, H. E.,** Photochemical smog systems: effect of dilution on ozone formation, *Science,* 188, 1113, 1975.
9. **Pitts, J. N., Jr.,** Keys to photochemical smog control, *Environ. Sci. Technol.,* 11, 456, 1977.
10. **Cicerone, R. J., Stolarski, R. S., and Walters, S.,** Stratospheric ozone destruction by chlorofluoromethanes, *Science,* 185, 1165, 1974.
11. *Chem. Eng. News,* 21, April 21, 1975.
12. *Chem. Eng. News,* 7, January 19, 1976.
13. National Academy of Sciences, *Degradation of Synthetic Organic Molecules in the Biosphere,* proceedings of a conference, San Francisco, Calif., June 12 and 13, 1971, National Academy of Sciences, Washington, D.C., 1972.
14. **Odum, E. P.,** *Fundamentals of Ecology,* 3rd ed., W. B. Saunders, Philadelphia, Pa., 1971, chapter 14.
15. **Woodwell, G. M., Craig, P. P., and Horton, A. J.,** DDT in the biosphere: where does it go? *Science,* 174, 1101, 1971.
16. **Hammond, A. L.,** Chemical pollution: polychlorinated biphenyls, *Science,* 175, 155, 1972.
17. *Chem. Eng. News,* 7, May 9, 1977.
18. **Bradford, G. R., Page, A. L., Lund, L. J., and Olmstead, W.,** Trace element concentrations of sewage treatment plant effluents and sludges – their interactions with soils and uptake by plants, *J. Environ. Qual.,* 4, 123, 1975.
19. **Kirkham, M. B.,** Trace elements in corn grown on long-term sludge disposal site, *Environ. Sci. Technol.,* 7, 765, 1975.
20. **Ganz, C., Liebert, C., Schulze, J., and Stensby, P. S.,** Removal of detergent fluorescent whitening agents from wastewater, *J. Water Pollut. Control Fed.,* 47, 2834, 1975.
21. **Dube, D. J., Veith, G. D., and Lee, G. F.,** Polychlorinated biphenyls in treatment plant effluents, *J. Water Pollut. Control Fed.,* 46, 966, 1974.
22. **Masselli, J. W., Masselli, N. W., and Burford, M. G.,** BOD?, COD?, TOD?, TOC?, *Tex. Ind.* (Atlanta), 53, September 1972.
23. **Glaze, W. H., Henderson, J. E., IV, Bell, J. E., and Wheeler, V. A.,** Analysis of organic materials in waste-water effluents after chlorination, *J. Chromatogr. Sci.,* 11, 580, 1973.
24. **Dowty, B. J., Carlisle, D. R., and Laseter, J. L.,** New Orleans drinking water sources tested by gas chromatography-mass spectrometry, *Environ. Sci. Technol.,* 9, 762, 1975.
25. **Marx, J. L.,** Drinking water: another source of carcinogens? *Science,* 186, 809, 1974.
26. Water contaminated throughout U.S., *Chem. Eng. News,* 18, April 28, 1975.
27. **Bellar, T. A. and Lichtenburg, J. J.,** The occurrence of organohalides in drinking water, *J. Am. Water Works Assoc.,* 66, 703, 1974.
28. **Odum, E. P.,** *Fundamentals of Ecology,* 3rd ed., W. B. Saunders, Philadelphia, Pa., 1971, chapter 3.
29. **Mitchell, D.,** Eutrophication of lake water microcosms: phosphate versus non-phosphate detergents, *Science,* 174, 827, 1971.
30. EPA Guidelines for Registering Pesticides, draft of May 1, 1972, v-38.
31. **Wood, J. M.,** Biological cycles for toxic elements in the environment, *Science,* 183, 1049, 1974.
32. **Hom, W., Risebrough, R. W., Soutar, A., and Young, D. R.,** Deposition of DDE and polychlorinated biphenyls in dated sediments of the Santa Barbara basin, *Science,* 184, 1197, 1974.

33. **Ledbetter, J. O.,** *Air Pollution, Part A: Analysis,* Marcel Dekker, New York, 1972, chapter 3.
34. **Hinkley, D. E.,** Air monitoring with tunable lasers, *Environ. Sci. Technol.,* 11, 564, 1977.
35. **Pitts, J. N., Jr., Finlayson-Pitts, B. J., and Winer, A. M.,** Optical systems unravel smog chemistry, *Environ. Sci. Technol.,* 11, 568, 1977.
36. **Hanst, P. L., Le Fohr, A. S., and Gay, B. W., Jr.,** Detection of atmospheric trace gases at parts-per-billion levels by infrared spectroscopy, *Appl. Spectrosc.,* 27, 188, 1973.
37. Methods of Air Sampling and Analysis, Intersociety Committee of the American Public Health Association, Washington, D.C., 1972, 31.
38. **Ledbetter, J. O.,** *Air Pollution, Part A: Analysis,* Marcel Dekker, New York, 1972, chapter 6.
39. **Bertsch, W., Chang, R. C., and Zlatkis, A.,** The determination of organic volatiles in air pollution studies; characterization of profiles, *J. Chromatogr. Sci.,* 12, 175, 1974.
40. **Russell, J. W.,** Analysis of air pollutants using sampling tubes and gas chromatography, *Environ. Sci. Technol.,* 9, 1175, 1975.
41. **Braman, R. S. and Johnson, D. L.,** Selective absorption tubes and emission technique for determination of ambient forms of mercury in air, *Environ. Sci. Technol.,* 8, 996, 1974.
42. **Hoffman, A. J., Curran, T. C., McMullen, T. B., Cox, W. M., and Hunt, W. F., Jr.,** EPA's role in ambient air quality monitoring, *Science,* 190, 124, 1975.
43. **Homolya, J. B.,** Coupling continuous gas monitors to emission sources, *Chem. Tech.,* 4, 426, 1974.
44. **Stevens, R. K. and Herget, W. F.,** *Analytical Methods Applied to Air Pollution Measurements,* Ann Arbor Science Publishers, Ann Arbor, Michigan, 1974.
45. **Lichtenstein, P. E., Schulz, K. R., and Fuhremann, T. W.,** Movement and fate of dyfonate in soils under leaching and non-leaching conditions, *J. Agric. Food Chem.,* 20, 831, 1972.
46. **Kilgore, W. W., Marei, N., and Winterlin, W.,** Parathion in plant tissues: new considerations, in *Degradation of Synthetic Organic Molecules in the Biosphere,* Proceedings of a Conference, San Francisco, Calif., June 12 to 13, 1971, National Academy of Sciences, Washington, D.C., 1972, 291.
47. **Walsh, G. E., Hollister, T. A., and Forester, J.,** Translocation of four organochlorine compounds by red mangrove (*Rhizophora mangle L.*) seedlings, *Bull. Environ. Contam. Toxicol.,* 12, 129, 1974.
48. **Bailey, G. W., Swank, P. R., Jr., and Nicholson, H. P.,** Predicting pesticide runoff from agricultural land: a conceptual model, *J. Environ. Qual.,* 3, 95, 1974.
49. **Crosby, D. G.,** Experimental approaches to pesticide decomposition, *Residue Rev.,* 25, 1, 1969.
50. *Fed. Regist.,* 41(232), 52780, December 1, 1976.
51. **Greenberg, A.,** *Standard Methods for the Examination of Water and Wastewater,* 14th ed., American Public Health Association, Washington, D.C., 1976.
52. **Ryckman, D. W., Prabhakara Rao, A. V. S., Buzzell, J. C., Jr.,** *Behavior of Organic Chemicals in the Aquatic Environment – A Literature Critique,* Manufacturing Chemists Association, Washington, D.C., 1966.
53. **Le Blanc, P. J.,** Review of rapid BOD test methods, *J. Water Pollut. Control Fed.,* 46, 2202, 1974.
54. **Caldwell, D. H. and Langlier, W. F.,** Manometric measurement of the biochemical oxygen demand of sewage, *Sewage Works, J.,* 20, 202, 1948.
55. **Thompson, J. E. and Duthie, J. R.,** The biodegradability and treatability of NTA, *J. Water Pollut. Control Fed.,* 40, 306, 1968.
56. **Vernimmen, A. P., Henken, E. R., and Lamb, J. C., III,** A short-term biochemical demand test, *J. Water Pollut. Control Fed.,* 39, 1006, 1967.
57. **Buzzell, J. C., Jr., Thompson, C. H., and Ryckman, D. W.,** *Behavior of Organic Chemicals in the Aquatic Environment. III. Behavior in Aerobic Treatment Systems (Activated Sludge),* Manufacturing Chemists Association, Washington, D.C., July 1969.
58. Test procedures and standards – ABS and LAS biodegradability, *Soap Deterg. Assoc. Sci. Tech. Rep.,* No. 3, January 1966.
59. **Eldib, I. A.,** Biodegradability of amphoteric detergents, *Soap Chem. Spec.,* May 1965.
60. Syndet testing could follow German lead, *Chem. Eng. News,* 65, February 18, 1963.
61. **Warren, C. B. and Malec, J. E.,** Biodegradation of nitilotriacetic acid and amino acids in river water, *Science,* 176, 277, 1972.
62. **Blankenship, F. A. and Piccolini, V. M.,** Biodegradation of nonionics, *Soap Chem. Spec.,* 75, December 1963.
63. **Kemp, H. T., Little, R. L., Holoman, V. L., and Darby, R. L.,** *Environmental Effects of Chemicals on Aquatic Life,* Water Quality Criteria Data Book, Vol. 5, Environmental Protection Agency, Washington, D.C., 1973.
64. **Brungs, W. A. and Mount, D. I.,** A device for continuous treatment of fish in holding chambers, *Trans. Am. Fish. Soc.,* 96, 55, 1967.
65. **Ganz, C. R., Schulze, J., Stensby, P. S., Lyman, F. L., and Macek, K.,** Accumulation and elimination studies of four detergent fluorescent whitening agents in bluegill (*Lepomis macrochirus*), *Environ. Sci. Technol.,* 9, 738, 1975.
66. **Sanborn, J. R. and Ching-Chieh, Y.,** The fate of dieldrin in a model ecosystem, *Bull. Environ. Contam. Toxicol.,* 10, 340, 1973.
67. **Bellar, T. A. and Lichtenberg, J. J.,** Determining volatile organics at microgram-per-litre levels by gas chromatography, *J. Am. Water Works Assoc.,* 739, December 1974.
68. **Bonelli, E. J., Taylor, P. A., and Morris, W. J.,** Mass fragmentography by GC/MS in the analysis of hazardous environmental chemicals, *Am. Lab.,* 29, July 1975.
69. Pesticide Analytical Manual, Food and Drug Administration, Rockville, Md., 1968, and subsequent revisions.
70. **Thompson, J. F. and Sherma, J.,** Manual of Analytical Quality Control for Pesticides and Related Compounds in Human and Environmental Samples, Environmental Protection Agency, Research Triangle Park, N.C., 1976.
71. **Junk, G. A., Richard, J. J., Grieser, M. D., Witiak, D., Witiak, J. L., Arguello, M. D., Vick, R., Svec, H. J., Fritz, J. S., and Calder, G. V.,** Use of macroreticular resins in the analysis of water for trace organic contaminants, *J. Chromatogr.,* 99, 745, 1974.
72. **Snyder, L. R. and Kirkland, J. J.,** *Introduction to Modern Liquid Chromatography,* John Wiley and Sons, New York, 1974.
73. **Scott, R. P. W., Scott, C. G., Munroe, M., and Hess, J., Jr.,** Interface for on-line liquid chromatography-mass spectroscopy analysis, *J. Chromatogr.,* 99, 395, 1974.

GENERAL REFERENCES

Benarde, M. A., *Our Precarious Habitat,* W. W. Norton and Co., New York, 1970.
Eckenfelder, W., Jr., *Water Quality Engineering,* Barnes and Noble, New York, 1970.
Artman, N. R., Safety considerations for detergents, *Prog. Water Technol.,* 3, 277, 1973.
Stokinger, H. E., Sanity in research and evaluation of environmental health, *Science,* 174, 662, 1971.
Cook, L. M., *Cleaning Our Environment – The Chemical Basis for Action,* American Chemical Society, Washington, D.C., 1969.
Kerri, K. D., Dendy, B. B., Brady J., and Crooks, W., Operation of Wastewater Treatment Plants – A Field Study Manual, EPA Office of Water Programs – Division of Manpower Training, Washington, D.C., 1970.

Section IX

Regulatory Review and Summary

TOXIC SUBSTANCES CONTROL

General Provisions and Inventory Reporting Requirements; Supplemental Notice; Public Meeting

Agency – Environmental Protection Agency.

Action – Proposed Rules; Notice of Public Meeting.

Summary – This notice reproposes the inventory reporting regulations first proposed on March 9, 1977 in the *Federal Register* and supplemented thereafter. Specifically, these reproposed regulations would require some manufacturers;

1. To report the identity of each chemical substance manufactured (or imported) for a commercial purpose and the site of such manufacture
2. To estimate the amount of each such chemical substance manufactured or imported at each site
3. To indicate whether each such chemical substance is manufactured and used only within one site
4. To indicate whether the respondent is a manufacturer, processor, and/or importer of each such chemical substance

In addition, these reproposed regulations would authorize certain other persons to report such information at their discretion.

Dates – Written comments must be received on or before September 16, 1977. EPA will hold a public meeting in Washington, D.C. on August 24, 1977 to provide an opportunity for oral comments. Details are provided below.

Address – Comments should be addressed to the Federal Register Section (WH-557), Office of Toxic Substances, Attention: Vicki Briggs, Environmental Protection Agency, 401 M Street SW., Washington, D.C., 20460. Comments should be filed in triplicate and bear the identifying notation OTS-081002. All written comments filed pursuant to this notice will be available for public inspection at that office from 8:30 a.m. to 4:00 p.m. Monday through Friday.

For further information contact – Mr. John Ritch, Office of Industry Assistance, Office of Toxic Substances (TS-788), Environmental Protection Agency, 401 M Street, SW., Washington, D.C., 20460, 202-755-0535.

Supplementary information – These regulations are proposed under the authority of Subsection 8(a) of the Toxic Substances Control Act (90 Stat. 2003; 15 U.S.C. 2601 et seq.; hereinafter referred to as TSCA).

On March 9, 1977, EPA first published in the *Federal Register* (42 FR 13130) proposed inventory reporting regulations to govern reporting of chemical substances for inclusion on an inventory of chemical substances required by Subsection 8(b) of TSCA. On April 12, 1977, EPA published a supplemental notice of proposed rulemaking in the *Federal Register* (42 FR 19298) providing additional information pertaining to the proposed inventory regulations. This notice set forth instructions for use of a Candidate List of Chemical Substances and specified minerals which EPA proposed to include in the inventory of chemical substances. On April 28, 1977, EPA published a notice of availability of the Candidate List of Chemical Substances for use in reporting chemicals for inclusion on the inventory (42 FR 21639). In addition, on July 8, 1977, the Agency published a notice to amend the procedures for securing a copy of the Candidate List on computer-readable tape (42 FR 35183).

On April 18, 1977, EPA held a public meeting in Washington, D.C. to provide interested persons an opportunity to comment publicly on the proposed regulations. In addition, approximately 200 persons have submitted written comments on the proposed regulations. Both the transcript of the public meeting and the written comments are available for inspection by the public in the Federal Register Office of the Office of Toxic Substances.

As a result of these comments, EPA has decided to repropose the inventory reporting regulations to require additional reporting by some persons and less reporting by others.

Participation in the Public Meeting

The public meeting on these proposed regulations will be on Wednesday, August 24, 1977 from 9:00 a.m. to 4:30 p.m. in the Thomas Jefferson Auditorium of the Department of Agriculture, 14th and Independence Avenue, SW., Washington, D.C. Persons who want to reserve time to present their comments at that meeting should contact Vicki Briggs at the address provided above or telephone 202–426–9819. Each person may request up to 15 min although less time may be allotted depending upon the number of participants. EPA will make a transcript of the proceedings for public inspection.

Status of Reproposal

The record of this rule making will include all comments received in response to the earlier notices of proposed rule making as well as the comments received in response to this notice. The public is encouraged to review the earlier notices of proposed rule making if any questions arise concerning the context of these reproposed regulations. While EPA would welcome comments on any aspect of these proposed regulations, persons are encouraged to direct their comments to the new provisions proposed here and not duplicate comments submitted earlier on other aspects of the proposed regulations. EPA will respond to all the comments submitted in response to the proposed rule making notices in the final inventory reporting regulations.

Modifications of Initial Reporting Requirements

The main purpose in revising the proposed approach is to use these initial reporting requirements not only to compile the inventory required by Section 8(b) but also to fulfill the Congressional intent, as stated in Section 2 of TSCA, that adequate data be developed for implementation of TSCA and other authorities directed to regulating risks associated with chemical substances. Although the regulations proposed on March 9, 1977 would have required manufacturers to report chemical substances manufactured for commercial purposes, the proposed approach would not have required reporting concerning production sites or the quantities produced.

In contrast to EPA's original proposal, the revised version published here would require certain manufacturers not only to identify the chemical substances in commerce but also to report where the chemical substances are manufactured and in what quantities. This information will be valuable for estimating the potential exposure to chemical substances for monitoring, control, and preventive actions. For example, plant site information would be useful in identifying possible sources of hazardous chemicals, especially in an emergency. Data on the quantities of chemical substances in commerce would enable EPA and other agencies to select substances for priority attention among the tens of thousands in commerce.

These amendments would expand the scope of the initial reporting requirements, but would limit the applicability of the requirements to those persons with establishments that are primarily engaged in the manufacture of chemical substances. Accordingly, only the approximately 20,000 establishments in the Standard Industrial Classification Major Group 28 (Basic Chemicals and Allied Products) and Group 2911 (Petroleum Refining) would be required to report each chemical substance manufactured at the production site and the volume of production. Manufacturers outside these groups would not be required to report. These latter persons could choose to report or could authorize a trade association to report to ensure chemical substances which they manufacture are included on the inventory. The hundreds of thousands of chemical processors may report during a limited period following publication of the initial inventory. EPA may require reporting by any of these manufacturers or processors as part of its phased reporting strategy under Section 8(a), discussed in the following section.

Other amendments to the March 9, 1977 proposal include a requirement that manufacturers indicate whether a chemical substance is manufactured and processed solely within one site and not distributed for a commercial purpose outside that site. EPA is considering specially designating these chemical substances on the inventory and providing under Section 5(a)(2) that any use of those substances for commercial purposes outside the manufacturing site would be considered a "significant new use." In addition, respondents would be required to indicate whether they manufacture, process, and/or import a chemical substance. Knowing which persons manufacture, import, or process a reported chemical substance would enable EPA to direct any future notice or requirement to appropriate persons and permit the Agency to estimate how much of a substance is manufactured domestically and how much is imported.

Various representatives of the Federal government and environmental groups have urged EPA to amend the initial reporting requirements to include reporting on uses of chemical substances. EPA recognizes the importance of obtaining use information in order to estimate exposure to a chemical substance. However, incorporating use reporting into the initial requirements would substantially delay the publication of the inventory, perhaps for more than a year after the statutory date. Premanufacture notification of new chemicals would be delayed accordingly.

For this and other reasons, EPA decided to postpone use reporting to the second phase of its reporting strategy, as described below.

Overall Strategy

By reproposing the inventory regulations, EPA recognizes that it will be unable to meet the statutory deadline for publication of the inventory in November 1977. Nonetheless, EPA believes that the proposed delay is warranted by the importance of the data base that would be generated as a foundation for implementation of TSCA. At the same time, EPA will not attempt to develop a comprehensive data base on all chemical substances through the initial reporting requirements. EPA has developed an overall strategy for data development under Section 8(a) of TSCA. These initial reporting requirements are the first of three phases.

The second phase of EPA's proposed strategy will be initiated after these regulations are final this fall. In this phase, EPA will address chemical substances selected because of their concern to EPA, the Occupational Safety and Health Administration (OSHA), the Consumer Product Safety Commission (CPSC), as well as to other agencies and interested parties. Manufacturers and processors of those chemical substances may be required to submit use information, including the estimated amounts of a chemical substance manufactured or processed for each use. In addition, EPA would consider asking for information on impurities, by-products, worker exposure, and other factors as needed for specific chemical substances or categories of chemical substances.

The third phase of EPA's reporting strategy would begin after the inventory is published in 1978. EPA would by regulation require reporting under Section 8(a) for additional chemical substances selected in part on the basis of their relative production volumes as reported under the initial reporting requirements. During this phase, EPA intends to develop the data base for a larger portion of chemical substances in commerce with respect to their use, exposure and other factors. Finally, in addition to such systematic reporting, EPA anticipates that it may ask for information on certain chemical substances as needed by the Department of Labor and others in emergency situations.

In determining what information to require in each of these phases, EPA will of course review alternative sources of data such as information available under Section 308 of the Federal Water Pollution Control Act Amendments of 1972 and other authorities, and will minimize duplicative reporting requirements.

Definitions of Small Manufactures for These Regulations Only

In proposing an expanded approach to the inventory reporting requirements, EPA would require certain manufacturers and importers to report information in addition to the identities of chemical substances in commerce. Paragraph 710.5(d) of these proposed regulations outlines this information. Although TSCA Section 8(a) provides broad authority to EPA to require information necessary for the administration of the Act, EPA may require "small manufacturers and processors" to submit only information required for compilation of the initial inventory or concerning a chemical substance which is subject to a proposed rule or order under TSCA Section 4, 5, or 6, or court action under Section 5 or 7.

Some of the additional information outlined in paragraph 710.5(d), such as production volume and the manufacturing sites of a chemical substance, may not be considered necessary for compilation of the initial inventory. Therefore, EPA may not be authorized to require submission of that information from "small manufacturers" under these regulations. Accordingly, EPA is proposing to define which persons qualify as "small manufacturers" for the purpose of these regulations and to exempt small manufacturers from certain of these reporting requirements.

The definition of "small manufacturer" proposed here is a one-time definition intended to apply solely to these regulations. Accordingly, it would only apply to manufacturers in SIC groups 28 and 2911 and to importers of chemical substances. Persons should not interpret this definition as indicative of future definitions which will be proposed for the purpose of subsequent regulations under Section 8(a) of TSCA. Those definitions for "small manufacturers" will take into account the burdens of complying with the future reporting and/or record keeping requirements.

Section 8(a)(3)(B) of TSCA provides that, after consulting with the Small Business Administration, the Administrator shall by rule prescribe standards for determining the manufacturers and processors which qualify as "small manufacturers and processors." The legislative history of TSCA shows that the Senate bill contained no exemption from the reporting requirements for small manufacturers and processors. The House bill first introduced this provision because reporting and record-keeping requirements "may impose a particularly heavy burden on small manufacturers and processors" (H.R. Rep. No. 94-1341, 94th Cong., 2d Sess. 42 (1976)). The Conference substitute retained the exemption of the House amendment in order to "protect small manufacturers and processors from *unreasonably* burdensome requirements" (italics added) (H.R. Rep. No. 94-1679, 94 Cong., 2d Sess. 80 (1976)).

In exempting "small manufacturers and processors" from certain reporting requirements, Congress intended that EPA balance its need for certain information with the burden imposed upon small manufacturers and processors in submitting that information. As discussed above, EPA believes that the information which would be required by these regulations is necessary to establish a data base for implementation of TSCA and other authorities directed to regulating risks associated with chemical substances. In developing the proposed exemption from these reporting requirements, EPA has consulted with the Small Business Administration (SBA) and others in order to assess the administrative and economic burdens for small manufacturers of complying with these reporting regulations.

As proposed in § 710.2 of these regulations, the term "small manufacturer or importer" means "a manufacturer who (1) has only a single manufacturing site, and either (2) has total sales of less that $100,000, based on the manufacturer's latest complete fiscal year, or (3) has no more than 2,000 pounds annual production (i.e., amount manufactured and imported) of each manufactured chemical substance. In the case of a company which is owned or controlled by another company, such factors would apply to the parent company and all companies owned or controlled by it taken together."

Manufacturers and importers which fall within this definition would be exempt from reporting production volume. They would not be exempt from reporting the following information, which is necessary for compilation of the inventory: the identities of the chemical substances they manufacture or import; the business address; whether a chemical substance is used solely within the manufacturing site; or whether they manufacture, process, and/or import the chemical substance. Any small manufacturer whose chemical substance is not included on the initial inventory would be subject to the premanufacturer notification requirements of TSCA Section 5.

In considering alternative definitions, EPA is evaluating the burden of complying with the expanded reporting requirements in light of the fact that manufacturers and importers would already be reporting the identities of chemical substances for the inventory. In promulgating these regulations, EPA will probably define "small" in terms of (1) plant site, and either (2) sales or (3) production levels, incorporating only two parameters in the final definition.

With respect to the alternative of defining the term "small manufacturer" in terms of clauses (1) and (2) above, the number of plant sites is indicative of the management structure of a company and the likelihood that the information required would already exist in a centralized form. Information on the total annual sales of all products in generally available to all manufacturers. It also is a measure of size in terms of dollars and therefore is relevant to the burden imposed by these reporting requirements. EPA considers $100,000 an appropriate level above which all manufacturers should be able to comply with the additional requirements of this regulation without undue economic burden. If the manufacturer is owned or controlled by another company, the manufacturer should compute both the total annual sales and the number of plant sites on the basis of the sales and number of plants in the U.S. for the company as a whole.

If "small" is defined in terms of (1) and (2), potentially as many as 20% of the firms in Standard Industrial Classification Major Group 28 (SIC 28), Chemicals and Allied Products, would be considered "small manufacturers." However, the exempted manufacturers contribute a very small fraction, less than 1% of the total sales within SIC 28.

Alternatively, if EPA chose to define "small manufacturer" in terms of clauses (1) and (3) above, one plant site and no more than 2,000 pounds annual production of any chemical substance, the number of establishments which would be exempted from reporting production would probably be far fewer. In fact, the effect of using the criteria in clauses (1) and (3) would be to exempt those persons who only manufacture chemical substances in less than 2,000 pounds from reporting the estimated production levels of those substances. EPA solicits comments on this alternative, especially with respect to the number of pounds selected for setting the exemption.

EPA also solicits comments on this proposed definition of "small manufacturer," including any quantitative data on the estimated costs of compliance, the number, sizes, and types of firms for which it may be a significant additional burden, or other information which describes the impact of these reporting requirements on small manufacturers. For example, EPA anticipates that reporting production may be burdensome for some manufacturers, particularly those who use batch processing to produce a variety of chemicals and keep records of production only on the basis of shipments or customer invoices. EPA would also appreciate any comments on other possible parameters for defining "small manufacturer," such as profits, market share, financial assets, or the number of employees.

Other Definitions

As indicated in § 710.2, EPA proposes to revise many of the definitions published in the March 9, 1977 proposed regulations. The definitions are included in § 710.2 rather than § 700.2 so that their applicability will be limited to these regulations and not automatically extend to subsequent regulations under TSCA. While minor proposed changes to the originally proposed terms are included in the new § 710.2, these changes are not discussed here as they will be addressed in the final regulations.

Several definitions included in the March 9, 1977, regulations were taken from other authorities. Specifically, the definitions of "food additive," "drug," "cosmetic," "device,"-"special nuclear material," "nuclear by-product material," "nuclear source material," and "pesticide" were incorporated without modification from other regulations. Instead of including these definitions in their entirety in these regulations, EPA would include them by reference. Thus any changes in the other statutes will automatically be reflected in these regulations.

Included in these proposed regulations are three additional terms. EPA is proposing to define "article" as a "manufactured item (1) which is formed to a specific shape or design during manufacture, (2) which has end use function(s) dependent in whole or in part upon its shape or design during end use, and (3) which is functional in its end use(s) without change of chemical composition during its end use; except that (4) fluids and particles are not considered articles regardless of shape or design." This definition is added to clarify proposed § 710.4(d)(6) which would exclude from the inventory, and from these reporting requirements, a chemical substance which is the result of a chemical reaction that occurs upon use of curable plastic molding compounds and other chemical substances to manufacture an article destined for the marketplace without further chemical change. Examples of this are chemical substances that form during the thermosetting process in forming plastic articles, firing pottery or enamel products, setting concrete sidewalks, or molding rubber products. This exclusion is discussed further below under the section "Chemical Substances Excluded From the Inventory."

Related to this is the proposed definition of "manufacture, process, or import "for commercial purposes" which means to manufacture, process, or import for use by the manufacturer, as well as for distribution in commerce, for use as a catalyst or an intermediate, and for test marketing purposes. This definition is intended to clarify that chemical substances that are used by the manufacturer, not only as an intermediate or catalyst in the manufacture of another chemical substance, but also in the manufacture of a mixture or an article or in any other way, would be subject to these regulations. Accordingly, chemical substances which are manufactured and then converted by the manufacturer into an article should be reported by that manufacturer and would not be excluded from reporting for the inventory under § 710.4(d)(6). For example, a person who manufactures a polymer and then converts the polymer into a synthetic fiber should report the polymer.

The term "establishment" is defined as "an economic unit, generally at a single site, as defined for purposes of the Standard Industrial Classification of Establishments. There may be more than one establishment at a single site." This term is necessary to clarify proposed § 710.3(a) which would provide that only manufacturers with establishments in certain Standard Industrial Classification (SIC) groups would be required to report.

EPA is proposing to define the term "site" as "each contiguous property unit where a chemical substance is maufactured or processed whether or not such site is independently owned or operated. Property divided only by a public right of way shall be considered one site. For the purposes of imported chemical substances, the site shall be the business address of the importer." While all persons are encouraged to report by site, § 710.5(a)(1) would require only those persons required to report under § 710.3(a) to report by site. EPA solicits comment on these definitions and the clarity of the reporting requirements with respect to articles, establishments, and sites.

Applicability: Who Must Report, Who May Report, Who May Not Report

Section 710.3 is intended to clarify who must report for the inventory under these regulations and for whom reporting is optional. In addition, the section states who may not report. Although these regulations expand the information obtained, EPA is proposing to limit the expanded reporting requirements primarily to those establishments which are the basic manufacturers of chemical substances.

Under the March 9, 1977 proposal, every person who currently manufactures a chemical substance for commercial purposes would have had to report that substance for the inventory. Many comments emphasized that this approach would require duplicative reporting by persons who are not generally recognized as part of the chemicals industry but who, for economic reasons or special purposes, manufacture a limited number of chemical substances essential to their processes. For example, in the pulp and paper industry, pulp mills manufacture sodium hydroxide and other chemical substances as part of their recovery processes. If EPA adopted the approach of the March 9, 1977 proposal, there may be more than 400 such establishments reporting that they manufacture sodium hydroxide or other chemical substances common to pulp manufacture.

As an alternative, these initial reporting requirements would focus on the chemical and allied products sector of the manufacturing industry and the petroleum refining sector, as defined by SIC Major Group 28 and Group 2911, respectively. Major Group 28 includes the manufacturers of chemicals such as acids, alkalies, and organics; synthetic fibers and plastics; dry colors and pigments; soaps and detergents; and paints and fertilizers. Group 2911, petroleum refining, is the basis of the organic chemicals industry.

Establishments subject to TSCA which fall outside these SIC groups are primarily involved in processing chemical substances, such as fabricating plastic and rubber products or treating articles such as textiles and metals, and would not be required to report. In most cases the chemical substances they manufacture would be reported by establishments in SIC Group 28. To the extent that they manufacture chemical substances for special purposes that may not otherwise be reported for the inventory, they would be responsible for ensuring, either through trade associations or individually, that those substances were included in the inventory. Otherwise, if these substances are not included in the inventory, any person who manufactured or imported the substances would be subject to premanufacture notification requirements under Section 5(a)(1)(A)of TSCA.

One advantage of limiting required reporting primarily to manufacturing establishments in the chemical and allied products sectors of industry would be that EPA would be able to direct the reporting requirements to 20,000 establishments rather than to 225,000 or more establishments, most of which primarily process chemical substances, as explained above. Those establishments in SIC group 28 represent approximately 95% of chemical production.

Further if someone identified a hazard associated with the processing of a chemical substance or wanted to know exactly what chemical substances may be manufactured or processed by establishments outside SIC groups 28 and 2911, EPA has authority under Section 8(a) to require such detailed reporting. As mentioned in the discussion of EPA's overall strategy, EPA intends to implement this general reporting authority to develop a data base on those substances for which there is significant human or environmental exposure or some other reason for concern.

EPA solicits comments on this proposal to limit required reporting to establishments in SIC groups 28 and 2911. Specifically, EPA assumes that there may be manufacturers who do not know in which SIC group their establishment appropriately belongs. Others may have been categorized in one group 5 years ago and have since changed their primary economic activity and belong in a different SIC group. EPA intends to notify each establishment which, to the best of EPA's knowledge, should be included in the SIC groups 28 or 2911. If for some reason a person has not been directly notified and would belong in SIC groups 28 or 2911, that person would still be required to report to EPA. Accordingly, it would be useful to EPA to know to what extent manufacturers are familiar with SIC groupings and whether the descriptions provided in the Standard Industrial Classification Manual published by the U.S. Government Printing Office would be adequate for manufacturers to determine in which group they belong.

Section 710.3(a)(2) would modify the original proposal and would only require importers to report those chemical substances imported into the U.S. for a commercial purpose since January 1, 1977. The March 9, 1977 proposed regulations would have required importers to report not only those chemical substances imported in bulk into the U.S. but also those chemical substances contained in the articles they import. Comments from industry and trade associations argued that it would be extremely burdensome for importers to identify the chemical substances contained in the articles they import. Moreover, they argued that the proposed regulations would have imposed a burden on importers of articles which was not imposed on domestic manufacturers of articles. The Administrator has decided to revise the original proposal to limit the reporting requirement to imported chemical substances. This included all chemical substances which are imported in cans, bottles, drums, barrels, packages, tanks, bags, and other devices which are used to contain the substances during importation. EPA solicits comments on this reproposal.

Aside from importers, under this proposal only establishments in SIC groups 28 and 2911 would be required to report the chemical substances they have manufactured since January 1, 1977. Section 710.3(b) provides that in addition to those required to report, any person who has manufactured, imported, or processed a chemical substance for a commercial purpose since January 1, 1975, may report that substance or authorize a trade association or other representative to report on his behalf.

As proposed in § 710.3(c), during a special reporting period, 120 days after the first publication of the inventory, any person who has processed or used a chemical substance (including the manufacture of a mixture or article containing that chemical substance) for a commercial purpose since January 1, 1975 may report that chemical substance if it was not included in the inventory. EPA would like to minimize duplicative reporting of chemical substances by processors during the initial reporting period to facilitate compilation of the initial inventory in a timely way. Many processors and users have expressed concern that the manufacturers of the chemical substances they process may fail to report. EPA hopes that this proposed provision would reduce the amount of duplicative reporting by processors seeking to ensure that chemical substances are included in the inventory.

As provided by Section 5(a)(1)(A) of TSCA, 30 days after publication of the initial inventory any person who manufactures or imports a "new chemical substance must submit premanufacture notification prior to manufacture or importation of the chemical substance. Processors are not subject to the provisions of Section 5(a)(1)(A). However, it is a prohibited act under Section 15(2) of TSCA for a person to use for commercial purposes a chemical substance which he had reason to know was manufactured in violation of Section 5. As a matter of Agency policy, the Agency will not enforce Section 15(2) with respect to processors and users of chemical substances (including manufacturers of a mixture or article containing that substance) during the 120-day period proposed in § 710.3(c). The Agency will, however, enforce Section 15(2) with respect to all manufacturers and importers of chemical substances during that period, and will enforce Sections 15(2) and (3) with respect to all persons after the period expires.

Section 710.3(d) would clarify who may not report chemical substances for the inventory, either because the chemical substances are automatically included in the inventory as provided in § 710.4(b), or because they are excluded from the inventory as provided in paragraphs (c) and (d) of § 710.4 a person should only report those substances which he knows and could verify are chemical substances as defined in the Act. In particular, chemical substances used exclusively as pesticides and drugs may not be reported. In addition, chemical substances manufactured solely in small quantities for research and development may not be reported. If a person does not know whether his customers may use the substance for a TSCA use and does not report, but discovers later that they do, he may add it to the inventory at that time. Any customer who uses the substance could add it to the inventory during the 120-day period provided in § 710.3(c).

Scope of the Inventory

In the March 9, 1977 proposal EPA relied upon certain definitions of terms such as "mixture," "manufacture or process for 'commercial purposes'," and "by-product" to clarify what chemical substances should be reported for the inventory. Because this approach was confusing to many, EPA has redrafted Section 710.4 to clarify what substances are eligible for inclusion on the inventory.

Basically, the Act provides in Sections 8(b) and (f) that any chemical substance may be included on the inventory if it has been manufactured or processed for a commercial purpose in the U.S. within 3 years of the effective date of these regulations. In the regulations proposed on March 9, 1977, EPA anticipated that the regulations would become effective by July 1, 1977, and that the 3-year period would date from July 1, 1974. The decision to repropose means that the final regulations will not be published until sometime in October. In order to define a reporting period that industry can rely upon, regardless of the actual date the final regulations are published, EPA is proposing that the 3-year period begin on January 1, 1975. This proposal would satisfy those who have urged that EPA define the period in terms of calendar years. EPA recognizes that some may still prefer to have the period include the full 3 years from whatever date the regulations are final. EPA specifically solicits comment on this matter.

Section 710.4(a) generally defines which chemical substances are manufactured, imported, or processed "for a commercial purpose." Because the term "manufacture" is defined to include "to import into the customs territory of the United States," chemical substances which are imported into the U.S. are subject to the same provisions as those which are manufactured in the U.S. The provisions in § 710.4(a) are consistent with the definition of the term "manufacture, process, or import 'for commercial purposes'" in § 710.2. Accordingly, any chemical substance manufactured, processed, or imported (1) for distribution in commerce, (2) for use as a catalyst or an intermediate, (3) for use by the manufacturer, or (4) for test market purposes, would be eligible for inclusion on the inventory.

Chemical Substances Automatically Included

Section 710.4(b) specifies that chemical substances which are naturally occurring and are unprocessed or processed only by manual, mechanical, or gravitational means; by dissolution in water; or by heating solely to remove water shall be automatically included on the inventory under the category "Naturally Occurring Chemical Substances." Examples of naturally occurring substances that would be included on the inventory are raw agricultural commodities, water, air, natural gas, crude oil, rocks, ores, and minerals.

In the March 9, 1977 regulations, EPA proposed to automatically include on the inventory the general category "raw agricultural commodities." The revised proposal would incorporate this category. Accordingly, as under the original proposal, manufacturers, importers, and processors of raw agricultural, horticultural, and silvicultural products, such as unprocessed cotton, wood, wool, straw, oat hulls, and raw hides, for example, would not report. For clarity, this proposal also cites water, air, and natural gas, and crude oil as examples of "naturally occurring chemical substances" that need not be reported. With respect to rocks, ores, and minerals, this revised approach would not attempt to list each automatically included mineral separately, but would include all rocks, ores, and minerals under the category "naturally occurring chemical substances."

In the supplement to the March 9, 1977 regulations published on April 12, 1977 (42 FR 19308), EPA proposed a list of minerals under consideration for inclusion on the inventory without reporting. Persons commenting on that proposal have emphasized the difficulty of knowing precisely all the minerals a company may be extracting from the earth and, accordingly, whether the minerals being mined were included on the inventory or not. Most rocks and ores contain many different substances of varying composition depending upon the geological formation of the mined area. With respect to creating an inventory of where each major mineral is mined, the U.S. Bureau of Mines already has such an inventory. For those minerals, such as asbestos, which may present a risk to human health or the environment, EPA will require information on uses, exposure, and other factors necessary in assessing that risk under Section 8(a) of the Act.

EPA solicits comments on the proposed approach to "naturally occurring chemical substances" and any suggestions for clarifying this category.

Chemical Substances Excluded by Definition or by TSCA Section 8(b)

Section 710.4(c) clarifies those substances which may not be reported for the inventory either because of the definition of "chemical substance" in Section 3(2) of the Act or the specific exemption for chemical substances manufactured, imported, or processed solely in small quantities for research in Section 8(b) of TSCA. In response to the March 9, 1977, proposal, EPA received extensive comments on the exclusion of chemical substances used in the manufacture of pesticides and drugs and will respond to the comments in the final regulations. Likewise, the final regulations will explain the exemptions for alloys, inorganic glasses, ceramics, frits and cements, including Portland cement.

A chemical substance used as a reagent in quality control testing, where the material tested is distributed in commerce, is itself considered to be distributed in commerce and should be reported to EPA for inclusion on the inventory unless it is known that the "small quantities for research and development" exemption applies. As mentioned above, Section 8(b) explicitly exempts from the inventory any chemical substance which is manufactured or processed only in small quantities solely for purposes of research, including analysis of another chemical substance. EPA would define "small quantities for research and development" as quantities that are not greater than reasonably necessary for such purposes and which, after the effective date of premanufacture notification requirements, are used for "research and development that is conducted by, or directly supervised by, a technically qualified individual(s)." Accordingly, unless the persons performing the quality control testing are themselves technically qualified persons, as defined in § 710.2, or are directly supervised by technically qualified individuals, the chemical substance would not be considered to be manufactured in a "small quantity for research and development" and would be subject to these reporting requirements or the premanufacture notification requirements.

Chemical Substances Excluded from the Inventory

Section 710.4 (d) clarifies that certain chemical substances which are not manufactured for distribution in commerce as chemical substances per se and have no commercial purpose separate from the mixture or article of which they may be a part are excluded from these reporting requirements. Specifically, impurities or chemical substances which are unintentionally present with another chemical substance are excluded. With respect to by-products, the proposed regulations of March 9, 1977, would have excluded by-products which have no commercial purpose. However, the proposal left unclear whether manufacturers should report by-products which are not manufactured for a commercial purpose but are used as a fuel or reprocessed. These regulations would provide that those by-products whose sole commercial value is to municipal or private organizations who (1) burn it as a fuel, (2) dispose of it as a waste, including as a landfill or for enriching soil or (3) extract component chemical substances which may have some commercial value, may be included on the inventory but that the reporting of such substances, insofar as they are by-products as defined in § 710.2, is optional.

In proposing to exempt from the reporting requirements such by-products which have some commercial purpose. EPA intends to encourage conservation and recycling of the energy and resources contained in the waste material that might otherwise be discarded because of reporting burdens under TSCA. Further, insofar as these wastes are hazardous, EPA intends to require reporting of them under the Resource Conservation and Recovery Act (RCRA) next spring or under TSCA Section 8(a)(2) during the second or subsequent phases of reporting. EPA explicitly solicits comments on this approach, particularly with respect to the proposed exemptions.

The exclusions in § 710.4(d)(3), (4), and (5) are for chemical substances which result from chemical reactions that occur incidental to exposure to environmental factors, or during storage or end use of a chemical substance or mixture. These exemptions clarify those provided in the March 9, 1977, proposed regulations and are consistent with the legislative history of the Act (H.R. Rep. 94-1341, 94th Cong., 2d Sess. 13 (1976)).

In § 710.4(6), EPA would exclude "any chemical substance which is the result of a chemical reaction that occurs upon use of curable plastic or rubber molding compounds, inks, drying oils, metal finishing compounds, adhesives, paints, or other chemical substances used to manufacture an article destined for the marketplace without further chemical change of the chemical substance except for those chemical changes that may occur as described elsewhere in this § 710.4(d). This provision expands upon the earlier proposal to exempt chemical substances formed upon use of curable plastic molding compounds. By providing a general exclusion for chemical substances manufactured as articles or parts of articles destined for the marketplace without further chemical change of the chemical substance, EPA intends to exclude from reporting persons who are primarily manufacturers of articles.

Section 710.2 proposes to define "article" as "a manufactured item (1) which is formed to a specific shape or design during manufacture, (2) which has end use function(s) dependent in whole or in part upon its shape or design during end use, and (3) which is functional in its end use(s) without change of chemical composition during its end use; except that (4) fluids and particles are not considered articles regardless of shape or design." Under this definition, fluids and particles such as dust, powders, dispersions, granules, lumps, and flakes would not be considered articles.

To illustrate, sodium hypochlorite and particles of titanium dioxide would not be considered "articles" in themselves. The function of sodium hypochlorite as a bleach, for example, depends upon a change of chemical composition during its end use. Similarly, although the distinctive shape of titanium dioxide particles may contribute to their ultimate usefulness, they are particles and would not be considered articles in themselves. Paints containing titanium dioxide particles would not be articles because they are liquids and do not have "shape or design" when manufactured. But automobiles coated with titanium dioxide containing paints are articles under this definition. Other examples of articles which need not be reported for the inventory include plastic films, synthetic fibers, leather goods, nails, iron bars, chrome plated bumpers, jewelry, paper, particle board, furniture, refrigerators, cloth, and clothing. EPA solicits comments on this provision and the clarity of the distinction between articles and chemical substances.

Further, precursors of the compositions covered by the exclusions in § 710.4(d)(5) and (6) may be supplied to users as two or more different products which need to be mixed as a first step in their use because of limited stability of the mixture. Chemical substances formed during such mixing would be excluded.

Finally, § 710.4(d)(7) provides an exemption for chemical substances that may occur as the result of a chemical reaction when a stabilizer, colorant, odorant, antioxidant, filler, solvent, carrier, surfactant, plasticizer, corrosion inhibitor, antifoamer or defoamer, dispersant, percipitation inhibitor, binder, emulsifer, deemulsifier, dewatering agent agglomerating agent, adhesion promoter, flow modifier, pH neutralizer, sequestrant, coagulant, flocculant, fire retardant, lubricant, chelating agent, quality control reagent, or a chemical substance which is solely intended to impart a specific physicochemical characteristic functions as intended. This provision expands upon the approach in the March 9, 1977, proposal which would have exempted as "mixtures" the result of chemical reactions that occur when certain chemical substances function as intended.

How to Report

Section 710.5(a) would provide general instructions for reporting chemical substances. As discussed in the earlier sections, only importers and manufacturers with establishments in SIC groups 28 and 2911 would be required to report, and "small manufacturers" in these SIC groups would not have to report certain information. The Agency would encourage any person not required to report all the information, to do so anyway because the information will establish the data base for future actions under TSCA.

Section 710.5(b) outlines how to report the name or specific identity of a chemical substance. In April 1977, EPA published and made available the TSCA Candidate List of Chemical Substances, and on April 12, 1977, published in the *Federal Register* a guide for using this list. EPA realizes that the Candidate List is not a complete list of chemical substances in commerce and does include some substances which would be excluded from the inventory. In addition there are some minor errors. On or before the date these reporting requirements are published in their final form in October 1977, EPA intends to revise the guide to the Candidate List and make necessary corrections and certain additions to the List itself.

Section 710.5(c) proposes that any person reporting a polymer for inclusion in the inventory must list in the description of the polymer composition at least those constituent monomers used at greater than two weight percent in the manufacture of the polymer. A person may include as part of the description of the polymer composition those monomers used at two weight percent or less in the manufacture of the polymer. Of course, all monomers themselves must be separate entries on the inventory. EPA received extensive comment on the issue of polymer reporting in response to the March 9, 1977, proposal and will respond to all the comments in the final regulations.

Additional information, which would be required according to the general instructions in § 710.5(a), is outlined in § 710.5(d). As mentioned earlier in this preamble, knowing who is manufacturing, importing, or processing a chemical substance would enable EPA to direct any future notice or requirement to the appropriate person. The reporting forms will provide three check boxes for persons to check any or all of them, as appropriate.

Manufacturers must report according to the site at which the chemical substance is manufactured. As explained earlier, the definition of "site" includes each contiguous property unit where a chemical substance is manufactured or processed whether or not such site is independently owned or operated. There may be more than one establishment, including subsidiaries or branches of a given company, at one site. The chemical substances manufactured at that site may all be reported on one form with the site address provided.

Paragraph (d)(3) of §, 710.5 would require manufacturers to designate whether they manufacture and process a chemical substance only within a site and do not distribute the chemical substance, or any mixture or article containing that substance, for commercial purposes outside that site. In most cases these chemical substances would be consumed by chemical reaction in the manufacture of another chemical substance. The exposure to such chemicals would be limited to persons involved in the manufacture, processing, and use at that site and immediate environs. Intermediates and catalysts would most likely form the greatest percentage of these chemical substances. However, this provision would also apply to any other chemical substances which are not distributed in commerce outside that site.

Section 710.5(d)(4) would require that manufacturers and importers report the amount of each chemical substance manufactured or imported in calendar year 1976. Alternatively, if the chemical substance was not manufactured or imported during 1976, a manufacturer or importer would either report the amount manufactured or imported during 1975 or the projected amount during 1977. If there has been no manufacture or importation since January 1, 1975 a manufacturer or importer should report the amount distributed to others for any purpose since that date. Processors would not be permitted to report amounts processed in order to avoid double-counting.

EPA is considering requiring that all production amounts be expressed in pounds. EPA would appreciate alternative suggestions for cases where conversion to pounds appears unreasonable.

As one alternative, persons would report amounts above 5,000 to only two significant figures. That is, only the first two figures of a six-figure number would be reported, such as 590,000 instead of 586,272. Instead, EPA could use a one-digit code to require reporting of the range of production volume. For example, production levels of 1,000 to 10,000 pounds would be reported by "1"; 10,000 to 50,000 reported by "2"; and so forth. While this approach may be easy to use, the results may not be as valuable since the imprecision of the ranges would be compounded when the production amounts are aggregated. AS a third alternative, EPA could require reporting production accurate to within 10% of the production. EPA would appreciate comments on these or other possible alternative ways of reporting production.

In addition, the reproposal at §, 710.4(e) would permit an importer to authorize his foreign supplier(s) to report on his behalf. For several reasons, including issues of confidentiality, importers often do not know exactly what they are importing. Because EPA's jurisdiction under TSCA extends only to importers, and not to their foreign suppliers, EPA will hold the importers liable for compliance with these reporting rules. However, the foreign suppliers will be permitted to act as agents for the importers with the latter remaining legally liable for their reports. To do so, the foreign suppliers must sign declarations on the reporting forms, and the importers must endorse these declarations. This approach is similar to that already followed by the U.S. Bureau of Customs in some of its 40 CFR Part 19 regulations.

Confidentiality

The expanded scope of these regulations would significantly increase the number of possible claims of confidentiality that persons reporting may make. Section 14 of TSCA provides that EPA must not disclose information which is exempt from mandatory disclosure under the Freedom of Information Act (5 U.S.C. 552(b)(4)). EPA has regulations dealing with the confidentiality of business information in Part 2, Subpart B of Title 40 of the Code of Federal Regulations (40 CFR Part 2; 41 FR 36906, September 1, 1976), which outline the general approach taken by EPA in dealing with confidentiality claims. EPA intends to add a new section to those regulations which will govern how the Agency will deal with claims of confidentiality with respect to information obtained under TSCA. A proposed version of this new section should be published in the *Federal Register* for public comment within the next several months.

With respect to these reporting regulations, §, 710.7(a) lists the items of information that may be claimed as confidential. EPA will design its reporting forms to allow all potential confidentiality claims to be asserted on the forms. EPA solicits comments on the various kinds of claims that might be asserted so that EPA can design its reporting forms accordingly.

Paragraphs 710.7 (b), (c), and (d) would provide that any claim of confidentiality must accompany the information at the time it is submitted to the Agency and that failure to make a claim on the reporting form could result in disclosure by EPA. EPA will consider only those claims that are asserted. Moreover, each company should take into account the possibility that EPA (or a court) might determine that some of the information should be released. Thus, if more than one claim applies concerning a particular chemical substance, the company should assert all those claims. For example, if a company believes that both the production volume of a chemical substance at a particular site or sites and the company-wide production of that substance are confidential, the company should specifically claim both of these items as confidential.

If a company makes a claim in the manner prescribed on the form, the information claimed to be confidential will be treated in accordance with EPA's confidentiality regulations and TSCA Section 14, including the 30-day notice prior to release. EPA is considering amending the proposed reporting forms to include statements which a company asserting a confidentiality claim could check to substantiate its claims. This would expedite the process of making final determinations by eliminating the need for obtaining that information at a later date. The success of this approach, however, is dependent upon businesses asserting only justifiable claims. If EPA determines that a particular claim or substantiation from a company is frivolous, EPA will take this into account in making other determinations concerning that company's other claims of confidentiality.

Section 710.7(e) of these proposed regulations modifies the March 9, 1977, proposal which would have required the submission of the following information from a person submitting a claim as to the specific name or identity of a chemical substance: (1) The confidential identity; (2) a proposed name which is only as generic as necessary to protect the substance's confidential identity; (3) a list of the elements of the chemical substance and its molecular weight; and (4) a bibliography identifying any published literature and summaries of any unpublished information concerning the health and ecological effects and environmental behavior of the chemical substance. These proposed regulations would require submission of only items (1) and (2) above at the time the person submits the claim of confidentiality.

Some comments suggested that EPA publish on the inventory only a noninformative code designation instead of a generic name and list of elements of the chemical substance and its molecular weight. As explained in the preamble to the March 9, 1977, proposed regulations, EPA originally proposed that the inventory include these items so that the public would have some indication of the undisclosed substance. Because of the likelihood that someone may be able to discern the identity of a confidential chemical substance from these data, EPA is proposing to publish only a generic name. EPA would either publish the generic name as proposed or, if EPA disagreed with the proposed generic name, consult with the person submitting it before publishing a revised generic name on the inventory.

Many comments recommended that EPA not require submission of a bibliography identifying any published literature because in many cases competitors could learn the identity of a chemical substance by reading the referenced literature. Others argued that without the specific identity of the chemical substance, the health and safety data would not be particularly useful to the public. Finally, some asserted that development of such a bibliography could be costly to prepare. Many suggested that EPA simply require a brief summary of known data on the health and environmental effects of the chemical substance. Others contended that a summary would not meaningfully contribute to the public understanding of the potential risks presented by the substance. Without the specific identity of the chemical substance, the public could not verify the information and, according to some, there would be a considerable possibility that the summaries would be misleading.

For these and other reasons, EPA is proposing to drop the requirement that persons submit a bibliography or summary of the health and safety studies pertaining to the confidential chemical substance at the time the claim is made. Under Section 8, EPA has authority to require submission of such information for any chemical substances in commerce and could request such information after the publication of the inventory. Moreover, TSCA Section 8(e) requires submission of information which supports the conclusion that a chemical substance presents a substantial risk of injury to health or the environment. EPA would appreciate comments on this proposal.

Alternatives for Handling Confidential Information

Of the information reported to EPA, only the identity of the chemical substances and perhaps designation of those chemical substances manufactured and used within a single site will be published. The remainder of the information will be used by EPA for various purposes under TSCA. EPA does not anticipate that this remaining information will be routinely released to the public. If EPA proposes to release confidential information, it will do so in accordance with EPA's confidentiality regulations.

EPA will be subject to disclosure requests under the Freedom of Information Act (FOIA, 5 U.S.C. 552). Under the Act, EPA must respond to any request for records by either releasing the records or denying the request because the information is exempted from disclosure. Records may be exempt from disclosure if they are "trade secrets and commercial or financial information obtained from a person and privileged or confidential." Section 14 of TSCA makes it clear that if EPA determines that information is exempt under 5 U.S.C. 552(b)(4), it must be kept confidential by EPA.

If a manufacturer claimed that the chemical name of a particular chemical substance is a trade secret, EPA would be confronted with conflicting statutory provisions. Section 8(b) apparently requires EPA to place the chemical name in the inventory. Section 14 appears to require EPA to keep the name confidential (at least temporarily, until a final determination is made by EPA or the courts). In addition, the Freedom of Information Act requires that EPA either release information in response to a request or provide reasons for any denial. EPA cannot refuse to answer the request. Normally, in responding to Freedom of Information requests, EPA would reply either: (1) "We have no records;" (2) "We have such records and are releasing them;" or (3) "We have such records but are denying them because they constitute a confidential trade secret and are exempt from disclosure." However, if the request concerned disclosure of the identity of a chemical substance which was allegedly a trade secret, to reply that EPA had a record but was refusing to release it would inform the requester that such a substance was being manufactured, imported, or processed for commercial purposes. The result would be to reveal the trade secret by denying the request.

Confidentiality assertions also pose a problem under Section 5 of TSCA. Section 5 requires anyone who proposes to manufacture a new chemical substance to furnish EPA with a 90-day premanufacture notice, during which time the person may not manufacture the new chemical substance. This delay may be even longer if a testing rule under Section 4 requires the manufacturer to develop and submit certain test data. However, if the chemical substance is on the Section 8(b) inventory, it is not a "new substance", and the Section 5 notice need not be given. If a company asserts that the name or specific identity of a chemical substance is confidential, EPA may not be able to list that substance on the inventory, and all other manufacturers would have to give premanufacture notification.

EPA has not decided how it will deal with these contradictory statutory requirements in the face of a claim that the identity of a chemical substance is confidential. Four issues arise that EPA must consider before deciding which approach to take. The issues are set forth below with a discussion of the options available under each:

1. What chemical substances' identities should be determined by EPA to be entitled to confidential treatment? EPA perceives three positions it could take:

a. No chemical identity is entitled to confidential treatment.

b. Only chemical identities of those chemical substances that are manufactured and used within one site and not distributed for a commercial purpose outside that site may be entitled to confidential treatment.

c. Any chemical identity may be entitled to confidential treatment.

If EPA decided that some or all of the chemical identities claimed as confidential were not entitled to confidential treatment, companies would be given the 30-day notice required in Section 14 of the Act. Before determining that a particular chemical identity was entitled to confidential treatment, EPA would have to make a specific determination, in accordance with 40 CFR Part 2, Subpart B, and the criteria in 40 CFR 2.208, that the particular chemical identity is entitled to confidential treatment.

2. Assuming that some chemical identities are temporarily (because no final determination has been made or because a court has enjoined EPA from disclosing the identity) or permanently (as a result of a final confidentiality determination by EPA or a court order) entitled to confidential treatment, how should EPA treat confidential chemical identities for purposes of the published inventory? EPA perceives four positions it could take:

a. The inventory could be published with only nonconfidential chemical substances in it. There would be no mention of confidential chemical identities. This approach would give the public no information concerning confidential chemicals.

b. The inventory could be published with nonconfidential chemical identities and generic names for those chemical substances that are entitled to confidential treatment. The use of generic names would inform the public in general terms about what types of confidential chemical substances had been reported.

c. The inventory could be published with nonconfidential chemical identities and random code numbers for those chemical substances that are entitled to confidential treatment. The use of a random code number would not give the public any information beyond that which would be available under A, except to acknowledge the existence of confidential chemical substances.

d. The inventory could be published with only nonconfidential chemical identities appearing on it and a notice that some chemical substances were reported that are confidential. This approach would give the public no more information than A or C.

3. Assuming that some chemical names are temporarily or permanently entitled to confidential treatment and that EPA is, therefore unable to publish the chemical identities on the Section 8 inventory list, how are present and future manufacturers to be treated under Section 5 of TSCA? EPA perceives four positions it could take:

a. If a manufacturer proposed to manufacture a chemical substance that did not appear on the inventory and asked EPA whether it was one of the reported confidential chemical substances, EPA could tell the manufacturer whether the chemical substance had been reported. If it had been reported, the manufacturer would be exempt from requirements of Section 5(a), as would the manufacturer that originally reported the chemical substance.

b. If a manufacturer proposed to manufacture a chemical substance that did not appear on the inventory and asked EPA whether it was one of the reported confidential chemical substances, EPA could refuse to answer the question. EPA would require the manufacturer to give premanufacture notification under Section 5(a), and the manufacturer would not be able to begin manufacturing for at least 90 days. The manufacturer that originally reported the chemical substance would be exempt from Section 5(a). This approach would treat the two manufacturers unequally allowing one to manufacture its chemical substance without delay or interruption while requiring the other to undergo premanufacture notification and a 90-day delay. Section 5 speaks in terms of "new chemical substances." This substance would not be "new" since it was already reported for the inventory.

c. If a manufacturer reported a confidential chemical substance for the inventory, EPA could require the manufacturer to give premanufacture notification under Section 5(a) and to cease manufacture for at least 90 days. If another manufacturer later proposed to manufacture the same chemical substance, EPA would require that the manufacturer give premanufacture notification and wait at least 90 days before starting manufacture. This approach would treat the two manufacturers equally. However, it would impose a burden on the first manufacturer reporting the chemical substance for the inventory in that it would have to stop manufacturing even though it may have been manufacturing the substance for some time before the inventory.

d. If a manufacturer reported a confidential chemical substance for the inventory, EPA could require the manufacturer to give premanufacture notification. However, EPA would not require the manufacturer to cease manufacture for 90 days. If another manufacturer later proposed to manufacture the same chemical substance, EPA would require that manufacturer to give premanufacture notification and wait at least 90 days before starting manufacture. This approach would give the first manufacturer reporting the chemical substance an advantage by allowing him to continue manufacture. Section 5(a) states that no person may manufacture a new chemical substance without giving notice at least 90 days before beginning manufacture. This approach would violate section 5.

4. Assuming that some chemical identities are temporarily or permanently entitled to confidential treatment, how is EPA to answer Freedom of Information requests for disclosure of records concerning confidential substances? EPA perceives two positions it could take:

a. If a request were made for disclosure of records concerning a reported chemical substance, the chemical identify of which was entitled to confidential treatment, EPA could reply: "EPA has such a record, but the request is denied because the record contains trade secrets that are exempt from disclosure by virtue of 5 U.S.C. 552(b)(4)." If a request were made for disclosure of records concerning a chemical substance that had not been reported, EPA would reply: "EPA has no such record." The result of this type of answer would be to confirm whether a particular chemical substance had been reported for the inventory and, therefore, was being manufactured, imported, or processed for commercial purposes.

b. If a request were made for disclosure of records concerning a particular chemical substance that did not appear on the inventory by chemical identity, EPA could reply: "The request is denied either because the record in question is exempt from mandatory disclosure by virtue of 5 U.S.C. 552(b)(4) or because EPA has no such record." This same answer would be given whether or not the chemical substance in question had been reported to EPA. In this way EPA would be able to give an answer that would allow the requester to pursue any judicial remedies under the Freedom of Information Act. At the same time, EPA would not have disclosed whether the particular chemical substance was being manufactured, imported, or processed for commercial purposes.

Any combination of the options under the four issues set forth above might be selected in the final approach taken by EPA. EPA intends to evaluate the advantages and disadvantages of these options before determining how to handle claims of confidentiality under the inventory reporting. EPA would appreciate comments concerning the various alternatives mentioned here and any other possible approaches.

Enforcement Liability

Because of the great importance of compiling a sound data base on the chemical substances in commerce, the Agency considers violation of these reporting requirements to be a serious violation of TSCA. Section 15(3)(B) of TSCA makes it "unlawful for any person to fail or refuse to submit reports, notices, or other information, as required by this Act or a rule thereunder." Section 16(a) provides for civil penalties of up to $25,000 for each violation of Section 15. Section 16(b) provides that criminal penalties of not more than $25,000 for each day of violation, or imprisonment for not more than one year, may be imposed on "any person who knowingly or willfully violates any provision of Section 15." Section 17(a) authorizes specific enforcement to restrain any violation of Section 15 and to compel the taking of any action required by or under this Act.

The Agency considers the most serious violations of these reporting requirements to include the following: (1) failure to report information required under the regulation; (2) falsification of information reported under the regulation; and (3) reporting of chemical substances which are specifically excluded from the inventory, such as chemical substances manufactured solely in small quantities for research and development.

Enforcement liability attaches to any person submitting a report for the inventory, including (1) those required to report, (2) manufacturers, importers, or processors who are not required to report, but voluntarily do so, (3) trade associations acting as agents for manufacturers, processors, or importers, and (4) those manufacturers, processors, or importers certifying to trade associations that a given chemical substance was manufactured, imported, or processed for a commercial purpose since January 1, 1975.

Economic Impact Analysis Statement

EPA has determined that the regulation does not require the compilation of an Economic Impact Analysis Statement as required by Executive Order 11821. This determination is based on the cost estimate for compilation of the inventory as originally proposed and an estimate of the additional burden created by the added reporting requirements. EPA has not completed an adequate analysis of the cost of complying with the requirements of this regulation because of the lack of time between the decision to repropose these regulations and the actual date for reproposal of the regulations. EPA is obtaining a better cost estimate at this time and will perform an Economic Impact Analysis if the cost exceeds the criteria for a major action, in general, $100 million annual cost. A discussion of the cost will accompany the final regulation.

The Environmental Protection Agency has determined that this document does not contain a major proposal requiring preparation of an Economic Impact Analysis Statement under Executive Order 11821 and OMB Circular A-107

Dated: July 27, 1977.

Douglas M. Costle
Administrator

Parts 700 and 710 as previously proposed are withdrawn, and it is proposed to establish a new Part 710 to read as follows:

PART 710 – INVENTORY REPORTING

§ 710.1 – Scope and Compliance

1. This part establishes regulations governing reporting by certain manufacturers, processors, and importers of chemical substances under Section 8(a) of the Toxic Substances Control Act (15 U.S.C. 2607). That subsection requires EPA to issue regulations for the purpose of compiling the inventory of chemical substances manufactured or processed for a commercial purpose, as required by Section 8(b) of the Act. In accordance with Section 8(b), EPA periodically will amend the inventory to include chemical substances which are manufactured, processed, or imported for commercial purposes; will revise the categories of chemical substances; and will make other amendments as appropriate.

2. Section 15(3) of TSCA makes it unlawful for any person to fail or refuse to submit information required under these reporting regulations. In addition, Section 15(3) makes it unlawful for any person to fail to keep, and permit access to, records required by these regulations. Section 16 provides that a violation of Section 15 renders a person liable to the U.S. for a civil penalty and possible criminal prosecution. Pursuant to Section 17, the Government may seek judicial relief to compel submittal of Section 8(a) information and to otherwise restrain any violation of Section 8(a).

3. Each person who reports under these regulations shall permit access to, and the copying of, records that document information reported under these regulations.

§ 710.2 – Definitions

For the purposes of this part, the following terms shall have the meaning contained in the Federal Food, Drug, and Cosmetic Act, 21 U.S.C. 321, and the regulations issued under such Act: "cosmetic," "device," "drug," "food," and "food additive." In addition, the term "food" includes poultry and poultry products, as defined in the Poultry

Products Inspection Act, 21 U.S.C. 453; meats and meat food products, as defined in the Federal Meat Inspection Act, 21 U.S.C. 60; and eggs and egg products, as defined in the Egg Products Inspection Act, 21 U.S.C. 1033. The term "pesticide" shall have the meaning contained in the Federal Insecticide, Fungicide, and Rodenticide Act, 7 U.S.C. 136, and the regulations issued thereunder. The following terms shall have the meaning contained in the Atomic Energy Act of 1954, 42 U.S.C. 2014, and the regulations issued thereunder: "nuclear by-product material," "nuclear source material," and "special nuclear material." In addition, "Act" means the Toxic Substances Control Act, 15 U.S.C. 2061, et seq.

"Administrator" means the Administrator of the U.S. Environmental Protection Agency or any employee of the Agency to whom the Administrator may either herein or by order delegate his authority to carry out his functions, or any person who shall by operation of law be authorized to carry out such functions.

An "article" is a manufactured item (1) which is formed to a specific shape or design during manufacture, (2) which has end use function(s) dependent in whole or in part upon its shape or design during end use, and (3) which is functional in its end use(s) without change of chemical composition during its end use; except that (d) fluids and particles are not considered articles regardless of shape or design.

"By-product" means a chemical substance produced without separate commercial intent during the manufacture of processing of other chemical substance(s) or mixture(s).

"Chemical substance" means any organic or inorganic substance of a particular molecular identity including (1) any combination of such substances occurring in whole or in part as a result of a chemical reaction or occurring in nature and (2) any chemical element or uncombined radical, and (3) except that "chemical substance" does not include:

1. Any mixture
2. Any pesticide when manufactured, processed, or distributed in commerce for use as a pesticide
3. Tobacco or any tobacco product, but not including any derivative products
4. Any nuclear source material, special nuclear material, or nuclear by-product material
5. Any pistol, firearm, revolver, shells, and cartridges
6. Any food, food additive, drug, cosmetic, or device, when manufactured, processed, or distributed in commerce for use as a food, food additive, drug, cosmetic, or device

"Commerce" means trade, traffic, transportation, or other commerce (1) between a place in a State and any place outside of such State, or (2) which affects trade, traffic, transportation, or commerce described in clause (1).

"Distribute in commerce" and "distribution in commerce" when used to describe an action taken with respect to a chemical substance or mixture or article containing a substance or mixture mean to sell or to transfer the ownership of the substance, mixture, or article in commerce, to introduce or deliver for introduction into commerce, or the introduction or delivery for introduction into commerce of the substance, mixture, or article; or to hold, or the holding of, the substance, mixture, or article after its introduction into commerce.

"EPA" means the U.S. Environmental Protection Agency.

"Establishment" means an economic unit, generally at a single site, as defined for purposes of the Standard Industrial Classification of Establishments. There may be more than one establishment at a single site.

"Importer" means any person who imports any chemical substance into the customs territory of the U.S. and includes: (1) the person primarily liable for the payment of any duties on the merchandise, or (2) an authorized agent acting on his behalf (as defined in 19 CFR 1.11).

"Impurity" means a chemical substance which is unintentionally present with another chemical substance.

"Intermediate" means any chemical substance (1) which is deliberately present in a chemical reaction sequence used to manufacture or process another chemical substance, (2) whose presence is known or reasonably ascertainable, and (3) which could be isolated and identified under conditions which are practically encountered in the environment.

"Manufacture" means to produce or manufacture in the U.S. or import into the customs territory of the U.S.

"Manufacture, process, or import 'for commercial purposes' " means to manufacture, process, or import: (1) for distribution in commerce, (2) for use as a catalyst or an intermediate. (3) for use by the manufacturer, or (4) for test marketing purposes.

"Mixture" means any combination of two or more chemical substances if the combination does not occur in nature and is not, in whole or in part, the result of a chemical reaction; except that "mixture" does include (1) any combination which occurs, in whole or in part, as a result of a chemical reaction if none of the chemical substances comprising the combination is a new chemical substance and if the combination could have been manufactured for commercial purposes without a chemical reaction at the time the chemical substances comprising the combination were combined, (2) hydrates of a chemical substance or hydrated ions formed by association of a chemical substance with water.

"New chemical substance" means any chemical substance which is not included in the inventory compiled and published under Subsection 8(b) of the Act.

"Person" means any natural or juridical person including any individual, corporation, partnership, or association, any State or political subdivision thereof, or any municipality, any interstate body and any department, agency, or instrumentality of the Federal Government.

"Process" means the preparation of a chemical substance

or mixture, after its manufacture, for distribution in commerce (1) in the same form or physical states as, or in a different form or physical state from, that in which it was received by the person so preparing such substance or mixture, or (2) as part of an article containing the chemical substance or mixture.

"Processor" means any person who processes a chemical substance or mixture.

"Site" means each contiguous property unit where a chemical substance is manufactured or processed whether or not such site is independently owned or operated. Property divided only by a public right-of-way shall be considered one site. For the purposes of imported chemical substances, the site shall be the business address of the importer.

"Small manufacturer of importer" means a manufacturer who (1) has only a single manufacturing site, and either (2) has total annual sales of less than $100,000 based on the manufacturer's latest complete fiscal year or (3) has no more than 2,000 pounds annual production (i.e., amount manufactured and imported) of each manufactured chemical substance. In the case of a company, which is owned or controlled by another company, such factors would apply to the parent company and all companies owned or controlled by it taken together.

"Small quantities for purposes of scientific experimentation or analysis or chemical research on, or analysis of, such substance or another substance, including any such research or analysis for the development of a product" (hereinafter sometimes shortened to "small quantities for research and development") means quantities of a chemical substance manufactured or processed or proposed to be manufactured or processed that (1) are no greater than reasonably necessary for such purposes and (2) after (the effective date of premanufacture notification requirements), are used by, or directly under the supervision of, a technically qualified individual(s).

"State" means any State of the U.S., the District of Columbia, the Commonwealth of Puerto Rico, the Virgin Islands, Guam, the Canal Zone, American Samoa, the Northern Mariana Islands, or any other territory or possession of the U.S.

"Technically qualified individual" means a person who, because of his education, training, or experience, or a combination of these factors, is capable of appreciating the health and environmental risks associated with exposure to the chemical substance which is used under his supervision, and who (1) is responsible for enforcing appropriate methods of conducting scientific experimentation, analysis, or chemical research in order to minimize such risks and (2) is responsible for the safety assessments and clearances related to the procurement, storage, use, and disposal of the chemical substance as may be appropriate or required within the scope of conducting the research and development activity.

"Test marketing" means the distribution of no more than a predetermined amount of a chemical substance, or mixture or article containing that chemical substance, by a manufacturer or processor to no more than a defined number of potential customers to explore market capability in a competitive situation during a predetermined testing period prior to the broader distribution in commerce.

"United States," when used in the geographic sense, means all of the States, territories, and possessions of the U.S.

§ 710.3 – Applicability: Who Must Report; Who May Report; Who May Not Report

Paragraphs (A), (B), and (C) of this section identify the persons subject to these requirements with respect to reporting chemical substances in accordance with § 710.4. Paragraph (D) of this section identifies the persons who may not report chemical substances for the inventory.

A. Who is required?

1. Manufacturers. Any person who manufactures, or has manufactured since January 1, 1977, a chemical substance(s) for a commercial purpose in an establishment included in the Chemical and Allied Products sector (as defined by Standard Industrial Classification (SIC) Major Group 28) or Petroleum Refining sector (as defined by SIC Group 2911) must report concerning the chemical substance(s) manufactured in that establishment.

2. Importers. Any person who imports, or has imported since January 1, 1977, a chemical substance(s) into the United States for a commercial purpose must report concerning that chemical substance(s).

B. Who may report?

1. In addition to those persons required to report by paragraph (A) of this section, any person who has manufactured, imported, or processed a chemical substance for a commercial purpose since January 1, 1975 may report concerning that chemical substance.

2. If a person manufactured or imported a chemical substance prior to January 1, 1975 but the substance was processed after that date, he may report that substance for the inventory if he certifies that the substance was processed after January 1, 1975.

3. A trade association may report on behalf of any person who would be permitted to report under paragraphs (B)(1) and (2) of this section. For every chemical substance reported by a trade association, at least one manufacturer, processor, or importer must have certified to that trade association, and be able to document to EPA in accordance with § 710.1(c), that the chemical substance was manufactured, imported, or processed for commercial purposes since January 1, 1975.

C. Who may report after publication of the inventory? During the 120-day period after the first publication of the inventory, any person who has processed or used a chemical substance (including the manufacture of a mixture or article containing that chemical substance) for a commercial purpose since January 1, 1975 may report that chemical substance if it was not included in the inventory.

Note – Premanufacture notification requirements under Section 5 for manufacturers and importers of new chemical substances will begin 30 days after the first publication of the inventory and will apply to all chemical substances not included on the first inventory.

Who may not report.

1. No person may report any chemical substance which is automatically included in the inventory under § 710.4(b).

2. No person may report any chemical substance which is excluded from the inventory under paragraphs (c) or (d) of § 710.4.

3. No person may report any chemical substance which has not been manufactured, processed, or imported for a commercial purpose since January 1, 1975.

§ 710.4 – Scope of the Inventory

Chemical substances Subject to These Regulations

The following chemical substances are manufactured, imported, or processed "for a commercial purpose".

Chemical substances which are manufactured, imported, or processed:

1. For distribution in commerce
2. For use as a catalyst or as an intermediate
3. For use by the manufacturer
4. For test marketing purposes

Naturally Occurring Chemical Substances Automatically Included

Any chemical substance which is naturally occurring and which is either unprocessed or processed only by manual, mechanical, or gravitational means; by dissolution in water; or by heating solely to remove water, shall be automatically included in the inventory under the category "Naturally Occurring Chemical Substances." Examples of such substances are (1) Raw agricultural commodities; (2) water, air, natural gas, and crude oil; and (3) rocks, ores, and minerals.

Substances Excluded by Definition or Section 8(b) of TSCA

The following substances are excluded from the inventory:

1. Any substance which is not considered a "chemical substance" as provided in Subsection 3(2)(B) of the Act and in the definition of "chemical substance" in § 710.2

2. Any mixture as defined in § 710.2. This term will include alloys, inorganic glasses, ceramics, frits, and cements, including Portland cement

3. Any chemical substance manufactured, imported, or processed solely in small quantities for research and development, as defined in § 710.2

Chemical Substances Excluded from the Inventory

The following chemical substances are excluded from the inventory insofar as they are not manufactured for distribution in commerce as chemical substances per se and have no commercial purpose separate from the mixture or article of which they may be a part.

Note – In addition, chemical substances excluded here would not be subject to premanufacture notification under Section 5 of the Act.

1. Any impurity.

2. Any by-product which has no commercial purpose.

Note – A by-product which has commercial value to municipal or private organizations who (i) burn it as a fuel, (ii) dispose of it as a waste, including in a landfill or for enriching soil, or (iii) extract component chemical substances which may have some commercial value, may be included on the inventory.

3. Any chemical substance which is the result of a chemical reaction that may occur incidental to exposure of another chemical substance, mixture, or article to environmental factors such as air, moisture, microbial organisms, or sunlight.

4. Any chemical substance which is the result of a chemical reaction incidental to storage of a chemical substance or mixture.

5. Any chemical substance which is the result of a chemical reaction that may occur upon end use of other chemical substances or mixtures such as adhesives, paints, miscellaneous cleansers or other housekeeping products, fuels and fuel additives, water softening and treatment agents, and which is not itself manufactured for distribution in commerce.

6. Any chemical substance which is the result of a chemical reaction that occurs upon use of curable plastic or rubber molding compounds, inks, drying oils, metal finishing compounds, adhesives, paints, or other chemical substances used to manufacture an article destined for the marketplace without further chemical change of the chemical substance except for those chemical changes that may occur as described elsewhere in this paragraph.

7. Any chemical substance which occurs as the result of a chemical reaction when a stabilizer, colorant, odorant, antioxidant, filler, solvent, carrier, surfactant plasticizer corrosion inhibitor, antifoamer or defoamer, dispersant, precipitation inhibitor, binder, emulsifier, deemulsifier, dewatering agent, agglomerating agent, adhesion promoter,

flow modifier, pH neutralizer, sequestrant, coagulant, flocculant, fire retardant, lubricant, chelating agent, quality control reagent, or a chemical substance which is solely intended to impart a specific physicochemical characteristic functions as intended.

§ 710.5 – How to Report

A. General instructions

1. Except for small manufacturers or importers, any person who is required to report under Section 710.3(a) shall follow the reporting procedures of paragraphs (B), (C), and (D) of this section.

2. Any person who chooses to report under § 710.3(b) shall follow the reporting procedures of paragraphs (B), (C), and (D) (3) of this section. In addition, the Agency encourages those persons to report in accordance with paragraphs (D)(1), (D)(2), and (D)(4) of this section. A trade association may report aggregated production data under paragraph (D)(4) of this section.

3. Any person who is required to report under § 710.3(a) and who is a small manufacturer or importer as defined in § 710.2 shall follow the reporting procedures of paragraphs (B), (C), and (D)(1) and (D)(3) of this section. In addition, the Agency encourages small manufacturers to report in accordance with paragraphs (D)(2) and (D)(4) of this section.

B. Reporting the identity of a chemical substance

1. To report a chemical substance, a person shall first consult the TSCA Candidate List of Chemical Substances and any amendment to the Candidate List.

2. To report a chemical substance found in the Candidate List, or in an amendment to the List, a person must complete, sign, and submit EPA inventory reporting Form A (EPA Form No. –).

3. To report a chemical substance not found in the Candidate List, or in an amendment to the List, a person must complete, sign and submit EPA inventory reporting Form B (EPA Form No. –).

4. For assistance in using the Candidate List or the reporting forms, consult "Guide to the Use of the TSCA Candidate List of Chemical Substances and Instructions for Reporting" published in Appendix A of these regulations.

C. Reporting polymers

1. To report a polymer a person must list in the description of the polymer composition at least those monomers used at greater than two weight percent in the manufacture of the polymer.

2. Those monomers used at two weight percent or less in the manufacture of the polymer may be included as part of the description of the polymer composition.

3. For purposes of of this paragraph, the "weight percent" of a monomer is the weight of the monomer expressed as a percentage of the weight of the polymeric chemical substance manufactured.

D. Reporting other information concerning a chemical substance

1. Designate whether the person manufactures, processes and/or imports the chemical substance.

2. Report the site(s) at which the person manufactures, processes, and/or imports the chemical substance.

3. Designate whether the person manufactures and processes the chemical substances only within a site and does not distribute the chemical substance, or any mixture or article containing that substance, for commercial purposes outside that site.

4. Report the amount of the chemical substance which the person manufactured at each site and/or imported during calendar year 1976. If the person did not manufacture or import the chemical substance during 1976, report the amount manufactured and/or imported during 1975 or projected for 1977. If there has been no manufacture or importation since January 1, 1975, report the amount distributed to others for any purpose since that date.

E. Importers

1. Any importer who is required to report or who chooses to report a chemical substance for the inventory may authorize the foreign supplier of an imported chemical substance(s) to report to EPA on behalf of the importer if both the foreign supplier and the importer sign the declarations provided on the reporting form.

2. The importer has the ultimate responsibility for reporting all information required by this part and for the completeness and truthfulness of such information. If certain information is not or cannot be provided by the foreign supplier, it must be provided by the importer.

§ 710.6 – When to Report

All reports concerning chemical substances manufacture, processed or imported for a commercial purpose during the period January 1, 1975 to (the effective date of these regulations) shall be submitted by (90 days after the effective date of these regulations).

All reports concerning chemical substances which are manufactured, processed, or imported for a commercial purpose for the first time during the period (the effective date of these regulations) to (the effective date of premanufacture notification regulations) shall be submitted when such manufacturing, processing, or importation begins.

§ 710.7 – Confidentiality

A. A manufacturer, importer, or processor may claim that for a particular chemical substance any or all of the following items of information submitted under this part are entitled to confidential treatment:

1. Company name
2. Site
3. The specific chemical name or identity
4. Whether the chemical substance is manufactured, imported, or processed
5. Whether the chemical substance is manufactured and processed only within one site and not distributed for commercial purposes outside that site
6. The quantity manufactured, imported, or processed

B. Any claims of confidentiality must accompany the information at the time it is submitted to EPA. The claims must appear on the form on which the information is submitted to EPA and in the manner described on the form.

C. Any information that is covered by a claim made as specified will be disclosed by EPA only to the extent permitted by, and by means of, the procedures set forth in Part 2 of this title (41 FR 36902).

D. If no claim accompanies the information at the time it is submitted to EPA, the information may be made public by EPA without further notice to the submitter.

E. If a claim of confidentiality is asserted concerning the specific chemical name or identity of a particular chemical substance, the person making the claim shall furnish EPA with (1) the specific chemical name and identity and (2) a proposed generic name which is only as generic as necessary to protect the confidential identity of the particular chemical substance.

[FR Doc. 77-22107 Filed 7-28-77; 11:02 a.m.]

Section X

References and Resources

ENVIRONMENTAL AND TOXICOLOGICAL TESTING LABORATORIES

The following list of laboratories was obtained from the Society of Toxicology, Toxicology Laboratory Survey, March 1976. Originally, the Technical Committee prepared a survey questionnaire that was mailed to all members of the Society of Toxicology and all other persons or institutions known to be engaged in toxicity testing or toxicology research. This listing is a product of that survey through a single mailing voluntary response.

Recently, CRC Press contacted each of the following laboratories to update the information included in the synopsis. Consequently, the listing constitutes the most current information available and an accessible compendium of laboratory services conducting environmental/toxicological testing. The following material is presented strictly as a resource listing and does not constitute certification or licensing of these laboratories. CRC Press gratefully acknowledges the Society of Toxicology for their cooperation in this endeavor.

Kettering Meyer
Southern Research Institute
Birmingham, Alabama

Individual to contact and questionnaire prepared by Donald L. Hill

Experience with:
Acute toxicity, irritation
mouse, rat, guinea pig, hamster
Subacute toxicity, oral, dermal, and parenteral
mouse, rat, guinea pig, hamster
Chronic toxicity, oral, dermal, and parenteral
mouse, rat, guinea pig, hamster
Carcinogenicity and Teratology:
mouse, rat, guinea pig, hamster
Metabolism, Distribution, Excretion
mouse, rat, guinea pig, hamster, rabbit, dog, cat

Routes of administration:
i.p., i.v., oral, and dermal

Classes of compounds studied:
Carcinogens, anticancer agents, hypoglycemic agents, urinary analgesics, antimicrobial agents, antimalarial agents, amebacides, amino acids, ascaricides, and commercial chemicals.

Experience with the **paint industry**

Capabilities available:
In-house pathology, clinical chemistry, data reduction analysis, and experience and capability in handling radioactive materials

Personnel:
253 persons in laboratory including 59 professionals: 3 toxicologists, 3 pathologists, 7 pharmacologists, 15 biochemists, and 14 chemists

Laboratory routinely performs contract research

University of Alabama
Department of Pharmacology
Birmingham, Alabama 35294

Individual to contact: Dr. Roy L. Mundy

Questionnaire prepared by Dr. Robert S. Teague

Experience with:
Metabolism, Distribution, Excretion
mouse, rat, guinea pig, hamster, rabbit, dog, cat
Subacute toxicity, oral, parenteral; Chronic toxicity, oral, parenteral
mouse, rat, guinea pig, hamster, rabbit, dog, cat
Teratology
chicken

Routes of administration:
oral, s.c., i.m., i.p., and i.v.

Classes of compounds studied and specific procedures experienced:
Drugs (chemical warfare agents, antiradiation compounds, antimalarials, hormones (estrogens), solvents, alcohols, endotoxins; LD 50's, ED 50's, various pharmacological screening procedures (smooth muscle baths, whole animal studies, pharmacokinetics, drug metabolism, organ pathology)

Experience with **government and pharmaceutical** industry

Experience in representation of data before regulatory agencies

Capabilities available:
In-house pathology, clinical chemistry, data reduction and analysis, pulmonary physiology, experience and capability to handle radioactive materials

Personnel:
6 persons in laboratory; 4 professionals, 1 toxicologist, 2 pharmacologists; 1 biochemist, 1 teratologist, 1 chemist

Laboratory frequently performs contract research

Arizona

University of Arizona College of Pharmacy
Department of Pharmacology and Toxicology
Tucson, Arizona 85721

Individual to contact: A. L. Picchioni, Ph.D.

Experience with:
- Acute toxicity, irritation
  - mouse, rat, dog
- Subacute toxicity, oral
  - rat, dog
- Teratology
  - rat
- Metabolism, Distribution, Excretion
  - rat

Routes of administration:
oral, i.v., i.p., s.c., and intracardiac

Classes of compounds studied:
Antiepileptic agents, tricyclic antidepressants, and other drugs, and heavy metals (cadmium)

Capable of clinical chemistry

Personnel:
4 persons in the laboratory including 3 pharmacologists

Laboratory frequently performs contract research

University of Arizona
Department of Entomology
Tucson, Arizona 85721

Individual to contact and questionnaire prepared by Dr. George W. Ware

Experience with:
- Acute toxicity, irritation
  - mouse, rat
- Metabolism, Distribution, Excretion
  - mouse, rat, guppy

Routes of administration:
oral, i.p., and i.v. (tail)

Classes of compounds studied:
Pesticides (chlorinated insecticides, organophosphates, and herbicides)

Experience with **pesticide manufacturers**

Ability and some experience in presentation of data before regulatory agencies

Capabilities available:
Data reduction and analysis, experience and capability to handle radioactive materials

Personnel:
6 in laboratory: 2 toxicologists, 2 biochemists, 2 chemists

Laboratory will consider contract research

Applied Biological Sciences Labs, Inc.
8320 San Fernando Road
Glendale, California 91201

Individual to contact and questionnaire prepared by Dr. J. B. Michaelson

Experience with:

Subacute toxicity, oral, parenteral; chronic toxicity, oral, parenteral;

Reproduction; Metabolism; Distribution; Excretion

mouse, rat, guinea pig, hamster, rabbit, dog, cat, chicken, duck, quail, other bird, trout, bluegill, other fish, aquatic organisms

Subacute toxicity, inhalation; Chronic toxicity, inhalation

mouse, rat, guinea pig, hamster, rabbit, dog, cat, chicken, duck, quail, other bird

Acute toxicity, irritation; Subacute toxicity, dermal; Chronic toxicity, dermal; Carcinogenicity; Teratology; Mutagenicity

mouse, rat, guinea pig, hamster, rabbit, dog, cat

Routes of administration:

oral, i.p., s.c., i.v., i.m., inhalation, dermal, intradermal, etc.

Classes of compounds studied:

Chemicals, drugs, food additives, plastics, agricultural products, chemical preservatives, cosmetics, detergents, and industrial cleaning compounds

Experience with **all types of industry**

Ability or experience in representation of data before regulatory agencies

Capabilities available:

In-house pathology, clinical chemistry, data reduction and analysis, pulmonary physiology, and inhalation experience

Personnel:

18 in laboratory: 11 professionals – 3 toxicologists, 1 pathologist, 2 pharmacologists, 1 biochemist, 1 chemist, 1 microbiologist, 2 M.D.s

Laboratory routinely performs contract research

University of California
Henderson Laboratory
Department of Pharmacology
School of Medicine
Davis, California

Individual to contact and questionnaire prepared by Dr. Gary L. Henderson

Experience with:

Acute toxicity, irritation; subacute toxicity, oral, parenteral; metabolism; distribution; excretion

mouse, rat, guinea pig, hamster, rabbit, dog, cat, other mammal (subhuman primates)

Routes of administration:

p.o. and i.v.

Classes of compounds studied and specific procedures:

Narcotic analgesics (methodone, acetyl methadol), gas chromatography, and radio tracer methodology

Experience with **federal government (National Institute on Drug Abuse) and pharmaceutical** industry

Ability or experience in representation of data before FDA and NIDA

Capabilities available:

In-house pathology, clinical chemistry, data reduction and analysis, experience and capability to handle radioactive materials

Personnel:

2 to 3 professionals in laboratory: 1 toxicologist, 1 pharmacologist, and 1 biochemist

Laboratory frequently performs contract research

University of San Francisco
Institute of Chemical Biology
San Francisco, California 94117

Individual to contact and questionnaire prepared by Dr. Arthur Furst, Director

Experience with:

Acute toxicity, irritation; subacute toxicity, oral, parenteral; chronic toxicity, oral, parenteral; carcinogenicity
  mouse, rat, guinea pig, hamster
Intratracheal
  mouse, guinea pig, hamster
Reproduction; teratology
  mouse, rat
Mutagenicity
  bacteria

Routes of administration:

oral (food, p.o.), i.m., intratracheal, skin painting, and intrathoracic

Classes of compounds studied:

Metal carcinogens (leading lab), i.m. and oral; hydrocarbons and asbestos – intratracheal, mice

Experience with **testing laboratories**

Ability or experience in representation of data before congressional committees

Capabilities available:

In-house pathology, clinical chemistry, data reduction and analysis, intratracheal experience, experience and capability to handle radioactive materials (^{14}C, ^{129}I, ^{131}I, ^{2}H, ^{63}Ni)

Personnel:

4 to 8 personnel in laboratory: 2 toxicologists, 1 pathologist, 1 pharmacologist, 2 biochemists, and 2 chemists

Laboratory frequently performs contract research

University of Connecticut School of Medicine
Toxicology-Carcinogenesis Division
Department of Laboratory Medicine
P.O. Box G
Farmington, Connecticut 06032

Individual to contact: Dr. F. William Sunderman, Jr.

Experience with:
Acute toxicity, irritation; subacute toxicity, oral and parenteral; chronic toxicity, oral and parenteral; carcinogenicity; metabolism, distribution, excretion
mouse, rat, guinea pig, hamster, rabbit, dog, cat
Reproduction and Teratology
mouse, rat, guinea pig, hamster, rabbit
Mutagenicity
mouse, rat, guinea pig, hamster, rabbit, dog

Routes of administration:
inhalation, intratracheal, intrarenal, intrahepatic, intrapleural, i.m., and i.p.

Classes of compounds studied:
Metals, asbestos, and chelating drugs

Experience with **metallurgical, chemical, and pharmaceutical industries**

Ability or experience in representation of data before regulatory agencies

Capabilities available:
In-house pathology, clinical chemistry, data reduction and analysis, pulmonary physiology, inhalation experience, and experience and capability to handle radioactive materials

Personnel:
16 persons in laboratory including 1 toxicologist, 1 pathologist, 1 pharmacologist, 1 biochemist, 1 teratologist, and 1 chemist

Laboratory routinely performs contract research

Florida

Dawson Research Corporation
P.O. Box 8272
Orlando, Florida 32806
305–851–3110

Individuals to contact: Mr. Gary A. Gardner, Vice President or Mr. Michael S. Kangiser, Director of Marketing

Experience with:
Acute toxicity (acute oral, oral LD_{50}, dermal LD_{50}, dose range, eye irritation, skin irritation, corrosivity, sensitization, hazardous substance toxicity); subacute toxicity; chronic toxicity; reproduction; teratology, efficacy; pesticides efficacy and toxicity on citrus, meat, milk and egg residue; C^{14} metabolism; simulated field toxicity and carcinogenesis studies. Also perform histology, clinicopathology, and pathology studies and consultant work

Animal models:
rat, mouse, rabbit, guinea pig, hamster, cat, cow, pig, sheep, goat, horse, quail, dog, primate

Routes of administration:
gastric intubation, intramuscular injection, intravenous injection, subcutaneous injection, intradermal injection, vitreal injection, dermal application, capsule (oral) and feed and water

Experience with **pharmaceutical, chemical, device, cosmetic, and food manufacturers**

Upon request, aids with the representation of data before regulatory agencies

Conducts all studies in compliance with the Good Laboratory Practices Act

Personnel:
66 professional; doctorate or equivalent 6; professional, bachelor or equivalent 23; technical 27; support 10

University of Miami
Research and Teaching Center of Toxicology
Department of Pharmacology
P.O. Box 24 8216
Coral Gables, Florida 33124

Individual to contact and questionnaire prepared by Dr. William B. Deichmann

Experience with:
Acute toxicity, irritation; subacute toxicity, oral
mouse, rat, guinea pig, rabbit, dog, pig, goat, sheep, chicken, duck, quail (new fencing required for pig, goat, sheep)
Subacute toxicity, inhalation
mouse, rat, guinea pig, rabbit, dog, chicken, duck, quail
Subacute toxicity, dermal
mouse, rat, guinea pig, rabbit, dog, pig
Subacute toxicity, parenteral
mouse, rat, guinea pig, rabbit, dog
Chronic toxicity, oral; carcinogenicity; reproduction; metabolism, distribution, excretion
mouse, rat, dog
Chronic toxicity, dermal, inhalation
mouse, rat

Routes of administration:
oral, dermal, inhalation, and parenteral

Classes of compounds studied and specific procedures:
Industrial chemicals and/or drugs for acute toxicity, eye, and skin irritation, hypersensitivity, subacute and chronic effects by inhalation, skin application, and feeding to determine carcinogenicity or other harmful effects; intensive study of phenolic compounds, nitroolefin air pollutants, organochlorine pesticides, aromatic amines

Experience with **chemical and pharmaceutical** industries

Ability or experience in representation of data before regulatory agencies

Capabilities available:
In-house pathology, clinical chemistry, data reduction and analysis, inhalation experience, experience and capability to handle radioactive materials

Personnel:
11 in laboratory: 4 professionals – 3 toxicologists, 2 pathologists, 3 pharmacologists, and 1 chemist

Laboratory routinely performs contract research

University of Miami School of Medicine
Department of Pharmacology
P.O. Box 520875, Biscayne Annex
Miami, Florida 33152

Individual to contact and questionnaire prepared by Dr. Jack L. Radomski

Experience with:

Carcinogenicity
 dog
Metabolism, distribution, excretion
 rat, hamster, rabbit, dog, cat

Routes of administration:

oral, i.p., s.c., i.m., and i.v.

Classes of compounds studied and specific procedures:

Drugs, agricultural chemicals, environmental substances, GC-mass spectrometry, gas chromatography, thin-layer chromatography, high-pressure liquid chromatography, column chromatography

Experience with **chemical** industry

Ability or experience in representation of data before regulatory agencies

Capabilities available:

Clinical chemistry, data reduction and analysis, experience and capability to handle radioactive materials

Personnel:

9 in laboratory: 2 professionals – 1 toxicologist and 1 chemist

Will consider contract research

Georgia

Poultry Disease Research Center
953 College Station Road
Athens, Georgia 30601

Individual to contact and questionnaire prepared by W. L. Ragland

Experience with:
Subacute toxicity, oral, parenteral; Chronic toxicity, oral, parenteral; Carcinogenicity; Metabolism, Distribution, Excretion
chicken, duck, quail, turkey

Routes of administration:
oral, i.v., and i.p.

Classes of compounds studied:
Pesticides, antibiotics, biologics, coccidiostats

Specific procedures:
GLC, tissue residues, enzyme assay and chemical composition, subcellular distribution, some of these coupled with immunologic detection, PAGE, fluorescence analyses

Experience with pharmaceutical and feed industries and instrument manufacturers

Capabilities available:
In-house pathology, clinical chemistry, data reduction and analysis, and experience and capability to handle radioactive materials

Personnel:
40 persons in laboratory including 9 professionals: 1 toxicologist, 2 pathologists, and 3 biochemists

Laboratory will consider performing contract research

IIT Research Institute
10 West 35th Street
Chicago, Illinois 60616

Individual to contact: Dr. Richard Ehrlich

Questionnaire prepared by Dr. Mary C. Henry

Experience with:

Acute toxicity, irritation; Subacute toxicity, oral, dermal, parenteral;
 mouse, rat, guinea pig, hamster, rabbit, dog, rhesus, squirrel monkey
Subacute toxicity, inhalation; Chronic toxicity, inhalation
 mouse, rat, guinea pig, hamster, rabbit, dog, rhesus, squirrel monkey
Chronic toxicity, oral, dermal, parenteral
 mouse, rat, hamster, rabbit, dog, rhesus, squirrel monkey
Behavioral toxicology, Developmental toxicology
 mouse, rat, monkey
Carcinogenicity
 mouse, rat, hamster
Reproduction
 mouse, rat, rabbit
Teratology
 rat
Metabolism, Distribution, Excretion
 rat
Mutagenicity
 dominant lethal, mouse, rat; Ames test
In vitro toxicity
 tissue culture, organ culture

Routes of administration:
Intragastric, i.v., i.p., s.c., skin painting, intratracheal, and inhalation

Classes of compounds studied and specific procedures:
Drugs, industrial chemicals, agricultural chemicals, foods, food additives, gaseous and particulate pollutants, heavy metals, and liquid aerosols

Experience with **government agencies and pharmaceutical, chemical and food industry**

Experience in presentation of data before FDA and EPA

Capabilities available:
In-house microscopic pathology, clinical chemistry, electron and scanning microscopy, data reduction and analysis, pulmonary physiology, inhalation, and experience and capability to handle radioactive materials

Personnel:
110 in laboratory: 65 professionals – 5 toxicologists; 2 pathologists; 1 laboratory animal veterinarian; 2 pharmacologists; 3 biochemists; 2 teratologists; 7 microbiologists, and more than 20 analytical and fine particle chemists

Laboratory routinely performs contract research

Industrial BIO-TEST Laboratories, Inc.
1810 Frontage Road
Northbrook, Illinois 60062

Individual to contact: Dr. M. L. Keplinger

Experience with:

- Acute toxicity, irritation
  - mouse, rat, guinea pig, hamster, rabbit, dog, cat, cow, pig, monkey, chicken, duck, quail, pigeon, pheasant, trout, bluegill, catfish, daphnia, shrimp, crayfish, crab
- Subacute toxicity, oral, dermal; chronic toxicity, oral, parenteral
  - mouse, rat, guinea pig, hamster, rabbit, dog, cow, pig, monkey, duck, chicken, quail, pigeon, pheasant
- Metabolism, distribution, excretion
  - mouse, rat, guinea pig, hamster, rabbit, dog, cow, pig, goat, sheep, chicken, duck, quail, catfish
- Reproduction
  - mouse, rat, guinea pig, hamster, rabbit, dog, pig, chicken, duck, quail, daphnia
- Chronic toxicity, dermal
  - mouse, rat, guinea pig, hamster, rabbit, dog, cow, pig, monkey, duck, chicken, quail, pigeon, pheasant
- Subacute toxicity, parenteral
  - mouse, rat, guinea pig, hamster, rabbit, dog, cow, pig, monkey
- Subacute toxicity, inhalation
  - mouse, rat, guinea pig, hamster, rabbit, dog, monkey, chicken
- Teratology
  - mouse, rat, guinea pig, hamster, rabbit, dog, pig, monkey
- Chronic toxicity, inhalation
  - mouse, rat, guinea pig, hamster, rabbit, dog, monkey
- Carcinogenicity
  - mouse, rat, guinea pig, hamster, rabbit, dog
- Mutagenicity
  - mouse, rat, hamster, dog

Routes of administration:
gavage, feeding, drinking water, i.v., i.p., subcutaneous, dermal, inhalation, immersion, eye instillation, implantation

Specific procedures:
Pharmacological/microbiological activity, carcinogenic, mutagenic, teratogenic, reproductive, acute and chronic oral, inhalation, dermal, eye irritation toxicity tests, clinical human evaluations pharmacological screens (cat, dog, mice, monkey)

Classes of compounds:
Medical devices (implants), agricultural chemicals, industrial solvents, industrial chemicals, food additives, cosmetics, soaps and detergents, and cleansers

Experience with **pharmaceutical, agricultural chemical, petroleum, chemical, food, cosmetic and toiletry industries**

Experience in representation of data before regulatory agencies

Capabilities available:
In-house pathology, clinical chemistry, data reduction and analysis, pulmonary physiology, inhalation experience, and experience and capability in handling radioactive materials

Personnel:
300 persons in laboratory include 45 professionals: 20 toxicologists, 5 pathologists, 3 pharmacologists, 6 biochemists, 5 teratologists, and 6 chemists

Laboratory routinely performs contract research

Travenol Laboratories, Inc.
6301 Lincoln Avenue
Morton Grove, Illinois

Individual to contact and questionnaire prepared by Dr. P. J. Garvin

Experience with:
- Acute toxicity, irritation
  - mouse, rat, guinea pig, rabbit, dog, cat
- Subacute toxicity, oral, parenteral; Chronic toxicity, oral
  - mouse, rabbit, dog
- Subacute toxicity, dermal
  - rat, guinea pig, rabbit
- Metabolism, Distribution, Excretion
  - mouse, rat, rabbit
- Reproduction; Teratology
  - rat, rabbit
- Subacute toxicity, inhalation
  - rat, guinea pig
- Chronic toxicity, parenteral
  - rat, dog
- Carcinogenicity
  - mouse, rat
- Mutagenicity
  - rat
- In vitro evaluation of Toxicity, Mutagenicity, and Carcinogenicity

Routes of adminstration:
oral, i.v., s.c., i.m., i.p., intrathecal, epidural, intradiscal, rectal, intravaginal, intraocular, intratracheal, and intravesicular

Classes of compounds studied and specific procedures:
Drugs, parenteral solutions, blood preservatives, biomedical materials; autoradiography, TEM, SEM, roentgenography

Experience with **pharmaceutical and biomedical devices** industries

Ability or experience in representation of data before regulatory agencies

Capabilities available:
Clinical chemistry, data reduction and analysis, pulmonary physiology (limited), inhalation experience (minimal), experience and capability to handle radioactive materials

Personnel:
21 in laboratory: 3 toxicologists; 3 cytologists; 1 pharmacologist; 3 biochemists; and 1 teratologist

Laboratory frequently performs government contract research

Indiana

Dow Chemical Company
Pathology-Toxicology Department
P.O. Box 68511
Indianapolis, Indiana 46268

Individual to contact and questionnaire prepared by Dr. J. E. LeBeau

Experience with:
- Subacute toxicity, oral
  - rat, guinea pig, hamster, rabbit, dog, cat, pig, chicken, duck
- Acute toxicity, irritation
  - mouse, rat, guinea pig, rabbit, dog, monkey
- Chronic toxicity, oral
  - mouse, rat, hamster, dog, monkey
- Pharmacokinetics
  - rat, monkey, dog
- Teratology
  - mouse, rat, rabbit, dog, chicken
- Subacute toxicity, parenteral
  - rat, rabbit, dog, monkey
- Carcinogenicity
  - mouse, rat, monkey
- Subacute toxicity, dermal
  - rat, guinea pig, rabbit
- Mutagenicity
  - mouse, rat
- Reproduction
  - rat

Routes of administration:
oral, i.v., i.p., topical, and i.m.

Classes of compounds studied:
Drugs, agricultural chemicals, industrial chemicals, medical devices, and biologics

Experience with **chemical, pharmaceutical, biological, and veterinary biologics/pharmaceutics industries**

Ability or experience in representation of data before regulatory agencies

Capabilities available:
In-house pathology, clinical chemistry, data reduction and analysis

Personnel:
30 in laboratory: 16 professionals – 8 toxicologists; 3 pathologists; 1 biochemist; 3 teratologists; 1 microbiologist; 1 electron microscopist

Laboratory frequently performs contract research

Indiana University
Section on Pharmacology
Medical Sciences Program
306 Myers Hall
Bloomington, Indiana 47401

Questionnaire prepared by R. P. Maickel

Experience with:
- Acute toxicity, irritation
  - mouse, rat, guinea pig
- Subacute toxicity, oral
  - mouse, rat
- Subacute toxicity, parenteral
  - mouse, rat, guinea pig
- Metabolism, Distribution, Excretion
  - mouse, rat, guinea pig

Routes of administration:
i.v., i.p., s.c., and i.m.

Experience with **pharmaceutical industry**

Ability or experience in representation of data before regulatory agencies (DEA, FDA)

Capabilities available:
Clinical chemistry, data reduction analysis, experience and capability to handle radioactive materials

Personnel:
10 persons in laboratory including 5 professionals with training in toxicology, pharmacology, biochemistry, and chemistry

Laboratory routinely performs contract research

Purdue University
West Lafayette
Indiana 47907

Individual to contact and questionnaire prepared by Dr. William W. Carlton and Dr. F. R. Robinson

Experience with:
Acute toxicity, irritation; Subacute toxicity, oral, dermal, parenteral
mouse, rat, guinea pig, hamster, rabbit, dog, cat, pig, chicken, duck
Chronic toxicity, oral, dermal, parenteral
mouse, rat, hamster, chicken
Carcinogenicity; Reproduction
mouse, rat, hamster

Classes of compounds studied:
Drugs, mycotoxins, agricultural chemicals, and carcinogens

Laboratory has experience with **pharmaceutical industry**

Capabilities available:
In-house pathology, inhalation experience

Personnel:
3 persons in laboratory – 1 toxicologist, 2 pathologists and 1 chemist

Laboratory will consider contract research

Purdue University
Department of Pharmacology and Toxicology
West Lafayette, Indiana

Individual to contact: Chairman

Questionnaire prepared by Gary P. Carlson and John H. Mennear

Experienced with:
Acute toxicity, irritation; Subacute toxicity, oral, parenteral; Chronic toxicity, oral, parenteral; Metabolism, Distribution, Excretion; Chemical interactions
mouse, rat, guinea pig, hamster, rabbit, dog, cat

Routes of administration:
oral, all parenteral, and inhalation

Classes of compounds studied and specific procedures:
Drugs, pesticides, organic solvents, heavy metals, and natural products; enzyme induction, drug metabolism, drug interactions, acute, subacute and chronic studies, isolated organs, and perfused organs

Experience with **pharmaceutical and consulting industries**

Capabilities available:
Clinical chemistry, data reduction and analysis, inhalation experience, experience and capability to handle radioactive materials

Personnel:
41 in laboratory (including 25 graduate students): 10 professionals – 2 toxicologists, 7 pharmacologists, and 1 biochemist

Laboratory will consider contract research

Iowa

Iowa State University
Veterinary Diagnostic Laboratory
Toxicology Section
College of Veterinary Medicine
Ames, Iowa 50010

Individual to contact and questionnaire prepared by William B. Buck

Experience with:
Acute toxicity, irritation; Subacute toxicity, oral, dermal, parenteral; Chronic toxicity, oral, dermal, parenteral
rat, hamster, rabbit, cow, pig, sheep, horse, squirrel monkey, chicken
Metabolism, Distribution, Excretion
rat, hamster, rabbit, dog, cat, cow, pig, sheep, horse, squirrel monkey, chicken
Reproduction; Teratology
rat, hamster, rabbit, dog, cat, cow, pig, sheep, horse, chicken
Forensic
dog, cat, cow, pig, sheep, horse

Routes of administration
oral, i.p., and dermal

Classes of compounds studied:
Organic iodine, organic arsenicals, lead insecticides, herbicides, fungicides

Field diagnostic services:
Rodenticides, insecticides, herbicides, fungicides, feed additives, biotoxins

Experience with **agricultural chemical producers and animal feed additive producers**

Experience in representation of data before regulatory agencies

Capabilities available:
In-house pathology, clinical chemistry, data reduction and analysis, experience and capability to handle radioactive materials

Personnel:
20 persons in laboratory; 10 professionals – 5 toxicologists; 3 pathologists; 2 chemists

Laboratory will consider contract research

University of Iowa
Department of Toxicology
Iowa City, Iowa 52240

Individual to contact and questionnaire prepared by R. J. Roberts

Experience with:
Acute toxicity
mouse, rat, guinea pig, hamster, rabbit, dog, pig
Subacute toxicity and Chronic toxicity (limited space)
Metabolism, Distribution, Excretion
mouse, rat, guinea pig, hamster, rabbit, dog, pig
Reproduction
mouse, rat, rabbit
Liver function; Lung toxicology
guinea pig, pig, rat, mouse

Classes of compounds studied:
Drugs (sed/hyp), antibiotics, salicylates, chemicals (CO, CS_2)

Experience with **drug industry**

Ability in representation of data before regulatory agencies

Capabilities available:
In-house pathology, clinical chemistry, data reduction and analysis, pulmonary physiology, inhalation experience, experience and capability to handle radioactive materials

Personnel:
8 in laboratory: 1 professional – 3 post-doctorals – clinicians – toxicologists; pharmacologists; biochemists

Laboratory will consider contract research

Kansas State University
Comparative Toxicology Laboratory
Manhattan, Kansas 66506

Individual to contact: Dr. F. W. Oehme

Experience with:

Acute toxicity, irritation; Subacute toxicity, parenteral; and Chronic toxicity, parenteral
- mouse, rat, guinea pig, hamster, rabbit, dog, cat, cow, pig, sheep, horse, chicken, duck, quail, trout, bluegill

Subacute toxicity, oral and dermal; and Chronic toxicity, oral and dermal
- mouse, rat, guinea pig, hamster, rabbit, dog, cat, cow, pig, sheep, horse, chicken, duck, quail

Carcinogenicity
- mouse

Reproduction
- mouse, rat, dog, cat, cow, pig, sheep, horse, gerbil, chicken, duck

Teratology
- mouse, rat, rabbit, dog, cat, pig, gerbil, chicken

Metabolism, Distribution, Excretion
- mouse, rat, rabbit, dog, cat, cow, pig, sheep, horse

Routes of administration:
dermal, ocular, i.v., s.c., and p.o.

Experience in **forensic chemical analysis** of all body fluids and tissues

Classes of compounds studied:
Drugs (most classes), pesticides, agricultural chemicals, environmental pollutants, heavy metals, naturally-occurring compounds (mycotoxins, plant extracts, venoms)

Specific procedures:
In vivo studies utilizing radioisotopes and the routine instrumentation for analysis of body fluids and tissues; AA; GC; various chromatography methods; spectrometry; ion-specific electrodes

Experience with **drug, floral, cosmetic, chemical, livestock, military, and aerospace industries**

Experience in representation of data before regulatory agencies

Capabilities available:
In-house pathology, clinical chemistry, data reduction and analysis, pulmonary physiology, and experience and capability in handling radioactive materials

Personnel:
13 in laboratory includes 8 professionals: 3 toxicologists, 1 pathologist, 1 pharmacologist, 1 biochemist, 1 chemist, and 1 physiologist

Laboratory frequently performs contract research

University of Kansas
Department of Pharmacology
Medical Center
Kansas City, Kansas

Individual to contact and questionnaire prepared by Dr. John Doull

Experience with:

Acute toxicity, irritation
 mouse, rat, guinea pig, hamster, rabbit, dog, cat, chicken, duck
Subacute toxicity, oral, inhalation, parenteral; Metabolism, distribution, excretion
 mouse, rat, guinea pig, hamster, rabbit, dog, cat
Subacute toxicity, dermal
 rat, rabbit

Classes of compounds studied:

Drugs, pesticides, solvents, metals

Experience with **pharmaceutical, pesticide, and agricultural industry**

Ability or experience in representation of data before FDA and EPA

Capabilities available:

In-house pathology, clinical chemistry, data reduction and analysis, inhalation experience, experience and capability to handle radioactive materials

Personnel:

4 clinical toxicologists, 6 toxicologists, 18 pharmacologists; in other departments – 20 pathologists, 14 biochemists, and 2 teratologists

Laboratory frequently performs contract research

Kem-Tech Laboratories
Division Borg-Warner Corp.
16550 Highland Road
Baton Rouge, Louisiana 70809

Individuals to contact: Dr. Leonard Nelms, Dr. Sham Sachdev, and Dr. John Budden

Experience with:
Acute toxicity, irritation
bluegill, *Ictalurus punctatus, Gambusia affinis, Pimephales* sp., *Micropterus salmoides, Fundulus grandis, Molliensia* sp., and *Notropis* sp.
Carcinogenicity
microbial organisms (*Salmonella typhimurium*)

Classes of compounds studied:
Metals, petro-chemical intermediate and final products, petro-chemical effluents

Experience with petro-chemical, wood product, metallic product, and inorganic chemical industries

Experience in:
Representation of data before regulatory agencies (state and federal air and water quality agencies

Personnel:
33 persons in laboratory including 18 professionals; 13 chemists

Laboratory frequently performs contract research

Kem-Tech Laboratories
1216 Port Neches
Port Neches, Texas 77651

Individual to contact: Dennis Mitchell

Same capabilities as listed above

Personnel:
8 persons in laboratory including 4 professionals; 4 chemists

Maine

Mount Desert Island Biological Laboratory
Salsbury Cove, ME 04672

Individual to contact: Dr. William B. Kinter

Experience with:
Acute toxicity, irritation; Subacute toxicity, oral and parenteral; Metabolism, distribution, excretion
duck, double crested cormorant, herring gull, black guillemot, killifish, eel, winter flounder, King O'Norway, dogfish shark
Subacute toxicity, dermal
killifish, eel, winter flounder, King O'Norway, dogfish shark

Routes of administration:
oral, i.v., intracoelomic and muscular, fish in vivo in water, and fish in vitro, studies with tissues

Classes of compounds studied:
Drugs (e.g., ouabain, phlorizin, probenecid), environmental pollutants (organochlorines such as DDT and PCB's, petroleum hydrocarbons, and heavy metals), cell membrane transport mechanisms and associated enzymes (as sites of action for drugs and pollutants, i.e., membrane toxicity)

Other test procedures:
Evaluation of cell membrane transport systems, e.g., (1) Na, K-ATPase activity and Na pump in gut, gill, nasal gland, and kidney (responsible for plasma osmoregulation); (2) Ca-ATPase activity and Ca pump in avian shell gland (responsible for DDT-induced eggshell thinning); (3) amino acid and sugar absorption in gut (nutrient absorption); and (4) phenol red-like organic acid pump in choroid plexus and renal tubule (excretory transport); transport measured chemically (e.g., photometry) and/or with radioactively labelled tracers (counting or autoradiography)

Capabilities available:
In-house pathology, clinical chemistry, data reduction and analysis, and experience in handling radioactive materials

Personnel:
6 persons in laboratory including 4 professionals: 1 toxicologist, 1 biochemist, 1 physiologist, and 1 pharmacologist

Laboratory will consider performing contract research

Huntingdon Research Center
Box 527
Brooklandville, Maryland 21022

Individual to contact: R. E. Wilsnack, D. V. M.

Questionnaire prepared by Dennis McKay

Experience with:

Acute toxicity, irritation; Subacute toxicity, oral
mouse, rat, guinea pig, hamster, rabbit, dog, cat, pig, sheep, horse, other mammal, chicken, duck, quail, other bird, trout, bluegill, other fish

Metabolism, distribution, excretion
mouse, rat, guinea pig, hamster, rabbit, dog, cat, cow, pig, sheep, horse, other mammal, chicken, duck, quail, other bird, trout, other fish

Reproduction
mouse, rat, guinea pig, rabbit, hamster, dog, cat, cow, pig, sheep, other mammal, chicken, quail, other bird, trout

Carcinogenicity
mouse, rat, guinea pig, hamster, dog, sheep, horse, other mammal, trout

Subacute toxicity, parenteral
mouse, rat, guinea pig, rabbit, dog, pig, chicken, quail

Chronic toxicity, oral
mouse, rat, hamster, rabbit, dog, cow, pig, sheep, horse

Teratology
mouse, rat, hamster, rabbit, dog, cat, sheep, chicken, duck

Subacute toxicity, inhalation
mouse, rat, guinea pig, hamster, rabbit, dog, chicken, quail, other bird

Subacute toxicity, dermal
mouse, rat, guinea pig, dog, cat, pig

Chronic toxicity, dermal
mouse, rat, rabbit, dog, pig

Chronic toxicity, inhalation
mouse, rat, rabbit, dog, sheep

Chronic toxicity, parenteral
dog, pig

Mutagenicity
mouse

Experience with all **routes of administration**

Classes of compounds studied:
Drugs, foods, food additives, food packaging materials, agricultural chemicals, cosmetics, detergents, toiletries, devices, aerosols, veterinary compounds, plastics, smoking materials, household, factory, and environmental materials

Experience with pharmaceutical, food, tobacco, cosmetic, detergent, chemical, agricultural, paper, petroleum industries

Experience in representation of data before almost all the world's leading regulatory agencies and advisory groups

Capabilities available:
In-house pathology, clinical chemistry, data reduction and analysis, pulmonary physiology, inhalation experience, and experience in handling radioactive materials

Personnel:
65 persons in laboratory including 20 professionals: 5 toxicologists, 1 pathologist, 1 pharmacologist, 1 teratologist, and 2 chemists

Laboratory routinely performs contract research

Maryland (continued)

Litton Bionetics, Inc.
5516 Nicholson Lane
Kensington, Maryland 20795

Individuals to contact: Dr. Robert J. Weir, Vice President or Mr. Harvey E. Giss, Manager, Program Development

Experience with:

Acute toxicity, irritation, oral, dermal, eye, inhalation, and parenteral
- mouse, rat, guinea pig, hamster, rabbit, dog, cat, cow, pig, sheep, horse, monkey, chimpanzee, chicken, duck

Subacute toxicity, oral and parenteral
- mouse, rat, guinea pig, hamster, rabbit, dog, cat, cow, pig, sheep, horse, monkey, chimpanzee, chicken, duck

Subacute toxicity, dermal
- mouse, rat, guinea pig, hamster, rabbit, dog, monkey

Subacute toxicity, inhalation
- mouse, rat, guinea pig, hamster, rabbit, dog, monkey, and limited experience with rabbit, dog, and cat

Chronic toxicity, oral and parenteral
- mouse, rat, guinea pig, hamster, rabbit, dog, cat, cow, pig, sheep, horse, monkey, chimpanzee, chicken, duck

Chronic toxicity, dermal
- mouse, rat, guinea pig, hamster, rabbit, dog, cat, cow, pig, sheep, horse, monkey

Chronic toxicity, inhalation
- limited experience with mouse, rat, guinea pig, hamster

Carcinogenicity
- mouse, rat, guinea pig, hamster, dog, monkey

Reproduction
- mouse, rat, guinea pig, hamster, rabbit, dog, cat, cow, pig, sheep, monkey, chicken

Teratology
- mouse, rat, hamster, rabbit, monkey

Mutagenicity
- mouse, rat, hamster

Metabolism, Distribution, Excretion
- mouse, rat, guinea pig, hamster, rabbit, dog, cat, monkey, chicken, duck

Chemistry
- method development, plant residue analysis, tissue residues, feed analysis

Routes of administration:
Oral (feed mix, intubation, water), dermal, inhalation, subcutaneous, eye instillation, implantation, immersion, i.p., and i.v.

Classes of compounds studied:
Drugs, pesticides, food additives, color additives, industrial chemicals, petroleum chemicals and products, cosmetics, smoking materials, soap products, medical devices (implants), and textile chemicals

Government experience includes FDA, EPA, U.S. Army, Navy, Air Force, NCI, NIDA, DOT, and DOA

Experience includes **pharmaceutical** (ethical and proprietary), **agricultural chemicals** (insecticide, weed control, growth regulator, fungicide, antimicrobial, etc.), **petroleum, bulk chemical producers, formulators, cosmetic, food, animal health,** etc.

Specific procedures and capabilities available:
In-house pathology, clinical chemistry, limited pulmonary physiology, inhalation experience, data reduction and analysis, and experience and capability in handling radioactive materials. This laboratory has the most complete commercial capability in mutagenesis and in vitro carcinogenicity available in the world today. Types of studies performed include:

Mutagenesis tests:
Microbial plate test (Ames); microbial suspension test; in vivo rat bone marrow cytogenetic assay; in vitro forward mutation cultured mammalian cells (L5178Y); host mediated assay; dominant lethal assay; heritable translocation assay; intrasanguineous host mediated assay using microbial indicator; DNA repair; detection of mutagenic metabolites in urine

In Vitro Carcinogenesis tests:
Microbial plate test; forward mutation cultured mammalian cells (L5178Y); in vitro transformation (DiPaolo); unscheduled DNA synthesis

Experience in representation of data before regulatory agencies such as FDA, EPA, Agriculture, OSHA, FTC, Transportation, etc.

Personnel:
1500 laboratory personnel (approximately 140 associated with safety testing) including: 4 Ph.D. toxicologists, 7 pathologists, 2 pharmacologists, 2 biochemists, 1 teratologist, 1 chemist, 3 geneticists, and 5 veterinarians

All studies are conducted according to Good Laboratory Practices as developed by FDA

Laboratory routinely performs contract research

Wildlife International Ltd.
Chestertown, Maryland 21620

Individual to contact: Dr. Robert Fink

Experience with:
Subacute toxicity, oral, dermal, parenteral; Chronic toxicity, oral, dermal, parenteral; Reproduction
chicken, duck, quail, goose

Routes of administration:
Feeding and oral

Services include:
Toxicology testing
Acute oral LD_{50}; subacute dietary LC_{50}; reproductive evaluation; cataract evaluation
Field capabilities
Simulated field studies; field monitoring; environmental impact analysis; bird rescue and rehabilitation
Consulting services
Relative to the wildlife requirements for pesticide registration; field problems relating to wildlife

Classes of compounds studied:
Drugs, agricultural chemicals, environmental chemicals

Specific procedures:
Protocols approved by EPA for pesticide registration

Experience with **agricultural chemical, pharmaceutical, forest products, and steel industries**

Extensive ability or experience in representation of data before regulatory agencies

Capabilities available:
Data reduction and analysis

Personnel:
17 professionals in laboratory including 2 toxicologists

Laboratory routinely performs contract research

Arthur D. Little, Incorporated
30 Memorial Drive
Cambridge, Massachusetts 02142

Individual to contact and questionnaire prepared by Dr. Paul E. Palm

Experience with:

Acute toxicity, irritation; Subacute toxicity, oral
mouse, rat, guinea pig, hamster, rabbit, dog, cat, monkey, chicken

Subacute toxicity, inhalation; Chronic toxicity inhalation
mouse, rat, guinea pig, hamster, rabbit, dog, monkey, chicken

Subacute toxicity, dermal, parenteral; Chronic toxicity, oral, dermal, parenteral; Metabolism, Distribution, Excretion
mouse, rat, guinea pig, hamster, rabbit, dog, monkey

Carcinogenicity
mouse, rat, guinea pig, hamster, rabbit, dog, cat, chicken

Chronic toxicity, oral
mouse, rat, guinea pig, hamster, rabbit

Reproduction; Teratology
mouse, rat, hamster, rabbit

Other toxicity studies, infusion
rat, mouse, rabbit, dog

Other toxicity studies, cardiovascular studies
rabbit, dog

Mutagenicity
mouse, rat

Other toxicity studies, check pouch microcirculation studies
hamster

Routes of administration:
oral, i.v., i.p., s.c., dermal, and inhalation

Classes of compounds studied:
Drugs, agricultural pesticides, flavorants, colorants, enzymes, tobacco products, air pollutants, industrial affluents, cosmetics, food products, etc.

Experience with **chemical, food, tobacco, cosmetic, packaging, soap, detergent, nonferrous smelting, steel, electric power, drug, etc. industries**

Experience in representation of data before FDA, Department of Agriculture, EPA, OSHA, CPSC

Capabilities available:
In-house pathology, clinical chemistry, data reduction and analysis, pulmonary physiology, inhalation experience, experience and capability to handle radioactive materials, drug metabolism and distribution

Personnel:
1,700 personnel in laboratory; 110 personnel in Life Sciences Section; 64 professionals in laboratory; 16 Ph.D., 48 M.S. or B.S.; Life Sciences Section only – 5 toxicologists; 2 pathologists; 3 pharmacologists; 6 biochemists; 2 teratologists; 6 chemists; 2 cytologists

Laboratory routinely performs contract research

Bio-Research Consultants
9 Commercial Avenue
Cambridge, Massachusetts 02141

Individual to contact and questionnaire prepared by Dr. Freddy Homburger

Experience with:
Subacute toxicity, dermal; Chronic toxicity, dermal
mouse
Subacute toxicity, inhalation; Chronic toxicity, inhalation; Reproduction
hamster
Chronic toxicity, oral; Chronic toxicity, parenteral; Carcinogenicity
mouse, rat, hamster

Routes of administration:
feeding, gavage, inhalation, s.c. injection, topical

Classes of compounds studied and specific procedures experienced:
Agricultural chemicals, drugs, lifetime feeding studies, lifetime inhalation studies, subcutaneous injection studies, gavage studies

Experience in representation of data before FDA, EPA, NCI

Capabilities available:
In-house pathology, clinical chemistry, data reduction and analysis, inhalation experience, experience and capability to handle radioactive materials, carcinogens, barrier-sustained controlled environment available

Personnel:
49 personnel in laboratory: 4 fulltime professionals; 1 pathologist; 1 pharmacologist; 2 biochemists; 1 immuno-pharmacologist; many consultants available

Laboratory performs contract research routinely

E G & G, Bionomics

Aquatic Toxicological Laboratory
790 Main Street
Wareham, Massachusetts 02571
617–295–2550

Individual to contact: Mr. Brouck Sleight

Marine Research Laboratory
Route 6, Box 1002
Pensacola, Florida 32507
904–453–4359

Mr. Rod Parrish

Questionnaire prepared by: Dr. S. R. Petrocelli

Experience with:
Acute toxicity, Chronic toxicity, Reproduction, Metabolism, Accumulation, Distribution, Excretion
fathead minnow, trout, bluegill, channel catfish, other fish, other aquatic organisms including algae and invertebrates (oysters, water flea, shrimp, crabs)

Classes of compounds studied:
Pesticides (DDT, Chlordane, Kepone), detergents, munitions compounds, industrial wastewaters, industrial chemicals, heavy metals, PCBs

Experience with **U.S. Army, EPA, Commonwealth of Massachusetts, U.S. Coast Guard, and major chemical companies**

Experience at EPA hearings on DDT, dieldrin, chlordane, heptachlor, and ocean dumping

Capabilities available:
Mobile laboratories for on-site testing, Mount & Brungs diluters, model aquatic ecosystems, culturing facilities, environmental chemistry, data reduction and analysis, experience and capability to handle radioactive materials, field surveys

Personnel:
50 persons: 25 professionals – 20 toxicologists, 5 chemists

Laboratory routinely performs contract research

Massachusetts (continued)

Mason Research Institute
23 Harvard Street
Worcester, Massachusetts 01608

Individual to contact: Mr. David G. Bennett

Experience with:

Acute toxicity, irritation; Subacute toxicity, oral, dermal, parenteral; Chronic toxicity, oral, dermal, parenteral; Carcinogenicity; Reproduction; Teratology; Metabolism, distribution, excretion

mouse, rat, guinea pig, hamster, rabbit, dog, cat, monkey

Routes of administration:

p.o., i.v., i.m., s.c., inhalation, drug delivery systems.

Classes of compounds studied:

Drugs, pollutants, food additives, pesticides, marihuana and constituents, tobacco, narcotic antagonists

Experience with government agencies and *drug and chemical industries*

Experience or ability in representation of data before regulatory agencies

Capabilities available:

In-house pathology, hematology, clinical chemistry, neurochemistry, data reduction and analysis, and experience and capability to handle radioactive materials.

Applied immunology:

Allergenicity testing; contact, immediate and delayed hypersensitivity
Immunogenicity of proteinaceous materials
Bacteriological sterility testing
Antiinflammatory assays
Plastics extractables and implantation tests

Bacterial mutagenesis assay:

Detection and evaluation of carcinogenic and mutagenic compounds

Personnel:

250 persons in laboratory including 15 professionals: 1 toxicologist, 5 pathologists, 1 pharmacologist, 3 biochemists, 1 teratologist, and 1 chemist

Laboratory routinely performs contract research

Massachusetts Institute Technology
Experimental Pathology
E18-611
Cambridge, Massachusetts

Individual to contact and questionnaire prepared by Dr. Paul M. Newberne

Experience with:

Subacute toxicity, oral, dermal, parenteral; Chronic toxicity, oral, dermal, parenteral

mouse, rat, guinea pig, hamster, rabbit, dog, cat, pig, chicken, duck, quail, other birds

Carcinogenicity; Reproduction; Teratology; Mutagenicity; Metabolism,
Distribution, Excretion

mouse, rat, guinea pig, hamster, rabbit, dog, cat

Routes of administration:

dermal, intratracheal, i.v., intragastric, and oral

Classes of compounds studied:

Chemical carcinogens (all classes), pesticides, and many drugs

Experience with **pharmaceutical, chemical, and food industry**

Ability or experience in representation of data before regulatory agencies

Capabilities available:

In-house pathology, clinical chemistry, data reduction and analysis, experience and capability to handle radioactive materials

Personnel:

20 in laboratory: 4 professionals – 1 toxicologist, 2 pathologists, 1 biochemist

Laboratory will consider contract research

The Dow Chemical Company
Biomedical Research
Midland, Michigan

Individual to contact: M. B. Chenoweth

Experience with:
Acute toxicity, irritation; Subacute toxicity, oral, dermal, and inhalation; Metabolism, Distribution, Excretion; and Noninvasive organ function tests
man

Routes of administration:
oral, inhalation, and cutaneous

Experience with **chemical/pharmaceutical industries**

Laboratory is a new unit planning to work with materials other than drugs

Ability or experience in representation of data before regulatory agencies (filed 10+ INDs during past 3 years)

Capabilities available:
In-house pathology, clinical chemistry, data reduction and analysis, pulmonary physiology, inhalation experience, and experience and capability in handling radioactive materials

Personnel:
3 persons in laboratory including 2 professionals (pharmacologists)

Dow Chemical U.S.A.
Toxicology Research Laboratory
1803 Building
Midland, Michigan 48640

Individual to contact and questionnaire prepared by Dr. P. J. Gehring

Experience with:
Acute toxicity, irritation
mouse, rat, guinea pig, hamster, rabbit, dog, monkey, chicken, duck, quail, trout, bluegill
Subacute toxicity, oral
mouse, rat, guinea pig, hamster, rabbit, dog, monkey, chicken, duck, quail
Subacute toxicity, dermal, inhalation, parenteral;
Chronic toxicity, oral, dermal, inhalation, parenteral
mouse, rat, guinea pig, hamster, rabbit, dog, monkey
Metabolism, Distribution, Excretion
mouse, rat, guinea pig, hamster, rabbit, dog, cat, monkey
Carcinogenicity
mouse, rat, guinea pig, hamster, rabbit, dog
Teratology
mouse, rat, rabbit, dog
Reproduction; Mutagenicity
mouse, rat, rabbit

Routes of administration:
i.v., i.m., s.q., oral, inhalation, intraocular, and dermal

Classes of compounds studied:
Drugs, chemicals, and agricultural chemicals

Experience with **chemical and pharmaceutical industry**

Ability or experience in representation of data before regulatory agencies

Capabilities available:
In-house pathology, clinical chemistry, data reduction and analysis, pulmonary physiology, inhalation experience, and experience and capability to handle radioactive materials

Personnel:
110 persons in laboratory: 54 professionals – 13 toxicologists; 5 pathologists; 5 pharmacologists; 6 biochemists; 4 teratologists; and 5 chemists

Laboratory will consider contract research

Michigan (continued)

International Research and Development Corporation
500 North Main Street
Mattawan, Michigan 49071

Individual to contact: Dr. F. Wazeter or Dr. E. Goldenthal

Experience with:

Acute toxicity, irritation
mouse, rat, guinea pig, hamster, rabbit, dog, cat, pig, nonhuman primates (rhesus, squirrel, cynomolgus, baboon), chicken, duck, quail, trout, bluegill

Subacute toxicity, oral
mouse, rat, guinea pig, rabbit, dog, cat, pig, chicken, duck, quail, trout, bluegill, rhesus, squirrel, cynomolgus, baboon

Chronic toxicity, oral
mouse, rat, hamster, rabbit, dog, cat, pig, rhesus, squirrel, cynomolgus, baboon

Subacute toxicity, parenteral
mouse, rat, rabbit, dog, cat, pig, rhesus, squirrel, cynomolgus, baboon

Subacute toxicity, inhalation
mouse, rat, guinea pig, rabbit, dog, rhesus, squirrel, cynomolgus, baboon

Chronic toxicity, parenteral
mouse, rat, rabbit, dog, pig, rhesus, squirrel, cynomolgus, baboon

Carcinogenicity
mouse, rat, hamster, dog, rhesus, squirrel, cynomolgus, baboon

Metabolism, Distribution, Excretion
mouse, rat, rabbit, cow, dog, rhesus, squirrel, cynomolgus, baboon

Subacute toxicity, dermal
mouse, guinea pig, rabbit, pig

Reproduction
mouse, rat, rabbit, dog

Chronic toxicity, dermal, inhalation; Teratology
mouse, rat, rabbit, monkey, dog

Eye toxicity
rabbit, rhesus monkey

Vaginal studies
dog, rhesus monkey

Intrauterine studies
baboon, swine

Routes of administration:
oral, parenteral, and dermal

Classes of compounds studied:
Human drugs, veterinary drugs, agricultural chemicals, food additives, pesticides, cosmetics, industrial chemicals, and colors

Experience with **drug, industrial chemical, cosmetic, pesticide, and food industries**

Experience in representation of data before regulatory agencies (FDA, USDA, EPA, and OSHA)

Capabilities available:
In-house pathology, clinical chemistry, data reduction and analysis, pulmonary physiology, inhalation experience, and experience and capability in handling radioactive materials

Personnel:
190 persons in laboratory including 25 professionals: 10 toxicologists, 5 pathologists, 2 pharmacologists, 1 biochemist, 1 teratologist, 5 chemists, and 1 statistician.

Laboratory routinely performs contract research

Michigan State University
Department of Pharmacology
East Lansing, Michigan

Individual to contact: Dr. J. B. Hook

Questionnaire prepared by S. Cagen

Experience with:
Acute toxicity, irritation; Subacute toxicity, oral, dermal, parenteral; Chronic toxicity, oral, dermal, parenteral; Reproduction; Teratology; Metabolism, Distribution, Excretion
mouse, rat, guinea pig, hamster, rabbit

Routes of administration:
oral and parenteral

Classes of compounds studied:
Drugs, pesticides, herbicides, environmental contaminants, and heavy metals

Experience or ability in representation of data before regulatory agencies

Capabilities available:
Data reduction and analysis and experience and capability to handle radioactive materials

Personnel:
8 persons in laboratory including 1 pharmacologist and 1 teratologist

Laboratory will consider performing contract research

The University of Michigan
Toxicology Laboratory
Ann Arbor, Michigan 48109

Individuals to contact: Dr. H. Cornish, Dr. R. Hartung, or Dr. R. Richardson

Experience with:
Acute, subacute, and chronic toxicity
mouse, rat, guinea pig, hamster, rabbit, quail
Inhalation chamber facilities
Continuous flow water system, acute and chronic studies
perch, fathead, aquatic insects
Delayed neurotoxicity testing
chickens

Classes of compounds studied:
Inhalation toxicology of solvents, biochemical interactions, metabolism, toxicology of aromatic nitro compounds, pesticides, heavy metals, and other organophosphorus compounds

Experience with automotive, plastics, aluminum, and chemical industry

Primarily interested in special problems that require an investigative or developmental approach

Capabilities available:
Inhalation chambers, experience and capability of handling radioactive materials, identification of metabolites, behavioral toxicology, data reduction and analysis, histopathology and chemical chemistry

Personnel:
10 to 20 persons in laboratory (variable including graduate students); 4 professionals – 4 toxicologists

While the laboratory does some regular toxicological testing which may not be readily available elsewhere, its major interest lies in research oriented problems

Warner-Lambert/Parke, Davis Pharmaceutical Research Division
Department of Toxicology
2800 Plymouth Road
Ann Arbor, Michigan 48106

2 Locations:
Ann Arbor, Michigan Laboratories
Sheridan Park, Ontario (Canada) Laboratories

Individual to contact and questionnaire prepared by F. A. de la Iglesia, M.D.

Experience with:
Acute toxicity, irritation; Subacute toxicity, oral, dermal, parenteral (ocular irritation and skin irritation studies for rabbit)
mouse, rat, rabbit, dog, rhesus monkey
Chronic, oral, dermal, parenteral
mouse, rat, dog, rhesus monkey
Metabolism, Distribution, Excretion
mouse, dog, rhesus monkey
Reproduction, Teratology
mouse, rat, rabbit
Carcinogenicity
mouse, rat
Mutagenicity
mouse
Others, including materials toxicology and electron microscopy

Routes of administration:
topical, oral, and parenteral

Classes of compounds studied:
Drugs, devices, cosmetics, OTC, consumer preparations and confectionary products

Specific procedures experienced:
Acute, subacute, chronic toxicity, reproduction and teratology, tumorigenesis, irritation, and inhalation

Experience in representation of data before regulatory agencies, domestic and international

Capabilities available:
In-house pathology, clinical chemistry, computerized data reduction and analysis, quantitative microscopy and histochemistry

Personnel:
77 persons including 37 professionals: 9 pathologists, 18 toxicologists, 3 teratologists, 2 data specialists, 3 biochemists and 3 clinical laboratory scientists

Laboratory infrequently performs contract research

Wayne State University
Department of Occupational and Environmental Health
625 Mullett
Detroit, Michigan 48226

Individual to contact: Andrew L. Reeves

Experience with:
Acute toxicity, irritation; Subacute toxicity, oral, dermal, inhalation, parenteral; Chronic toxicity, oral, dermal, inhalation, parenteral; Carcinogenicity; Metabolism, distribution, excretion
mouse, rat, guinea pig, hamster, rabbit

Classes of compounds studied:
Inorganic compounds and asbestos

Experience with beryllium mining and refining and with asbestos cement manufacturing

Experience in representation of data before regulatory agencies

Capabilities available:
In-house pathology, clinical chemistry, data reduction and analysis, pulmonary physiology, inhalation experience, and experience and capability to handle radioactive materials

Personnel:
10 persons in laboratory including 3 professionals: 1 toxicologist, 1 biochemist, and 1 chemist

Laboratory frequently performs contract research

University of Mississippi
Research Institute of Pharmaceutical Sciences
School of Pharmacy
University, Mississippi 38677

Individual to Contact: Dr. Coy W. Waller

Questionnaire prepared by Dr. Wallace L. Guess

Experience with:
Acute toxicity, irritation
mouse, rat, guinea pig, rabbit, dog, cat, pig
Subacute toxicity, oral
mouse, rat, rabbit, dog
Subacute toxicity, dermal
guinea pig, rabbit, pig
Subacute toxicity, parenteral
mouse, rat, rabbit, dog, cat, pig
Chronic toxicity, oral and parenteral
mouse, rat, rabbit, dog
Chronic toxicity, dermal
guinea pig, rabbit, pig
Reproduction and Teratology
mouse, rat, hamster
Metabolism, Distribution, Excretion
mouse, rat, guinea pig, hamster, rabbit, dog, pig, monkey

Routes of administration:
oral, gavage, dosed feed, parenteral, and dermal

Classes of compounds studied:
Drugs, cosmetics, agricultural chemicals, rodenticides, and food additives

Experience with **chemical, pharmaceutical, agricultural, cosmetic, and nutritional industrial products**

Ability or experience in representation of data before regulatory agencies – staff has worked with and provided data for the FDA, CPSC, and EPA

Capabilities available:
Clinical chemistry, data reduction and analysis, and experience and capability in handling radioactive materials

Personnel:
80 persons in laboratory including 50 professionals: 3 toxicologists, 6 pharmacologists, 3 biochemists, 2 teratologists, and 12 chemists

Laboratory frequently performs contract research

University of Mississippi Medical Center
Department of Pharmacology and Toxicology
Jackson, Mississippi 39216

Individual to contact: Dr. A. Wallace Hayes

Experience with:
Acute toxicity, irritation; Subacute toxicity, oral, dermal, parenteral; Chronic toxicity, oral, dermal, parenteral
mouse, rat, guinea pig, hamster, rabbit, dog, cat, cow, sheep, horse, monkey
Carcinogenicity; Reproduction; Teratology
mouse, rat, guinea pig, hamster, rabbit, dog, cat, monkey
Metabolism, Distribution, Excretion
mouse, rat, dog, cat, monkey
Mutagenicity
mouse, microorganism

Routes of administration:
oral, s.c., intracutaneous, i.m., i.p., i.v., and dermal

Classes of compounds studied:
Drugs, agricultural chemicals, mycotoxins, industrial chemicals, and drugs of abuse

Specific procedures:
Whole animal techniques – metabolism, distribution, excretion; whole organ perfusion techniques – lung, liver, kidney; tissue slice techniques – lung, liver, kidney, brain, etc.; subcellular preparations – all major tissues; microbiological techniques; routine histopathological techniques and electron and scanning microscope techniques; neuro and behavioral toxicology; enzyme kinetics, e.g., microsomal enzymes; full forensic capabilities; full clinical laboratory capabilities; routine acute and chronic toxicology including reproductive and teratogenic procedures

Experience with **environmental and regulatory industries**

Experience or ability in representation of data before regulatory agencies

Capabilities available:
In-house pathology, clinical chemistry, data reduction and analysis, pulmonary physiology, and experience in handling radioactive materials

Personnel:
22 persons in the laboratory including 8 professionals – 2 toxicologists, 1 pathologist, 2 pharmacologists, 1 biochemist, 1 teratologist, and 1 chemist

Laboratory would like to perform contract research

Missouri

Midwest Research Institute
425 Volker Boulevard
Kansas City, Missouri 64110

Individual to contact: Dr. C. C. Lee or Dr. Wm. B. House

Questionnaire prepared by Drs. T. R. Castles and C. C. Lee

Experience with:
- Acute toxicity, irritation
  - mouse, rat, guinea pig, hamster, rabbit, dog, monkey
- Subacute toxicity, oral; Chronic toxicity, oral
  - mouse, rat, guinea pig, hamster, rabbit, dog, monkey
- Subacute toxicity, dermal
  - rabbit
- Subacute toxicity, inhalation and parenteral; Chronic toxicity, inhalation and parenteral
  - mouse, rat, guinea pig, hamster, rabbit, dog, monkey
- Carcinogenicity
  - mouse, rat, guinea pig, hamster
- Reproduction and Mutagenicity
  - mouse, rat, guinea pig, hamster, rabbit, dog
- Teratology
  - mouse, rat, guinea pig, hamster, rabbit
- Metabolism, distribution, excretion
  - mouse, rat, guinea pig, hamster, rabbit, dog, pig, monkey, chicken, duck

Routes of administration:
p.o., s.q., i.m., i.p., and i.v.

Classes of compounds studied:
Drugs, agricultural chemicals, inorganic metals, munitions, GRAS substances, industrial chemicals

Experience with **pharmaceutical, agricultural chemicals, paint, petroleum, and industrial chemical industries**

Special capabilities:
In-house pathology, clinical chemistry, data reduction and analysis, and radiobiology

Personnel:
4 toxicologists, 2 pathologists, 1 biochemist, 2 pharmacologists, 2 teratologists and over 200 technical staff, including chemists in the Institute

Mobay Chemical Corporation
Chemagro Agricultural Division
Box 4913
Hawthorn Road
Kansas City, Missouri 64120

Individual to contact: D. W. Lamb

Experience with:
- Acute toxicity, irritation
  - mouse, rat, guinea pig, rabbit, dog, cat, cow, pig, sheep, horse, chicken, duck, quail, trout, bluegill, channel catfish, crayfish
- Subacute toxicity, oral
  - mouse, rat, dog, cat, cow, pig, sheep, horse, chicken, duck, quail
- Subacute toxicity, dermal
  - rat, rabbit
- Chronic toxicity, oral
  - mouse, rat, dog
- Carcinogenicity
  - mouse, rat
- Metabolism, distribution, excretion
  - rat, rabbit, dog, cow, pig, goat, chicken, quail, trout, bluegill, channel catfish

Routes of administration:
oral and dermal

Classes of compounds studied:
Agricultural chemicals, plastic intermediates, and dye chemicals

Has experience in representation of data before regulatory agencies

Capabilities available:
Clinical chemistry, data reduction and analysis, inhalation experience, and experience in handling radioactive materials

Personnel:
14 persons in laboratory including 2 toxicologists, 1 pathologist

Laboratory will consider performing contract research

University of Missouri
College of Veterinary Medicine
Columbia, Missouri 65201

Individual to contact: Dr. Gary Osweiler

Questionnaire prepared by Dr. Gary Van Gelder

Experience with:

Acute toxicity, irritation; Subacute toxicity, oral, dermal, parenteral; Chronic toxicity, oral, dermal, parenteral; Reproduction; Metabolism, Distribution, Excretion

mouse, rat, guinea pig, hamster, rabbit, dog, cat, cow, pig, sheep, horse

Routes of administration:

oral, parenteral, and dermal

Forensic testing:

Laboratory maintains a full service veterinary diagnostic laboratory including diagnostic toxicology, analytical toxicology, pathology, bacteriology, and virology; field investigations can be performed as well as conducting necessary tests on submitted animals, tissues, fluids, feed, soil, or plant samples to establish a diagnosis

Analytical chemistry capabilities:

Atomic absorption spectrophotometry, gas-liquid chromatograph, thin layer chromatography and light, UV and infrared spectrophotometry

Laboratory can perform behavioral toxicology tests ranging from simple schedules to complex sensory and learning paradigms; exposures can be prenatal or postnatal in young or adult subjects

Classes of compounds studied:

Agricultural chemicals (herbicides, insecticides), metals and chemical elements, feed additives, toxic plants, rodenticides, fungicides

Experience with agricultural chemical and feed industry

Capabilities available:

In-house pathology, clinical chemistry, data reduction and analysis, pulmonary physiology, inhalation experience, and experience and capability to handle radioactive materials

Personnel:

9 persons in laboratory including 8 professionals: 4 toxicologists, 1 pathologist, 1 pharmacologist, 1 biochemist, and 1 chemist

Laboratory will consider performing contract research

Nebraska

Harris Laboratories, Inc.
P.O. Box 80837
Lincoln, Nebraska 68501

Individual to contact: Dr. James McClurg

Questionnaire prepared by Dr. Lewis E. Harris

Experience with:

Acute toxicity, irritation; Subacute toxicity, oral
  mouse, rat, guinea pig, hamster, rabbit, dog, cat, cow, pig, sheep, horse, chicken, duck, quail, pheasant, trout, bluegill
Chronic toxicity, oral
  mouse, rat, guinea pig, hamster, rabbit, dog, cat, cow, pig, sheep, horse, chicken, duck, quail, pheasant
Chronic toxicity, dermal
  mouse, rat, guinea pig, hamster, rabbit, dog, cow, pig, sheep, horse, chicken
Subacute toxicity, parenteral
  mouse, rat, guinea pig, hamster, rabbit, dog, cow, pig, sheep, horse
Subacute toxicity, dermal
  mouse, rat, guinea pig, hamster, rabbit, dog, cat, cow, pig, sheep, horse
Reproduction
  rat, guinea pig, rabbit, dog, cow, pig, sheep, chicken
Teratology
  rat, guinea pig, rabbit, cow, pig, sheep
Subacute toxicity, inhalation
  mouse, rat, guinea pig, hamster, rabbit, dog
Metabolism, Distribution, Excretion
  dog, cow, pig, sheep
Carcinogenicity
  mouse, rat
Mutagenicity
  rat

Routes of administration:
oral and parenteral

Classes of compounds studied and specific procedures:
Drugs – toxicity studies and clinical pharmacology; agricultural chemicals – toxicity studies, particularly those required by EPA, including residue assays; food additives – toxicity studies and residue analyses; cosmetics – toxicity studies (animal and human)

Experience with **human pharmaceutical, animal health products, agricultural chemicals, cosmetics, household products, and food additives industries**

Ability or experience in representation of data before FDA, USDA, and EPA

Capabilities available:
Clinical chemistry, data reduction and analysis, inhalation experience, and limited experience and capability to handle radioactive materials

Personnel:
75 persons in laboratory; 25 professionals – 1 toxicologist; 3 pathologists (part-time); 1 pharmacologist; 3 biochemists; and 7 chemists

Laboratory routinely performs contract research

AMR Biological Research, Inc.
P.O. Box 5700
Princeton, New Jersey 08540

Individual to contact: S. Margolin, Ph.D., President

Experience with:
- Acute toxicity, irritation
  - mouse, rat, guinea pig, hamster, rabbit, dog, cat, cow, pig, sheep, horse, mini-pig, goat, rhesus monkey, squirrel monkey, chicken, duck, quail, trout, bluegill
- Subacute toxicity, oral
  - mouse, rat, guinea pig, hamster, rabbit, dog, cat, cow, pig, sheep, mini-pig, goat, rhesus monkey, squirrel monkey
- Subacute toxicity, dermal, parenteral
  - mouse, rat, guinea pig, hamster, rabbit, dog, cat, cow, pig, sheep, horse, monkeys, mini-pig, goats
- Subacute toxicity, inhalation
  - mouse, rat, guinea pig, hamster, rabbit, dog, monkeys
- Chronic toxicity, dermal
  - mouse, rat, guinea pig, hamster, rabbit, dog, monkey, mini-pig
- Chronic toxicity, oral
  - mouse, rat, guinea pig, hamster, rabbit, dog
- Chronic toxicity, inhalation
  - mouse, rat, guinea pig
- Chronic toxicity, parenteral
  - mouse, rat, guinea pig, hamster, rabbit, dog, cat, cow, sheep, horse, mini-pig, goat, monkeys, chicken
- Carcinogenicity
  - mouse, rat, hamster, dog
- Reproduction and Teratology
  - mouse, rat, rabbit, dog
- Mutagenicity
  - mouse, rat, hamster
- Metabolism, Distribution, Excretion
  - mouse, rat, guinea pig, hamster, rabbit, dog, cat, mini-pig, monkey, man, chicken, duck, quail
- Pharmacology – all phases, by classic techniques screening for new compounds using classic pharmacologic and microbiologic techniques
  - dog, cat, mini-pig

Classes of compounds studied:
Drugs (human and veterinary), cosmetics, agricultural chemicals (herbicides and pesticides), industrial chemicals, and consumer products

Experience with industries producing **drugs, agricultural chemicals, industrial chemicals, cosmetics, and consumer products**

Experience in representation of data before regulatory agencies (FDA, EPA, Department of Agriculture, CPSC, OSHA, DOT)

Capabilities available:
In-house pathology, clinical chemistry, data reduction and analysis, pulmonary physiology, inhalation experience, experience and capability to handle radioactive materials

Personnel:
20 persons in laboratory including 4 professionals: 1 toxicologist, 1 pharmacologist, 1 biochemist, 1 chemist, 1 microbiologist

Laboratory routinely performs contract research, and consultation on regulatory affairs including preparation of petitions; all studies in compliance with regulatory requirements

New Jersey (continued)

Bio/dynamics, Inc.
Mettlers Road
East Millstone, New Jersey 08873

Individual to contact: Dr. Thomas J. Russell

Questionnaire prepared by Carol S. Auletta

Experience with:
- Acute. Subacute and Chronic Toxicity
  - mouse, rat, guinea pig, hamster, rabbit, dog, cat, cow, pig, sheep, chicken, monkey (rhesus, cynomalogus, squirrel)
- Metabolism, distribution and excretion
  - mouse, rat, guinea pig, hamster, rabbit, dog, cat, cow, pig, sheep, chicken, monkey (rhesus, cynomalogus, squirrel)
- Carcinogenicity
  - mouse, rat, guinea pig, hamster, rabbit, dog, cat
- Reproduction/Teratology
  - mouse, rat, hamster, rabbit, dog

Routes of administration:
oral, dermal, parenteral, inhalation

Classes of compounds studied and specific procedures experienced:
Drugs (human and veterinary), agricultural chemicals, cosmetics, soaps and soap additives, industrial chemicals, bone implants and devices

Experience with **pharmaceutical, chemical, food and cosmetic industries**

Experience in representation of data before FDA and EPA

Capabilities available:
In-house pathology, clinical chemistry, inhalation, data reduction and analysis, experience and capability to handle radioactive materials

Personnel:
200 persons in laboratory; 24 professionals – 6 toxicologists; 5 pathologists; 1 pharmacologist; 3 biochemists; 1 teratologist; 6 chemists

Laboratory routinely performs contract research

College of Medicine and Dentistry of New Jersey
Department of Preventive Medicine
Newark, New Jersey 07103

Individual to contact: Morris M. Joselow, Ph.D.

Experience with:
- Acute, subacute, and chronic toxicity; oral
  - mouse, rat, guinea pig, rabbit, dog
- Subacute toxicity, dermal
  - rabbit
- Metabolism, absorption, and excretion
  - mouse, rat, guinea pig, hamster, rabbit, goldfish
- Carcinogenicity, reproduction, mutagenicity
  - mouse, rat, guinea pig, dog

Experience with collection and analyses of environmental samples (air, water, waste)

Classes of compounds studied:
Industrial chemicals, heavy metals (e.g., lead, mercury, cadmium, thallium, bismuth), trace substances, drugs

Experience with chemical industry and in making presentations before regulatory agencies (EPA, FDA)

Capabilities available:
Chemical laboratory for trace analyses; animal facilities; computers for data reduction and analyses

Laboratory will consider contract research

Cooper Laboratories
SMP Division
Cedar Knolls, New Jersey

Individual to contact: J. Loux

Experience with:
Acute toxicity, irritation
mouse, rat, guinea pig, hamster, rabbit, dog, gerbil
Subacute toxicity, oral and parenteral
mouse, rat, guinea pig, hamster, gerbil
Subacute toxicity, dermal, ocular
guinea pig, rabbit
Chronic toxicity, dermal
rabbit
Reproduction; Teratology
mouse, rat
Metabolism, Distribution, Excretion
mouse, rat, rabbit

Routes of administration:
oral, i.p., topical, and i.v.

Classes of compounds studied:
Ethical drugs and proprietary preparations

Laboratory has experience with **pharmaceutical industry**

Experience in representation of data before regulatory agencies (FDA)

Capabilities available:
Clinical chemistry, data reduction and analysis, and experience in handling radioactive materials

Personnel:
4 persons in laboratory including 1 toxicologist

Leberco Laboratories
123 Hawthorne Street
Roselle Park, New Jersey 07204

Individual to contact and questionnaire prepared by Dr. Irving Levenstein.

Experience with:
Acute toxicity, irritation
mouse, rat, guinea pig, hamster, rabbit, dog, cat
Subacute toxicity, oral, dermal, parenteral
mouse, rat, guinea pig, hamster, rabbit, dog, cat
Chronic toxicity, oral, dermal, parenteral
mouse, rat, guinea pig, hamster, rabbit, dog, cat
Carcinogenicity, Reproduction, and Teratology
mouse, rat

Routes of administration:
oral, dermal, s.c., and i.v.

Classes of compounds studied and specific procedures:
Drugs, chemicals, cosmetics, and colors

Laboratory has experience with **industrial products in areas of food, drugs, and cosmetics**

Ability or experience in representation of data before regulatory agencies

Capabilities available:
Clinical chemistry – gas chromatography, electron capture, high-pressure liquid chromatography, spectrophotofluorometer, and atomic absorption

Personnel:
3 toxicologists and 2 chemists

Laboratory routinely performs contract research

Lever Bros. Research
Toxicology Section
45 River Road
Edgewater, New Jersey 07020

Individual to contact and questionnaire prepared by: Dr. Edward J. Singer

Experience with:
- Acute toxicity, irritation
  - mouse, rat, guinea pig, hamster, rabbit
- Subacute toxicity, oral, dermal, inhalation
  - mouse rat, guinea pig, hamster, rabbit
- Chronic toxicity, oral, dermal, inhalation
  - mouse, rat, guinea pig, hamster
- Carcinogenicity and Mutagenicity
  - mouse, rat
- Reproduction and Teratology
  - mouse, rat, hamster, rabbit
- Metabolism, distribution, excretion
  - mouse, rat, guinea pig, hamster, rabbit, man

Routes of administration:
oral, dermal, i.v., i.p.

Classes of compounds studied and specific procedures experienced:
Surfactants, sequestrants, antimicrobials, siliceous/nonsiliceous dusts, enzymes (detergent use), perfumes, flavors, food additives (acidulant, emulsifier); percutaneous absorption/localization, dermal toxicology, sensitization, fibrogenic response

Experience with industrial products in areas of soap/detergent, food, and cosmetics

Experience in representation of data before regulatory agencies (OTC Panel, CPSC, FDA, Florida Pollution Control)

Capabilities available:
Clinical chemistry, data reduction and analysis, inhalation experience, and experience and capability to handle radioactive materials

Personnel:
21 persons in laboratory – 13 professionals, 8 toxicologists, 1 pharmacologist, 2 biochemists, teratologist, and 1 chemist

Laboratory will consider performing contract research

Inhalation Toxicology Research Institute
Lovelace Biomedical and Environmental Research Institute
P. O. Box 5890
Albuquerque, New Mexico 87115

Individual to contact: Dr. Roger O. McClellan

Questionnaire prepared by Dr. Robert K. Jones

Experience with:
Subacute toxicity, inhalation; chronic toxicity, inhalation
Metabolism, distribution, excretion following inhalation
mouse, rat, guinea pig, hamster, dog, primate, pony
Chronic toxicity, inhalation; carcinogenicity; mutagenicity
mouse, rat, hamster, dog, primate

Classes of compounds studied:
Radioactive isotopes (beta, gamma, and alpha emitters), effluents from fossil fuel energy sources, and chemical particulates from aerosolized consumer products

Experience in representation of data before regulatory agencies

Capabilities available:
In-house pathology, clinical chemistry, pulmonary physiology, analytical chemistry, radiochemistry, inhalation exposure, and experience in handling radioactive materials, animal breeding
rat, mouse, dog

Personnel:
262 persons in laboratory including 49 professionals: 5 toxicologists, 5 pathologists, 4 biochemists, 6 chemists, 4 physiologists, 8 biophysicists, 6 aerosol physicists, 3 bioengineers, and 5 biologists

Laboratory routinely performs contract research for government agencies only

Albert Einstein College of Medicine
Neurotoxicology Unit
Rose F. Kennedy Center (502)
1410 Pelham Parkway
Bronx, New York 10461

Individuals to contact: Dr. P. S. Spencer and Dr. H. H. Schaumburg

Experience with:
Acute (neuro) toxicity, oral cat, rat
Acute (neuro) toxicity, parenteral cat, rat
Acute (neuro) toxicity, direct organotypic tissue cultures*
Chronic (neuro) toxicity, oral monkey, cat, rat
Chronic (neuro) toxicity, parenteral cat, rat
Chronic (neuro) toxicity, direct organotypic tissue cultures*

Classes of compounds studied:
Drugs (human and experimental), cosmetics, agricultural chemicals (insecticides), industrial chemicals (e.g., acrylamide, methyl n-butyl ketone and other hexacarbons), heavy metals

Experience with chemical industry, NIOSH, and NIH, and in representation of data before NIOSH and EPA

Capabilities available:
In-house neuropathology including electron microscopy, neurometabolism studies, data reduction and analysis, experience and capability to handle radioactive chemicals, tissue culture laboratory (Dr. M. Bornstein)

Personnel:
23 persons in neurotoxicology and tissue culture laboratories including 5 professionals: 2 neurologist/neuropathologists, 2 neuroscientist/neurotoxicologists, 1 neurochemist

Unit routinely conducts contract research

*Composed of central or peripheral nerve tissue grown in a Maximow slide assembly and fed with nutrient fluid in which controlled concentrations of toxins may be introduced, toxic metabolites identified by g.c-m.s, and structural changes determined by direct visualization of the living tissue

American Health Foundation
Naylor Dana Institute
Valhalla, New York 10595

Individual to contact and questionnaire prepared by Dr. John H. Weisburger

Experience with:
Acute toxicity; Subacute toxicity, oral, dermal, parenteral; Chronic toxicity, oral, dermal, parenteral; Carcinogenicity; Metabolism, Distribution, Excretion
mouse, rat, guinea pig, hamster
Subacute toxicity, inhalation
mouse (limited), rat, guinea pig, hamster

Routes of administration:
oral (food and gavage), s.c., i.p., i.v., i.m., inhalation, intratracheal, intralaryngeal, intrarectal

Classes of compounds studied and specific procedures experienced:
chemicals, drugs, food additives, food constituents in rodent systems; rapid bioassays such as cell transformation and mutagenicity, metabolism in vitro and in vivo, interpretation and evaluation of data, epidemiology and significance to man, microanalysis of mixtures, identification of chemicals

Experience with **government labs and regulatory agencies, chemical, petrochemical, food, drug, and energy industries**

Experience in representation for data generated in labs under their supervision before regulatory agencies

Capabilities available:
In-house pathology, clinical chemistry, data reduction and analysis, pulmonary physiology (limited), inhalation experience (limited), experience and capability to handle radioactive materials, germ-free animals, analytic chemistry and instrumentation

Personnel:
127 persons in laboratory: 60 professionals (M.D., D.V.M., Ph.D., or M.S.) – 18 epidemiologists and nutritionists, 2 toxicologists, 41 pathologists, 2 pharmacologists, 15 biochemists, 1 veterinarian, 17 chemists, 2 statisticians, 1 microbiologist

Laboratory frequently performs contract research

City of New York Department of Health
455 Fifth Avenue
New York, New York

Individual to contact and questionnaire prepared by Dr. Bernard Davidow

Experience with:
Acute toxicity, irritation; Subacute toxicity, oral, parenteral; Chronic toxicity, oral, parenteral; Carcinogenicity; Reproduction; Teratology; Mutagenicity; Metabolism, Distribution, Excretion
mouse, rat, guinea pig, hamster, rabbit, dog, cow, pig, sheep, horse, chicken, duck, quail
Subacute toxicity, dermal
dog
Forensic and diagnostic services
Chemistry – anodic stripping voltammetry; atomic absorption spectrophotometry; clinical chemistry; gas chromatography; immunoassays (RIA, emit, hemagglutination inhibition); thin layer chromatography, ultra violet, infrared, and fluorimetric spectrophotometry
Cytology
Hemotology – blood grouping, coagulation, complete blood counts
Microbiology – bacteriology, mycology, parasitology
Serology
Virology

Routes of administration:
oral, parenteral, and dermal

Classes of compounds studied:
Drugs, agricultural chemicals, food additives, and cosmetics

Experience with **federal government and New York state industry**

Ability or experience in representation of data before regulatory agencies

Capabilities available:
In-house pathology, clinical chemistry, and experience and capability to handle radioactive materials

Personnel:
426 in laboratory; professionals – 16 Ph.D.s, 16 M.S., and 124 B.S.; 3 toxicologists; 1 pathologist; 2 pharmacologists; 4 biochemists; and 3 chemists

Laboratory will consider federal contracts

General Foods Technical Center
555 South Broadway
Tarrytown, New York

Mailing address:
250 North Street
White Plains, New York 10625

Individual to contact: Dr. James Scala

Questionnaire prepared by Dr. John C. Kirschman

Experience with:
Acute toxicity, irritation; Subacute toxicity, oral, dermal, and parenteral; Chronic toxicity, oral and parenteral; Carcinogenicity; Reproduction; and Metabolism, distribution, excretion
mouse, rat

Classes of compounds studied:
Food ingredients and food additives

Experience in representation of data before regulatory agencies (FDA)

Capabilities available:
Clinical chemistry, data reduction and analysis, and experience and capability in handling radioactive materials

Personnel:
3 toxicologists, 1 pharmacologist, and 1 biochemist

Health, Safety, and Human Factors Laboratory
Kodak Park
Rochester, New York 14650

Individual to contact and questionnaire prepared by: C. J. Terhaar

Experience with:

Acute toxicity, irritation
mouse, rat, guinea pig, rabbit, dog, cat, chicken, trout, fathead minnow, *Daphnia* algae spp., flatworms, snails
Subacute toxicity, oral
mouse, rat, guinea pig, rabbit, dog, cat, chicken, duck, fathead minnow
Subacute toxicity, dermal
mouse, rat, guinea pig, rabbit
Subacute toxicity, inhalation
mouse, rat, guinea pig, rabbit, dog, cat
Subacute toxicity, parenteral
mouse, rat, guinea pig, rabbit, dog, cat
Chronic toxicity, oral
rat,
Chronic toxicity, dermal
guinea pig, rabbit
Chronic toxicity, inhalation
rat, cat
Chronic toxicity, parenteral
rat, dog
Carcinogenicity
mouse, rat
Reproduction
rat, *Daphnia*
Teratology
rat
Mutagenicity
mouse, rat
Metabolism, Distribution, Excretion
rat, dog

Routes of administration:
p.o. and i.p.

Classes of compounds studied and specific procedures:
Photographic, industrial, plastics, food additives, and dyes

Experience or ability in representation of data before regulatory agencies

Capabilities available:
In-house pathology, clinical chemistry, inhalation experience, and experience and capability to handle radioactive materials; data reduction and analysis

Personnel:
21 persons in toxicology laboratory and 10 in biochemistry laboratory: 8 toxicologists, 1 pathologist, and 1 teratologist

Huntingdon Research Center
216 Congers Road
New City, New York 10956

Individual to contact and questionnaire prepared by Dr. Charles O. Ward, Ph.D.

Experience with:
- Acute toxicity, irritation
  - mouse, rat, guinea pig, hamster, rabbit, dog, cat, primate
- Subacute toxicity, parenteral; Chronic toxicity, oral
  - mouse, rat, guinea pig, hamster, rabbit, dog, cat, primate
- Subacute toxicity, oral
  - mouse, rat, guinea pig, hamster, rabbit, dog, primate
- Subacute toxicity, inhalation
  - mouse, rat, guinea pig, hamster, rabbit, dog, cat, primate
- Chronic toxicity, parenteral
  - mouse, rat, guinea pig, hamster, dog, cat, primate
- Chronic toxicity, inhalation
  - mouse, rat, guinea pig, hamster, dog, primate
- Subacute toxicity, dermal
  - mouse, rat, guinea pig, rabbit, dog
- Carcinogenicity
  - mouse, rat, dog, cat, primate
- Metabolism, Distribution, Excretion
  - mouse, rat, rabbit, dog, chicken
- Reproduction; Teratology
  - mouse, rat, hamster, rabbit, dog
- Chronic toxicity, dermal
  - mouse, rat, guinea pig, rabbit
- Mutagenicity
  - mouse, rat

Routes of administration:
oral, dermal, parenteral and inhalation

Classes of compounds studied:
Drugs, agricultural chemicals, industrial chemicals, cosmetics, food additives, devices, household products

Experience with **government, pharmaceutical, chemical, cosmetic, and food industry**

Experience in representation of data before regulatory agencies

Capabilities available:
In-house pathology, data reduction and analysis, pulmonary physiology, inhalation experience, experience and capability to handle radioactive materials

Personnel:
35 persons in laboratory: 8 professionals – 6 toxicologists, 1 pathologist, 1 chemist

Laboratory routinely performs contract research

New York (continued)

South Shore Laboratory, Inc.
148 Islip Avenue
Islip, New York 11751

Individual to contact and questionnaire prepared by Dr. N. Rakieten

Experience with:
Acute toxicity, irritation
mouse, rat, guinea pig, hamster, rabbit, dog, cat, chicken, duck
Subacute toxicity, oral, dermal, parenteral; Chronic toxicity, oral, dermal, parenteral
mouse, rat, guinea pig, hamster, rabbit, dog, cat
Carcinogenicity; Reproduction; Teratology; Mutagenicity
mouse, rat
Metabolism, Distribution, Excretion
dog, cat

Routes of administration:
p.o., i.m., i.v., i.p., and s.c.

Classes of compounds studied:
Drugs and chemicals

Experience with **NIH, pharmaceutical, and cosmetic industries**

Ability or experience in representation of data before regulatory agencies

Capabilities available:
In-house pathology, clinical chemistry, data reduction and analysis, pulmonary physiology, inhalation experience

Personnel:
7 in laboratory: 2 toxicologists; 1 pathologist; 2 pharmacologists; 1 biochemist; and 1 chemist

Laboratory routinely performs contract research

University of Rochester
Department of Radiation Biology and Biophysics
Medical Center
Rochester, New York 14642

Individual to contact and questionnaire prepared by P. E. Morrow

Experience with:
Subacute toxicity, oral
rat, rabbit
Subacute toxicity, inhalation
rat, guinea pig, hamster, rabbit, dog
Subacute toxicity, parenteral
mouse, rat
Chronic toxicity, oral
mouse, rat
Chronic toxicity, inhalation
rat
Carcinogenicity
rat, hamster
Reproduction
rat, dog
Teratology
mouse
Mutagenicity
mouse, hamster
Metabolism, distribution, excretion
rabbit, dog

Experience with all routes of administration

Classes of compounds studied and specific procedures experienced:
Drugs, heavy metals, atmospheric pollutants, food additives and radionuclides, as well as all listed procedures

Experience with products in the **chemical industry**

Capabilities available:
In-house pathology, clinical chemistry, data reduction analysis, pulmonary physiology, inhalation experience, and experience and capability to handle radioactive materials

Personnel:
250 persons in laboratory including 50 professionals: 20 toxicologists, 2 pathologists, 6 pharmacologists, 18 biochemists, 2 teratologists, and 2 chemists

Laboratory will consider peforming contract research

EN-CAS Analytical Laboratories
807 Brookstown Ave.
Winston-Salem, North Carolina

Individual to contact: Dr. Charles Ganz

Experience with:
Product environmental screening
Environmental impact studies
Wastewater analyses
Drinking water studies
Waste treatability studies
Activated sludge metabolism studies
Fish toxicity bioassays
Workplace and ambient air analyses
Industrial hygiene monitoring
Trace organic chemicals analysis
Trace metals analysis
Pesticide analysis
Characterization of organic substances
Studies with radiolabeled (^{14}C) materials
Photochemical studies

Facilities and equipment:
Present facilities consist of two laboratories totaling approx. 3000 square feet; in addition to general laboratory ware and a wide array of sample preparation equipment, the following major instrumentation is also available: total organic carbon analyzer, infrared spectrometer, UV-visible spectrophotometer, atomic absorption spectrometer, ion selective electrode system, gas chromatographs fitted with the following detectors: flame ionization, electron capture, flame photometric, hall electrolytic conductivity, bendix flasher (for thermal desorption of trapped organic vapors), personnel and area air sampling and analysis equipment, UV-phase contrast microscope, liquid scintillation counter

Personnel:
Staff includes as many as 9 experienced technical personnel: 1 Ph.D. organic-analytical chemist, 1 B.S. + biochemist, 1 B.S. + chemist, 1 B.S. technologist, 1 AAS equivalent technologist, and 2 to 4 technicians; a number of specialized consultants are also available

Environmental Toxicology
NIEHS
P.O. Box 12233
Research Triangle Park, North Carolina 27709

Individual to contact: Dr. R. L. Dixon

Experience with:
Acute toxicity, irritation; Subacute toxicity, oral, inhalation, parenteral; Chronic toxicity, oral, inhalation; Reproduction; Metabolism, Distribution, Excretion
mouse, rat, guinea pig, hamster, rabbit
Carcinogenicity; Mutagenicity
mouse, rat
Teratology
mouse, rat, rabbit

Routes of administration:
oral, parenteral, and inhalation

Classes of compounds studied:
Pesticides, food additives, growth stimulants (livestock), metals, and occupational chemicals

Experience with **chemical industry**

Experience or ability in representation of data before regulatory agencies

Capabilities available:
In-house pathology, clinical chemistry, data reduction and analysis, pulmonary physiology, inhalation experience, and experience and capability to handle radioactive materials

Personnel:
60 persons in laboratory including 15 professionals: 4 toxicologists, 1 pathologist, 4 pharmacologists, 3 biochemists, and 3 teratologists

National Institute of Environmental Health Sciences
Research Triangle Park, North Carolina

Individual to contact: R. E. Staples

Experience with:
- Teratology
  - mouse, rat, hamster, rabbit
- Embryo culture and transfer
  - mouse, rat, rabbit

Routes of administration:
gavage, diet, s.c., and i.p.

Classes of compounds studied:
Environmental agents (pesticides, solvents, asbestos, microwave, embryo freezing, uterine proteins)

Ability in representation of data before regulatory agencies

Other than for post doctoral scientists, future studies will be conducted by contract

Agents studied:
add noise, PCB components and isomers

Aerospace Medical Laboratory
Wright Patterson Air Force Base
Toxic Hazards Division
Dayton, Ohio

Individuals to contact: Dr. A. A. Thomas, Dr. K. C. Back, Dr. V. Carter

Questionnaire prepared by Dr. A. A. Thomas

Experience with:
Acute toxicity, irritation
mouse, rat, guinea pig, hamster, rabbit, dog, guppies, fathead minnows, algae, soil bacteria
Subacute toxicity, oral
mouse, rat, guinea pig, hamster, rabbit, dog
Subacute toxicity, inhalation; Chronic toxicity, inhalation
mouse, rat, guinea pig, hamster, rabbit
Subacute toxicity, parenteral
mouse, rat
Metabolism, Distribution, Excretion
rat, dog
Subacute toxicity, dermal
rabbit

Routes of administration:
Chronic toxicity, inhalation; Carcinogenicity inhalation and feeding

Classes of compounds studied and specific procedures:
Military chemicals (fuels, missile propellants, lubes and additives, etc.), materials (plastics, thermal degradation), and environmental pollutants

Experience with **defense** industry

Ability or experience in representation of data before regulatory agencies

Capabilities available:
In-house pathology, clinical chemistry, data reduction and analysis, vivarium laminar air-flow housing, pulmonary physiology, experience and capability to handle radioactive materials, altitude chambers (8 Thomas domes), 4 Rochester, and 2 Longley chambers

Personnel:
110 in laboratory: 40 professionals – 8 toxicologists; 4 pathologists; 1 pharmacologist; 3 biochemists; 12 chemists; 5 veterinarians; 1 M.D.

Laboratory frequently performs contract research and will consider contract research only for other government agencies

Consolidated Biomedical Laboratories, Inc.
P.O. Box 2289
Columbus, Ohio 43216

Individual to contact and questionnaire prepared by George V. Foster, Ph.D.

Experience with:
Clinical chemistry tests, Enzymes, BUN, glucose, electrolytes, etc. on most animals, toxicology tests, drug screens, anti convulsants, trace metals, therapeutic drug monitoring, development of new methods, environmental air monitoring, histology

Capabilities available:
In-house pathology, clinical chemistry, toxicology, endocrinology, histology, microbiology, hematology, serology, cytogenetics, data reduction

Techniques available:
Radioassay, gas chromatography, liquid chromatography, automated chemistry analyses, atomic absorption, anodic stripping voltametry, fluorescent antibody, thin layer chromatography fluorometry

Personnel:
10 pathologists, 5 biochemists

Laboratory does contract research

Ohio State University
Division of Toxicology
1645 Neil Avenue
Columbus, Ohio 43210

Individual to contact and questionnaire prepared by Dr. Daniel Couri

Experience with:
Acute toxicity, irritation; Subacute toxicity, oral; Metabolism, Distribution, Excretion
mouse, rat, guinea pig, hamster, rabbit, dog, cat, chicken, quail
Subacute toxicity, inhalation, parenteral; Chronic toxicity, oral, inhalation, parenteral; Carcinogenicity; Reproduction
mouse, rat, guinea pig, hamster, rabbit, chicken, quail
Subacute toxicity, dermal
mouse, rat, guinea pig, hamster, rabbit, dog, cat
Chronic toxicity, dermal
mouse, rat, guinea pig, hamster, rabbit
Forensic toxicology
human autopsy
Clinical toxicology, body fluids and tissue analyses

Routes of administration:
oral, dermal, parenteral, and inhalation

Classes of compounds studied and specific procedures:
Acid, basic, neutral, industrial solvents, heavy metals, pesticides; spectrophotometric (UV-vis), ir, fluorescence, atomic absorption, immunologic RIA, EMIT, chromatographic (columns, TLC, GLC, GS), GC-MS

Experience with plastic coating, rendering plant, food processing, paint and lacquers, and foundry industry

Experience in representation before NIOSH, NIDA, NHLBI, EPA

Capabilities available:
In-house pathology, clinical chemistry, data reduction and analysis, inhalation experience, experience and capability to handle radioactive materials, analytical toxicology

Personnel:
15 persons in laboratory; 2 faculty; 7 graduate students; 6 toxicology specialists (B.S. or M.S. with 2 to 5 years experience)

Laboratory will consider contract research

Oregon

Oregon State University
Oak Creek Laboratories of Biology
Department of Fisheries and Wildlife
Corvallis, Oregon 97331

Individual to contact and questionnaire prepared by Dr. Lavern J. Weber or Dr. Charles Warren

Experience with:

Acute toxicity, irritation; Subacute toxicity, oral, inhalation (aquatic fish), parenteral; Reproduction; Metabolism, distribution, excretion
  trout, bluegill, Coho salmon, bass, guppies
Acute toxicity, irritation
  aquatic insects
Subacute toxicity, inhalation
  aquatic fish

Classes of compounds studied:

Carbamates, organophosphates, chlorinated insecticides, heavy metals, nitrogenens, phosphate agricultural fertilizers and chlorine

Experience with **Rare Earth Refinery and paper** industry

Ability or experience in representation of data before regulatory agencies; some members of lab but not contact

Capabilities available:

Clinical chemistry, data reduction and analysis, gill transport physiology uptake, experience and capability to handle radioactive materials

Personnel:

19 in laboratory: 6 professionals – 3 toxicologists, 2 pharmacologists, and 6 basic biologists (ecology)

Laboratory will consider some types of contract research

Pennsylvania

Biosearch Incorporated
P.O. Box 8598
Philadelphia, Pennsylvania 19101

Individual to contact and questionnaire prepared by Dr. Karl L. Gabriel

Experience with:

Acute toxicity, irritation
mouse, rat, guinea pig, hamster, rabbit, dog, cat, cow, pig, sheep, horse, human, chicken, duck, quail, trout, bluegill, other fish

Subacute toxicity, oral, dermal, parenteral; Chronic toxicity, oral, dermal, parenteral
mouse, rat, guinea pig, hamster, rabbit, dog, cat, cow, pig, sheep, horse, chicken, duck, quail

Metabolism, Distribution, Excretion
mouse, rat, rabbit, dog, cat, cow, pig, sheep, horse, human, chicken, duck, quail

Reproduction
mouse, rat, rabbit, dog, cat

Carcinogenicity
mouse, rat, dog, cat

Subacute toxicity, inhalation; Chronic toxicity, inhalation
mouse, rat, guinea pig

Teratology
mouse, rat, rabbit

Routes of administration:
Oral, percutaneous, i.v., s.c., i.p., and inhalation

Classes of compounds studied and specific procedures:
Agricultural chemicals, drugs, household products, industrial chemicals; human patch test, usage studies

Experience with **drug, chemical, food, etc. industries**

Ability or experience in representation of data before FDA, EPA, etc.

Capabilities available:
In-house pathology consultant, clinical chemistry, data reduction and analysis, inhalation experience, experience and capability to handle radioactive materials

Personnel:
5 in laboratory: 3 professionals – 1 toxicologist; 1 pathologist; and 1 pharmacologist

Laboratory routinely performs contract research

Carnegie-Mellon University
Chemical Hygiene Fellowship
Carnegie-Mellon Institute of Research
4400 Fifth Avenue
Pittsburgh, Pennsylvania 15213

Individual to contact: E. F. Cox or C. S. Weil

Questionnaire prepared by E. F. Cox

Experience with:

Metabolism, Distribution, Excretion
mouse, rat, guinea pig, hamster, rabbit, dog, cat, cow, pig, sheep, horse, other mammal, chicken, duck, quail, other bird, trout, bluegill, other fish, aquatic organisms (in vivo and in vitro, including man)

Inhalation
mouse, rat, guinea pig, hamster, rabbit, dog

Acute toxicity, irritation
mouse, rat, guinea pig (skin sensitization), rabbit, dog

Chronic toxicity, inhalation
mouse, rat, hamster

Carcinogenicity; Subacute toxicity, oral; Chronic toxicity, oral
mouse, rat

Reproduction; Teratology; Mutagenicity
rat

Subacute toxicity, dermal
rabbit

Classes of compounds studied and specific procedures experienced:
Agricultural and other industrial chemicals; all phase of industrial toxicology, including experimental design and statistical analysis of data

Experience with **chemical, petroleum,** and **pharmaceutical** industries

Experience in representation of data before EPA, FDA, NIOSH, some government groups outside the U.S.A., eg., Canada and Holland

Capabilities available:
In-house pathology, clinical chemistry, data reduction and analysis, inhalation experience, experience and capability to handle radioactive materials

Personnel:
70 persons in laboratory; 41 professionals – 18 toxicologists/pharmacologists, 13 biologists/biochemists, 1 pathologist, 6 chemists

Laboratory frequently performs and will consider contract research

Hahnemann Medical College and Hospital
230 North Broad Street
Philadelphia, Pennsylvania 19102

Individual to contact and questionnaire prepared by Dr. Benjamin Calesnick

Experience with:
Acute toxicity, irritation; Subacute toxicity, oral, dermal, inhalation, parenteral; Chronic toxicity, oral, dermal, inhalation, parenteral; Reproduction; Teratology; Mutagenicity; Metabolism, Distribution, Excretion; Forensic
mouse, rat, guinea pig, hamster, rabbit, dog, cat, cow, pig, sheep, horse, nonhuman primates, chicken, duck, quail, pigeon, trout, bluegill, bass – no salt water fish

Routes of administration:
oral, dermal, parenteral, and inhalation

Classes of compounds studied:
Drugs, agricultural chemicals, industrial pollutants, and food additives

Experience with **U.S. Army, FDA, pharmaceutical and cosmetic industries**

Ability or experience in representation of data before FDA and U.S. Army

Capabilities available:
In-house pathology, clinical chemistry, data reduction and analysis, pulmonary physiology, inhalation experience, experience and capability to handle radioactive materials

Personnel:
32 in laboratory: 5 animal care; 1 toxicologist; 2 pathologists; 20 pharmacologists; 2 biochemists; 1 teratologist; and 1 chemist

Laboratory frequently performs contract research

Jefferson Medical College
Department of Pharmacology
1620 Locust Street
Philadelphia, Pennsylvania 19107

Individual to contact and questionnaire prepared by Dr. A. J. Triolo

Experience with:
Acute toxicity, Subacute toxicity, oral, dermal, parenteral
Chronic toxicity, oral
mouse, rat, dog, cat
Biochemical mechanism of toxicity
Chemical induction of tumors in laboratory animals

Routes of administration:
oral, i.p., i.v., and topical

Classes of compounds studied:
Pesticides, therapeutic drugs

Experience with **EPA and pharmaceutical** industry (EPA Pesticide Grant)

Experience in representation of data before regulatory agencies

Capabilities available:
In-house pathology, clinical chemistry, analysis, radioactive materials

Personnel:
4 persons in laboratory: 1 professional – toxicologist, pharmacologist

Laboratory will consider contract research

M. B. Research Laboratories, Incorporated
Spinnerstown, Pennsylvania 19868

Individual to contact and questionnaire prepared by Dr. Oscar M. Moreno

Experienced with:
Acute toxicity, irritation; Subacute toxicity, oral, dermal, parenteral; Chronic toxicity, oral, dermal, parenteral
mouse, rat, guinea pig, hamster, rabbit, dog, cat
Carcinogenicity
mouse, rat, guinea pig, hamster, rabbit, dog
Reproduction; Teratology
rat, rabbit

Routes of administration:
oral (food or intubation), dermal, and parenteral

Classes of compounds studied:
Drugs, agricultural chemicals, cosmetics, detergents, and petroleum products

Experience with **chemical, pharmaceutical, cosmetics and personal products, petroleum, and dental industries**

Ability in representation of data before regulatory agencies

Capabilities available:
Data reduction and analysis, inhalation experience (limited)

Personnel:
7 in laboratory: 3 professionals – 2 toxicologists and 1 pharmacologist

Laboratory routinely performs contract research

Pharmakon Laboratories
1140 Quincy Avenue
Scranton, Pennsylvania 18510

Individual to contact and questionnaire prepared by Dr. Richard J. Matthews

Experience with:
Acute toxicity, irritation; Subacute toxicity, oral, dermal, parenteral;
Chronic toxicity, oral, dermal, parenteral
mouse, rat, guinea pig, hamster, rabbit, dog, cat
Carcinogenicity
mouse, rat
Teratology
rat, rabbit
Reproduction; Mutagenicity
rat

Routes of administration:
oral, i.p., i.v., s.c., and topical

Classes of compounds studied and specific procedures:
Preclinical and clinical drug candidates (tranquilizers, antihypertensives, anti-inflammatory, analgesics, and hemastatic agents); evaluation of chemicals and pesticides for mammalian toxicity only and evaluation of chemicals and cosmetics for skin and eye irritation

Experience with **pharmaceutical, agricultural, chemical, cosmetics, and toilet goods industries**

Ability or experience in representation of data before regulatory agencies

Capabilities available:
In-house pathology, clinical chemistry, data reduction and analysis

Personnel:
20 in laboratory: 7 professionals – 2 toxicologists; 1 pathologist; 2 pharmacologists; 1 biochemist; and 1 teratologist

Laboratory routinely performs contract research

The Skin & Cancer Hospital TUHSC
3322 North Broad Street
Philadelphia, Pennsylvania 19140

Individual to contact: Dr. P. D. Forbes

Experience with:
Acute toxicity, irritation
mouse, rat, guinea pig, hamster, rabbit, pig
Subacute toxicity, oral, dermal
mouse, pig
Subacute toxicity, parenteral; Chronic toxicity, oral, dermal, parenteral
Photocarcinogenicity
mouse
Phototoxicity, photoallergy, and photocarcinogenesis

Classes of compounds studied:
Antibiotics, food additives, fragrance materials, and detergent additives

Experience with **drug, nutrition, perfume, and cleaning industries**

Experience in representation of data before regulatory agencies (FDA)

Capabilities available:
In-house pathology, clinical chemistry, data reduction and analysis, and experience in handling radioactive materials

Personnel:
12 persons in laboratory including 5 professionals

Laboratory will consider performing contract research

University of Pittsburgh
Pittsburgh, Pennsylvania 16251

Individual to contact and questionnaire prepared by Y. Alarie

Experience with:
Subacute toxicity, inhalation; Acute toxicity, irritation
mouse, rat, guinea pig, rabbit
Sensory and pulmonary irritation
mouse, guinea pig, rabbit
Chronic toxicity, inhalation
mouse

Route of administration:
inhalation

Classes of compounds studied:
Industrial chemicals (gases, vapors, and inhalation hazard), polymer combustion products, and inhalation drugs

Capabilities available:
In-house pathology, data reduction and analysis, pulmonary physiology, inhalation experience, experience and capability to handle radioactive materials

Personnel:
11 in laboratory: 3 professionals – 1 toxicologist; 1 pathologist; and 1 biochemist

Laboratory may consider contract research

W. H. Rorer, Inc.
Research Division
Fort Washington, Pennsylvania 19034

Individual to contact: Dr. G. N. Mir

Experience with:
Acute toxicity, irritation; Subacute toxicity, oral
mouse, rat, guinea pig, rabbit, dog, cat
Subacute toxicity, dermal
guinea pig, rabbit
Subacute toxicity, inhalation
guinea pig
Subacute toxicity, parenteral
rabbit, dog
Metabolism, Distribution, Excretion
rat, dog, monkey

Routes of administration:
oral, i.p., s.c., i.v., and i.d.

Classes of compounds studied:
Cardiovascular, CNS depressants, antidiarrheals, antisecretory (gastric), anti-inflammatory, and local anesthetics

Experience with **pharmaceutical industry**

Laboratory regularly represents data to FDA in form of INDs and NDAs

Capabilities available:
Clinical chemistry, data reduction and analysis, inhalation experience, and experience and capability to handle radioactive materials

Personnel:
110 persons in laboratory including 22 professionals: 5 toxicologists, 13 pharmacologists, 10 biochemists, and 10 chemists

Bio Tox Laboratories
Route 91
Woodriver Junction, Rhode Island 02894
(401)364-7731

Individual to contact and questionnaire prepared by Dr. H. Lal

Experience with:
Subacute toxicity, parenteral
mouse, rat, guinea pig, hamster, rabbit, dog, cat
Acute toxicity; irritation; subacute toxicity, oral, dermal
mouse, rat, guinea pig, rabbit, cat
Subacute toxicity, inhalation; Chronic toxicity, oral, dermal, inhalation, parenteral
mouse, rat

Classes of compounds studied:
Insecticides, neuroleptics, volatile hydrocarbons, SH enzyme inhibitors, chemical intermediates

Experience with **drug** and **chemical** industry

Experience in representation of data before FDA

Capabilities available:
In-house data reduction and analysis, inhalation experience, experience and capability to handle radioactive materials

Personnel:
20 personnel in laboratory; 3 professionals – 1 toxicologist; and 1 consultant pathologist

University of Rhode Island
Department of Pharmacology and Toxicology
Kingston, Rhode Island

Individual to contact: Dr. J. DeFeo

Questionnaire prepared by Dr. G. Fuller

Experience with:
Acute toxicity, irritation; Subacute toxicity, oral
mouse, rat, rabbit
Metabolism, Distribution, Excretion
mouse, rat, rabbit
Subacute toxicity, dermal, parenteral
rabbit

Classes of compounds studied:
Drugs and pesticides, especially acetylcholine inhibitors and connective tissue response to chemicals (FIBROSIS)

Experience with **drug industry**

Ability or experience in representation of data before regulatory agencies

Capabilities available:
Clinical chemistry, data reduction and analysis, inhalation experience, experience and capability to handle radioactive materials

Personnel:
32 in laboratory: 6 professionals – 2 toxicologists; and 4 pharmacologists

Laboratory frequently performs contract research

University of Tennessee Center for the Health Sciences
Materials Science Toxicology Laboratories
26 South Dunlap
Memphis, Tennessee 38163

Individual to contact and questionnaire prepared by Dr. W. H. Lawrence

Experience with:
- Acute toxicity, irritation
  - rabbit
- Subacute and Chronic toxicity, oral
  - mouse, rat, guinea pig, rabbit, dog
- Subacute and Chronic toxicity, dermal
  - rabbit
- Subacute and Chronic toxicity, inhalation
  - mouse, rat
- Subacute and Chronic toxicity, parenteral
  - mouse, rabbit, dog
- Carcinogenicity and Teratology
  - rat
- Reproduction
  - mouse, rat
- Mutagenicity
  - mouse
- Metabolism, distribution excretion
  - mouse, rat, guinea pig, rabbit

Routes of administration:
oral, i.p., i.v., s.c., and inhalation

Experience with tissue culture procedures (agar-overlay and inhibition of cell growth); hemolysis tests (*in vitro*); inhalation; pyrolysis-inhalation; isolated, perfused rabbit heart; isolated intestine; C-V function in anesthetized dogs

Classes of compounds studied:
Biomedical polymeric materials; monomers (e.g., acrylates, methacrylates, etc.); plasticizers (e.g., phthalates, adipates, etc.); polymeric iniators, etc.; and related chemicals (e.g., 2-chloroethanol, chloracetaldehyde, etc.)

Industrial laboratory experience with mostly biomedical device manufacturing fabricators; some chemical manufacturing companies

Ability or experience in representation of data before regulatory agencies

Capabilities available:
In-house pathology; some clinical chemistry; inhalation experience; data reduction and analysis; and experience and capability to handle radioactive materials

Personnel:
20 persons in laboratory – 5 professionals (+ 3 postdoctorates = 8 toxicologists) 1 pathologist, 3 pharmacologists, 1 biochemist, 1 teratologist, and 2 physical chemists

Laboratory frequently performs contract research

Vanderbilt University
Center in Toxicology
Department of Biochemistry
Nashville, Tennessee 37232

Individual to contact and questionnaire prepared by Dr. Robert A. Neal

Experience with:
Acute toxicity, irritation; Subacute toxicity, oral, dermal, parenteral; Chronic toxicity, oral, dermal, parenteral
mouse, rat, guinea pig, hamster, rabbit, chicken

Classes of compounds studied:
Drugs, pesticides, and industrial chemicals

Capabilities available:
In-house pathology, clinical chemistry, data reduction and analysis, pulmonary physiology, and experience and capability in handling radioactive materials

Personnel:
24 persons in laboratory including 8 toxicologists, 1 pathologist, 1 pharmacologist, 2 biochemists, 1 teratologist, and 2 chemists

Laboratory will consider performing contract research

Vanderbilt University
Naturally Occurring Toxicants Laboratory
School of Medicine
Nashville, Tennessee 37232

Individual to contact and questionnaire prepared by Dr. Benjamin J. Wilson

Experience with:
Acute toxicity, irritation; Mutagenicity, Carcinogenicity; Teratogenicity
mouse, rat, guinea pig, hamster, rabbit, cow, horse, chicken, duck

Classes of compounds studied and specific procedures:
Mycotoxins (selected ones) – qualitative analysis; sweet potato toxins – currently developing qualitative and quantitative procedures; higher plant toxins

Experience with veterinary diagnostic laboratories

Ability or experience in representation of data before regulatory agencies

Capabilities available:
In-house pathology, inhalation experience, experience and capability to handle radioactive materials

Personnel:
4+ in laboratory: 2 professionals – 1 toxicologist; 1 part-time pathologist; 1 medicinal chemist; and 1 microbiologist

Laboratory frequently performs contract research; currently have 2-year contract to develop analytical methods for pulmonary toxins in sweet potatoes

Vanderbilt University Medical Center
Department of Pharmacology and
Center in Toxicology, Department of Biochemistry
Nashville, Tennessee 37232

Individual to contact: Dr. Raymond D. Harbison

Experience with:

Acute toxicity, irritation; Subacute toxicity, oral, dermal, parenteral
Chronic toxicity, oral, dermal, parenteral; Teratology;
Metabolism, distribution, excretion
mouse, rat, guinea pig, hamster, rabbit, dog, cat, sheep
Subacute toxicity, inhalation
mouse, rat, guinea pig, hamster, rabbit
Chronic toxicity, inhalation; Mutagenicity
mouse, rat
Carcinogenicity; Reproduction
mouse, rat, guinea pig, hamster, rabbit, dog, cat
Analytical toxicology and pharmacology

Routes of administration:
i.p., oral, s.c., i.v., inhalation, and dermal

Classes of compounds studied:
Pesticides, drugs, industrial chemicals, and environmental contaminants

Experience with chemical and drug industries and government

Experience in representation of data before regulatory agencies

Capabilities available:
In-house pathology, clinical chemistry, data reduction and analysis, inhalation experience and experience in handling radioactive materials

Personnel:
9 persons in laboratory including 6 professionals: 1 toxicologist, 1 pathologist, 1 pharmacologist, 1 biochemist, 1 teratologist, and 1 chemist

Laboratory frequently performs contract research

Research Laboratory, ARS, USDA
Veterinary Toxicology and Entomology
P.O. Drawer GE
College Station, Texas 77840

Individual to contact and questionnaire prepared by Dr. Harry E. Smalley, Director

Experience with:
Acute toxicity, irritation; Subacute toxicity, oral, parenteral; Chronic toxicity, oral, parenteral; Metabolism, Distribution, Excretion
mouse, rat, guinea pig, hamster, rabbit, cow, pig, sheep, horse, goat, chicken, duck, quail
Reproduction; Teratology; Radioimmunoassay
mouse, rat, guinea pig, hamster, rabbit, cow, pig, sheep, chicken, duck, quail
Amino acid analysis; Clinical chemistry; Hematology; Hormones; Photodegradation; Telemetry; Radioimmunoassay
hamster, rabbit, cow, pig, sheep, horse, goat, chicken, duck, quail
Subacute toxicity, dermal; Chronic toxicity, dermal
rabbit, cow, pig, sheep, horse, goat

Classes of compounds studied:
Agricultural chemicals, feeds, feed additives, and toxic plants

Experience with **agricultural chemicals industry**

Capabilities available:
In-house pathology, clinical chemistry, data reduction and analysis, pulmonary physiology, experience and capability to handle radioactive materials

Personnel:
130 in laboratory: 24 professionals – 8 toxicologists; 1 pathologist; 1 pharmacologist; 4 biochemists; 1 teratologist; 5 chemists; and 4 entomologists

Laboratory performs government research

Syscon Incorporated
P.O. Box 3486
Bryan, Texas 77801

Individual to contact and questionnaire prepared by Dr. B. J. Camp

Experience with:
Acute toxicity, irritation; Subacute toxicity, oral, dermal, parenteral; Chronic toxicity, oral, dermal, parenteral; Carcinogenicity; Metabolism, Distribution, Excretion
catfish

Routes of administration:
oral, parenteral, and water bath-gills

Classes of compounds studied:
Heavy metals, PCBs, PBBS and pesticides

Capabilities available:
In-house pathology, clinical chemistry, data reduction and analysis

Personnel:
4 in laboratory: 4 professionals – toxicologists, pathologists, and biochemists

Laboratory will consider contract research

Utah

Utah State University

Dept. Vety. Science
UMC 56, Utah State University
Logan, Utah 84322

Individual to contact: R. P. Sharma, D.V.M., Ph.D.

Experience with:

Acute toxicity, irritation; Subacute toxicity, oral and parenteral; and Chronic toxicity, oral and parenteral
- mouse, rat, guinea pig, hamster, rabbit, chicken

Metabolism, distribution, excretion
- mouse, rat, guinea pig, hamster, rabbit, dog, cow, pig, sheep, chicken

Immune Suppression
- mouse, rat, guinea pig, hamster, rabbit

Routes of administration:
p.o., i.v., i.p., i.d., and s.c.

Classes of compounds studied:
Heavy metals, fluoride, and agricultural chemicals

Has ability, but no prior experience, in representation of data before regulatory agencies

Capabilities available:
In-house pathology, clinical chemistry, data reduction and analysis, and experience and capability to handle radioactive materials

Personnel:
8 persons in laboratory including 3 professionals: 1 toxicologist, 1 pathologist, and 1 pharmacologist

Laboratory will consider performing contract research

Experimental Pathology Laboratories, Inc.
P.O. Box 474
Herndon, Virginia 22070

Individual to contact:
William M. Busey, D.V.M., Ph.D.

Classes of compounds studied:
Pharmaceuticals, agricultural chemicals, industrial chemicals, aerosol cosmetics, topical cosmetics, and air pollutants

Experience or ability to represent data before regulatory agencies

Capabilities available:
General toxicologic pathology, carcinogenesis pathology, pulmonary pathology, electron microscopy, electron probe analysis, necropsy services and consultation in the design and interpretation of toxicology studies

Personnel:
45 persons in laboratory including 6 professionals, all pathologists

Laboratory routinely performs contract research

No animal facilities

Hazleton Laboratories
9200 Leesburg Turnpike
Vienna, Virginia

Individual to contact: Dr. F. E. Reno

Experience with:
Acute toxicity, irritation
mouse, rat, guinea pig, hamster, rabbit, dog, monkey, cow, pig, sheep, duck, quail
Subacute toxicity, oral, dermal, and parenteral
mouse, rat, guinea pig, hamster, rabbit, dog, monkey, cow, pig, sheep
Subacute toxicity, inhalation; Chronic toxicity, oral, dermal, inhalation, and parenteral
mouse, rat, guinea pig, hamster, rabbit, dog, monkey
Carcinogenicity
mouse, rat, hamster
Reproduction
mouse, rat, hamster, rabbit, dog, monkey
Teratology
mouse, rat, hamster, rabbit
Mutagenicity
mouse, rat, in vitro systems
Metabolism, Distribution, Excretion
mouse, rat, guinea pig, hamster, rabbit, dog, monkey, cat, cow, chicken

Routes of administration:
Experience in all

Laboratory has virtually worked with all classes of compounds, employing all of the procedures used commonly in safety evaluation

Laboratory has worked with all types of industry

Experience and ability in representation of data before regulatory agencies

Capabilities available:
In-house pathology, clinical chemistry, data reduction and analysis, pulmonary physiology, inhalation experience, and experience and capability to handle radioactive materials

Personnel:
450 persons in laboratory (approximately 220 associated directly with safety evaluation program) including 10 toxicologists, 5 pathologists, 1 pharmacologist, 2 biochemists, 1 teratologist, and 3 chemists

Laboratory routinely performs contract research

MCV/VCU/HSD, Box 726
Department of Pharmacology
Richmond, Virginia 23298

Individual to contact and questionnaire prepared by Dr. Joseph F. Borzelleca

Experience with:
Acute toxicity, irritation; Subacute toxicity, oral, dermal, parenteral; Chronic toxicity, oral, dermal, parenteral
mouse, rat, guinea pig, hamster, rabbit, dog, cat, cow, pig, sheep, chicken, duck, monkey (Rhesus, squirrel)
Biotransformation, Distribution, Excretion
mouse, rat, guinea pig, hamster, rabbit, dog, cat, cow, pig, sheep, chicken, monkey (Rhesus, squirrel)
Reproduction
mouse, rat, cow, pig
Carcinogenicity
mouse, rat
Teratology
mouse, rat, guinea pig, rabbit, pig

Experience with all **routes of administration**

Classes of compounds studied:
Pesticides, cosmetics, drugs, food additives, tobacco constituents, including smoke

Experience in representation of data before FDA, EPA, USDA

Capabilities available:
In-house pathology, clinical chemistry, data reduction and analysis, experience and capability to handle radioactive materials, analytical chemistry

Personnel:
7 toxicologists, 17 pharmacologists, 6 biochemists, 1 teratologist, 2 chemists

Laboratory performs routinely and frequently and will consider contract research

Battelle Pacific Northwest Laboratory
Biology and Ecosystems Department
P.O. Box 999
Richland, Washington 99352

Individual to contact: Dr. W. J. Bair

Questionnaire prepared by Dr. D. D. Mahlum

Experience with:
Acute toxicity, inhalation, parenteral; Subacute toxicity, inhalation, parenteral, oral; Chronic toxicity, inhalation, oral, parenteral; Carcinogenicity; Reproduction; Metabolism, Distribution, Excretion
mouse, rat, hamster, rabbit, dog, pig, sheep, trout, salmon and various other species
Teratology
mouse, rat, hamster, rabbit, dog, pig, trout, salmon and various other species
Subacute toxicity, parenteral
mouse, rat, hamster, dog, pig, sheep, trout, salmon and various other species
Subacute toxicity, inhalation; Chronic toxicity, inhalation
mouse, rat, hamster, dog, pig
Chronic toxicity, dermal
mouse, rat, hamster, rabbit, pig
Subacute toxicity, dermal
mouse, rat, hamster, pig

Routes of administration:
oral, inhalation, parenteral, and dermal

Classes of compounds studied and specific procedures:
Inhalation – radionuclides including transuranics, chelating agents, asbestos, automotive exhausts, fly ash, cigarette smoke; teratologic and reproductive – radionuclides, heavy metals, and trypan blue; carcinogenicity – radionuclides, dimethylaminoazobenzene, DAF, ethionine, narcotic antagonistics, marihuana

Experience with **nuclear, mining, and drug industry**

Capabilities available:
In-house pathology, clinical chemistry, data reduction and analysis, pulmonary physiology, inhalation experience, experience and capability to handle radioactive materials

Personnel:
225 in laboratory: 60 professionals – 10 toxicologists; 3 pathologists; 1 pharmacologist; 6 biochemists; 3 teratologists; and 5 chemists

Laboratory frequently performs contract research

University of Washington
Toxicology Laboratory
Department of Pharmacology [SJ-30]
School of Medicine
Seattle, Washington 98195

Individual to contact and questionnaire prepared by Dr. Ted A. Loomis

Experience with:
Acute toxicity, irritation
mouse, rat, rabbit, dog
Subacute, oral; Chronic, oral; Chronic, parenteral
mouse, rat, dog
Subacute, dermal
rat, rabbit, dog
Subacute, inhalation; Reproduction; Teratology
rat
Subacute, parenteral
mouse, rat
Chronic, dermal
rabbit
Metabolism, Distribution, Excretion
rat, dog
Forensic services
Analytical – GC, GC-MS, IR, UV, thin layer chromatography, high-pressure LC, RIA
Consultation

Routes of administration:
oral, i.p., and i.v.

Classes of compounds studied:
Agricultural chemicals and drugs

Experience with **forensic and agricultural chemistry industry**

Has small amount of ability or experience in representation of data before regulatory agencies

Capabilities available:
Clinical chemistry, experience and capability to handle radioactive materials

Personnel:
3 in laboratory: 2 professionals; 1 toxicologist; and 1 chemist

Laboratory will consider contract research

Washington State University
College Hall – 5B
Pullman, Washington 99113

Individual to contact and questionnaire prepared by James Way

Experience with:
- Acute toxicity, irritation
  - mouse, rat, guinea pig, hamster, rabbit, dog, cat, cow, pig, sheep, horse, monkey, chicken, duck, trout, bluegill
- Subacute toxicity, oral, dermal, and parenteral
  - mouse, rat, guinea pig, hamster, rabbit, dog, cat, cow, pig, sheep, horse, monkey, chicken, duck, trout, bluegill
- Chronic toxicity, oral
  - mouse, rat, guinea pig, hamster, rabbit, dog, cat, cow, pig, sheep, horse, monkey, chicken, duck, trout, bluegill
- Chronic toxicity, dermal
  - mouse, rat
- Chronic toxicity, parenteral
  - mouse, rat, guinea pig, hamster, rabbit, dog, cat, cow, pig, sheep, horse, monkey, chicken, duck
- Metabolism, Distribution, Excretion
  - mouse, rat, guinea pig, hamster, rabbit, dog, cat, cow, pig, sheep, horse

Classes of compounds studied:
Alkylphosphates, cyanide, nitrites, nitrates, purine and pyrimidine, and antineoplastic cpd

Experience with the **pharmaceutical industry**

Ability or experience in representation of data before regulatory agencies

Capability and experience in handling radioactive materials

Personnel:
4 persons in laboratory include 2 professionals; 1 toxicologist and 1 pharmacologist

Laboratory will consider performing contract research

Howard University Medical School
Department of Pharmacology
Washington, D.C. 20059

Individual to contact: Dr. Frederick Sperling

Experience with:
Acute toxicity, irritation
mouse, rat, guinea pig, hamster, rabbit, dog, cat
Subacute toxicity, oral, dermal, and parenteral
mouse, rat, guinea pig, hamster, rabbit
Subacute toxicity, inhalation; Chronic toxicity, inhalation
mouse, rat, hamster
Chronic toxicity, oral, dermal, and parenteral
mouse, rat, guinea pig, hamster
Reproduction; Teratology; Metabolism, Distribution, Excretion; Forensic
mouse, rat, guinea pig, hamster, rabbit

Routes of administration:
oral, subcutaneous, all parenteral, inhalation, and inunction

Classes of compounds studied:
Chlorinated hydrocarbon and anticholinesterase pesticides, carcinogenic agents, and solvents

Specific procedures:
Acute, subacute, and chronic testing by various routes and evaluating metabolic, pharmacodynamic, and pathological reactions and effect

Experience in representation of data before regulatory agencies; has presented material to FDA and Department of Agriculture (FIFRA)

Capabilities available:
In-house pathology, clinical chemistry, data reduction and analysis, pulmonary physiology, inhalation experience, and experience and capability to handle radioactive materials

Personnel:
Laboratory includes 2 toxicologists, 9 pharmacologists, and 1 teratologist

Laboratory will consider performing contract research

West Virginia

West Virginia University Medical Center
Department of Pharmacology
Morgantown, West Virginia 26506

Individual to contact and questionnaire prepared by Dr. John A. Thomas

Experience with:

- Acute toxicity, irritation
  - mouse, rat, guinea pig, hamster, dog
- Subacute toxicity, oral
  - mouse, rat, guinea pig, hamster
- Subacute toxicity, parenteral
  - mouse, rat, guinea pig, hamster, dog
- Chronic toxicity, oral and parenteral
  - mouse, rat, guinea pig, hamster
- Carcinogenicity, Reproduction, and Metabolism, Distribution, Excretion
  - mouse, rat, guinea pig, hamster, rabbit, dog
- Spermatogenesis
  - mouse

Routes of administration:
oral, i.v., i.m., and s.c.

Classes of compounds studied:
Pesticides (all types) and herbicides; synthetic hormones and most major classes of drugs; and carcinogens (MCA, ENU, DMBA, etc.)

Ability or experience in representation of data before regulatory agencies (Witness for EPA hearings, occasional consultant on carcinogenesis)

Capabilities available:
In-house pathology, experience and capability with radioactive materials, clinical chemistry, and data reduction and analysis

Personnel:
20 persons in laboratory including 6 professionals in the areas of toxicology, pathology, pharmacology, biochemistry, and chemistry

Laboratory frequently performs contract research

Endocrine Laboratories of Madison, Inc.
3301 Kinsman Blvd.
P.O. Box 7546
Madison, Wisconsin 53707

Individual to contact and questionnaire prepared by G. N. Rao, Vice President

Experience with:
Acute toxicity, irritation; Subacute toxicity, oral, parenteral
mouse, rat, guinea pig, hamster, rabbit, dog, cat, rhesus monkey
Chronic toxicity, oral, parenteral
mouse, rat, hamster, rabbit, dog, cat, rhesus monkey
Subacute toxicity, dermal
guinea pig, rabbit, dog, cat, rhesus monkey
Metabolism, Distribution, Excretion
mouse, rat, dog, cat, rhesus monkey
Carcinogenicity; Reproduction; Teratology
mouse, rat, dog, rhesus monkey
Experienced in primate research, maintains permanent closed colony of 200 rhesus monkeys

Routes of administration:
oral, parenteral, suppository, and dermal

Classes of compounds studied:
Hormones, contraceptives, endocrines, regulators, drugs, cosmetics, agrichemicals, environmental chemicals, growth regulators

Specific procedure experience:
Endocrine bioassays, radioimmunoassays, reproduction studies, antifertility studies, immunology techniques, metabolic and drug distribution studies

Experience with **pharmaceutical, chemical manufacturing, and cosmetic** industry

Capabilities available:
In-house pathology, clinical chemistry, data reduction and analysis, experience and capability to handle radioactive materials

Personnel:
25 persons in laboratory including 6 professionals – 1 toxicologist, 1 pharmacologist, 2 biochemists, and 1 chemist

Laboratory routinely performs contract research

The Medical College of Wisconsin
Department of Environmental Medicine
8700 West Wisconsin Avenue
Milwaukee, Wisconsin 53226

Individuals to contact: Richard D. Stewart, M.D. and Carl L. Hake, Ph.D.

Questionnaire prepared by Dr. Carl L. Hake

Experience with:
Metabolism, distribution, excretion
mouse, rat, guinea pig, hamster, rabbit, dog, cat, monkey, human

Routes of administration:
oral and inhalation

Classes of compounds studied:
Industrial solvents, aerosol propellants, and air pollutants, alone and in combination with drugs or alcohol in controlled environment chambers using procedures to measure physiological, behavioral, and health effects, and metabolic products

Experience with products of the chemical petroleum, automobile, consumer products, and armed services industries

Experience in representation of data before regulatory agencies

Capabilities available:
Clinical chemistry, data reduction and analysis, pulmonary physiology, inhalation experience, and experience and capability in handling radioactive materials; facilities and experience for analyses for ingestions or overexposure of various drugs and compounds including alcohols, aromatic solvents, carbon monoxide, chlorinated solvents

Personnel:
16 persons in laboratory including 12 professionals: 1 toxicologist, 4 chemists, 2 physicians, 1 physiologist, 2 bioengineers, 1 psychologist, and pathologists, pharmacologists, and biochemists are available for consultation

Laboratory routinely performs contract research

WARF Institute, Inc.
Box 7545
Madison, Wisconsin 53707

Individual to contact: Dr. D. J. Aulik, Director Client Services

Questionnaire prepared by Paul O. Nees, DVM, Client Service Executive

Experience with:
- Acute toxicity, oral
  - mouse, rat, guinea pig, hamster, rabbit, dog, cat, cow, pig, sheep, subhuman primates, chicken, duck, quail, sparrow, pheasant, pigeon, trout, bluegill, carp, tropical daphnia, algae
- Subacute toxicity, parenteral
  - mouse, rat, guinea pig, hamster, rabbit, dog, cat, cow, pig, sheep, subhuman primates, chicken, duck, quail, sparrow, pheasant, pigeon, trout, bluegill, carp
- Subacute toxicity, oral
  - mouse, rat, guinea pig, hamster, rabbit, dog, cat, pig, sheep, subhuman primates, duck, quail, sparrow, pheasant, pigeon, trout, bluegill, carp
- Metabolism, Distribution, Excretion
  - mouse, rat, rabbit, dog, cat, cow, pig, sheep, subhuman primates, chicken, duck, quail, trout, bluegill
- Chronic toxicity, oral, parenteral
  - mouse, rat, guinea pig, hamster, rabbit, dog, cat, pig, sheep, subhuman primates, chicken, duck, quail
- Reproduction
  - mouse, rat, hamster, rabbit, dog, pig, subhuman primates, chicken, duck, quail
- Subacute toxicity, dermal
  - mouse, rat, guinea pig, rabbit, dog, cat, pig
- Subacute toxicity, inhalation
  - mouse, rat, guinea pig, hamster, rabbit, dog, cat
- Chronic toxicity
  - rat, guinea pig, rabbit, dog, cat, pig
- Teratology
  - mouse, rat, hamster, rabbit, pig, subhuman primates
- Chronic toxicity, inhalation
  - mouse, rat, guinea pig, hamster, rabbit
- Carcinogenicity
  - mouse, rat, hamster, dog, subhuman primates, chicken
- Mutagenicity
  - mouse, rat, drosophila

Classes of compounds studied:
Drugs, agricultural chemicals, general chemicals, cosmetics, food and feed additives, natural toxins, nutritional ingredients, household products

Specific procedures:
Acute, subacute, and chronic toxicology, metabolism studies, tissue residue analysis and method development, carcinogenicity, teratogenicity, and reproductive studies, clinical chemistry, pathology, chemical and biochemical analytical support, nutritional effects

Routes of administration:
oral (intubation and add mix in feed), parenteral, dermal, and inhalation

Experience with **pharmaceutical, agrichemical, chemical specialties, cosmetics, food and feed production and processing industries**

Capabilities available:
In-house pathology, clinical chemistry, data reduction and analysis, inhalation experience, and experience and capability to handle radioactive materials

Personnel:
260 persons in laboratory including toxicologist, pathologists, pharmacologist, biochemists, teratologists, and chemists

Laboratory routinely performs contract research

Bio-Research Laboratories, Ltd.
265 Hymus Boulevard
Pointe Claire, Quebec, Canada H9R 1G6

Individual to contact: F. Fried, Ph.D.

Questionnaire prepared by: B. G. Procter, D.V.M., M.Sc.

Experience with:
- Acute toxicity, irritation
  - mouse, rat, guinea pig, hamster, rabbit, dog, cat, cow, pig, sheep, horse, primates, chicken
- Subacute toxicity, oral
  - mouse, rat, guinea pig, hamster, rabbit, dog, cow, pig, sheep, horse, primates, chicken
- Subacute toxicity, parenteral
  - mouse, rat, rabbit, dog, cow, pig, sheep, horse, primates
- Chronic toxicity, oral
  - mouse, rat, guinea pig, hamster, dog, primates
- Metabolism, Distribution, Excretion
  - mouse, rat, hamster, rabbit, dog, primates
- Subacute toxicity, dermal
  - mouse, rat, hamster, guinea pig, rabbit
- Carcinogenicity
  - mouse, rat, hamster, dog, primates
- Reproduction; Teratology
  - mouse, rat, hamster, rabbit
- Subacute toxicity, inhalation
  - rat, dog, primates
- Chronic toxicity, dermal
  - mouse, rat, dog, guinea pig, hamster
- Chronic toxicity, inhalation
  - rat, dog, primates
- Mutangenicity
  - mouse, rat, hamster
- Chronic toxicity, parenteral
  - primates, dogs

Routes of administration:
oral, dermal, i.v., s.c., i.p., i.m., and inhalation

Experience in representation of data before regulatory agencies in Ottawa and Washington

Classes of compounds studied:
Pharmaceuticals, agricultural chemicals, foods and food additives, and some industrial chemicals

Tests conducted:
Range from LD_{50} to lifetime feeding studies

Experience with **pharmaceutical, food, chemical and cosmetic** industry

Capabilities available:
In-house pathology, clinical chemistry, data reduction and analysis, pulmonary physiology, inhalation experience and experience and capability in handling radioactive materials

Personnel:
80 persons in laboratory include 12 professionals: 2 toxicologists, 3 pathologists, 1 pharmacologist, 1 biochemist, 1 analytical chemist, 2 microbiologists, 1 pharmacokineticist and 1 ophthalmologist

Laboratory routinely performs contract research

Canada (continued)

Health Protection Branch, Health & Welfare
Toxicology Research Division
Food Directorate
Tunney's Pasture, Ottawa K1A OL2
Ontario, Canada

Individual to contact and questionnaire prepared by Drs. I. C. Munro & K. S. Khera

Experience with:
Acute toxicity; Subacute toxicity, oral, parenteral;
Chronic toxicity, oral, parenteral
mouse, rat, guinea pig, hamster, rabbit, dog, cat, primate
Metabolism, Distribution, Excretion
mouse, rat, cat, primate
Teratology
mouse, rat, rabbit, cat
Carcinogenicity; Reproduction; Mutagenicity

Routes of administration:
oral, gavage, dietary feeding, i.p., and s.c.

Classes of compounds studied:
Food additives, environmental chemicals, veterinary drugs, and pesticides

Capabilities available:
In-house pathology, clinical chemistry, data reduction and analysis, and experience and capability in handling radioactive materials

Personnel:
40 persons in laboratory including 15 professionals: 7 toxicologists, 3 pathologists, 1 pharmacologist, 2 biochemists, 1 teratologist, and 1 chemist

Huntingdon Research Centre
Huntingdon, Cambs.
PE18 6ES
England

Individual to contact: Dr. D. L. G. Rowlands

Questionnaire prepared by Dr. Alastair N. Worden

Experience with:

Metabolism, Distribution, Excretion
: mouse, rat, guinea pig, hamster, rabbit, dog, cat, cow, pig, sheep horse, other mammal, chicken, duck, quail, other bird, trout, other fish, plants (pesticide metabolism)

Acute toxicity, irritation
: mouse, rat, guinea pig, hamster, rabbit, dog, cat, pig, sheep, other mammal, chicken, duck, quail, other bird, trout, bluegill, other fish

Subacute toxicity, oral
: mouse, rat, guinea pig, hamster, rabbit, dog, cat, cow, pig, sheep, horse, chicken, duck, quail, other bird, trout, bluegill, other fish

Reproduction
: mouse, rat, guinea pig, hamster, rabbit, dog, cat, cow, pig, sheep, other mammal, chicken, quail, other bird, trout

Subacute toxicity, inhalation
: mouse, rat, guinea pig, hamster, rabbit, dog, chicken, quail, other bird

Subacute toxicity, parenteral
: mouse, rat, guinea pig, rabbit, dog, pig, chicken, quail

Chronic toxicity, oral
: mouse, rat, hamster, rabbit, dog, cow, pig, sheep, horse

Carcinogenicity
: mouse, rat, guinea pig, hamster, dog, pig, sheep, other mammal, trout

Teratology
: mouse, rat, hamster, rabbit, dog, pig, other mammal, chicken

Subacute toxicity, dermal
: mouse, rat, guinea pig, rabbit, dog, pig

Chronic toxicity, dermal
: mouse, rat, rabbit, dog, pig

Chronic toxicity, inhalation
: mouse, rat, dog, sheep

Parenteral toxicity
: dog, pig

Mutagenicity
: mouse

Other mammals:
5 species of primate, mink, other hamsters, chinchilla

Other birds:
Starling, sparrow, greenfinch, budgerigars, canaries

Other fish:
Goldfish, harlequins

Aquatic organisms:
Daphnia magna

Test procedures:
All procedures relating to safety evaluation and preclinical testing required or advised by regulatory agencies and advisory bodies throughout the world

Services:
All toxicological and safety evaluation procedures, all testing for harm to wildlife, human metabolism, pharmacokinetics, and bioavailability (in conjunction with clinical pharmacology units), specialized procedures relating to nutrition, investigational surgery and medicine

Classes of compounds studied:
Drugs, foods, food additives, food packaging materials, agricultural chemicals, cosmetics, detergents, toiletries, devices, aerosols, veterinary compounds, plastics, smoking materials, household, factory, and environmental materials

Experience with **pharmaceutical, food, tobacco, cosmetic, detergent, general chemical, agricultural, paper, petroleum, and heavy chemical industry**

Ability or experience in representation of data before almost all the world's leading regulatory agencies and advisory groups

Capabilities available:
In-house pathology, clinical chemistry, data reduction and analysis, pulmonary physiology, inhalation experience, experience and capability to handle radioactive materials

Personnel:
820 in laboratory: 320 professionals – 140 toxicologists, 28 pathologists; 16 pharmacologists; 36 biochemists; 15 teratologists; 36 chemists + environmental scientists; botanists; physicists; engineers; statisticians; experimental surgeons; clinicians; registration, library, and editorial staff

Laboratory routinely performs contract research

Life Science Research
Stock, Essex CM4 9PE
England

Individual to contact and questionnaire prepared by Dr. K. H. Harper

Experience with

Acute toxicity, irritation; Subacute toxicity, oral, inhalation, parenteral; Chronic toxicity, oral, dermal, inhalation, parenteral; Reproduction; and Metabolism, Distribution, Excretion
  mouse, rat, guinea pig, hamster, rabbit, dog, primate, chicken, duck, quail, pheasant
Teratology
  mouse, rat, guinea pig, hamster, rabbit, dog, primate, chicken
Subacute toxicity, dermal
  mouse, rat, guinea pig, hamster, rabbit, dog, primate
Carcinogenicity
  mouse, rat, hamster, dog, primate
Mutagenicity
  mouse, rat, hamster, rabbit, in vitro systems

Classes of compounds studied:
All types, along with a full range of procedures

Experience with all types of industries

Experience in representation of data before regulatory agencies

Capabilities available:
In-house pathology, clinical chemistry, data reduction and analysis, pulmonary physiology, inhalation experience, and experience in handling radioactive materials

Personnel:
200 persons in laboratory including 38 toxicologists, 2 pathologists, 2 pharmacologists, 10 biochemists, 8 teratologists, and 6 chemists

Laboratory routinely performs contract research

Pfizer Research Centre
37400, Amboise, France

Individual to contact: Dr. A. M. Monro

Experience with:
Acute toxicity, irritation; Subacute toxicity, oral
mouse, rat, guinea pig, hamster, rabbit, dog, cat
Subacute toxicity, parenteral
mouse, rat, dog
Subacute toxicity, topical (dermal, vaginal, rectal, ocular)
rat, dog, rabbit, guinea pig
Reproduction; Teratology, Fertility, Prenatal
mouse, rat, rabbit, dog, cat
Chronic toxicity, 3-generation studies, oral
mouse, rat, hamster, dog
Carcinogenicity
mouse, rat, hamster
Metabolism, distribution, excretion
mouse, rat, rabbit, dog

Routes of administration:
oral, i.m., topical, and i.v.

Classes of compounds studied:
Drugs for human and veterinary use, and chemicals (mainly food additives)

Experience with **pharmaceutical and chemical industries**

Experience in representation of data before regulatory agencies (FDA, CSM-Britain)

Capabilities available:
In-house pathology, hematology, clinical chemistry, data reduction and analysis, pulmonary physiology, and embryology

Personnel:
95 persons in laboratory including 22 professionals: 4 toxicologists, 5 pathologists, 1 pharmacologist, 3 biochemists, 3 teratologists, and 1 statistician

Italy

Consorzio Ricerca Farmaceutica
Via Tito Speri
14–Pomezia
Italy

Individual to contact and questionnaire prepared by Professor Giulio Perri

Experience with:

Acute toxicity, irritation
 mouse, rat, guinea pig, hamster, rabbit, dog, cat, pig, chicken
Subacute toxicity, oral, dermal, inhalation, parenteral
 mouse, rat, guinea pig, hamster, rabbit, dog, cat, pig
Chronic toxicity, oral, dermal, inhalation, parenteral
 mouse, rat, guinea pig, hamster, rabbit, dog, pig
Carcinogenicity; Reproduction; Teratology; Mutagenicity
 mouse, rat, guinea pig, rabbit, dog
Bacterial mutagenic tests: Ames, Spot tests, Aspergillus tests, with and without microsomal activation
Studies on structure – activities for analogues series with mutagenic tests
Farmacokinetics studies with labeled compound (C_{14} and tritium
Metabolism, Distribution, Excretion
 rat, guinea pig, rabbit, dog

Classes of compounds studied:
Drugs, agricultural chemicals, insecticides, pesticides, food additives, dyes

Experience with **pharmaceuticals, dairy, cosmetics, and chemical** industries

Laboratory has authorization by the Italian Ministry of Health for Toxicology-Pharmacology

Capabilities available:
In-house pathology, clinical chemistry, data reduction and analysis, inhalation experience, experience and capability to handle radioactive materials

Personnel:
30 persons in laboratory including 12 professionals – 1 toxicologist, 1 pathologist, 1 pharmacologist, 2 biochemists, 1 teratologist, 4 chemists

Laboratory routinely performs contract research

University of Puerto Rico
Toxicology Laboratories
School of Medicine
P. O. Box 5067
San Juan, Puerto Rico 00936

Individual to contact and questionnaire prepared by Dr. Sidney Kaye

Experience with:
Evaluation, Metabolism, Distribution, Excretion; Other;
Forensic, Industrial ($Cl_{2\,9}$, Hg, As, Pb, Co, etc.)
other mammal, man, acute and chronic

Routes of administration:
oral, i.v., and inhalation

Experience with **drug and chemical** industries

Experience in representation of data before regulatory agencies

Scotland

Inveresk Research International
Edinburgh, EH21 7UB, Scotland

Individual to contact and questionnaire prepared by Dr. H. Reinert

Experience with:

Acute toxicity, irritation
- mouse, rat, guinea pig, hamster, rabbit, dog, cat, cow, pig, sheep, horse, baboon, cynomolgus, rhesus

Chronic toxicity, oral
- mouse, rat, hamster, dog, pig, sheep, baboon, cynomolgus, rhesus

Subacute toxicity, oral
- mouse, rat, guinea pig, hamster, rabbit, dog, pig, sheep, baboon, cynomolgus, rhesus

Metabolism, Distribution, Excretion
- mouse, rat, guinea pig, hamster, rabbit, dog, cat, cow, pig, sheep, horse, baboon, cynomolgus, rhesus

Subacute toxicity, dermal; Chronic toxicity, dermal
- mouse, rat, guinea pig, hamster, rabbit, dog, pig, sheep, baboon, cynomulgus, rhesus

Subacute toxicity, inhalation, parenteral; Chronic toxicity, inhalation, parenteral
- mouse, rat, guinea pig, hamster, rabbit, dog, baboon, cynomolgus, rhesus

Teratology
- mouse, rat, rabbit, baboon, cynomolgus, rhesus

Reproduction
- mouse, rat, baboon, rhesus, cynomolgus, baboon

Carcinogenicity
- mouse, rat, hamster, rhesus, cynomolgus, baboon, dog

Mutagenicity
- mouse, hamster, rats

Classes of compounds studied:
Drugs, agrochemicals, industrial chemicals, tobaccos, synthetic tobaccos, cosmetics, household chemicals, veterinary chemicals

Experience with all industries

Experience in representation of data before regulatory agencies in most countries of the world

Capabilities available:
In-house pathology, clinical chemistry, haematology, pulmonary physiology, pharmacology, dominant lethal test, heritable translocation, cytogenetic analysis, host mediated assay, in vitro bacteria and yeasts mutation, mammalian cell toxicity and transformation, peri- and postnatal studies including behaviour, male and female fertility studies in animals, teratology, pharmacology, drug dependence studies in baboons, pharmacokinetics in animals and man, local tolerance testing in man, inhalation experience, experience and capability to handle radioactive materials

Personnel:
234 persons in laboratory including 78 professionals: 11 toxicologists; 4 pathologists; 9 pharmacologists; 7 biochemists; 4 teratologists; 4 mutageneticists; 11 chemists

Laboratory routinely performs contract research

SUPPLEMENTAL LISTING OF TESTING LABORATORIES

The following represents a listing of additional laboratories offering environmental/toxicity testing services. This list was provided through the courtesy of the U.S. government.

CALIFORNIA

Northrop Services, Inc.
500 East Orangethrope Avenue
Anaheim, California 92801
(714) 871-5000

Sponsored Projects Information Services
Stanford University
Stanford, California 94305
(415) 497-2883

Stanford Research Institute
333 Ravenswood Avenue
Menlo Park, California 94025

University of Southern California
Office of the Executive Vice President
University Park
Los Angeles, California 90007
(213) 746-2311

FLORIDA

Papanicolaou Cancer Research Institute at Miami
1155 N.W. 14th Street
P.O. Box 236188
Miami, Florida 33123

ILLINOIS

Industrial Bio-Test Laboratories, Inc.
1810 Frontage Road
Northbrook, Illinois 60062

Rosner-Hixson Laboratories
3570 North Avondale
Chicago, Illinois 60618
(312) 588-8500

The University of Chicago
Department of Radiology
950 East 59th Street
Chicago, Illinois 60637
(312) 947-6868

University of Health Sciences
2020 West Ogden Avenue
Chicago, Illinois 60612
(312) 226-4100

INDIANA

T. P. S., Inc.
P.O. Box 333
Mount Vernon, Indiana 47620
(812) 985-3832

KENTUCKY

The University of Kentucky Research Foundation
East Wing, Kinkead Hall
Lexington, Kentucky 40506
(606) 258-4666

LOUISIANA

Gulf South Research Institute
7700 GSRI Avenue
Baton Rouge, Louisiana 70808

MARYLAND

Andrulis Research Corporation
7420 Montgomery Lane
Bethesda, Maryland 20014
Dr. Robert Katz

Frederick Cancer Research Center
Bioassay Research
Box B
Frederick, Maryland 21701

Microbiological Associates
5221 River Road
Bethesda, Maryland 20016
(301) 654-3400

Pharmacopathics Research Laboratories, Inc.
9705 N. Washington Boulevard
Laurel, Maryland 20810
(301) 776-8036

Preventive Health Programs, Inc.
2301 Research Boulevard
Rockville, Maryland 20850
(301) 948-4160

U.S. Army Environmental Hygiene Agency
Aberdeen Proving Ground, Maryland 21010

MASSACHUSETTS

Harvard Medical School
Youville Hospital
1575 Cambridge Street
Cambridge, Massachusetts 02138
ATTN: Dr. William A. Skornik

MICHIGAN

Dow Chemical Company
Contract Research, Development, and Engineering
Building 566
Midland, Michigan 48640

MISSOURI

Ralston Purina Company
Research 900
Checkerboard Square Plaza
St. Louis, Missouri 63188
(314) 982-3393

NEW JERSEY

Biometric Testing, Incorporated
156 Algonquin Parkway
Whippang, New Jersey 07981

Ethicon Research Foundation
Somerville, New Jersey 08876

Food and Drug Research Laboratories, Inc.
60 Evergreen Place
East Orange, New Jersey 07018
(201) 677-9500

Foster D. Snell, Inc.
General Laboratories
Hanover Road
Florham Park, New Jersey 07932
(201) 377-6700

NEW MEXICO

Albany Medical College
International Center of Environmental Safety
P.O. Box 1027
Holloman AFB, New Mexico 88330
ATTN: Dr. Travis Griffin

Rutgers University
Cook College – NJAES
Food Science Bldg., Room 107
New Brunswick, New Jersey 08903
(201) 932-1766

University Laboratories, Inc.
810 North Second Avenue
Highland Park, New Jersey 08904
(201) 246-1146

Wells Laboratories, Inc.
25-27 Lewis Avenue
Jersey City, New Jersey 07306
ATTN: Dr. Arthur F. Peterson
Director of Laboratories

NEW YORK

American Health Foundation
1370 Avenue of the Americas
New York, New York 10019
(212) 489-8700

State University of N.Y. at Buffalo
Department of Microbiology
333 Sherman Hall
Buffalo, New York 14214
(716) 831-3816

NORTH CAROLINA

U.S. Environmental Protection Agency
Environmental Toxicology Division
Research Triangle Park, North Carolina 27711

OHIO

Battelle – Columbus Laboratories
505 King Avenue
Columbus, Ohio 43201
(614) 424-6424

Bio/Tox Research Laboratories, Inc.
553 North Broadway
Spencerville, Ohio 45887
(419) 647-4196

Consolidated Biomedical Laboratories
P.O. Box 2289
Columbus, Ohio 43216
(614) 889-1061

Kettering Laboratory
3223 Eden Avenue
Cincinnati, Ohio 45267

OREGON

Oregon State University
Environmental and Health Science Center
Corvallis, Oregon 97331

University of Oregon
Kresge Hearing Research Laboratory
Health Sciences Center
Portland, Oregon 97201

PENNSYLVANIA

Cannon Laboratories, Inc.
P.O. Box 3627
Reading, Pennsylvania 19605
(215) 375-4536

TEXAS

Environmental Sciences Division
Southwest Foundation for Research & Education
8848 West Commerce Street
San Antonio, Texas 78284
(512) 674-1410

Southwest Foundation for Research and Education
Environmental Sciences Department
P.O. Box 28147
San Antonio, Texas 78284

Technology Incorporated
Life Sciences Division
8531 North New Braunfels Avenue
San Antonio, Texas 78217
(513) 426-2405

UTAH

Intermountain Laboratories, Inc.
P.O. Box 10633
870 East 7145 South
Midvale, Utah 84047
(801) 561-2223

University of Utah Research Institute
Utah Biomedical Test Laboratory (UBTL)
520 Wakarra Way
Salt Lake City, Utah 84108
(801) 581-7236

Utah Biomedical Test Laboratories
520 Walkara Way
Salt Lake City, Utah 84108

Weber State College
3750 Harrison Boulevard
Ogden, Utah 84403

VIRGINIA

Environmental Consultants, Inc.
4807 Colley Avenue
Norfolk, Virginia 23508
(804) 423-1858

Flow Research Animals, Inc.
P.O. Box 1065
Dublin, Virginia 24084
(301) 881-2900

CANADA

University of Guelph
Toxicology
Department of Biomedical Sciences
Guelph, Ontario, Canada

GERMANY

Dr. med. Horst Brune
c/o Vaselinwerk
2 Hamburg 11
Worthdamm 13-27
78 13 21-27

PUBLICATION LIST FROM THE OFFICE OF TOXIC SUBSTANCES

Activities of Federal Agencies Concerning Selected High Volume Chemicals – PB240133/AS
Papercopy: $3.75 Microfiche: $2.25

Asbestos in the Drinking Water of the Ten Regional Cities. Part I. PB-252-620
Papercopy: $4.50

Assessment of Liquid Siloxanes (Silicones). PB-247-778
Papercopy: $5.50 Microfiche: $2.25

Assessment of Wastewater Management Treatment Technology and Associated Costs for Abatement of PCBs Concentrations in Industrial Effluents. Task II. PB-251-433
Papercopy: $9.25

Benzene Environmental Sources of Contamination, Ambient Levels, and Fate – PB244139/AS
Papercopy: $4.25 Microfiche: $2.25

Chemical Technology and Economics in Environmental Perspectives. Task I – Technical Alternatives to Selected Chlorofluorocarbon Uses. PB-251-146
Papercopy: $8.00

Compilation of State Data for Eight Selected Toxic Substances: Arsenic, Beryllium, Cadmium, Chromium, Cyanide, Lead, Mercury, Polychlorinated Biphenyls, Volume I: Final Report. PB-248-660
Papercopy: $6.75 Microfiche: $2.25

Compilation of State Data for Eight Selected Toxic Substances. Volume II: Directory of State Toxic Substances Monitoring Agencies. PB-248-661
Papercopy: $4.00 Microfiche: $2.25

Compilation of State Data for Eight Selected Toxic Substances. Volume III: Annotated Bibliography of State Data and Information Sources. PB-248-662
Papercopy: $4.00 Microfiche: $2.25

Compilation of State Data for Eight Selected Toxic Substances. Volume IV: Compilation of Summaries and Analyses of State Data. PB-248-663
Papercopy: $16.25 Microfiche: $2.25

Compilation of State Data for Eight Selected Toxic Substances. Volume V: Monitoring Program Capability Descriptor Tables. PB-248-664
Papercopy: $9.75 Microfiche: $2.25

Current Awareness of Toxic Substances. Part I. PB-250-074
Papercopy: $12.50

Current Awareness of Toxic Substances. Part II. PB-250-075
Papercopy: $12.75

Development of a Study for Definition of PCBs Usage, Wastes, and Potential Substitution in the Investment Casting Industry. Task III. PB-251-842
Papercopy: $4.00

Draft Economic Impact Assessment for the Proposed Toxic Substances Control Act, S.776-PB242826/AS
Papercopy: $3.75 Microfiche: $2.25

An Ecological Study of Hexachlorobenzene. PB-252-651
Papercopy: $4.50

Ecological Study of Hexachlorobutadiene. PB-252-671
Papercopy: $4.00

Environmental Aspects of Chemical Use in Rubber Processing Operations (Conference Report) – PB244172/AS
Papercopy: $11.25 Microfiche: $2.25

Environmental Aspects of Chemical Use in Well-Drilling Operations. PB-246-947
Papercopy: $13.75 Microfiche: $2.25

Environmental Aspects of Chemicals Used in Printing Operations. Conference Proceedings. PB-251-406
Papercopy: $12.00

Environmental Contamination from Hexachlorobenzene PB-251-874
Papercopy: $4.00

Environmental Hazard Assessment of One and Two Carbon Fluorocarbons. PB-246-419
Papercopy: $9.00 Microfiche: $2.25

Environmental Hazard Assessment Report: Chlorinated Naphthalenes. PB-248-834
Papercopy: $4.00 Microfiche: $2.25

Environmental Hazard Assessment Report: Higher Benzene Polycarboxylates. PB-248-835
Papercopy: $3.50 Microfiche: $2.25

Epidemiology Studies: Vinyl Chloride. PB-248-426
Papercopy: $5.50 Microfiche: $2.25

All publications listed may be purchased from the National Technical Information Service, U.S. Department of Commerce, P.O. Box 1553, Springfield, VA 22151, (804)321-8500.

Estimating Limiting Risk Levels from Orally Ingested DDT and Dieldrin, Using Updated Version of Mantel-Bryan Procedure – PB243009/AS
Papercopy: $3.75 Microfiche: $2.25

Framework for the Control of Toxic Substances (Compilation of Speeches). PB243459/AS
Papercopy: $4.25 Microfiche: $2.25

Hexachlorobenzene and Hexachlorobutadiene Pollution from Chlorocarbon Processing – PB243641/AS
Papercopy: $7.00 Microfiche: $2.25

Identification of Organic Compounds in Effluents from Industrial Sources – Drinking Water – PB241641
Papercopy: $7.25 Microfiche: $2.25

Identification Systems for Selecting Chemicals or Chemical Classes as Candidates for Evaluation – PB238196/AS
Papercopy: $6.25 Microfiche: $2.25

Impact of Intensive Application of Pesticides and Fertilizers on Underground Water Recharge Areas Which May Contribute to Drinking Water Supplies. PB-251-181
Papercopy: $5.50

Industry Survey of Test Methods of Potential Health Hazards. PB-239-840
Papercopy: $4.75

Investigation of Selected Potential Environmental Contaminants: Chlorinated Paraffins. PB-248-634
Papercopy: $5.50 Microfiche: $2.25

Investigation of Selected Potential Environmental Contaminants: Haloethers. PB-246-359
Papercopy: $7.25 Microfiche: $2.25

Investigation of Selected Potential Environmental Contaminants: Ketonic Solvents. PB-252-970
Papercopy: $10.00

Laboratory Test Methods to Assess the Effects of Chemicals on Terrestrial Animal Species – PB241505/AS
Papercopy: $12.00 Microfiche: $2.25

Literature Study of Selected Potential Environmental Contaminants: Antimony and Its Compounds. PB-251-433
Papercopy: $6.75

Literature Study of Selected Potential Environmental Contaminants – Titanium Dioxide – PB242293
Papercopy: $5.75 Microfiche: $2.25

Manufacture and Uses of Selected Alkyltin Compounds. Task II. PB-251-819
Papercopy: $6.00

Manufacture and Uses of Selected Aryl and Alkyl Aryl Phosphate Esters. Task I. PB-251-678
Papercopy: $6.00

Manufacture and Use of Selected Inorganic Cyanides. PB-251-820
Papercopy: $8.00

Materials Balance and Technology Assessment of Mercury and Its Compounds on National and Regional Bases. PB-247-000
Papercopy: $11.75 Microfiche: $2.25

An Ordering of the NIOSH Suspected Carcinogens List Based Only on Data Contained in the List. PB-251-851
Papercopy: $16.25

Papers of a Seminar on Early Warning Systems for Toxic Substances (Conference Report) – PB244412/AS
Papercopy: $7.25 Microfiche: $2.25

PCB's in the United States: Industrial Uses and Environmental Distribution. Task I. PB-252-012
Papercopy: $12.50

Pollution Potential of Polybrominated Biphenyls – PB243690/AS
Papercopy: $4.25 Microfiche: $2.25

Preliminary Assessment of the Environmental Problems Associated with Vinyl Chloride and Polyvinyl Chloride – PB239110/AS
Papercopy: $5.25 Microfiche: $2.25

Preliminary Assessment of Suspected Carcinogens in Drinking Water – PB244415/AS
Papercopy: $3.75 Microfiche: $2.25

Preliminary Assessment of Suspected Carcinogens in Drinking Water (Appendices) – PB244416/AS
Papercopy: $7.25 Microfiche: $2.25

Preliminary Assessment of Suspected Carcinogens in Drinking Water, December 1975 Report to Congress. PB-250-961
Papercopy: $5.50

Preliminary Environmental Hazard Assessment of Chlorinated Naphthalenes, Silicones, Fluorocarbons, Benzenepolycarboxylates, and Chlorophenols – PB238074/AS
Papercopy: $8.75 Microfiche: $2.25

Preliminary Investigation of Effects on the Environment of Boron. PB-245-984
Papercopy: $5.25 Microfiche: $2.25

Preliminary Investigation of Effects on the Environment of Indium. PB-245-985
Papercopy: $3.75 Microfiche: $2.25

Preliminary Investigation of Effects on the Environment of Nickel. PB-245-986
Papercopy: $4.75 Microfiche: $2.25

Preliminary Investigation of Effects on the Environment of Selenium. PB-245-987
Papercopy: $5.25 Microfiche: $2.25

Preliminary Investigation of Effects on the Environment of Tin. PB-245-988
Papercopy: $5.25 Microfiche: $2.25

Preliminary Investigation of Effects on the Environment of Vanadium. PB-245-989
Papercopy: $4.75 Microfiche: $2.25

Preliminary Study of Selected Potential Environmental Contaminants – Optical Brighteners, Methyl Chloroform, Trichloroethylene, Tetrachloroethylene, Ion Exchange Resins – PB243910/AS
Papercopy: $8.75 Microfiche: $2.25

Review and Evaluation of Available Techniques for Determining Persistence and Routes of Degradation of Chemical Substances in the Environment – PB243825/AS
Papercopy: $13.00 Microfiche: $2.25

A Review of Concentration Techniques for Trace Chemicals in the Environment. PB-247-946
Papercopy: $12.50 Microfiche: $2.25

Review of the Environment Fate of Selected Chemicals – PB238908
Papercopy: $3.75 Microfiche: $2.25

Review of the Environmental Fate of Selected Chemicals. Task II – Polynuclear Aromatic Hydrocarbons. PB-250-948
Papercopy: $6.00

Sampling and Analysis of Selected Toxic Substances: Ethylene Dibromide. PB-246-213
Papercopy: $3.75 Microfiche: $2.25

Sampling and Analysis of Selected Toxic Substances. Task III – Vinyl Chloride, Secondary Sources. PB-252-966
Papercopy: $4.00

Structure – Activity Correlation – PB240658
Papercopy: $4.25 Microfiche: $2.25

A Study of Flame Retardants for Textiles. PB-251-441
Papercopy: $7.50

Summary of Office Requirements Resulting from Toxic Substances Control Act and a Preliminary Specification for a Data Management System – PB238088/AS
Papercopy: $5.75 Microfiche: $2.25

Technical and Microeconomic Analysis: Cadmium and Its Compounds. PB-244-625
Papercopy: $7.25 Microfiche: $2.25

Technical and Microeconomic Analysis of Cadmium and Its Compounds – PB244625/AS
Papercopy: $7.25 Microfiche: $2.25

Test Methods for Assessing the Effects of Chemicals on Plants. PB-248-198
Papercopy: $8.00 Microfiche: $2.25

OTHER PUBLICATIONS WITH ENVIRONMENTAL/TOXICITY DATA

CRC Handbook of Analytical Toxicology, Sunshine, I., Ed., CRC Press

CRC Handbook of Chemistry and Physics, Weast, R., Ed., CRC Press

CRC Handbook of Environmental Control, Bond, R. G. and Straub, C., Eds., CRC Press

CRC Handbook of Laboratory Safety, Steere, N. V., Ed., CRC Press

Chemistry of Pesticides, Melnikov, N., Ed., Springer-Verlag

Dangerous Properties of Industrial Materials, Sax, N. I., Reinhold

Guidelines for Analytical Toxicology Programs, Thoma, J., Bondo, P., and Sunshine, I., Eds., CRC Press

Handbook of Analytical Chemistry, Meites, L., Ed., McGraw-Hill

Industrial Hygiene and Toxicology, Patty, F. A., Ed., Wiley-Interscience

Lange's Handbook of Chemistry, Dean, J. A., Ed., McGraw-Hill

The Merck Index, Windholz, M., Ed., Merck & Co.

Methodology for Analytical Toxicology, Sunshine, I., Ed., CRC Press

Physicians' Desk Reference, Baker, C. E., Jr. (Publ.), Medical Economics Co.

Practical Toxicology of Plastics, LeFaux, R., Ed., CRC Press

Registry of Toxic Effects of Chemical Substances, Christensen, H. E. and Luginbyhl, T. T., Eds., NIOSH, U.S. Department of Health, Education, and Welfare

Toxicity Metabolism of Industrial Solvents, Browning, E., Ed., Elsevier

PUBLISHERS' INDEX

CRC Press, Inc.
18901 Cranwood Parkway
Cleveland, Ohio 44128

Elsevier Scientific Publishing Co.
52 Vanderbilt Ave.
New York, New York 10017

McGraw-Hill Book Co.
1221 Ave. of the Americas
New York, New York 10036

Medical Economics Co.
Litton Industries
Oradell, New Jersey 07649

Merck & Co., Inc.
Rahway, New Jersey 07065

NIOSH, U.S. Department of Health, Education, and Welfare
Public Health Service
Center for Disease Control
National Institute for Occupational Safety and Health
Rockville, Maryland 20852

Springer-Verlag New York Inc.
175 Fifth Ave.
New York, New York 10010

Van Nostrand Reinhold Co.
Division of Litton Educational Publishing, Inc.
450 W. 33rd Street
New York, New York 10001

Wiley-Interscience
John Wiley & Sons, Inc.
605 Third Ave.
New York, New York 10016

Section XI

Appendix

THE WHITE HOUSE

STATEMENT BY THE PRESIDENT*

I have signed S. 3149, the "Toxic Substances Control Act." I believe this legislation may be one of the most important pieces of environmental legislation that has been enacted by the Congress.

This toxic substances control legislation provides broad authority to regulate any of the tens of thousands of chemicals in commerce. Only a few of these chemicals have been tested for their long-term effects on human health or the environment. Through the testing and reporting requirements of the law, our understanding of these chemicals should be greatly enhanced. If a chemical is found to present a danger to health or the environment, appropriate regulatory action can be taken before it is too late to undo the damage.

The legislation provides that the Federal Government through the Environmental Protection Agency may require the testing of selected new chemicals prior to their production to determine if they will pose a risk to health or the environment. Manufacturers of all selected new chemicals will be required to notify the Agency at least 90 days before commencing commercial production. The Agency may promulgate regulations or go into court to restrict the production or use of a chemical or to even ban it if such drastic action is necessary.

The bill closes a gap in our current array of laws to protect the health of our people and the environment. The Clean Air Act and the Water Pollution Control Act protect the air and water from toxic contaminants. The Food and Drug Act and the Safe Drinking Water Act are used to protect the food we eat and the water we drink against hazardous contaminants. Other provisions of existing laws protect the health and the environment against other polluting contaminants such as pesticides and radiation. However, none of the existing statutes provide comprehensive protection.

This bill provides broad discretionary authority to protect the health and environment. It is critical, however, that the legislation be administered in a manner so as not to duplicate existing regulatory and enforcement authorities.

In addition, I am certain that the Environmental Protection Agency realizes that it must carefully exercise its discretionary authority so as to minimize the regulatory burden consistent with the effective protection of the health and environment.

The Administration, the majority and minority members of the Congress, the chemical industry, labor, consumer, environmental and other groups all have contributed to the bill as it has finally been enacted. It is a strong bill and will be administered in a way which focuses on the most critical environmental problems not covered by existing legislation while not overburdening either the regulatory agency, the regulated industry, or the American people.

*From the Office of the White House Press Secretary, October 12, 1976.

THE WHITE HOUSE

FACT SHEET

S. 3149 – THE TOXIC SUBSTANCES CONTROL ACT*

The President today signed S. 3149 – The Toxic Substances Control Act. This Act provides, for the first time, comprehensive authority for the Federal Government to regulate all substances or the use of all substances that may produce toxic effects.

Highlights

This new law will better enable us to minimize the risk of unknown hazards to health of the environment from toxic substances while permitting us to continue to reap the benefits which these substances can contribute. The bill contains some 53 pages of intricate regulatory material.

Generally speaking, the bill gives authority to the EPA Administrator to:

– require private industry to provide test data and supply detailed information on specified substances;
– prevent, or place limitations on, the marketing of new substances which the Administrator believes harmful; and,
– ban or limit continued marketing of existing substances.

The Toxic Substances Control Act is designed to prevent problems. By allowing early and selective regulation of only those uses that are likely to be hazardous, the Act minimized adverse regulatory impacts on the chemical industry. In addition, this preventative approach should help reduce the need for regulations under other laws which hurt important industries such as fishing, food processing, and the many other manufacturers who rely on chemical products.

Background

New chemical substances are being formulated rapidly and new commercial applications are being found almost daily. The production of metals, metal compounds and synthetic organics, which has been growing at a rate of 10 to 15% over the past 20 years, will continue to provide many new benefits to our society. For example, organic chemicals, which can be tailored in structure and properties to fit almost any imaginable need, are being used in ever-increasing quantities to produce dyes, pigments, flavors, perfumes, plastics, rubber products, detergents, pharmaceuticals, and so on. Yet, substances which in some applications have been extremely useful have been found in other applications to cause unanticipated and undesirable side effects on the environment and human health. Examples are vinyl chloride, polychlorinated and polybrominated biphenyls, kepone, fluorocarbons, and lead. There presently exists a number of statutory authorities to regulate toxic substances. Among these are the:

– Federal Food, Drug, and Cosmetic Act which regulates substances which are used as foods, drugs, or cosmetics;
– Occupational Safety and Health Act which regulates contact with substances in the work place;
– Consumer Product Safety Act regulates dangers from consumer products;
– Federal Insecticide, Fungicide, and Rodenticide Act (FIFRA) which regulates substances used as pesticides;
– Safe Drinking Water Act which regulates the level of toxic substances that can be present in drinking water supplies; and
– The Federal Water Pollution Control Act provides for State and Federal regulation over industrial discharges of toxic pollutants into the Nation's waters.

However, there are certain important gaps in the regulatory framework. For example, there is presently no effective way to regulate PCB's until and unless their dispersion into the environment affects water supply. This type of situation would be subject to control under various provisions of the bill.

*From the Office of the White House Press Secretary, October 12, 1976.

ENVIRONMENTAL PROTECTION AGENCY – IMPLEMENTATION OF TOXIC SUBSTANCES CONTROL ACT – PUBLIC MEETING [FRL 654-6]

On November 23, 1976, the Environmental Protection Agency (EPA) published in the *Federal Register*, 41 FR 51648, a notice of public meeting to be held December 14 and 15, 1976, in Washington, D.C., on the implementation of the Toxic Substances Control Act (TSCA) Public Law 94-469. The purpose of this meeting is to receive comments from all interested parties on issues relevant to the Act and its implementation. A draft list of issues representing various points of view identified so far is included in this notice following the meeting agenda. The Agency is interested in having comments on any of these issues and/or on other aspects of this statute.

The morning of the first day will be a plenary session during which there will be a presentation by EPA on the principal provisions of TSCA and Agency actions to date followed by an opportunity for general comments or questions from attendees. The afternoon of the first day will be divided into four concurrent sessions to provide an opportunity for comments on more specific aspects of the legislation and its implementation. EPA staff members will attend each session to receive comments. The morning of the second day will again be a plenary session at which the moderators will present reports on the subsessions. An opportunity will also be provided for general observations and comments by attendees.

The proceedings will be recorded and a transcript available for public inspection after January 1.

Subject to time limitations, all those wishing to make brief oral presentations will be given the opportunity to do so and are asked to notify, in writing or by telephone, Mr. Richard Somoskey, Office of Toxic Substances (WH-557), Room 711, East Tower, U.S. Environmental Protection Agency, 401 "M" Street, S.W., Washington, D.C. 20460, telephone number (202)755-4880, prior to December 13. Written comments are welcome and may be submitted to the same address.

Dated: December 3, 1976.

Kenneth L. Johnson,
Acting Assistant Administrator
for Toxic Substances.

Agenda

December 14

Thomas Jefferson Auditorium, U.S. Department of Agriculture, Room 1072, South Agriculture Building, 14th and Independence Ave., S.W., Washington, D.C. (enter Wing Four on Independence Ave.).

9–10:30 a.m. – General overview; review of principal provisions of TSCA; EPA actions to date; TSCA strategy group, integrated toxic strategy group, TSCA implementation activities.

10:30–12 p.m. – Questions and discussion.

2–5 p.m. – (Concurrent sessions.)
General policy issues, Department of Agriculture Auditorium.
Testing issues, Pierre Room, L'Enfant Plaza Hotel, 480 L'Enfant Plaza East, S.W., Washington, D.C.
Premarket/initial inventory issues, Hall of States A, Skyline Inn, South Capitol and I Streets, S.W., Washington, D.C.
Other issues including research/data, systems/discretionary reporting/confidentiality/state programs, Hall of States B. Skyline Inn, South Capitol and I Streets, S.W., Washington, D.C.

December 15

Department of Agriculture Auditorium
9–10:30 a.m. – Reports on concurrent sessions.
10:30–12 p.m. – Observations by attendees.

List of Issues Identified Relevant to Implementation of the Toxic Substances Control Act (Public Law 94-469)

General Issues

Strategic Policy Issues

1. EPA spokesmen have repeatedly stated that the overall objective of TSCA is to reduce the probability of chemical incidents harmful to man or the environment without unnecessarily raising the costs of products, retarding R & D, distorting the configuration of U.S. industry, or jeopardizing our international competitive position. Should this objective be reaffirmed, modified, or changed? If so, how? Can more precise near-term objectives be specified which will be useful in measuring the favorable and unfavorable impacts of the legislation on societal interests?

2. Implementation activities will impact on industry in at least three ways, namely, imposition of specific requirements under TSCA; support of specific requirements under other authorities; and stimulation of new patterns of industrial R & D of different industrial approaches to toxic chemicals in general, and of modified buying, selling, and diversification strategies. How important is strict enforcement of specific TSCA regulations in the context of the broader impacts that the legislation will undoubtedly have?

3. In the absence of any TSCA implementation experience, should differing emphases be given to different sections of the law and to different activity areas at the

outset? If so, what should be the considerations in determining emphases? How can a mid-course correction capability to adjust these emphases be built into implementation? How should the emphases be manifested (e.g., resource allocations, management attention, public statements). Among the areas of concern are:

a. Assessment and regulation of old chemicals vs. new chemicals.

b. Attention to many chemicals vs. more detailed attention to a few.

c. Preliminary premarket screening of many new chemicals as an "alert" device vs. detailed premarket assessment of a few new chemicals to trigger regulation during the 90-day period.

d. Chemical-by-chemical approach vs. category-by-category approach.

e. Information acquisition to support other agencies vs. information acquisition to support EPA.

f. Information acquisition vs. regulatory actions.

g. Acquisition of exposure data vs. acquisition of toxicity data.

h. Developing sound TSCA data system vs. reorientation of Government-wide data systems.

i. Use of existing test methods vs. development of new test methods.

j. Use of information acquisition authority to assess specific plants vs. to assess industrial trends.

k. Systematized approach to chemical prioritization vs. "ad hoc" priority decisions at time resources are committed.

4. How aggressively should EPA at the very outset develop regulations under Sections 4 and 6 taking into account (a) limitations on available resources and uncertainties concerning the most important problem chemicals on the one hand, and (b) at the same time the backlog of unattended known problems, the Congressional interest, and the need for EPA to exert early leadership in setting priorities?

5. To what extent, and how, should flexibility be built into the implementation approach to accommodate unanticipated crisis chemicals and citizen's petitions? What external parties should be involved in determinations as to the priority to be given to such unanticipated developments? In this regard should citizens petitions be subjected to "peer" review?

6. To what extent and how should Government-sponsored research on effects of specific chemicals be reoriented in view of the policy declaration that such testing is the responsibility of industry? Under what circumstances should the Government conduct testing of effects of specific chemicals (e.g., validate new methods, avoid delays in requiring industry tests, confirm suspect data).

7. What types of actions should require formal Environmental Impact Statements? What should be the purpose of these statements? To what extent should the statements be relied upon as the device to gain meaningful public participation? To what extent will they be redundant of the analyses inherent in actions under TSCA? Will they slow down the rulemaking process?

8. Should rules of procedures for hearings for all relevant sections be developed through rulemaking at the outset?

9. How aggressively and through what mechanisms should the United States seek international consistency in the testing and regulation of toxic substances? In the absence of such consistency, what steps can be taken to help insure equal treatment of U.S. manufacturers, importers, and exporters?

Linkages Among Sections

10. Should a Section 8 reporting requirement be imposed on all existing chemicals for which testing under Section 4 is required? for which a regulatory action short of a ban is taken under Section 6? for all new chemicals reported under Section 5 subsequent to the 90-day period?

11. Should a "significant new use" premarket notification requirement under Section 5 be required for all new uses of chemicals partially regulated by Section 6?

12. Should all new chemicals (note: Probably chemical classes) subject to testing under Section 4 also be put on the "risk" list under Section 5?

13. Should the "use" categories developed for the routine premarket notification form under Section 5 and for the reporting form under Section 8 be the same "use" categories as for reporting significant new uses under Section 5?

14. To what extent, if at all, should the activities of the Technical Assistance Office be related to inspection activities? to evaluation of the overall impact of the legislation?

15. Should the list of "risk" chemicals under Section 5 include all chemical classes subject to testing under Section 4?

16. Are there any linkages between Section 6 (regulatory actions) and Section 9 (relationship to other laws) that should be addressed in a general sense or is a case-by-case approach more appropriate?

17. To what extent, if at all, should the division of responsibility for research between EPA (Section 10) and HEW (Section 27) be clarified?

18. To what extent, if at all, should the division of responsibility for data systems between EPA (Section 10) and CEQ (Section 25) be clarified?

19. Several definitional questions may affect more than one section of TSCA. In some instances, consistency would seem appropriate. Among the issues are

a. What type of chemical intermediates should be

included in the initial inventory and be subject to premarket notification (e.g., those that leave plant property, leave a building, leave a totally enclosed pipe and kettle system)? Should the same definition delimit the scope of Section 4 (i.e., testing)? What are the environmental implications of delimiting coverage of intermediates? the operational implications? the trade secret implications?

b. Should there be a minimum poundage requirement for chemicals on the inventory and for those subject to premarket notification?

If so, what should the cutoff be? If not, how will research chemicals be defined to exclude them?

c. Should there be guidelines to determine when the constituents of a mixture react "incidentally"?

d. Is there a need to clarify what is regulable under FIFRA and the Atomic Energy Act and therefore not regulable under TSCA?

e. Should there be general ground rules for defining chemical classes or should the definition be addressed on a case-by-case basis?

f. Can chemical "equivalency" be clarified in the abstract?

Section-By-Section Issues

Testing

20. What should be the purpose of developing test data? To provide a basis for near-term regulatory decisions? By EPA? By other agencies? To identify potential problem chemicals for further investigations? To delimit the world of problem chemicals? To stimulate additional and improved industry test activities?

21. To what extent should test requirements for specific chemicals be tied with concomitant efforts to develop exposure and economic information so that when test data are received, decisions can be made on appropriate regulatory action?

22. Should more detailed criteria be developed for prioritizing chemicals for testing? What would be the legal and operational implications of such criteria?

23. How can EPA assure that the test data are credible? Can the FDA inspection program service EPA's needs? Will this cover foreign testing labs?

24. Should test requirements be structured largely around chemical classes? How broad can a class be and still be legally defensible?

25. Should test requirements consist of general guidance under which industry must develop and submit detailed protocols for EPA approval? or should EPA initially promulgate definitive requirements?

26. Will testing be required on the pure chemical or the technical grade chemical? or will this be decided on a case-by-case basis?

27. How will OSHA's needs for toxicology data based on 40 hr/week exposure be reconciled with EPA's need for data based on 24 hr/day?

28. How can test data submitted by industry be best formatted to assist in the subsequent decision-making process? (Does this presuppose development of a decision-making framework and/or procedures for evaluating the results of the required tests?)

29. Is any useful purpose served by attempting to define "health risk" in the abstract or should this be more appropriately addressed on a case-by-case basis?

30. What cost-sharing formulae will be used to provide reimbursement? For data which are yet to be developed vs. existing or in development data? Small business considerations?

31. What is the role of the Testing Committee and priority? What reliance should be placed on its recommendations? What level of rationale will be attached to the list of 50? Should EPA propose testing requirements prior to receiving the Committee's initial recommendations? Should the Committee be considered as a source of recommendations for regulatory as well as testing recommendations?

32. Can the Test Committee list for priority attention 50 chemical "classes?"

33. To what extent should TSCA test requirements for specific effects be consistent with FDA and FIFRA requirements for the same effects? How much of a delay can be tolerated in seeking this consistency among the three programs?

34. Should there be a relationship between the extent of testing that is required and (a) the likely exposure, and (b) the value of the chemical?

35. With regard to test petitions concerning new chemicals, what should be the character and product of informal discussions that are required to attempt to avoid the necessity for the petitions?

36. When is it appropriate for the processor rather than the manufacturer to conduct tests?

37. How are the availability of facilities and personnel to be determined? Should test requirements be limited by "existing" facilities or should they be used to stimulate development of additional facilities?

Premarket Notification

38. What can be done to discourage/avoid premarket 'false alarms'? Charge a notification fee? Require subsequent reporting under Section 8? Put a time limit on the validity of the notification?

39. How should premarket notifications be formatted to help ensure that the 5-day deadline for *Federal Register* publication will be met? How will claims of confidentiality be handled?

40. Should EPA determine and specify in the abstract what data shall be considered as adequate or adverse and possible EPA responses for the several categories of submissions (e.g., if a testing rule exists; if there is no testing rule, but a chemical is on the list; if there is no

testing rule and a chemical is not on the list)? Or is this a case-by-case determination?

41. Should criteria be developed to provide guidance as to when (and what type of) action should be taken under (e) and (f) (e.g., when a court injunction should be sought to prohibit manufacture of a chemical)? What EPA or industry (or other) actions will follow? What would be the legal and operational implications of such criteria?

42. To what extent, under what circumstances, and for what purposes should EPA require premarket notification of significant new uses of existing chemicals? What will be defined as a "significant new use"? How can "uses" be categorized in an environmentally meaningful way?

43. To what extent, under what circumstances, and for what purposes should EPA place chemicals on the pre-market "risk" list? How do chemicals on this list relate to chemicals for which testing is required under Section 4? Should a test protocol be developed for each chemical on the list? How important is this list? What will be the review system (is there a viable matrix?) against which to predict behavior/risk associated with new chemicals?

44. How will exemptions be defined and handled? Should EPA publish guidance for industry on this?

45. When are intermediates subject to premarket notification? If they leave plant property? If they leave a building? If they leave a closed system?

46. What general factors could trigger a justified extension of the notice period? Should these be set forth in the abstract or must they be case-by-case determinations?

47. What on-call capability should EPA have to confirm questionable premarket notification data? Effects data? Other data?

48. Should guidelines be developed to clarify "test marketing"?

49. How will the premarket notification "gap" from July to December 1977 be handled?

50. How will "small quantities" be clarified?

51. In the event that the molecular identity of a new chemical is confidential, will publication in the *Federal Register* of the generic class allow meaningful public input into premarket deliberations? How will "for commercial purposes" be defined? This delimits the coverage of Section 5.

Regulation and Imminent Hazard

52. Is there a need to clarify in the abstract how the authorities in Section 6 relate to other authorities or should this be done on a case-by-case basis? For example:

a. "Quality control" authority vs. OSHA authority.

b. "Labelling" authority vs. authority of OSHA, CPSC, and DOT.

c. "Disposal" authority vs. authority of Resource Conservation and Recovery Act.

d. "Handling" vs. in-transit storage authority of DOT.

53. Is it the case that for the required PCB regulations, EPA is "not" required to consider and publish a statement on the risks, benefits and costs?

54. Does the authority to regulate chemicals which are contained in their original state in products – including imported products – need clarification or should this be addressed on a chemical-by-chemical basis?

55. Should there be a general policy concerning the desirability of having labelling and disposal requirements apply retroactively to existing products, or should this be addressed on a case-by-case basis?

56. Should criteria be developed in the abstract for invoking the different authorities of Sections 6 and 7? Or should this be a case-by-case decision? What are the legal implications of such criteria?

Reporting and Record Keeping

57. A number of questions are important with regard to the initial inventory, including:

a. Should the initial "straw" list be based on available data and industry required only to supplement the list? Should only Government data be used in this approach or should "unconfirmed" data from SRI and other sources also be used?

b. What criteria should be used for "basket categories" in the list? How detailed should the description of these categories be?

c. What happens in terms of publishing the inventory if a manufacturer can show that the molecular structure of an existing chemical, which is not a basket category, is confidential?

d. Should data in addition to the chemical structure (e.g., use, by-products) be collected when compiling the inventory? What would be the operational implications? What evidence is there that anyone really wants such data?

e. How will natural products be handled?

58. What evidence should there be that someone is ready to use the data collected under discretionary reporting prior to requiring such data?

59. Should there be general discretionary reporting procedures established by rulemaking with the specific reports on specific chemicals or specific facilities required by subsequent FR notices or by letters comparable to the FWPCA 308 procedure?

60. How can record-keeping requirements best serve the interests of other Federal agencies (e.g., OSHA) and state authorities?

61. Does notification (Section 8e) overlap the spill reporting authority of FWPCA? How should this be sorted out? Is it necessary to clarify "substantial risk"? Does this have implications for other sections?

62. Will "health and safety studies" need clarification?

63. Should "environmental" criteria influence the definition of "small business"?

Relationship to Other Laws

64. If action has been transferred to another agency, but that agency has not acted, should the action be retrieved by EPA if new data are developed?

65. Does reducing the number of agencies involved in regulating a chemical, and presumably improving efficiency of regulation, serve "the public interest"?

66. How can EPA minimize distorting the priorities of other agencies when transferring actions?

67. How strong a case of risk should be made before action is transferred to another agency?

Research and Data Systems

68. How should research priorities be set? for EPA? for the Government? Is general guidance on priorities in addition to that set forth in TSCA desirable?

69. What are the training needs? Who pays the bill?

70. How can access to confidential information collected under TSCA be facilitated without losing secure control of the data?

Inspections

71. Is inspection of a plant appropriate if the plant is not required to comply with a TSCA regulation? What would be the purpose of the inspection?

72. Should the purpose of inspection be to scout problems or insure compliance?

73. Can EPA inspections serve OSHA purposes and vice versa? The needs of States? What are the operational implications?

74. Should chemical samples be taken during inspection?

Exports and Imports

75. How will EPA know that a new chemical poses a risk in the workplace until after the damage is done if exports are exempt from premarket notification? Can Section 8 be used to require reporting of all exports at the time manufacture commences?

76. Should the notification of foreign governments about hazardous exports be "pro forma" or should more vigorous steps be taken?

77. Many imported products contain natural products. Where will the line be drawn as to definition of "chemical substance"?

78. Many imported products contain chemicals. Should customs inspection be changed in light of TSCA?

Disclosure of Data

79. Should "confidential" data be segregated from "nonconfidential" in industry submissions? What are the operational implications? What penalties should there be if industry does not segregate the data?

80. What will be the mechanism for resolving conflicts over claims of confidentiality prior to judicial review?

81. Should a scheme for aggregating confidential production data into releasable gross production figures be developed along the lines of the ITC scheme?

82. Should "confidential" premarket test data automatically become nonconfidential at the end of the 90-day period? When manufacture commences?

Prohibited Acts, Penalties, Enforcement, and Seizure

83. What might be the impact of current and future case law on interpretations of potential violations? Are there general guidelines in this regard or is this a case-by-case matter?

84. If importers attempt to pass responsibility to foreign manufacturers, is there a jurisdictional problem for the courts?

Preemption

85. Should EPA promulgate a "General Rule," developed in the abstract to handle all applications for exemptions, or simply publish general guidance and handle each application as a separate rulemaking in its own right?

86. How could EPA admit that the conditions for an exemption are met, under (b) (1) and (2) without tacitly admitting that its own rule was inadequate?

87. Should "similar" tests or regulations be defined in the abstract or should this be a case-by-case matter?

88. What are the enforcement aspects of States taking inappropriate actions without seeking an exemption?

Judicial Review

89. How can "all" relevant documentation be kept in an easily retrievable form for each type of TSCA action from its inception?

Citizens Civil Action

90. Is clarification needed as to whether any (and if so, which ones) imminent hazard actions are discretionary?

Administration of the Act

91. Should there be a general policy on "fees"? If so, what should it be? What should be the purpose of fees? Collect revenue? Reduce false alarms? Improve credibility of data?

92. To what extent, if at all, should the Office of Technical Assistance be decentralized? What should be the scope of the Office's activities?

93. Are general guidelines for "categories" desirable?

State Programs

94. Is it expected that there will be a significant

State-level program? What responsibilities could or should States assume? Should EPA grant funds at 75% of the establishment and operation costs, as allowed by the legislation, or at some lower percentage to (1) identify those programs for which the local commitment is substantial, and (2) get the maximum mileage out of the very limited authorization?

95. Should the objective of State programs be to "fill in the gaps" in the Federally directed program(s)? What are these gaps?

96. Should the grant allocation be made on an "institutional" or "project" basis? That is, should EPA develop a nationwide "allocation formula" under which the funds would be allocated among all the States, to be run primarily by the Regions and designed to provide supplemental funding to ongoing programs? Or, alternatively, should each grant application be evaluated on a competitive basis in a program run from Headquarters designed primarily as a source of funding for demonstration projects – with the money divided among relatively few States?

97. How should any such State program(s) under TSCA tie in with other EPA programs, e.g., 208?

98. How will grant applications be evaluated, and by whom? Will there be external review?

99. To what extent should the criteria set forth in Section 28 be amplified or elaborated?

[FR Doc. 76-36096 Filed 12-6-76; 8:45 am]

ASSESSMENT AND CONTROL OF CHEMICAL PROBLEMS: AN APPROACH TO IMPLEMENTING THE TOXIC SUBSTANCES CONTROL ACT*

TABLE OF CONTENTS

*Reprinted from the U.S. Environmental Protection Agency Office of Toxic Substances, Washington, D.C., February 17, 1977, Draft No. 3.

PREFACE

The Toxic Substances Control Act of 1976 establishes a major new area of government responsibility to protect public health and the environment against the hazards of chemicals and other toxic substances. The huge numbers of chemicals currently used within our society, the difficulties of evaluating potential health effects attributable to individual substances, and the general lack of adequate data in this field all combine to make the implementation of this statute an imposing challenge to EPA.

One fact is clear. The complex questions presented by toxic substances will not be settled quickly. We are still in the early stages of developing a scientific understanding of these matters, and the translation of scientific knowledge into regulatory action will continue over a period of many years or even decades. The first need of EPA in addressing this new statute therefore is to lay as solid a foundation as possible for a successful long-term program. The Agency is anxious to obtain the broadest possible consensus among all concerned groups as to the basic policies and procedures established in the initial phases of this work.

As a first step we have prepared the attached draft document which describes a possible approach to the implementation of the Toxic Substances Control Act. The purpose of this draft is to stimulate comments from all interested parties to help the Agency develop an overall strategy for carrying out the Act. Your comments in response to this document are needed now.

A public meeting, for the purpose of soliciting comments on the strategy, has been scheduled for March 22–23 in Washington, D.C. The meeting will begin each day at 9:00 a.m. in the Jefferson Auditorium, U.S. Department of Agriculture. If you cannot attend that meeting, you may send written comments prior to April 15 to John B. Ritch, Office of Toxic Substances; Room 1041, West Tower; U.S. Environmental Protection Agency; 401 M Street, S.W.; Washington, D.C. 20460.

In reviewing the draft document, you will note that there is no discussion of resources needed for carrying out the Agency's responsibilities under the Act, nor is there a discussion of alternative implementation schedules which might be followed. A serious examination of these items will be conducted by the Agency over the next several weeks, concurrent with public review of the implementation approach draft as it presently stands. Subsequent to the March 22–23 meeting and based upon comments made, the results of the resource examination, and a continued assessment of basic issues, the Agency will produce a final strategy. Your contribution to that strategy is essential.

John R. Quarles, Jr.
Acting Administrator
U.S. Environmental Protection Agency

II. INTRODUCTION

This draft document describes a possible approach to the implementation of the Toxic Substances Control Act (TSCA). It represents the first step in the development of an overall strategy for implementation of the new law.

The purpose of this draft document is to stimulate comments from the many parties interested in implementation of the Act. These comments, together with estimates currently being developed concerning the resources required by EPA to carry out alternative levels of TSCA activities, will be considered in preparing a strategy document later this spring.

In an effort to help focus public discussion, the draft document proposes a general goal and two purposes in implementing TSCA. It suggests the initial emphasis and priority to be given to different program activities and addresses some of the major policy issues. The first 3 years of implementation are highlighted, and timetables for program activities during this period are proposed.

Comments on the draft document are invited. They would be particularly helpful if received by April 15, 1977. They should be sent to Mrs. Vickie Briggs, Environmental Protection Agency, Office of Toxic Substances, Room 715 East Tower (WH-557), 401 "M" Street, SW, Washington, D.C. 20460. (To facilitate handling and disposition of comments, please refer to OTS-000002 in the cover letter.)

II. EXECUTIVE SUMMARY

The goal of the Toxic Substances Control Act (TSCA) is to protect human health and the environment from unreasonable risks – now and in future generations. To achieve this goal, TSCA implementation activities will emphasize not only control of specific problems under TSCA regulatory provisions, but also use of TSCA authorities to support other governmental and nongovernmental programs to control toxic substances. These programs include other activities of EPA and other Federal and State agencies, activities of environmental and public interest groups and of professional societies, policies of the financial and investment communities, and efforts of individual companies and trade associations.

During the first several years, EPA will give top priority to the following implementation activities:

1. Establishment and implementation of a premarket review system – The system will emphasize a. the responsibility of industry to develop adequate data for meaningful chemical assessments, b. categorization of new chemicals by broad chemical classes and broad uses with particular attention to those categories of greatest environmental concern, and c. procedures for rapid decisions and adequate documentation of these decisions.

2. Establishment of initial testing requirements – Testing will be required in a hierarchical manner on selected categories of both new and existing chemicals. Industry will be required to develop data concerning both toxicity and exposure and to conduct risk assessments of the data. Quality assurance of the data that are developed will be stressed.

3. Regulatory actions to control a limited number of environmental problems associated with existing chemicals – In addition to early action on PCBs and selected chlorofluorocarbons, a limited number of serious chemical problems for which adequate data are currently available will be selected for intensive review and for regulatory action as appropriate. Concurrently, a systemmatized approach for identifying, characterizing, and controlling toxic substances under TSCA will be developed and implemented as rapidly as possible.

4. Assessment and control of unanticipated problems of urgent concern – Unexpected problems will inevitably arise and provisions will be made to respond to such problems without unduly disrupting other priority activities.

In these four priority areas, as well as in other areas, continuing attention will be directed to several overarching concerns. Activities will be oriented to serve the interests of both EPA and other organizations, particularly with regard to data dissemination. Data will be gathered on a highly selective basis to serve specific purposes. Confidentiality aspects will be a major factor influencing data collection, use, and dissemination strategies and activities.

III. BACKGROUND

A. The Problem and TSCA

An estimated 1,000 new chemical substances* are introduced into commerce each year in addition to the 30,000 or more which are currently used in many ways. The problems posed by the presence of some of these chemicals in the environment are too well known. Assessing and dealing with chemical problems involves the complex tasks of measuring their presence, estimating their effects, and evaluating the economic and social costs and benefits of their use and control. At the same time, there exists a wide variety of often overlapping and uncoordinated regulatory authorities and support activities directed to toxic substances at the national, state, and local levels. TSCA is designed to help reduce scientific uncertainties concerning toxic substances and to add coherence to the national effort to protect man and the environment from unreasonable risks without unnecessarily blunting a dynamic sector of our economy.

*"Chemicals" and "chemical substances" are used interchangeably in this document in addressing those substances subject to TSCA. In general, pesticides, foods, food additives, drugs, and cosmetics are excluded from TSCA coverage. The problems associated with mixtures and formulations are given special attention in the law.

B. The Congressional Interest

1. Summary of the Legislative Deliberations

Early in 1971, the Administration transmitted to Congress a proposal for toxic substances legislation. The report of the Council on Environmental Quality accompanying the proposal, *Toxic Substances,* documented some of the environmental problems which such legislation should address. The two Houses of the 92nd Congress passed different versions of the legislation late in the session. However, there was not time to appoint a conference committee to resolve the differences prior to adjournment.

Five different bills were introduced during the 93rd Congress, and by July 1973, the two Houses had passed different versions. However, the conference committee could not resolve several of the key differences between the two bills, including provisions concerning premarket screening and the relationship of TSCA to other laws.

In March 1976, in the 94th Congress, the Senate passed a somewhat revised version of the bill it had considered previously. In August 1976, the House passed still a different version. A conference committee met in early September, and the compromise bill that it developed was passed by both the House and the Senate in September 1976, with an effective date of January 1, 1977. The President signed the legislation on October 11, 1976.

During the deliberations of both the House and Senate in the 94th Congress, a number of technical aspects of the legislation were considered in detail. Thus, the committee reports of both houses, together with the report of the conference committee, provide considerable guidance concerning Congressional intent.

2. Findings

In proposing toxic substances legislation, the Congress found that human beings and the environment are exposed to many chemicals, some of which may present unreasonable risks due to their manufacture, processing, distribution, use, or disposal and that to address this problem, regulation of both interstate and intrastate commerce is necessary [2(a)].

3. Policy and Intent

The Policy of the United States enunciated in the Toxic Substances Control Act is that a. adequate data on the effects of chemical substances should be developed as the responsibility of those who manufacture and process them; b. authority should exist to regulate such chemicals which pose unreasonable risks and act on those which are imminent hazards; and c. exercising this authority should assure that chemical substances will not present unreasonable risks yet not unduly impede technological innovation [2(b)]. The Congress intends that the Act be carried out in a reasonable manner, taking into account the environmental, economic, and social impact of actions taken under the law [2(c)].

C. Highlights of TSCA

TSCA authorizes EPA to obtain information about existing and new chemicals and take appropriate action against those which may present unreasonable risks. Manufacturers or processors of chemicals may be required to conduct tests and submit to EPA data on the effects and behavior of chemicals. EPA must be notified 90 days in advance of the manufacture of new chemicals and supplied with information necessary to evaluate their effects. When necessary, EPA is authorized to take steps to limit manufacturing, processing, use, or disposal of a chemical substance which may present an unreasonable risk.

TSCA contains several explicit authorities to promote better coordination among federal agencies in identifying, assessing, and controlling toxic substances. If the Administrator determines that an unreasonable risk may be prevented or sufficiently reduced under a law not administered by EPA, he will request the relevant agency to evaluate the problem and take appropriate action. Similarly, other laws administered by EPA will be used in preference to TSCA when these authorities can adequately address the problems.

IV. OVERALL PROGRAM DIRECTION

A. General Goal in Implementing TSCA

The general goal in implementing TSCA can be stated as follows: To protect human health and the environment from unreasonable risks presented by chemical substances.

The program to be implemented pursuant to TSCA is one of a number of efforts, within and outside government, directed toward this common goal. These other efforts include not only other programs of EPA and other Federal agencies, but also programs of the states, activities of environmental and public interest groups and of professional societies, policies of the financial and investment communities, and efforts of individual companies and trade associations.

However, the TSCA implementation program is somewhat unique in view of its breadth of coverage and its wide ranging authorities as contrasted to the narrower scope of other programs in the toxic substances area. Thus, there are greater expectations that the program will be able to work toward the common goal on a far broader basis than has been possible in the past.

The number of adverse chemical incidents can and must be reduced. But accidents will continue to occur, and chemicals posing environmental problems will undoubtedly slip through the net of assessment and control. Development of efficient and effective means to minimize the likelihood that such adverse incidents will occur – now and in the future – is the challenge of this legislation.

B. Purposes of TSCA Implementation Activities

The purposes in implementing the regulatory and non-regulatory provisions of TSCA are twofold, namely 1. To control toxic substances directly, and 2. To support other governmental and nongovernmental programs to control toxic substances.

Inherent in the concept of control is the necessity to match the control requirements with the problems. In many cases, chemicals do not pose a threat to health or the environment, and no control is warranted. At the other extreme, there may be a necessity to ban a chemical altogether. In each case, judicious use of control alternatives must undergird successful implementation.

The dual purposes of direct control through the regulatory authorities of TSCA on the one hand and, on the other hand, indirect control through the use of TSCA authorities to support the efforts of other programs in a position to control toxic chemicals reflect the clear intent of Congress. Implementation of this legislation must rely on achieving a "multiplier effect" through the efforts of many organizations if a significant number of commercial chemicals are to be addressed in the near future. Furthermore, the dual purposes recognize the many common interests among a large variety of public and private organizations and the necessity to share and conserve resources whenever possible in addressing similar problems. Indeed, more than any other environmental legislation, TSCA should be considered as a mechanism to serve the interests of many organizations.

Several principles will guide implementation activities directed to these dual purposes:

1. The Environmental Protection Agency will assume a position of national leadership in the field of toxic substances control. At the same time, other organizations will be encouraged to utilize the full range of their authorities and capabilities to control toxic substances. In those cases when there are overlapping authorities or capabilities, EPA will encourage the use of the most expeditious and effective approach for addressing urgent environmental problems.

2. Information obtained under the law will be made available as promptly and as widely as feasible to enable the expertise of other federal, state, and local agencies and of the private sector to be utilized as fully as practical in meeting the purposes of the Act. Similarly, relevant information will be actively solicited from all available sources and used appropriately. International exchange of experiences and of data concerning toxic substances assessment and control will be strongly supported.

3. All interested parties will have adequate opportunity to participate in development of regulatory requirements. The basis for regulatory decisions will be clearly stated at the time the decisions are made. To the extent possible, such statements will distinguish among scientific facts and uncertainties, scientific judgments, and value judgments.

Operationally, it may be desirable to establish formal agreements with other Federal agencies, and perhaps with other governments and with international organizations. Also, systematic approaches to regularly obtaining up-to-date views on the priorities of other organizations will be developed.

C. Major Functional Areas

The activities to be conducted under TSCA have been divided into four major functional areas and several supporting areas (see Figure 1), recognizing that all of these areas are interrelated as discussed below. In particular, the acquisition and assessment of information provides the basis for regulatory and related decisions by both EPA and other organizations.

1. Major Functional Area 1: Acquire Information and Assess Risks to Health and the Environment

TSCA provides several new authorities by which EPA may require industry to develop and provide to EPA information concerning chemical substances. These activities include submittal of information concerning the production, by-products of production, uses, and effects of chemicals. In some cases, such data may be readily available to industry. Sometimes, it will be necessary for industry to develop the data.

EPA will obtain information from other sources as well, including other Governmental organizations and private institutions. Also, EPA will, as necessary, carry out laboratory and field studies to supplement or to confirm data available from external sources. However, the emphasis will be on requiring industry to develop data necessary to assess the environmental acceptability of chemicals since TSCA explicitly places this responsibility squarely on industry [2(b)(1)].

The Agency considers that industry has the responsibility not only of gathering and assembling data but also of assessing the data to determine possible environmental risks. This responsibility is particularly important with regard to test data. Industry will be expected to prepare risk assessments in accordance with guidelines issued by EPA on data submitted pursuant to premarket notification and testing requirements. Obviously, the Agency will review such assessments and also conduct independent assessments when necessary.

In some cases when EPA is using TSCA authorities to acquire for other agencies data which are not of priority interest to EPA, the assessment process will be left to the other agencies. More often, there will probably be a congruence among the interests of EPA and other agencies in the same environmental problems and joint risk assessments among agencies will be in order.

2. Major Functional Area 2: Use of TSCA Regulatory Authorities when Necessary to Control New Chemicals

TSCA calls for establishment from scratch by the end of

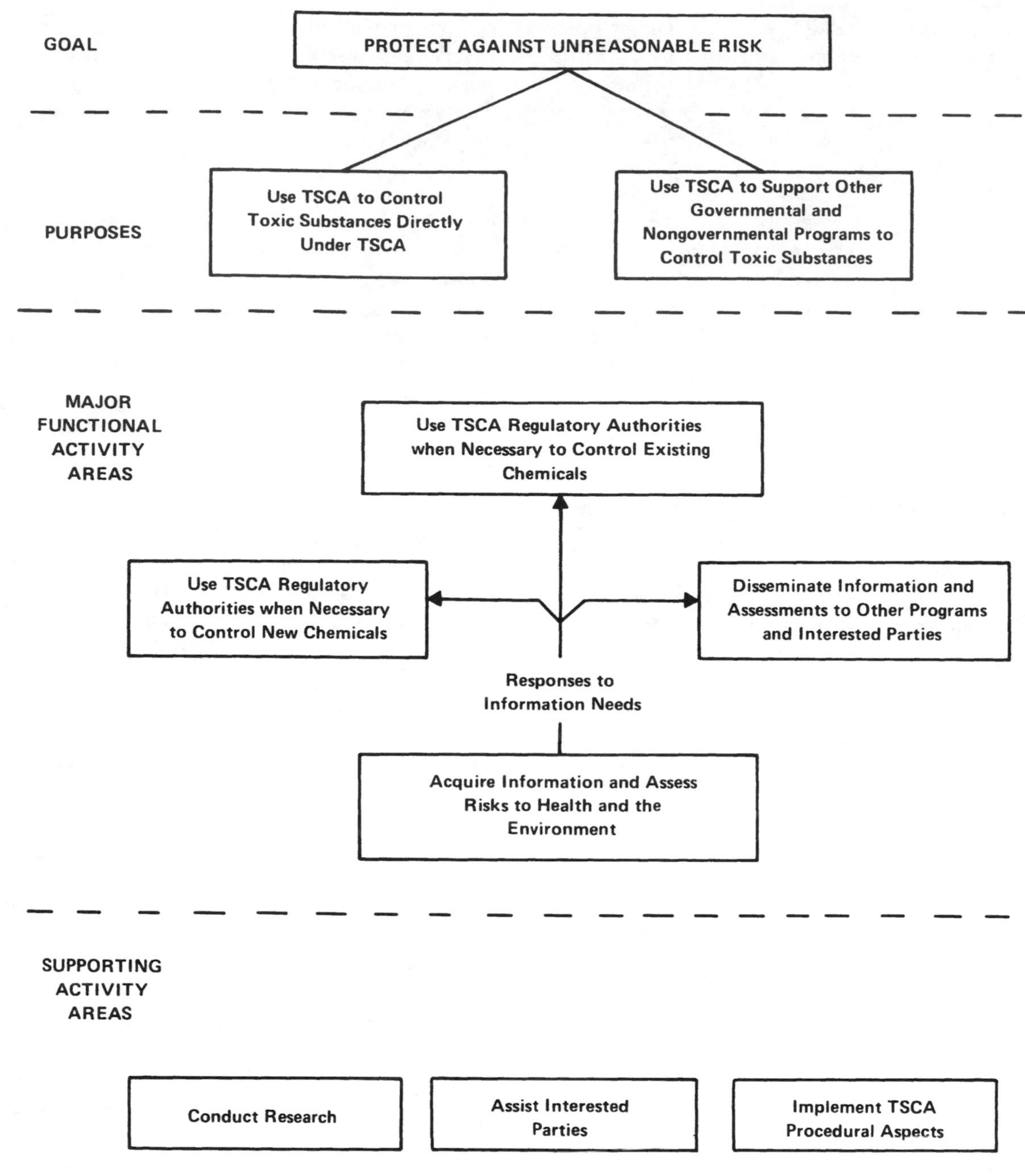

FIGURE 1. Goal, purposes, and activity areas.

1977 of a premarket review system for considering the environmental acceptability of all new commercial chemicals. The system includes a. final regulations clarifying which chemicals are subject to premarket notification and the notification requirements, b. an appropriately trained staff to receive and assess the notifications, and c. an internal EPA policy and procedural framework to help insure consistency, efficiency, and objectivity of the reviews.

The following regulatory mechanisms are available to control new chemicals when appropriate:

1. Premarket notifications will not be accepted by EPA unless they include all information required by regulations [5(a) and 5(b)].

2. If additional information is needed to conduct an adequate assessment, and such information is not available, the manufacture of the chemical will be delayed pending development of the information [5(e)].

3. Should a proposed new chemical be determined to present a significant risk, regulatory steps may be taken to control the manufacture, use, or disposal of the chemical [5(f)].

3. Major Functional Area 3: Use of TSCA Regulatory Authorities when Necessary to Control Existing Chemicals

Among the types of regulatory actions provided for in TSCA are [6(a) and 6(b)]

1. Banning or limiting manufacture, processing, distribution, or use of a chemical
2. Requiring warning labels
3. Requiring specified disposal methods
4. Requiring specified quality control measures during the manufacturing process.

There are many other statutes available for controlling toxic chemical problems. The rather comprehensive authority of TSCA is intended to be used when necessary to fill the gaps among these other authorities. Thus, the starting point in determining when and how TSCA regulatory authorities are to be used is an overall assessment of environmental problems associated with a chemical or group of chemicals and a determination as to which regulatory approach will most effectively reduce the problems.

If the most appropriate statute is administered by another Agency, EPA may request the other Agency to take action or explain the basis for nonaction within a specified period of time [9(a)]. Close collaboration with the other regulatory agencies beginning with the initial discovery of a chemical problem is essential to insure that this system of formal "referrals" does not unnecessarily disrupt the priority activities of these agencies.

4. Major Functional Area 4: Disseminate Information and Assessments to Other Programs and Interested Parties

A key to achieving "indirect" control of toxic chemicals through other programs is an effective system to aggressively disseminate to the broadest possible audience information obtained through TSCA. To the extent possible the quality and reliability of the data should be clear, and information should be in a format that will be most meaningful to the users [10(b) and 25(b)].

An effective dissemination program is intimately linked to an earlier determination of the information needs of the users and the timing of these needs so that the appropriate data can be generated in the first place. Thus, the first step is a continuing program to assess user interests and needs and, to the extent possible, to shape the data collection efforts to meet their needs.

Much of the data collected for one user, including EPA, will undoubtedly be of interest to many other users as well. Thus, information will be placed on the public domain as rapidly as possible after receipt by EPA.

Some of the data collected under TSCA possibly will be subject to claims of confidentiality. Rapid and efficient mechanisms for segregating data which are confidential from other data are essential if the user community is to be adequately serviced.

5. Interrelationships Among the Major Functional Areas

As shown in Figure 2, the four major functional areas are integral components of the overall system of assessing and controlling toxic substances. Figure 3 shows in more detail crosswalks among specific provisions of TSCA and indicates the reinforcing character of these provisions.

In general, the information acquisition activities (Area #1) are determined by the needs for data to a. provide a basis for decisions concerning the control of new or existing chemicals under TSCA (Areas #2 and #3) and b. service the interests of others (Area #4). In some cases, the data lead directly to risk assessments within EPA; in other cases, the raw data are forwarded for assessment to other programs.

D. Initial Priorities

During the initial implementation phase, it will be necessary to establish priorities among the many possible TSCA activities. These priorities should reflect continuing concerns of the Congress, the Agency, and the public that there may be a large number of currently unattended environmental problems which should be addressed very promptly. At the same time, early action must be taken to establish the policy and procedural framework for a long-term program, with the groundrules clearly understood by all interested parties. This is particularly true with regard to the premarket review system which should be environmentally meaningful and administratively efficient from the outset. And, of course, the legislative deadlines should be met.

With regard to regulatory activities, newly identified problems that present a significant risk to health or the environment must be dealt with immediately. Also, the chemical problems cited during the five years of legislative hearings should be promptly reviewed to determine whether regulatory action is needed. Finally, selection of early regulatory activities should include consideration of utilizing a variety of approaches that will not only reduce environmental risks but also clarify the dimensions, the strengths, and the weaknesses of TSCA.

Information is the lifeblood of the overall system. Early development of policies and procedures to establish a broadly based and technically sound data base, drawing on domestic and foreign sources and readily accessible to all interested parties, is essential if future regulatory actions are to be soundly conceived. Given the lead time necessary to develop health and environmental data, steps should be promptly initiated to begin to develop such data on those chemicals which could present serious problems. The information on the health and environmental aspects of commercial chemicals which is currently available only to individual agencies or companies should be promptly made more widely available.

In view of the unusual opportunity provided by TSCA to respond to these general concerns, EPA will give top priority to the following operational activities during the initial 3 years of TSCA implementation:

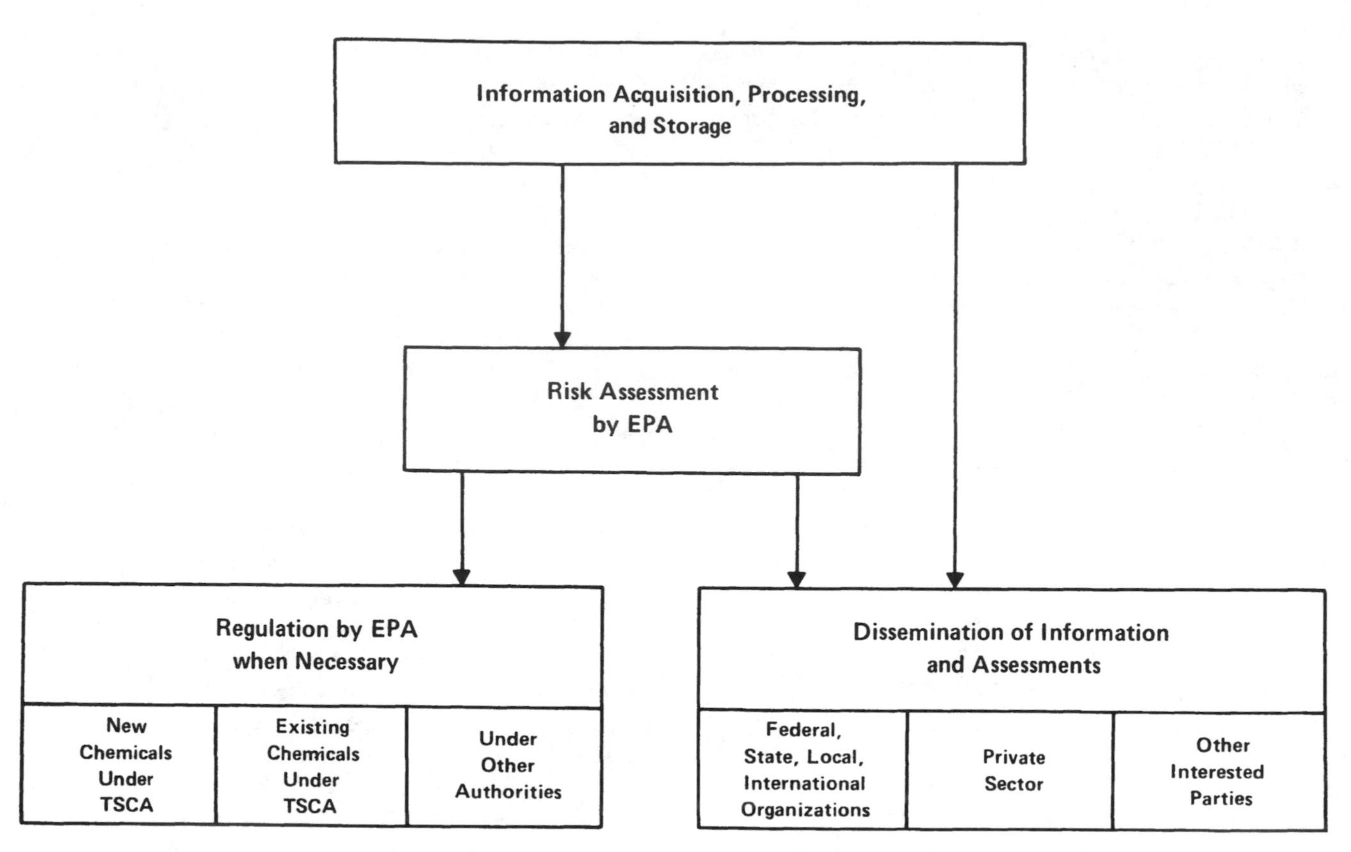

FIGURE 2. Information flow among functional areas.

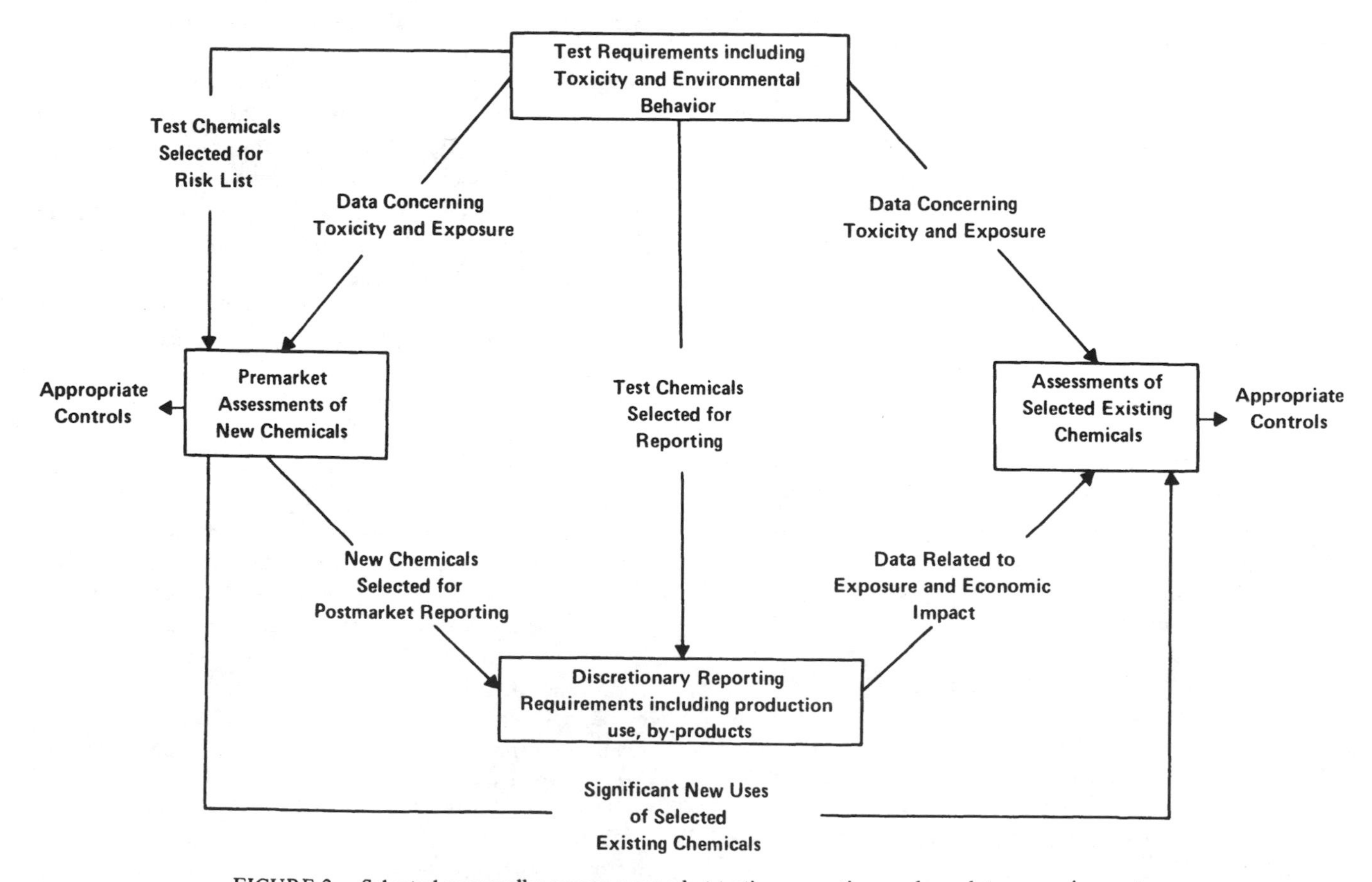

FIGURE 3. Selected crosswalks among premarket testing, reporting, and regulatory requirements.

1. Establishment and implementation of a premarket review system – The system will emphasize the a. responsibility of industry to develop adequate data for meaningful chemical assessments, b. categorization of new chemicals by broad chemical classes and broad uses with particular attention to those categories of greatest environmental concern, and c. procedures for rapid decisions and adequate documentation of these decisions.

2. Establishment of initial testing requirements – Testing will be required in a hierarchical manner on selected categories of both new and existing chemicals. Industry will be required to develop data concerning both toxicity and exposure and to conduct risk assessments of the data. Quality assurance of the data that are developed will be stressed.

3. Regulatory actions to control a limited number of environmental problems associated with existing chemicals – In addition to early action on PCB's and selected chlorofluorocarbons, a limited number of serious chemical problems for which adequate data are currently available will be selected for intensive review and for regulatory action as appropriate. Concurrently, a systemmatized approach for identifying, characterizing, and controlling toxic substances under TSCA will be developed and implemented as rapidly as possible.

4. Assessment and control of unanticipated problems of urgent concern – Unexpected problems will inevitably arise and provisions will be made to respond to such problems without unduly disrupting other priority programs.

In these four priority activity areas, as well as in other areas, continuing attention will be directed to several overarching concerns. Activities will be oriented to serve the interests of both EPA and other organizations, particularly with regard to data dissemination. Data will be gathered on a highly selective basis to serve specific purposes. Confidentiality aspects will be a major factor influencing data collection, use, and dissemination strategies and activities.

There are many uncertainties as to the number of chemicals subject to premarket notifications and the availability of data concerning these chemicals, the number and extent of unanticipated problems associated with existing chemicals that will emerge, and the difficulties in gaining a scientific consensus concerning the most appropriate approaches to testing. Thus, there is little basis for estimating at this time the relative emphases that should be placed on each of these four types of activities. However, in setting priorities among the many types of potential environmental problems falling within these activities, the following principles will be considered:

1. Toxic substance problems of national or global dimensions, and particularly those problems affecting many people or extensive ecological resources, will receive priority over localized problems which affect smaller populations.

2. Special attention will be given to the effects of toxic substances on human health, recognizing that ecological impacts can also directly and indirectly affect human health.

3. Toxic chemicals which are discharged into the environment in significant quantities and which persist and/or bioaccumulate will be of particular concern.

E. Protecting Against Unreasonable Risk

The concept of unreasonable risk is the central element in relating TSCA implementation activities to the overall goal of TSCA. This concept recognizes that some level of risk must be accepted in our activities involving chemicals. Also, the welfare of future generations as well as our present population must be of concern.

In general, the most severe risks trigger the imminent hazard provision of TSCA [7(a)], and the possibility of less severe risks may trigger testing or other information gathering activities. In some instances the term "unreasonable risk" is used in the Act, in one case "substantial risk" is used [8(e)], and in other cases elaborations of "unreasonable risk" are presented.

The burden of proof for establishing the degree of risk varies in different sections. For example, the proponent of use must establish that selected new chemicals which are on a "risk" list [5(b)(4)(A)] do not present an unreasonable risk. On the other hand, EPA must make explicit findings on risk to take regulatory actions [6(a)]. Meanwhile, there is an overarching requirement on EPA to consider social and economic impacts, as well as environmental impacts, in taking action under any provision of TSCA [2(c)].

The most detailed elaboration in the Act as to the types of considerations involved in assessing risk is set forth in the guidance provided to the Interagency Testing Committee, but this guidance is still very general [4(e)(1)(A)]. Several other provisions of the law, as well as the legislative history, also emphasize the importance of considering carcinogenesis, teratogenesis, and mutagenesis when evaluating risks. While there are no explicit references in the law to acute toxicity concerns, there is no reason for not considering such risks.

Recent court opinions, such as those concerning lead in gasoline and asbestos in drinking water, provide some indication as to the important factors in risk assessment. The 1975 NAS report *Decision Making for Regulating Chemicals in the Environment* also makes some observations concerning those factors that should be considered in assessing risks and in balancing costs, risk, and benefits in decision-making.

To foster a degree of consistency in the approach to risk assessments, the following steps will be taken:

1. Minimum data requirements for conducting risk assessments will be developed for selected categories of chemicals (e.g., chemical classes, use categories).

2. Guidelines for use by industry and the agency concerning risk assessments and the factors involved in cost-risk-benefit decisions will be developed and used.

3. Risk assessments of the same chemical or same environmental problems by different programs and different agencies will to the extent feasible be consistent even though the determination as to whether regulatory action is warranted may vary depending on the differing requirements of different statutes.

4. Summaries of relevant court opinions related to unreasonable risk will be prepared periodically and made available to risk assessors and decision makers.

5. The feasibility of developing criteria for triggering different types of TSCA actions when the chemicals meet the criteria will be explored in detail.

F. Supporting Functional Activities

1. Research

TSCA authorizes a broad range of research activities [10 and 27], and explicitly calls for EPA to direct attention to:

1. Development of screening techniques to help assess health and ecological effects [10(c)]

2. Development of monitoring techniques and instruments [10(d)]

3. Basic research to provide the scientific basis for screening and monitoring developments [10(e)]

4. Training of federal personnel to utilize the new techniques [10(f)]

HEW and other agencies are expected to intensify their research efforts directed to toxic substances as well.

EPA will be developing a more detailed framework for its initial 5-year research program during the next few months. EPA intends to give high priority to research both in its own programs and in encouraging a more broadly based and more effectively coordinated national effort involving research by many organizations. Also, the Agency will encourage efforts to expand the pool of technical manpower needed by many organizations to carry out TSCA requirements.

2. Assistance to Interested Organizations

Given the limited authorization of $1.5 million for each of the first 3 years of TSCA for support of a program of state grants (28), the initial grants will be limited to programs in only a few states. The states will be selected on the basis of the likelihood that the proposed programs will significantly upgrade local capabilities to address environmental problems. Should the initial grants prove to be successful in providing an important new dimension to toxic substances control, EPA will consider seeking additional funding in later years and will assess the desirability of incorporating the program into the broader concept of bloc grants.

The Agency will not be in a good position for some time to provide responses to petitions from industry requesting specification of test requirements for individual new chemicals [4(g)]. Within a few years the Agency should have available testing guidelines covering a broad range of chemicals, effects, environmental behavior patterns, and routes of exposure. At that time, authoritative responses to such petitions should be relatively easy to provide. In the interim, heavy reliance will be placed on using, whenever possible, the relevant portions of testing guidelines already prepared by EPA and other agencies for other programs such as those directed to pesticides, food additives, and drugs, taking into account the different types of exposures involved with industrial chemicals.

With regard to EPA assistance in defraying the costs of certain attorneys and witnesses in regulatory proceedings [6(c)(4)(A)], only limited funding will be available in fiscal year (FY) 1977, but significant additional funding will be sought in FY 1978. Meanwhile, the criteria for determining eligibility are being developed. Available resources will be distributed on an equitable basis among all bona fide claimants at the end of each fiscal quarter for services rendered during that quarter.

An area of Congressional concern has been the capability of small and medium industrial firms to understand the legal and policy complexities of toxic substances control in relation to specific chemical concerns. Thus, in addition to exempting small business from certain reporting requirements, TSCA calls for an EPA Assistance Office to provide clarification on regulatory and related requirements [26(d)]. The Office has been established. In addition, each of the ten EPA Regional Offices will have an appropriate contact point to help clarify TSCA requirements for the small businessman in particular, and the public in general. The Assistance Office will not be in a position in the near future to provide advice on the environmental acceptability of specific commercial chemicals of interest to individual parties. However, the Office will help identify some of the parameters that should be considered in such assessments and will of course assist in obtaining available information relevant to the particular chemicals of interest.

3. Procedural Aspects

A number of TSCA provisions call for development of procedural rules. Also, more explicit guidance will help facilitate implementation of other provisions. Some of EPA's earliest activities will be directed to:

1. Procedures governing the hearings to be conducted under TSCA and particularly with regard to regulatory actions [6(a)]

2. The requirements for selected EPA and HEW employees to disclose their financial interests [26(e)]

3. Clarification of the requirements concerning claims of confidentiality and requests for information under the Freedom of Information Act [14]

4. Preemption of state laws [18]

Several TSCA provisions are of special interest to another federal agency and will receive attention in the near future. For example:

1. National defense waivers (Department of Defense) [22]

2. Employee protection (Department of Labor) [23]

3. Definition of small manufacturers and processors (Small Business Administration) [8(a)(3)(B)]

4. Notice to foreign governments of exports (Department of State) [12]

5. Imports (Department of Treasury) [13]

6. Reimbursement for use of test data generated by another party (FTC/Department of Justice) [4(b)(3) and 4(c)]

Finally, special studies and reports are called for as follows:

1. Study of the identification aspects of all laws administered by EPA [25(a)]

2. Annual reports to Congress on TSCA actions [30]

V. ACQUISITION OF INFORMATION AND ASSESSMENT OF RISKS TO HEALTH AND THE ENVIRONMENT

A. General Policy Framework

The gathering, processing, and storing of information will be designed to support specific requirements of EPA and other organizations and will not be considered an end in itself. Data requirements may be very specific or quite broad. For example, information activities may be oriented to:

1. Development of specific regulations to control new or existing chemicals under TSCA or other EPA authorities

2. Identification and prioritization of problem chemicals for attention under TSCA or other authorities

3. Supporting specific activities of other agencies and other organizations to assess and control problem chemicals

4. Informing the public of chemical activities and associated problems

The intended use of the information will determine the type and extent of data that are needed, the timing, the sources, and the most appropriate mechanisms for acquiring the data. When appropriate, data already available in government files will be used. When industrial data are required, data already available to industry should be used to the extent possible. However, in a number of cases, and particularly with regard to new chemicals, additional data must be developed by industry on a routine basis.

Given the interests of many parties in data that can be developed under TSCA, and the multiplicity of data acquisition efforts already in place, the development of new TSCA data requirements will be coordinated widely within and outside the Government. Such coordination should assist in a. reducing the likelihood of gaps in information needed for regulatory purposes, b. conserving resources of EPA and other agencies, and c. limiting the total reporting burden on industry.

Particular attention will be given to assuring that information that is acquired, and particularly industrial data, is accurate, complete, and current. Among the steps to be considered are

1. Clarity and precision of record keeping and data submission requirements

2. Procedures for assuring appropriate quality control in the acquisition and reporting of information

3. Interim reports of progress in fulfilling long-term data requirements to avoid discovery only at a very late date that unacceptable data have been generated

4. Spot checks and audits of data generation activities conducted in the U.S. or abroad

5. Development of confirmatory data to check on accuracy and reliability of submissions

6. Prompt and vigorous enforcement against violators of data requirements

7. Encouragement of open communication with industry to discuss implementation problems in satisfying data requirements

B. Disclosure of Data

Assertions of trade secrecy and related confidentiality matters could cause many implementation problems and must be addressed promptly. Prior to gathering information which is likely to include confidential data, the need for the information should be very clear. Submitters will be afforded the opportunity to make confidentiality claims at the time of submission. A clear explanation of the consequences of failing to assert confidentiality and the procedure for resolving disputes will be a part of the request for information.

TSCA specifically excludes from claims of confidentiality health and safety studies on chemicals offered for commercial distribution and on chemicals subject to premarket notification and/or testing requirements [14(b)]. Other data which the manufacturer, processor, or distributor consider confidential may be so designated but must be segregated from other nonconfidential data. Thus, EPA

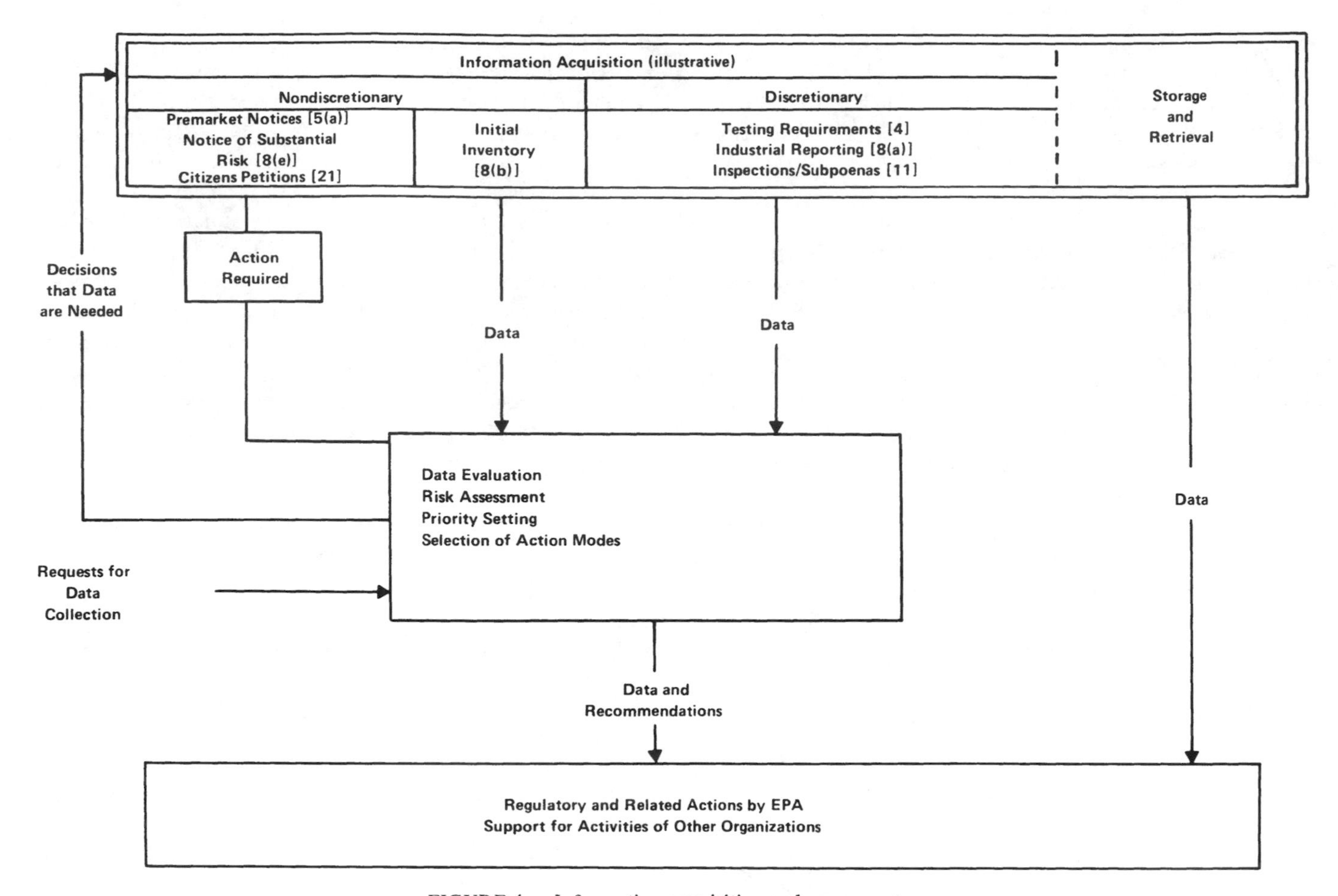

FIGURE 4. Information acquisition and assessment.

reporting forms will require confidentiality designations for individual items, and a general confidential stamp for an entire form will not be accepted.

If there is a Freedom of Information request or other action concerning release of data designated as confidential, the originator of the data will be given an opportunity to justify the claim of confidentiality in detail prior to the Administrator's decision on the release of the data. Also, he will have an opportunity to seek judicial relief if there is disagreement with the Administrator's determination. However, if the release of data is necessary to protect health or the environment against an imminent, unreasonable risk, the advance notice of release can be as short as 24 hr [14(c)].

The information system for receiving and storing TSCA data will insure appropriate protection of confidential information.

C. TSCA Information Gathering Authorities

Figure 4 identifies some of the information that will be received as a result of both nondiscretionary and discretionary provisions of TSCA. Nondiscretionary provisions include:

1. Reporting of chemicals to be included on the initial inventory of existing chemicals [8(a) and 8(b)]
2. Premarket notifications [5(a)]
3. Industrial reporting of substantial risks associated with commercial chemicals [8(e)]
4. Citizen's petitions [21]

Some of the provisions to be implemented at the discretion of EPA are

1. Testing requirements [4(a) and 4(b)]
2. Industrial reporting of production and related activities [8(a)]
3. Industrial submission of records of adverse reactions to health or the environment [8(c)]
4. Industrial submission of health and safety studies [8(d)]
5. Inspections and subpoenas [11]
6. Acquisitions of information from other Agencies [10(b)(2)]

Figure 5 sets forth the implementation timetable for the principal activities related to TSCA testing provisions and

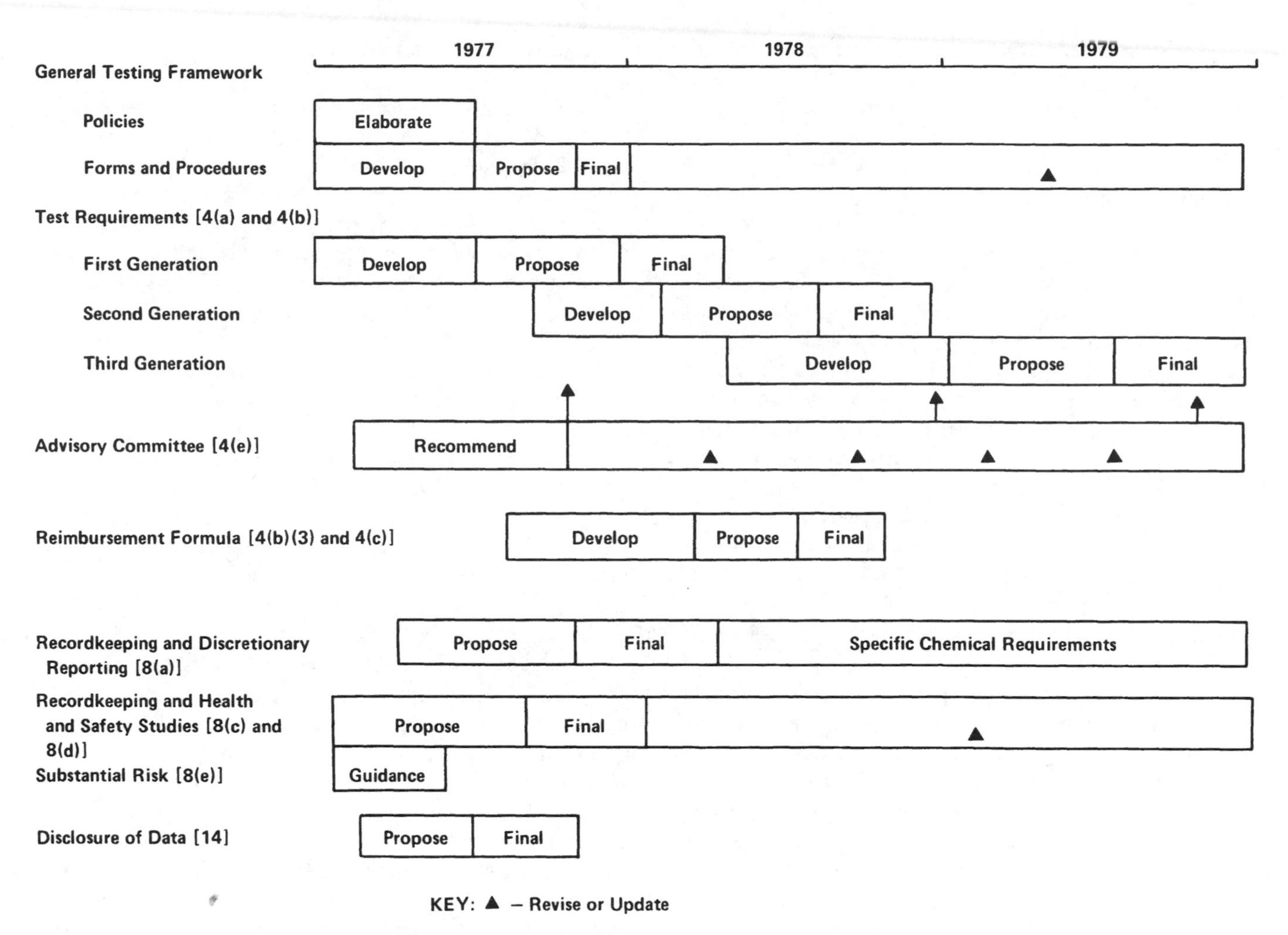

FIGURE 5. Implementation timetable: testing and reporting.

the provisions concerning industrial record keeping and reporting. These activities are also discussed below.

D. Risk Assessments

Risk assessments are required under a number of TSCA provisions. Risk assessments are inevitably conducted with less than optimal data – either in terms of quantity or quality. However, a minimum level of data is essential if assessments are to be meaningful, and EPA will develop guidelines concerning such minimum data requirements.

Manufacturers will be expected to conduct and submit risk assessments, prepared in accordance with EPA guidelines, when developing data under TSCA testing regulations and when submitting premarket notices. Also, risk assessments may be required in connection with information provided under other TSCA provisions. The assessments will utilize not only data developed by the manufacturer but also other available data on the chemicals, and when appropriate, on closely related chemicals. In addition to providing a better basis for EPA evaluation of the likely environmental problems associated with toxic substances, this requirement should stimulate a much broader industrial effort to better understand the environmental acceptability of many chemicals being produced and used.

While much remains to be done to clarify, even in a preliminary way, what is involved in risk assessments, the following types of considerations are important:

1. Chemical structure, physical properties, contaminants
2. Source assessment and exposure potential: quantities produced, production processes, uses, reactions involved in uses, and types of frequency of environmental discharges
3. Environmental behavior and fate, including degradation rates and products, chemical reactions in the environment and inadvertent products, bioaccumulation and biomagnification potential, and possible synergistic effects
4. Acute, chronic, and subacute effects on man; absorption, excretion, and metabolism

5. Effects on vertebrates, invertebrates, microorganisms, and plants

6. Effects on inanimate objects and structures

E. Testing Requirements

Almost all agencies and organizations in a position to assess and control toxic substances have stressed the need for more timely and more reliable test data on a wide range of chemicals. Test data will be used to support specific regulatory actions and to identify problems that need attention. TSCA test requirements that are developed can be pacesetters for the entire field of environmental assessment techniques. Thus, priority will be given to implementation of the testing provisions of TSCA, recognizing that scientific uncertainties and overlapping organizational interests will complicate rapid progress in this area.

The initial implementation activities related to testing will involve a. establishing the procedures and format for submitting test data under a variety of TSCA provisions, including premarket notifications which will begin to arrive in December 1977, b. determining general policies which will provide the framework for TSCA test requirements for determining specific effects of specific chemicals, including procedures for assuring the quality of test data, and c. developing the initial TSCA test requirements for selected categories of both new and existing chemicals.

In view of the number of chemicals of potential interest, test requirements directed to categories of chemicals is the most appropriate approach. The categories will be based on similarities in chemical structure, chemical use, and/or levels and routes of likely exposure, recognizing that production volume may often be an appropriate surrogate for exposure potential. Not only will such an approach simplify the sorting of priorities, but it will also provide a means for anticipating problems with new chemicals and new uses. In general, EPA will develop general testing requirements for each selected category of chemicals and the manufacturer will in turn propose a detailed protocol for each chemical for EPA approval.

Given the current limitations on the availability of testing facilities and personnel and the scientific uncertainties concerning some types of test methods, the testing requirements during the first several years will be developed on a selective basis directed to some of those chemical effects, types of chemical behavior, and routes of exposure of immediate concern. At the same time, a more comprehensive effort will be undertaken to develop testing approaches on a much wider range of chemicals, effects, behavior, and exposure routes. The recommendations of the Interagency Testing Committee and of other interested parties will be considered within this framework. In short, the categorization scheme can be "tuned", in terms of number and breadth of categories and types of effects and exposure, in accordance with the capability to generate sound data and to use the data effectively.

In general, TSCA testing requirements will:

1. Require only such data as are necessary to reach regulatory and related decisions.
2. Provide the flexibility to enable initial state-of-the-art protocols to be effected immediately while permitting modifications and improvements to be made as the science of testing progresses. EPA will update protocols when appropriate without jeopardizing the acceptability of testing already underway in accordance with earlier protocols.
3. Document the laboratory analytical procedures and the requirements for estimating risks so that industry will clearly know what is required and how the information will be assessed. Industry should be able to conduct the same assessments of their data as EPA.
4. Provide an approach for developing test protocols in a hierarchical manner so that the degree of testing and evaluation is related to the likely exposure and the projected economic and social benefit.

Consistency of test requirements issued by different programs directed to the same classes of chemicals is important. Also, compatability with recommendations of international organizations and with requirements of other governments is important. However, two pitfalls must be avoided. First, overstandardization of requirements could stifle advances in the state-of-the-art. Secondly, the types and levels of exposures associated with industrial chemicals usually differ significantly with the exposures associated with drugs, food additives, and pesticides. Therefore, the types and extent of testing of industrial chemicals may differ greatly from the more comprehensive approaches used in these other programs.

In developing the near-term requirements, as well as the more comprehensive long term approach, the Congressional concern with the following aspects will be kept in mind:

1. Carcinogenicity, mutagenicity, teratogenicity
2. Levels of environmental exposure
3. Behavioral effects
4. Synergistic and cumulative effects

With regard to assuring the quality of data that are submitted by industry, EPA will explore the feasibility of expanding the FDA laboratory inspection program and the use of the FDA Good Laboratory Practices manual to encompass TSCA concerns. A counterpart system for environmental testing will be explored within EPA. In the longer term, other approaches may be more appropriate to assure that data are obtained and presented in a technically credible manner [3(12)(B)].

During 1977 the Interagency Advisory Committee will designate up to 50 priority chemicals for testing. The Agency will have up to 1 year to respond to these

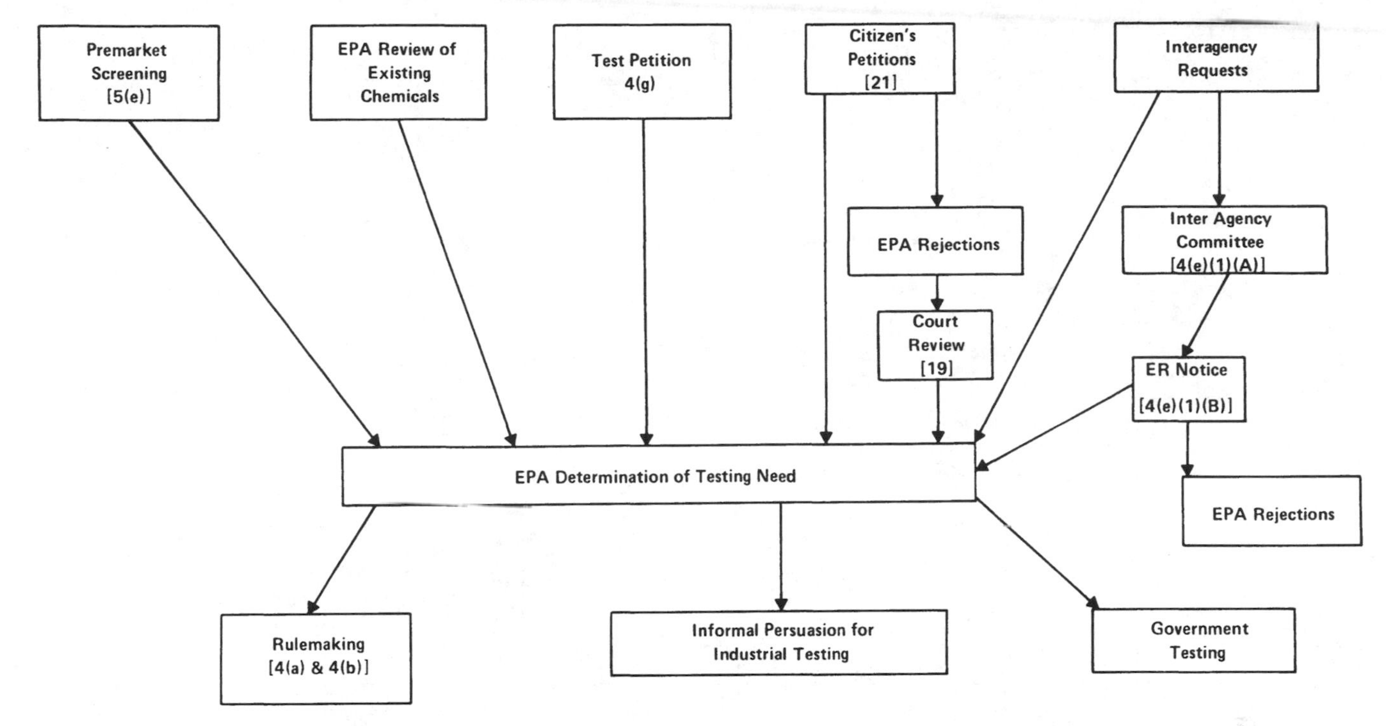

FIGURE 6. Origin of test requirements.

recommendations [4(e)(1)]. Meanwhile, as indicated in Figure 6, other testing recommendations will be received from a variety of sources. A sorting of the various recommendations, taking into account the availability of testing facilities, will be a major activity during 1977–78. While such prioritizing is important, promulgation of initial test requirements will not be delayed pending development of an elaborate sorting system which may never be feasible. Early EPA action will demonstrate the Agency's intent to consider all suggestions but then to move in the direction of the priorities as viewed by the Agency.

Policies and procedures will be prepared for enabling the developer of test data to receive fair reimbursement from another party who wishes to use the data [4(b)(3) and 4(c)].

F. Industrial Reporting and Retention of Information

The record-keeping and reporting provisions of TSCA are illustrated in Figure 7.

Discretionary acquisition of industrial data concerning production, by-products, and uses [8(a)] will be used for a number of purposes, including:

1. To keep track of the commercial development of new chemicals after they have been marketed for the first time; although the initial production and use patterns might not warrant regulatory intervention in the premarket period, changes in these patterns could become of environmental concern
2. To provide information related to likely types of exposure on existing chemicals which are subject to testing requirements
3. To provide data related to exposure and also related to economic impact of controls on those chemicals which are being considered for regulatory action
4. To provide information on the impact of controls which limit but do not ban chemicals
5. To help pinpoint the types and locations of exposure potential when unexpected chemical problems of urgent concern arise
6. To provide trend data concerning newly emerging chemical technologies which should be of special concern

The initial emphasis will be on requirements for the establishment within individual industrial firms of complete, up-to-date, and easily accessible records concerning each of the individual chemicals which are used by the firm. Such records will include information concerning the chemical identity, production levels, uses, by-products, health and safety studies, alleged adverse reactions, and other factors of environmental significance [8(a) and 8(c)]. These records will be available for on-site reviews which might trigger requests for the data to be reported to EPA. Also, as individual chemicals are being analyzed in detail by EPA and other agencies, information concerning these chemicals will be requested. In order to know which manufacturers to approach concerning specific chemicals, EPA will require each manufacturer to identify all the

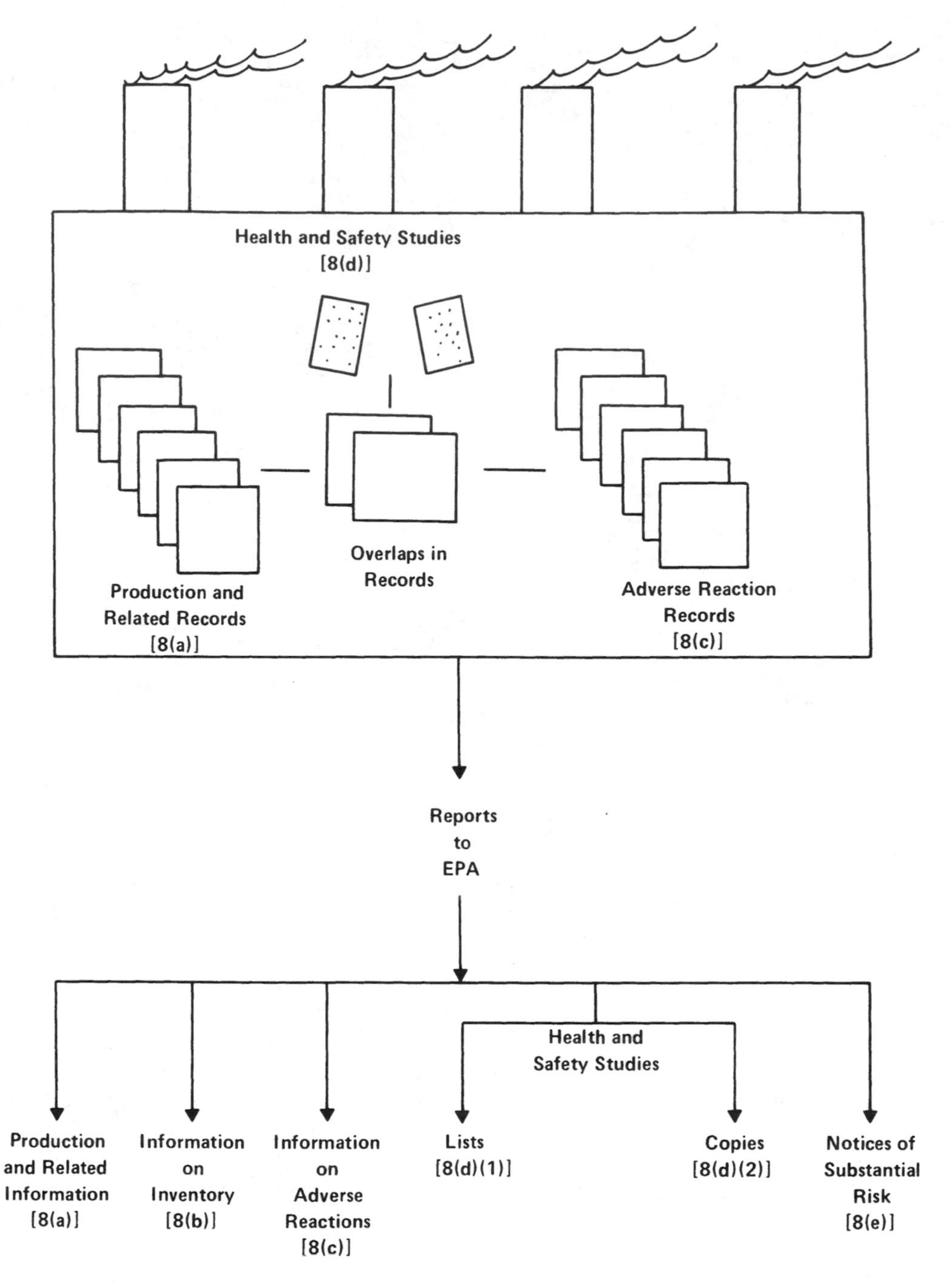

FIGURE 7. Industrial record keeping and reporting for existing chemicals.

chemicals he manufactures in connection with compiling the initial inventory of existing chemicals [8(b)].

The responsibilities of importers and their suppliers concerning record keeping require special attention.

The general procedural requirements for reporting will also be established promptly with the specific chemicals to be subject to reporting identified subsequently [8(a)]. Data on specific chemicals could then be obtained with a minimum of delay. In the early years, this authority will be used selectively to help assess chemical problems under active investigation. At a later date, it may be desirable to build up within EPA a broader base of industrial data. However, before comprehensive reporting requirements are developed, more thorough investigations of the availability of comparable data through other governmental and nongovernmental mechanisms will be carried out to avoid duplicative reporting.

During the Congressional hearings, concern was repeatedly voiced that considerable amounts of data related to the health and environmental acceptability of commercial chemicals were available in industry files. While the value of such data are unknown, the extent of the backlog of data will be promptly ascertained [8(d)]. Should a manufacturer become aware of data generated by himself or others concerning the likelihood of substantial risks due to chemical exposures [8(e)], notifications to EPA of this data will trigger the activities described under the response section below.

VI. USE OF TSCA REGULATORY AUTHORITIES WHEN NECESSARY TO CONTROL NEW CHEMICALS

A. TSCA Authorities to Control New Chemicals

Priority will be given to establishing the program for premarket screening which must be in place by the end of 1977. TSCA requires notification of all new chemicals at least 90 days prior to their manufacture to provide the Agency an opportunity for determining whether such chemicals pose a risk. If the Agency believes that a new chemical poses an unreasonable risk, or if information is lacking to make an evaluation, the Agency may initiate action leading to the following:

1. Delay in the manufacture of the new chemical pending development of additional information needed to evaluate the risk [5(e)]
2. Ban or marketing limitation on the new chemical to protect against an unreasonable risk [5(f)]
3. Requirements to track the commercial development of the new chemical after initial marketing through periodic reporting by industry of its production levels, uses, by-products, and related aspects [8(a)].

Also linked to the premarket review are several TSCA authorities which can be used to require certain types of information to be submitted at the time of notification, namely:

1. Production, use, by-product and related data [5(a)(1)]
2. Data required by testing regulations [5(b)(1)]
3. Health and environmental data on chemicals on a "risk" list [5(b)(4)]

If required data are not submitted, the notification will not be accepted.

B. Preparation of the Inventory of Existing Chemicals

The first order of business to implement premarket screening for new chemicals is the compilation of an inventory of existing chemicals. This inventory will be regularly updated and will serve as the baseline for determining which chemicals to be manufactured after publication of the inventory are "new" and, therefore, subject to premarket notification.

Each manufacturer will be required to report for inclusion in the inventory all chemicals produced after January 1, 1977. Manufacturers may report other chemicals produced between July 1, 1974, and January 1, 1977, if they wish to have these chemicals included; chemicals which are not reported will be subject to premarket notification requirements. To assist in standardizing the nomenclature used in reporting for the inventory, the Agency will publish a candidate list of about 30,000 chemicals, with appropriate names and CAS registry numbers. Also, instructions will be provided for reporting other chemicals not on the candidate list.

The initial inventory will not differentiate among various technical grades of chemicals. The problems related to impurities will be addressed at a later date. The initial inventory will include some categories of chemicals since it is not practical in the short time available, and it may not be desirable, to list every variation of all existing chemicals. Raw agricultural products, for example, will be considered as a single category. Also, minerals will be included on the basis of relatively broad categories.

The exemption from the inventory and from premarket notification of "small" quantities for research purposes will include those amounts no greater than what is reasonably necessary for scientific experimentation, testing, analysis, or research, including such research or analysis necessary for the development of a product [5(h)(3)]. "Test marketing" will be limited to distribution of a chemical to a defined number of potential customers for purposes of evaluating particular uses of that chemical during a predetermined evaluation period [5(h)(1)].

Many intermediate chemicals are often involved in the synthesis of industrial chemicals and are subject to TSCA. Only those intermediate chemicals which cannot be isolated in a practical sense from the immediate vicinity of the reaction process will be exempt from inclusion on the initial inventory and from premarket notification. Thus, for example, those chemicals which normally exist in only a pipeline would be included.

The Agency is considering requiring that all premarket notifications be accompanied by a fee [26(b)]. This requirement would not only help defray costs of administration but would also help insure that Agency efforts are directed to environmental assessments of serious commercial endeavors and are not diverted to address theoretical curiosities submitted by parties who have no intention to manufacture the chemicals commercially. Also, the Agency is considering placing a limit of perhaps 1 year on the time between premarket notification and initiation of manufacture of the chemical. If the time is exceeded, a second notification would be required.

C. Data Requirements for the Review of New Chemicals

A meaningful review of new chemicals requires adequate data to review. Thus, EPA reviewers will be provided with internal guidance on minimum data requirements for new chemicals based on chemical categorization considerations. Data requirements will vary among categories. This guidance will be made publicly available.

The guidance for each category will identify the types of data which should normally accompany notification of a new chemical in that category including in some instances data not explicitly required by regulations. A notification submitted without the data will be a candidate for possible action to delay its commercialization. In that event, the Administrator may issue a proposed order to delay manufacture, based on insufficiency of information to "permit a reasoned evaluation of the health and environmental effects" of the new chemical [5(e)]. The guidance will merely serve to alert Agency reviewers to possible problems, and the absence of data will not automatically cause the Agency to issue a proposed order. The Agency is prepared, however, to issue proposed orders when appropriate.

In some cases, it may be appropriate to convert portions of the internal guidance into formal testing requirements. In that event, premarket notifications would not be accepted in the absence of such data. As the guidance is developed, the desirability of such testing requirements will be actively explored.

The Agency does not plan to establish a risk list of new chemicals which pose "an unreasonable risk of injury to health or the environment" at the outset [5(b)(4)]. The internal guidance by chemical categories will help insure a meaningful review of premarket notices, and in a sense perform much of the same function as a risk list. Also, the Agency does not plan to activate reporting of significant new uses of existing chemicals in the short term because of the urgency and complexity of the task of instituting premarket notification for new chemicals and the difficulty of preparing a meaningful categorization of uses of environmental significance [5(a)(2)].

D. Limiting Manufacture or Marketing of New Chemicals

The internal guidance on data needs will also assist in determining the key factors underlying decisions as to whether to restrict new chemicals [5(f)]. However, each regulatory decision must be made on a case-by-case basis, and decision-making criteria beyond a check list of the factors to be considered does not appear feasible at this time.

There are often many uncertainties about the commercial viability of new chemicals. In some cases, the likelihood that the potential environmental problems uncovered during premarket review will become actual problems may be questionable. Therefore, alternatives to the resource-intensive process of formal rulemaking and, when necessary, court intervention to limit manufacture or marketing will be considered. For example, simply releasing adverse data to interested parties may in certain instances result in limitations on the type of commercialization which poses environmental problems. However, when necessary, EPA will intervene in a regulatory mode. If worker exposure is the principal concern, referral to OSHA might be the most appropriate course [9(a)].

E. Review Procedures for New Chemicals

As indicated in Figure 8, the Agency will initially determine if the premarket notification contains the information required by regulations. If such information is missing, the notification will not be accepted. A *Federal Register* notice indicating receipt of a complete notification will be published within 5 days [5(d)]. The notification will be accepted even if it is not accompanied by the additional data specified in the internal EPA guidance for the new chemical's category.

The internal guidance will also indicate a. those categories of chemicals which should in all cases be subjected to in-depth reviews, and b. those categories which should be screened to determine if in-depth reviews are warranted. In the latter case, technical reviewers will identify the specific chemicals for which no action appears warranted and those that deserve in-depth review, along with all the new chemicals in the first group.

The in-depth review will routinely be completed in 30 days and will include consideration of:

1. Physical and chemical properties of the chemical and its by-products
2. Health and ecological effects
3. Environmental behavior and fate including persistance and bioaccumulation
4. Likely sources of environmental discharges and exposed populations
5. Technological and economic factors
6. Industry's risk assessment

In addition to the data, included in the premarket notice, other data on the same chemical and, when appropriate, on chemical relatives which may not have been available to the manufacturer will be reviewed. Criteria will be developed for triggering extensions of up to 90 days of the in-depth review period. These criteria might include the necessity to validate questionable data.

At the completion of the in-depth review, including the supplementary review when appropriate, recommendations will be prepared as to the action that should be taken on each chemical. The recommendations and supporting justification will be concisely summarized in a short decision-making document.

A policy decision body including representatives from appropriate Agency offices will meet on a regular basis and

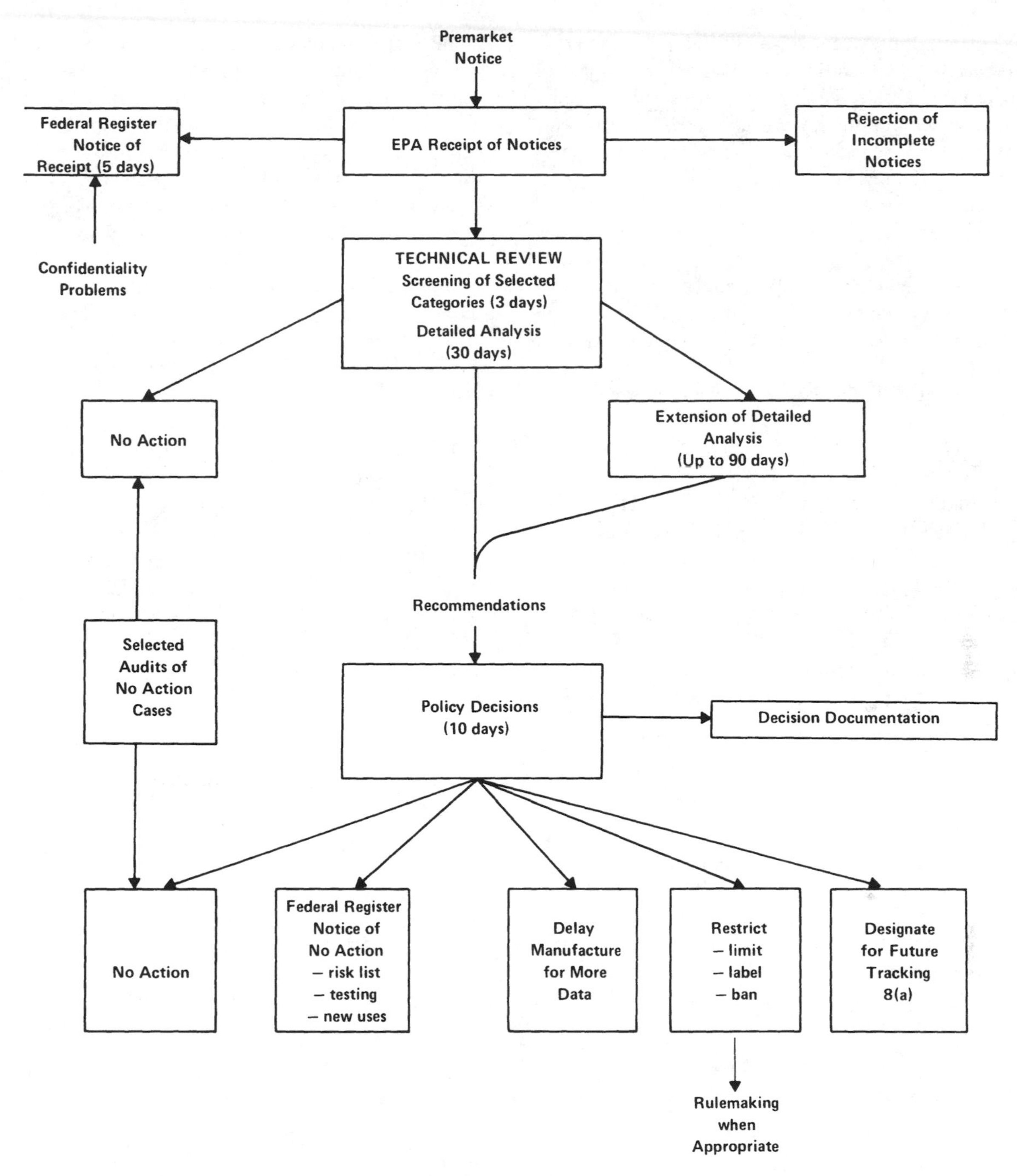

FIGURE 8. Premarket review system.

decide the disposition of each of the recommendations of the technical reviewers. The decision will be appropriately documented so that at a later date the basis for the decision is clear. The following types of decisions are envisaged:

1. No action is warranted.

2. No action is warranted but a *Federal Register* notice of the decision is required if the chemical is subject to testing requirements. Also, at a later data a *Federal Register* notice will be required if the chemical is on the risk list or is subject to significant new use reporting [5(g)].

3. Manufacture will be delayed pending development of additional data [5(e)].

4. Manufacturing and marketing will be regulated [5(f)].

5. While no action is taken during the premarket period, the chemical will be ear-marked for future reporting after it becomes commercialized [8(a)].

"No action" can be decided by technical screeners on

chemicals in selected categories and by the policy body on other chemicals. All "no action" decisions will be subject to a later audit on a selective basis to help insure that the necessity to screen out quickly the less worrisome problems does not result in inappropriate chemicals inadvertently slipping through the review process.

The Agency will give prompt attention to development of a data recovery system for premarket submissions. This will allow use of the Agency reviewers' comments and prior analysis of one new chemical when later reviewing a different, although similar new chemical. The data system will also permit use of premarket data for other purposes under TSCA.

The premarket screening implementation timetable is set forth in Figure 9.

VII. USE OF TSCA REGULATORY AUTHORITIES WHEN NECESSARY TO CONTROL EXISTING CHEMICALS

A. General Regulatory Approach

TSCA is designed to fill the void in the regulatory span of other authorities. It is intended both to prevent problems and to correct problems.

As indicated in Figure 10, EPA plans to initiate a limited number of regulatory actions directed to existing chemicals in the near future. In addition to addressing important environmental problems, such early actions should stimulate a greater degree of introspection within industry concerning preventive and corrective measures in anticipation of future environmental problems or additional regulatory actions. Also, this early attention to specific problem chemicals should enable EPA to play a lead role in establishing regulatory priorities rather than having external developments become the dominant factor in determining the Agency's priorities. Finally, initial actions directed to a variety of environmental situations should help clarify the scope and limitations of TSCA and its interfaces with other regulatory programs.

The regulatory strategy will emphasize overall assessments of the causes of the environmental problems associated with chemicals and the most effective measures to address these underlying causes. A key factor will be a determination of the most effective regulatory approach using TSCA, another statute, or a combination of statutes. In many cases, authorities other than TSCA will provide the appropriate means for reducing the problems. If the most appropriate statute is administered by another Agency, EPA may request the other Agency to take action or explain nonaction within a specified time [9(a)]. However,

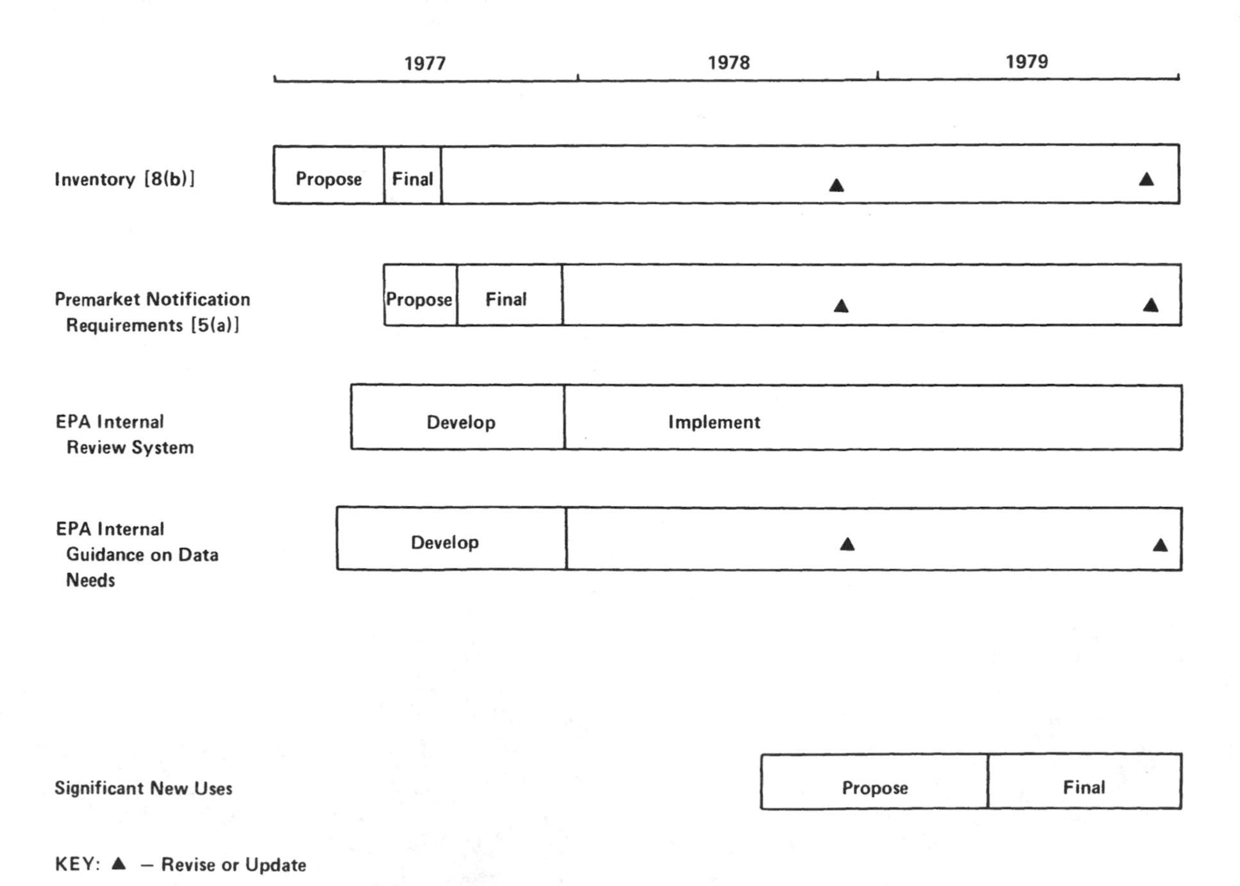

FIGURE 9. Premarket screening implementation timetable.

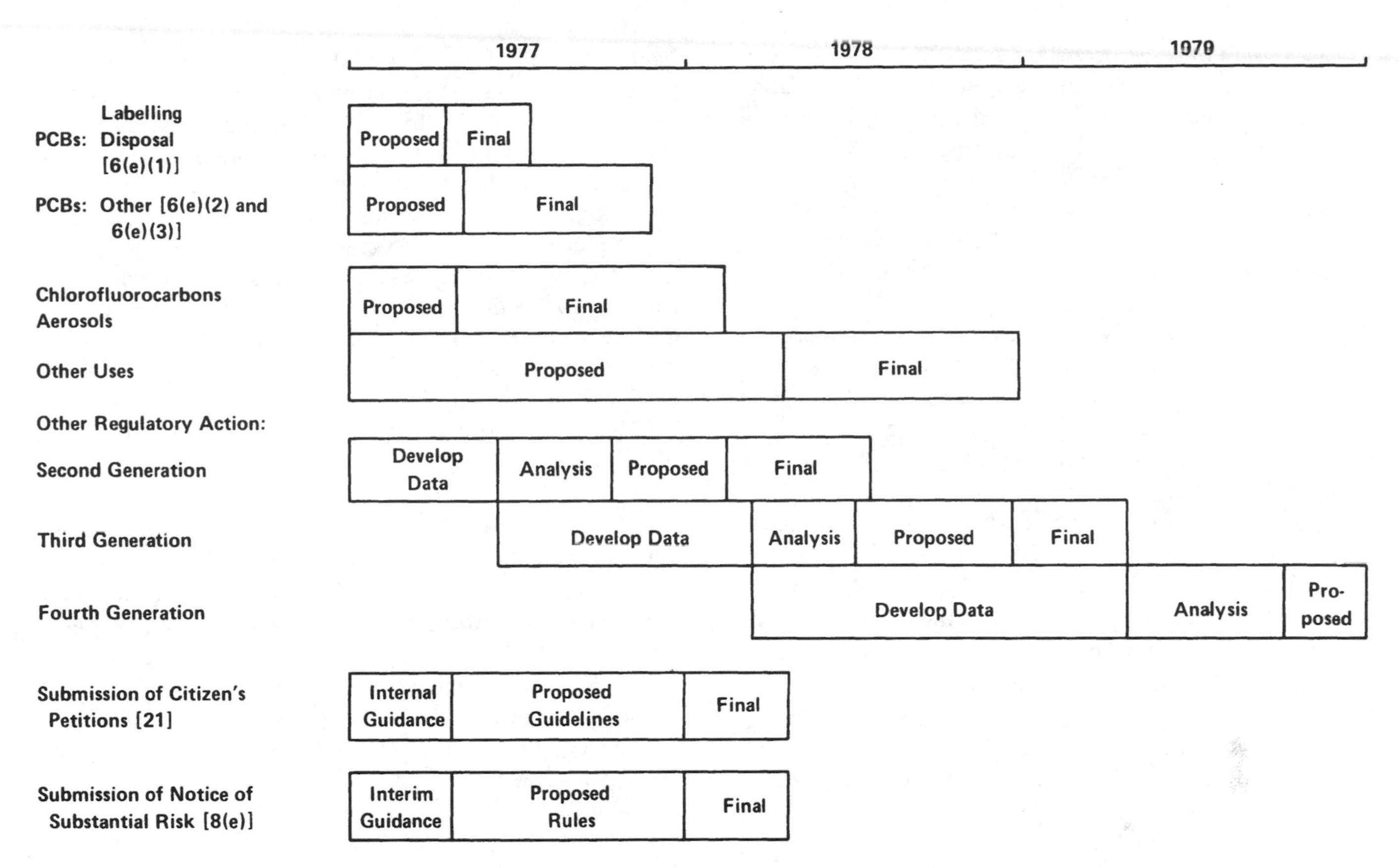

FIGURE 10. Use of Toxic Substances Control Act regulatory authorities to control existing chemicals.

a collaborative rather than adversarial interagency approach is essential, and close interaction with other agencies beginning with the initial discovery of a chemical problem will help insure that this system of formal referrals does not unnecessarily disrupt the priority activities of these agencies.

When formally referring chemical problems to other agencies for action, EPA will provide all available information concerning potential risk. While EPA may not have in hand complete documentation concerning all aspects of the risk, there should be a reasonably good indication that preventive or corrective action in the near term deserves serious attention. EPA will follow up with the other agencies. If additional data are subsequently developed, and the follow through by the other agency appears inadequate, EPA will not hesitate to review again the situation and to determine appropriate additional actions.

B. Policy Considerations

The basic consideration for all regulatory decisions is the risk to man or the environment of the problems being addressed. The appropriateness and character of the regulation must then take into account the possibility of actions under other laws [6(c) and 9(a)], the least burdensome approach under TSCA [6(a)], and the overarching requirement to take economics and social factors, as well as environmental concerns, into consideration [2(c)]. The environmental acceptability of the substitute chemicals or alternative technologies that are likely to emerge as the result of the regulatory action will be given special attention. Sometimes, the likely alternatives may be less than optimal from the environmental viewpoint and the incremental environmental gains in adopting the substitutes will be considered.

Often the regulatory action taken under TSCA will be in addition to regulatory actions that have been or could be taken by other programs. Different programs must by law give different weights to the economic and to some of the other factors concerning regulations, and there may be some differences in the orientation of the regulatory decisions made by different programs.

Regulatory actions directed to existing chemicals will not be limited only to correcting problems which are known or believed to be having adverse health and ecological effects but will also be directed to preventing such problems in the future. On occasion, TSCA action may address only a portion of the environmental problem posed by a chemical. Nevertheless, such limited steps can be important, and TSCA action to correct a piece of the problem need not be delayed until the entire problem can be effectively addressed.

C. Orientation of Initial Activities

Polychlorinated biphenyls and selected chlorofluorocarbons have been of special concern to the Congress for several years, and work is well underway to develop appropriate regulations for limiting their discharge into the environment. The other unattended chemical problems cited during the Congressional hearings are also being examined to determine whether additional regulatory actions are needed immediately.

Many agencies already have lengthy lists of chemicals of particular environmental concern. These lists are being reviewed before reaching a judgment as to whether they can provide useful guidance or whether they should be supplemented with still additional lists. Also, the list to be developed by the Interagency Testing Committee on priorities for testing may be useful in determining regulatory priorities as well.

Expanded efforts will be directed to more systematic procedures for screening and establishing priorities among chemical problems for regulatory attention. While establishing priorities will inevitably involve a number of judgmental decisions, it may be possible to develop improved techniques for assisting in the setting or priorities. Also, the utilization of chemical classes or categories of chemical use might assist in narrowing the vast array of chemicals to a more manageable number.

Underpinning the entire regulatory effect must be a technically sound program of chemical assessment. Chemical assessments have traditionally involved a. hazard assessment, b. source assessment, c. identification of substitutes and alternative technologies, d. development of control options, and e. evaluation of the environmental, economic, and related impacts of controls. In some cases, much of the needed data is at hand; more often, supplementary data must be developed. The assessment program will include analyses of individual chemicals, categories of chemicals, chemical technologies, and geographic problems.

D. Responses to Urgent Problems

Response to the uncovering of chemical problems which might pose urgent risks to health or ecological resources will receive the highest priority, and an on-call response capability will help minimize diversion of resources from other priority activities. In the past such problems have usually come to light as the result of new toxicity test results, new monitoring data, or identification of human or ecological victims of chemical exposure. The discovery of real or alleged urgent problems often results from the conduct of Government programs (e.g., PCBs in the milk of nursing mothers), findings of the scientific community (e.g., nitrosamines in the atmosphere), industrial revelations (e.g., worker deaths from vinyl chloride), and press investigations (e.g., cancer death rates in the Little Elk Valley of Maryland).

TSCA provides additional mechanisms for bringing to light urgent problems. Two of these mechanisms – citizens' petitions [21] and risks uncovered by test data [4(f)] – require a response within a specified time limit. A third principal TSCA mechanism, notification by industry of substantial risk [8(e)], does not have a mandated response time. Regardless of the source of the discovery, the urgency of the problem must drive the response timing.

TSCA also provides several new regulatory mechanisms for limiting chemical exposures quickly if warranted (e.g., imminent hazard [7], regulatory action immediately effective [6(d)(2)], referral to other agency with short deadline for action by the other agency [9(a)]. In the past, local agencies and industry itself have often been willing to take immediate corrective steps in the face of chemical crises. In any event, prompt and effective action to prevent additional damage using TSCA or other programs will be the immediate objective.

The EPA response capability will include available technical specialists to assess the discovery, on-call field and laboratory capability to confirm and supplement the data concerning the discovery, and coordinative and organizational mechanisms, involving a variety of programs within and outside EPA, for implementing prompt and effective corrective actions.

While each new problem will have its own pecularities, there are usually some common concerns relating to toxicity, exposure, chemical behavior, related commercial activities, and routes of environmental discharges. Generalized checklists of typical concerns are being developed to help insure that technical and policy analyses do not overlook important factors in the face of tight timetables. Also, steps will be taken to insure that interested parties are continuously informed of developments, and particularly acquisition of additional data, in view of the broad political interests in these types of problems.

A very general framework for the response activity is

1. Identification of problems associated with chemical activities as a result of:
   - Systematic screening of available information
   - Monitoring, toxicological, and epidemiological screening programs
   - Ad hoc environmental incidents, research findings, and allegations
   - Discoveries submitted under TSCA
2. Characterization of the problems with particular attention to:
   - Health and ecological effects and environmental behavior
   - Current and projected sources, environmental levels, and exposed population
   - Substitutes, control technology, and related cost and economic factors
   - Actions to date and actions underway to clarify and control the problems

3. Development and stimulation of preventive and corrective approaches including consideration of:

- Role of relevant authorities of EPA and other agencies
- Alternative approaches to voluntary or regulatory redress
- Environmental and economic impact of approaches
- Implementation of appropriate approach

VIII. DISSEMINATION OF INFORMATION AND ASSESSMENTS TO OTHER PROGRAMS AND INTERESTED PARTIES

A. The Broad Interests in Toxic Substances

Throughout the Congressional consideration of TSCA, there was a recognition of a TSCA role for many of the regulatory and nonregulatory programs of a number of organizations that were in place directed to the assessment and control of toxic substances. Since the enactment of the new law, such programs have already increased in number and in the scope of their interests.

At the federal level, for example, a large number of chemicals will be explicitly regulated under the Federal Water Pollution Control Act. Many more will be affected by more general standards under that authority and also under the Clean Air Act. The National Academy of Sciences will recommend a large number of chemicals to be considered for possible regulation under the Safe Drinking Water Act. NIOSH has several hundred criteria documents completed or in preparation which will add to the list of chemicals already regulated by OSHA. The Department of Transportation and the Mining Enforcement and Safety Administration similarly have regulations in place or under development affecting many toxic substances. All of these activities must be based on assessments of environmental and related data.

A number of states have taken steps in the toxic substances area. Virginia and Illinois, for example, are particularly interested in data reporting systems. Several states in the Great Lakes area have taken steps concerning PCBs and phosphates. Several states are concerned with chlorofluorocarbons. New Jersey, Texas, and California have very broad concerns over the heavy concentrations of the chemical industry and can be expected to expand efforts in the near future.

Central to the way industry does business are the policies and the attitudes of the financial community concerning investment capital. TSCA, in effect, adds one more risk dimension in the investment world. As this community begins to focus on toxic substances, it needs access to reliable and timely data. The information must be packaged in an understandable and usable form. Meanwhile, the insurance industry is rapidly expanding its interests in toxic substances, particularly with regard to product liability. The issue of substitutes for PCBs also sharpened concern over hazards associated with acceptability of substitutes. Both access to data and early awareness of possible TSCA regulatory actions are important to this side of the commercial community.

Other forces influencing the future directions of the chemical industry are the labor unions, environmental and public interest groups, and the consumer. In all of these cases the specter of possible harmful effects of chemicals can have a direct impact on industrial behavior. TSCA can provide important "early warnings" to these groups who in turn can provide the government with other early warning signals.

A handful of the larger chemical companies who are responsible for a large proportion of chemical sales have for a number of years conducted sizeable programs to assess the environmental aspects of industrial chemicals. Although there have been many soft spots in these efforts, they nevertheless provide a good foundation for expanded activities. Complimentary efforts have also been supported by industry through a number of trade associations and most recently through the Chemical Industry Institute of Toxicology.

However, relatively few companies have adequate data available to conduct environmental assessments of the chemicals they buy and sell. Despite these shortcomings of the past, the greatest potential impact resulting from TSCA in terms of the number of chemicals that are addressed may lie in the expanded internal assessments and procedures of individual companies, activities that must rest on a solid base of environmental data.

Finally, as the Congress and the courts deepen their involvement in this area, the availability of experts that they can call upon and the credibility of scientific data take on added importance. TSCA will be an important tool for developing the information base which will undergird many major decisions of the future.

B. Policy Considerations

Several explicit TSCA mechanisms enable interested parties to obtain information. Perhaps the most far-reaching mechanism is the Interagency Testing Committee which provides for seven agencies, in addition to EPA, to set forth the highest priority needs for testing [4(e)]. The provision for citizen's petitions allows any interested party to seek information under the testing and reporting sections of the law [21]. The requirement to place in the *Federal Register* notices of receipt of premarket data will alert parties concerned with new chemicals reaching the marketplace [5(e)]. The provision concerning disclosure of data clarifies some of the uncertainties concerning the public access to health and safety studies [14(b)]. Of course the overarching requirements of the Freedom of Information Act are designed to enable all interested parties to obtain information collected under TSCA and other laws. All of these explicit provisions of TSCA underscore the clear intent of the Congress that this legislation service the interests of many organizations in a variety of ways, and particularly with regard to acquisition and dissemination of data.

Therefore, EPA will emphasize coordinated approaches with other federal agencies to the assessment and control of toxic substances, with particular attention to identifying common information requirements that can be best satisfied through TSCA. The Agency will actively solicit the views of other interested parties as well as to information needs. Information acquired under TSCA will be made available as widely and as promptly as possible. In this regard the interests of the international community are particularly important as we consider the possibility of an international convention to deal on a global basis with the control of toxic substances.

A major effort will be made at the outset to insure that sound procedures are in place which will facilitate a flow of data to all interested parties. Since establishment of these procedural aspects will take priority, it will not be possible to respond to all requests for information to be collected under TSCA until adequate systems are in place to handle the flow. Even in the long run, it will continue to be necessary to prioritize the competing claims from many parties for information services under TSCA.

C. The Establishment and Operation of Data Systems

1. System for Receiving and Storing TSCA Data

Data received pursuant to TSCA will be retained in a discrete data system module with entry into and withdrawal from the system controlled, at least initially, by the Office of Toxic Substances. The TSCA data module obviously will be but one additional component to the much larger governmental data systems covering toxic substances. To the extent feasible, the TSCA module will be made compatible with and linked to the other existing systems. Figure 11 sets forth the implementation timetable for data systems.

While the volume of data received under TSCA may be relatively small at the outset, this volume will probably grow rapidly. Therefore, from the outset the system will incorporate automated components as rapidly as possible. The system will be designed and time-phased so that data received in the initial implementation stages are compatible with data received several years into the future, thus avoiding the potential problem of recoding. Development in submodules is envisaged, with components (personnel, hardware, and software) added as implementation progresses and data volumes grow. At the outset, however, methods for coding, cataloging, and retrieving data will be established so that consistent ground rules will guide data requirements resulting from rulemaking activities.

The chemicals included in the initial inventory of existing chemicals [8(b)] will provide a core index for the system which will be expanded as the number of chemicals in the system grows. In general, production and use data obtained pursuant to premarket notification [5(a)] and discretionary reporting [8(a)] requirements will be entered from the outset into an automated system. Initially, test data will be retained in hard copy with a locator tag incorporated into the automated system. A test data summary format which could be automated is currently being explored with the National Library of Medicine.

With regard to coding, CAS numbers and manufacturer identification will provide two keys. A chemical use coding system is currently under development to provide another key.

Access to confidential data will be strictly controlled and initially limited to the examination by authorized officials of available information *in situ*.

2. Making TSCA Data Available

EPA plans to aggressively develop and carry out procedures for disseminating to the public information obtained under TSCA. For example, test data will be submitted to EPA in sufficient copies so that one copy can be made publicly available without delay through the National Technical Information Service or other appropriate mechanism. Studies of the feasibility of establishing on-line terminal access to nonconfidential data and of periodically publishing such data will be soon initiated.

3. Retrieval of Data from Other Systems

There are many sources of external technical, scientific, and economic data which should be utilized in implementing TSCA. These sources are located both within and outside the federal government. CEQ is presently conducting a survey of such data bases. Meanwhile, some states are initiating additional toxic materials data banks. These data activities in particular will be examined in some detail with an eye toward reducing reporting requirements, on the one hand, and avoiding duplicate data searches during examination of chemical for potential hazards, on the other.

4. Dissemination of Non-TSCA Data to Interested Parties

Many parties are interested in improved ways for tapping the multiplicity of Governmental data banks concerning chemicals. As part of the overall data system effort, practical means for facilitating such access will be explored. Initially, an inventory of existing data bases is being conducted. Hopefully, meaningful road maps to governmental data can be provided to all interested users in the future.

5. Standardization of Government-wide Data Activities

Data standardization is particularly important when considering interlocking data systems. Data required by TSCA rulemaking should be requested in a format which can be used readily by a multiplicity of interested parties. As an initial step towards improved standardization of government-wide approaches, emphasis will be given to standardizing the formatting and dissemination of newly

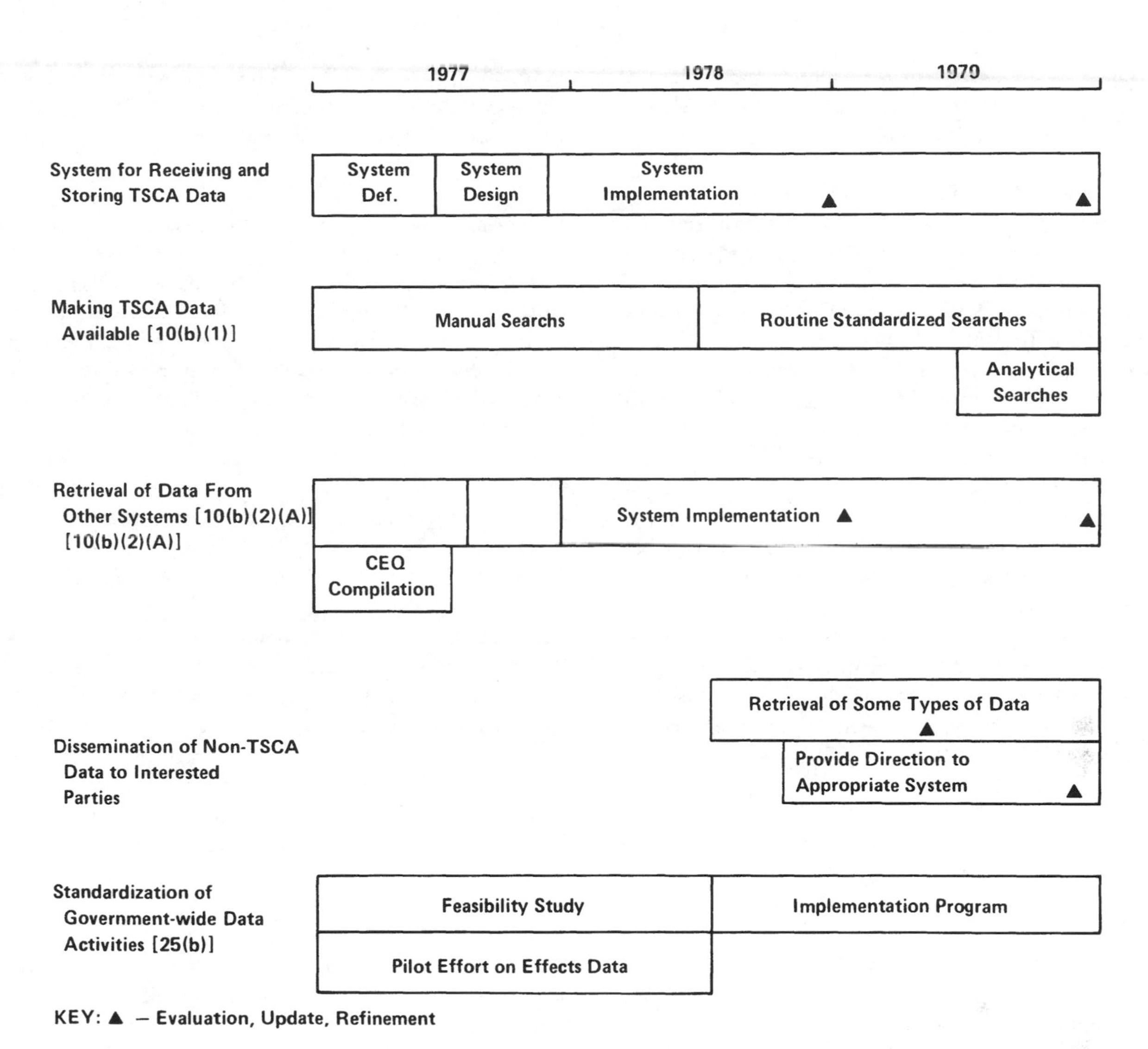

FIGURE 11. Data systems implementation.

collected data on health and ecological effects. At a later date, the effort will be expanded to include other types of data, and as time and resources permit, efforts will be directed to achieving compatibility of data already in the files of many agencies with the newly collected and standardized data.

IX. TYPES OF ANTICIPATED IMPACTS FROM IMPLEMENTATION ACTIVITIES

The types and extent of the impacts that will result from implementation activities are speculative at best, given the many uncertainties concerning the effects of chemicals and their behavior in the environment, the large number and variety of chemical products, and the continuing rapid growth of the chemical industry. The five years of Congressional testimony included many general statements concerning the environmental benefits that are likely to ensue. A number of unfortunate chemical incidents were cited as examples of the types of problems that can be avoided. On the economic side, the EPA report *Draft Economic Assessment of the Impact of the Toxic Substances Control Act* of June 1975, as well as economic assessments prepared by industry and by the General Accounting Office, attempted to identify some of the types of direct impact that will undoubtedly occur. However, those discussions were very limited, and little effort was made to address indirect impacts.

The following discussion provides but a very superficial framework for the impact evaluation effort that will accompany implementation activities.

A. Health and Environment

A primary legislative concern is reducing adverse health effects, and particularly chronic effects. The legislation should offer an opportunity to clarify the health effects of many chemicals and, over time, to reduce the number of deaths and the disease rates attributable to such effects. Also, the possibility of preventing acute effects will be addressed.

Data obtained under TSCA should enhance the capability of OSHA, MESA, and state authorities to reduce the incidence of worker deaths and diseases. Other agencies and organizations which obtain information and support under this legislation should also assist in preventing adverse health impacts at the national and local levels.

The likely impact of TSCA on reducing ecological damage is even more difficult to predict. The early experience in addressing PCBs (i.e., destruction of aquatic resources) and chlorofluorocarbons (i.e., depletion of the ozone layer) clearly demonstrates the importance of the legislation in this regard. A number of toxic chemicals end up in the aquatic environment where ecological damage can be extensive. Prevention of such damage will largely depend on the specific regulatory actions. Ecological test requirements should assist in clarifying the impact of chemicals on the ecosystem and in setting quality standards for water and other media under other authorities. This aspect of environmental assessment has been largely neglected by industry in the past.

B. Commerce and the Economy

Implementation requirements will add a new dimension in financial planning within industry for the development, manufacture, and marketing of chemicals. For example, there will be far greater reluctance to expand commercial investments in chemicals of questionable toxicity, and the search for broader applications of chemicals which are environmentally acceptable will intensify. Some marginal products for which testing is required may give way to substitutes which become commercially competitive. Many firms will be far more cautious in purchasing or selling products of unknown chemical composition. The ripple effect of such adjustments in current marketing practices will impact on a broad range of downstream processors and users.

There will probably be a tendency among some of the large companies toward greater self reliance on in-house chemical assessments of old and new chemicals and on conducting their own synthesis of small batches of highly reactive chemicals previously purchased from small suppliers. Given the concern over quality control of test data and the shortage of laboratory facilities, in-house toxicological and ecological testing laboratories should become more commonplace. Meanwhile, small firms may tend to move away from product lines that become targets for TSCA attention.

The international development and marketing strategies of multinational firms will also be impacted. Test marketing may be more heavily concentrated in countries where premarket requirements are minimal. Also, there may be a surge of new chemical imports into the United States to establish them as "existing" chemicals and, thus, eliminate the notification period for future imports. In general, more careful planning of international shipment of chemicals will be required.

C. Industrial Research and Development

Premarket notification requirements and premarket testing requirements will cause some adjustments in the research and development cycle. In a few cases, the testing requirements may in large measure codify existing industrial practices. In most cases, however, the new requirements will alter substantially the time phasing, the types of expertise, and the review processes involved in developing new chemicals. These adjustments will in turn impact on a) the decisions as to which new chemicals and products should be explored and then developed, b) the criteria for investing in research and development when there is an increased risk for commercial introductions, and c) the efforts to "design around" potentially troublesome chemicals from the environmental viewpoint.

The number and quality of environmental assessments conducted by industry should increase markedly. More qualified technical personnel will be attracted to the field, methodological approaches will be significantly upgraded, and the quality assurance procedures will be improved. However, should governmental requirements "overstandardize" assessment techniques, there is a danger that industrial creativity in improving the state-of-the-art of environmental assessment could be stifled.

In the early years of implementation, the number of new chemicals reaching the marketplace may decline due to uncertainties as to future regulatory requirements, technical and financial difficulties in adjusting to the new procedural requirements, and the increased research and development costs and lead times for some products with limited market potential. However, in the longer run innovation in introducing new products need not be stifled. More intensive investigations of environmentally acceptable chemicals, coupled with incorporation of premarket requirements into a routine research and development cycle, should continue to allow ever expanding benefits to the consumer from the uses of chemicals.

D. The Scientific Base

The heavy emphasis in the legislation on improving the scientific methodologies undergirding chemical assessments, and the attendant implications for strengthening the technical manpower base, should have a major impact on the chemical and biological sciences. Not only will the legislation give impetus to the advancement of these individual disciplines, but it should stimulate a closer coupling of these disciplines with engineering, economics, and other areas of importance to toxic substances control. This general impetus to the broad spectrum of sciences may far overshadow the scientific impact of individual regulatory actions.

The inadequacy of science to provide clear answers for regulatory decisions will probably be subjected to frequent criticism by many impatient parties. However, carefully documented scientific investigations will be a key to many actions. There is no doubt that the importance of credible technical data, albeit inconclusive, will be widely recognized.

All should benefit from the expanded sharing of scientific data. The importance of common procedures and common formats in carrying out the reporting scientific investigations will take on added importance as the problems involved in exchanging noncompatible data bases become clear.

E. Social Concerns

Many of the impacts cited above are tied to social concerns, such as employment effects, increased costs of products, and rights of inspection. However, there are even more fundamental social concerns which will be affected by TSCA implementation such as:

1. How much effort should be directed to protecting the welfare of future generations?
2. To what extent should the public participate in decision making that has previously been the exclusive domain of private industry?
3. How are health concerns to be balanced with economic costs in determining "unreasonable risk"?

There is little experience in measuring the types of social impact that could far outweigh in importance the other more narrow impacts of this legislation. A continuing effort to identify and understand these impacts is the key to determining the value of TSCA as an instrument of public policy.

Section XII

Index

INDEX

A

B

C

D

E

F

G

H

I

J

K

L

M

N

O

P

Q

R

S

T

U